Encyclopaedia Anatomica

Encyclopaedia Anatomica

A Collection of Anatomical Waxes
Sammlung anatomischer Wachse
Collection des cires anatomiques

With contributions by | Mit Beiträgen von | Textes de
Monika v. Düring & Marta Poggesi

Photographs by | Fotografien von | Photographies de
Saulo Bambi

Museo di Storia Naturale dell'Università di Firenze,
sezione di zoologia La Specola

TASCHEN

KÖLN LONDON LOS ANGELES MADRID PARIS TOKYO

The figures in square brackets below the captions give information about the locations of the works shown here from the La Specola Museum. The Roman numerals indicate the room, the Arabic numerals the display cabinet.

Die Angaben in den eckigen Klammern unter den Bildunterschriften geben Auskunft über den Standort der abgebildeten Werke im Museum La Specola: Die römischen Ziffern bezeichnen den Saal, die arabischen die Vitrine.

Les chiffres entre crochets apparaissant sous les légendes désignent l'emplacement des œuvres au musée La Specola : les chiffres romains indiquent les salles, les chiffres arabes les vitrines.

To stay informed about upcoming TASCHEN titles, please request our magazine at www.taschen.com/magazine or write to TASCHEN, Hohenzollernring 53, D–50672 Cologne, Germany, contact@taschen.com, Fax: +49-221-254919. We will be happy to send you a free copy of our magazine which is filled with information about all of our books.

Hohenzollernring 53, D–50672 Köln
www.taschen.com

Original edition: © 1999 Benedikt Taschen Verlag GmbH
Concept by Volker Gebhardt, Cologne
Edited by Petra Lamers-Schütze, Yvonne Havertz, Cologne
Design by Claudia Frey, Cologne
Cover design by Sense/Net, Andy Disl and Birgit Reber, Cologne
Production by Ute Wachendorf, Cologne

Anatomical texts by Monika v. Düring
Anatomical captions by Monika v. Düring, Volker Gebhardt, Jürgen Goldschmidt
English translation by Fiona Elliott (Marta Poggesi), Francis Steel and Britta Fricke (Monika v. Düring et al.)
German translation Daniele dell'Agli (Marta Poggesi)
French translation by Thérèse Chatelain-Südkamp (Marta Poggesi), Henri Sick and Jean-Marie Le Minor (Monika v. Düring et al.)

Printed in Singapore
ISBN 3–8228–5039–X

Contents
Inhalt
Sommaire

The Wax Figure Collection in 'La Specola' in Florence

Marta Poggesi

The History of the Collection

When Peter Leopold of Habsburg-Lotharingen (1747–1792), Grand Duke of Tuscany (ill. p. 7), decided in 1771 to bring together all the 'scientific' collections from the various galleries in the Grand Duchy, the result was an innovation that was unique not only in Europe but anywhere in the world.

Much earlier than other potentates this enlightened ruler – an enthusiastic student of the natural sciences – had understood the importance of the sciences for the cultural advancement of any society. The first thing he did was to examine the various ways in which the findings of science could be made accessible to all those who were interested.

And indeed the Imperial Regio Museo di Fisica e Storia Naturale (The Imperial-Royal Museum for Physics and Natural History, later widely known as 'La Specola', which means 'observatory' in Italian) was the first of its kind, in that from the day of its opening on 21 February 1775 it admitted the general public to its collections. It is true that there were separate opening hours for the upper and the lower classes: the latter – "provided they were cleanly clothed" – were allowed to visit between 8.00 and 10.00 in the morning, which then left enough time before the "intelligent and well-educated people" were admitted at 1.00 in the afternoon. But however discriminatory this distinction may seem to us today, nevertheless one can still sense how innovative it was to open the museum's doors to this broader public.

These collections had originally been started by the Medici family; immensely important as patrons and connoisseurs of the arts they had also done much to promote the sciences. Ample evidence of this may be seen in the Accademia del Cimento (1657–1667) which had amongst its staff such famous scientists as Redi, Magalotti and Galileo's favourite student, Viviani. After the death of Giangastone, the last descendant of the family, the Grand Duchy of Tuscany went to Francis III of Habsburg-Lotharingen, who decided to have an inventory made of all the treasures in his residence. This task was entrusted to the physician and natural historian Giovanni Targioni-Tozzetti (1712–1783), who completed the work in just under a year in 1763/64.

When Peter Leopold succeeded his father to the throne in 1765 –

after the latter had become Emperor of Austria – he therefore found that the groundwork for a re-organisation of the collection had already been done. The work involved in this major undertaking fell to Felice Fontana (1730–1805; ill. p. 29), by profession a teacher of logic at the University of Pisa, but also an anatomist, physicist, chemist and, above all, an internationally renowned physiologist. He devoted himself to restructuring the buildings with such passion that by the end of 1771 the first items could already be moved into the new rooms. As early as 1771 the Grand Duke had already bought the Palace of the Torrigiani in the via Romana, very close to the Palazzo Pitti. He had also bought a number of neighbouring houses and had commissioned the Abbot Felice Fontana of Rovereto to draw up plans for alterations to these buildings in order to create a home for his scientific collections. As director of the new museum, in the early years Fontana travelled throughout Europe acquiring books and collections and establishing contacts in different countries. As a result the museum in Florence became one of the most important museums of its day, not least for its rich scientific library.

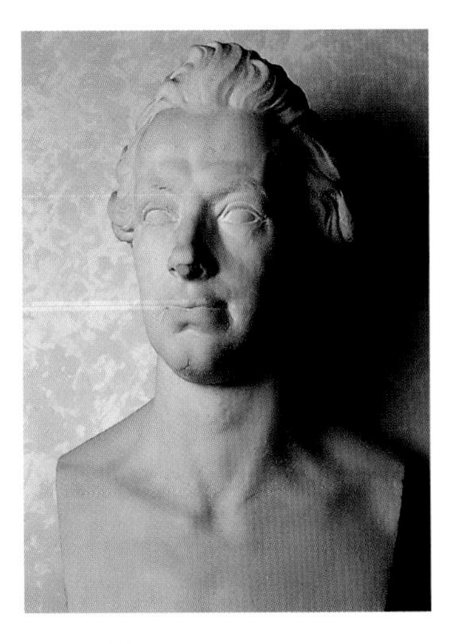

Bust of Peter Leopold of Habsburg-Lotharingen

Büste von Peter Leopold von Habsburg-Lothringen

Buste de Pierre Léopold de Habsbourg-Lorraine

Fontana directed the museum until his death. Throughout this time Giovanni Fabbroni was at his side – at first as his assistant (and constant antagonist) and from 1784 onwards as the museum's deputy director, accompanying Fontana on numerous journeys. In 1805 he took over as Director but only for one year.

It is interesting to note that the money to construct the museum for physics and natural history (and many other projects instigated by Peter Leopold) was raised by the sale of valuable objects once owned by the Medici family – despite a testament left by the Palatinate Electress Anna Maria Luisa (1688–1743), Giangastone's sister, which decreed that the estate of the Medici family should in its entirety and irrevocably remain in the ownership of the city of Florence.

The core holdings of the museum, which came principally from the Uffizi, consisted of collections (minerals and shells, for instance), natural history curiosities from the era of the Medici family, Galileo's instruments and equipment from the Accademia del Cimento as well as four wax figures by the Sicilian sculptor Giulio Gaetano Zumbo.

In 1771 a ceroplastic studio was set up together with other workshops essential to the running of the museum (a carpentry shop, a glaziery and a taxidermy studio): these will be discussed in more detail later in this essay. Keen to have astronomy and meteorology included in the museum as well, in 1780 Peter Leopold commissioned the architect Gaspare Paoletti (who had earlier been involved in the restructuring of the palace) to build the Osservatorio Astronomico (the so-called 'little tower'). This was later to be the source of the name 'Specola', which means observatory. The work turned into a major project which led to the building's being much larger than originally planned. In fact many experts had advised against it and had advocated building on the Acetri hills instead. In 1789, the year of the completion, a part of the Boboli Park was incorporated as the museum's Botanic Garden (ill. p. 8).

After the death of his brother Josef in 1790, Peter Leopold became Emperor of Austria and put Tuscany into the hands of his second son Ferdinand III (1769–1824), who had neither his father's vision nor his skill in matters of government. But the times were also against him: Napoleonic expansionism forced the Lotharingians to give up Tuscany which, after various complications, went to the Bourbons of Parma. During this period the museum started to organise the teaching of

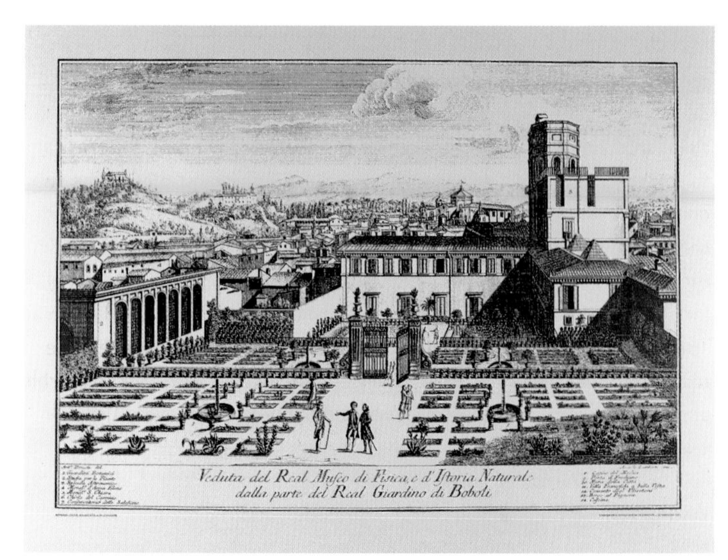

Veduta del Real Museo di Fisica, e d'Istoria Naturale
dalla parte del Real Giardino di Boboli

'La Specola' seen from the Boboli Garden; engraving from the late 17th century

Die Specola vom Boboli-Garten aus gesehen; Kupferstich vom Ende des 17. Jahrhunderts

La Specola vue du jardin Boboli ; gravure de la fin du XVIIe siècle scientific subjects, which was continued in 1814 after the restoration of the Lotharingians.

Under Ferdinand's rule the museum had to survive a period of unrest and lost a great deal of the scientific importance attached to it in its early years when it enjoyed a reputation as one of the most important centres of learning in Europe according to illustrious foreign authors such as Goethe and Bernouilli. Ferdinand also had major structural alterations made to the palace: in 1820, under the direction of the architect Pasquale Poccianti a corridor was constructed joining the Specola with the *Meridiana* wing of the Palazzo Pitti, so that the Poccianti Corridor now extends the Vasari Corridor – which leads from the Palazzo Vecchio via the Uffizi to the Palazzo Pitti – as far as the Specola.

After his death in 1824 Ferdinand was succeeded by his son Leopold II (1797–1870), affectionately nicknamed 'Canapone' by the Florentines on account of his blond hair. It was thanks to him that the teaching and lectures in the sciences were reinstated, with a particular emphasis on the applied sciences relevant to agriculture and the cultivation of reclaimed land. During his time as regent the Tribuna di Galileo was constructed. Dedicated to the memory of one of the greatest scientists, it was inaugurated in 1841 on the occasion of the third congress of Italian scientists in Florence. The Tribuna is a large room on the first floor of the building and was partially rebuilt by the architect Giuseppe Marelli. Work was started in 1830, the initial plan being simply to add an apse to the existing room, but the scheme was then altered in accordance with the wishes of the Grand Duke: in order to lend more weight to his intended celebration of Galileo and his work, the whole room was to be dedicated to the renowned scientist. As well as a statue, it was to contain all the surviving Galileo memorabilia and the instruments from the Accademia del Cimento. The building of the Tribuna, one of the few examples of late Neo-Classicism in Florence, led to a number of important alterations to the palace, most particularly to the floors below the Tribuna. It also meant that a part of the courtyard was roofed over. The room itself was decorated specifically and exclusively with Tuscan marble and with works by artists – sculptors and painters – who were also exclusively from Tuscany.

After the fall of Leopold II – the last Grand Duke of Tuscany – in 1859 the Istituto di Studi Superiori e di Perfezionamento (Institute of Higher Study and Further Education) was founded. In 1923 this became

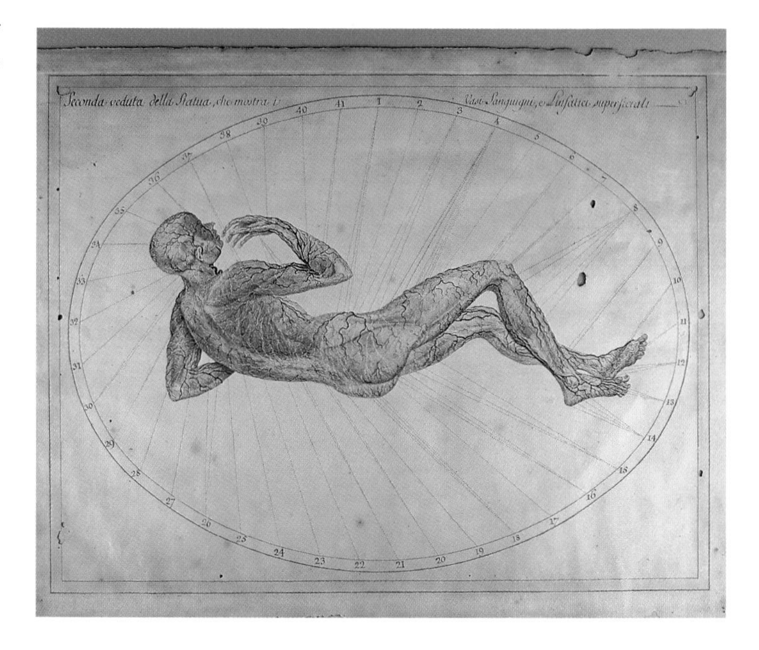

Original drawing of a recumbent figure, called *lo spellato* (Room XXVIII, display cabinet no. 740)

Originalzeichnung einer liegenden Figur, genannt *lo spellato* (Saal XXVIII, Vitrine Nr. 740)

Dessin original d'une figure couchée appelée *lo spellato* (salle XXVIII, vitrine n° 740)

the University of Florence. The museum buildings became part of the Institute and now contain the University's Department of Physics and Natural Sciences.

While the founding of the Institute was of inestimable importance it also led to the breaking up of the museum into its various disciplines, which – along with the relevant collections and libraries – had progressively to be relocated at other sites in view of the constantly increasing numbers of students. Even the resistance of many scientists at the time could not hinder this process, with the result that the historical buildings in the via Romana now only contain the zoological collections and the bulk of the anatomical ceroplastics.

The Early Days of Anatomical Wax Models

Wax has been used to make models since time immemorial; it has always been a popular material for artists, both for aesthetic and for technical reasons, because it is easy to work and allows the finished piece to be cast in metal afterwards. It was used widely for religious motifs such as statues of saints or cribs and votive panels. At the time of the Renaissance and throughout 17th century huge numbers of wax

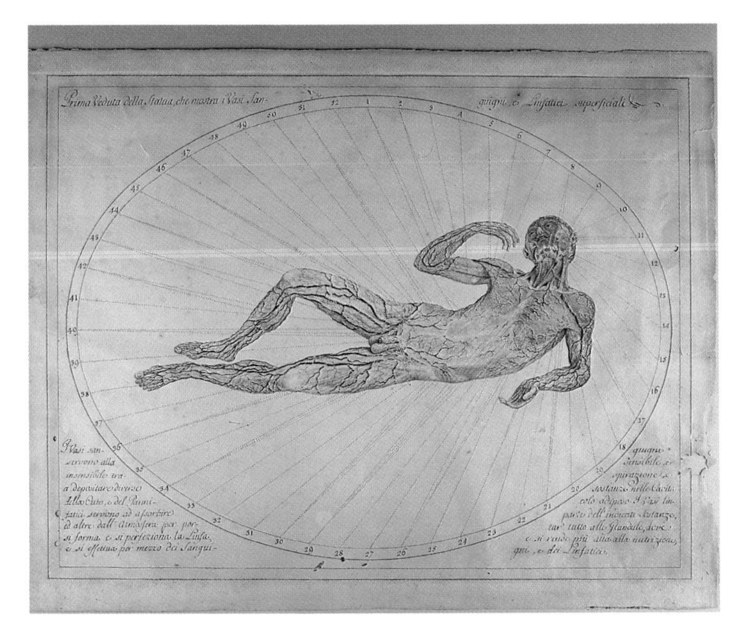

Original drawing of a recumbent figure, called *lo spellato* (Room XXVIII, display cabinet no. 740)

Originalzeichnung einer liegenden Figur, genannt *lo spellato* (Saal XXVIII, Vitrine Nr. 740)

Dessin original d'une figure couchée appelée *lo spellato* (salle XXVIII, vitrine n° 740)

figures could be found in churches, particularly in Orsammichele and in Santissima Annunziata. These included limbs, organs or parts of organs as well as portraits, objects and statues – often even life-sized. In the 17th century, wax models were also increasingly used for scientific purposes.

Although the state – and even more so the Church – did its best to prevent corpses being used for anatomical studies and penalised anyone who attempted to do so, by the mid-15th century the first anatomical drawings and treatises were beginning to emerge, generally drawn by great artists and sculptors such as Leonardo da Vinci, Michelangelo, Raphael and Titian, to name but a few. Renowned anatomists such as Fallopio, Cesalpino and Vesalius used drawings by these masters to illustrate their treatises.

Towards the end of the 17th century Gaetano Giulio Zumbo, working in Bologna where there was a famous school of anatomy, was the first to make anatomical models using coloured waxes. Zumbo came from Syracuse and his works, which are illustrated here, include two male heads (one now in Florence and one in Paris), a female figure (now unfortunately lost) and some other smaller works. Bologna was also the first place to have a proper school of ceroplastics and it was here that the wax modellers working in the museum in Florence during the second half of the 18th century were trained. The painter, sculptor and architect Ercole Lelli (1702–1766) is credited with founding the Bolognese school and, like Zumbo before him, he too made wax models of muscles and inner organs by studying real corpses. He was able to obtain access to the corpses he needed for this work through the support of Cardinal Prospero Lambertini, who became Pope Benedetto XIV in 1740. As an amateur scientist and a man of learning the Cardinal encouraged the study of anatomy at the school in Bologna and bought all the models Lelli made: they are magnificent pieces, now preserved in the Istituto d'Anatomia Umana Normale at the University of Bologna.

Amongst Lelli's assistants were Giovanni Manzolini (1700–1755), who with his wife Anna Morandi (1716–1774) also made numerous models that can likewise still be seen at Bologna University.

At the time, the study of anatomy was progressing throughout Europe but the school of ceroplastics in Florence was in fact a direct offshoot of the school in Bologna. The link was made by the surgeon and child-birth specialist Giuseppe Galletti. Having been much impressed by models made by Lelli and Manzolini, the physician Galetti – together with the wax modeller Giuseppe Ferrini – himself made a number of models in wax and terracotta (now in the Museo di Storia della Scienza in Florence) in order to demonstrate different birth procedures, both normal and with complications.

The Wax Figures in the Specola: The School, Works, Techniques

Although the ceroplastic workshop existed for more or less a century – from 1771 until the mid 19th century – the majority of the wax models in the museum, like the commissioned works for other institutions in Florence and elsewhere, were made during the first 50–60 years of the workshop's activities.

The founding of the school goes back to Felice Fontana. Supported by Grand Duke Peter Leopold, Fontana devoted himself with great energy and determination to setting up a wax workshop, particularly since – as an anatomist and pathologist – he was personally involved in the creation of the models. At first there was just the wax modeller Giuseppe Ferrini working under Fontana's guidance in the workshop, but subsequently the anatomist Antonio Matteucci and the very young

Clemente Susini were also employed. Immensely talented and hugely productive, Susini was later to become the most important and famous of the wax modellers of the Florentine School (ill. p. 47).

Page · Seite · Page 13:

G. G. Zumbo: Specimen of a head (cf. pp. 18, 345, 348)

Präparat eines Kopfes (vgl. S. 18, 345, 348)

Préparation anatomique d'une tête (cf. p. 18, 345, 348)

We have no precise information today as to where the wax was worked, although it seems most likely that this would have been in studios on the ground floor in the south wing of the palace which is divided into a number of small courtyards and has windows looking out on to the via Romana. Practically none of the original equipment used has survived, but archive records indicate that the following at least were purchased: copper vats in various sizes for melting the wax; modelling tools (ill. p. 33); metal wire in different thicknesses; marble slabs for pressing the wax into thin sheets; scales; tripods for heating substances; slates for making notes and drawings during dissections; crates with handles for transporting the corpses, wooden crates with carrying poles for transporting the wax figures; containers, vases and bottles made from clay or glass for pigments and other substances that were added to the wax. Many of the latter were found with their contents still intact in the museum's store-rooms (ill. p. 32).

Egisto Tortori: Small clay bust, model for a plaster bust of C. Susini

Kleine Tonbüste, Modell für eine Gipsbüste von C. Susini

Petit buste en argile, modèle d'un buste en plâtre de C. Susini

Furthermore, it is also clear from the archives how many corpses or parts of a corpse were necessary to make a wax model: astoundingly, over two hundred for a single figure. This startlingly large number is a direct consequence of the fact that there was no way of preserving or freezing the corpses. Consequently there was a constant need for fresh subjects if a dissection were to be accurate and thorough. In order to keep an exact record of all the corpses or parts of corpses – which came from the Santa Maria Nuova hospital about two kilometres away – a register was put up at the door where all corpses being admitted or dispatched to the cemetery were listed: "[...] with regard to the corpses from Santa Maria Nuova, of which it is not known how many there are and because one may not have people die purely for the benefit of the wax collection, a register was put up at the entrance door on which all admissions were noted as well as the number of corpses sent to the cemetery." (from a document written in 1763).

Despite the small number of employees the work progressed very rapidly, when one considers that in 1790 – twenty years after the workshop was set up – the models already filled eight rooms, and this did not even include the models that were made specifically for the hospital in Florence and for other institutions in Italy and abroad (cf. ill. p. 48). Most outstanding amongst the wax figures from the Specola workshop that were soon to be found all over Europe was the collection commissioned in 1781 by the Austrian Emperor Josef, Peter Leopold's older brother, for the military school of medicine in Vienna (called the Josephinum after the Emperor). This collection contains 1200 items made in 1786 and transported to Vienna in two deliveries on the backs of mules. Other models were destined for Pavia, Cagliari, Bologna, Budapest, Paris (today in Montpellier), Uppsala, London, Leiden and elsewhere. These outside commissions occasionally caused problems since the Grand Duke was determined that they should not hold up the production of works for the museum. Thus in the case of such commissions the stipulation was that the modellers and anatomists – Fontana himself worked on the collection for Vienna – were allowed to use the museum's moulds and tools, but that they had to procure the materials themselves and carry out the work with the assistance of craftsmen not in the employ of the museum.

It was Fontana'a ambition to produce as many wax figures as possible in order to create a teaching resource which would in future obviate the need to exhume corpses for the study of the human anatomy. To this end he included in his collections a number of tempera drawings (ill. pp. 10, 11) showing individual parts with numbers around the drawn figure linked by fine lines to the different organs, which could then be identified using these numbers. The relevant written explanations would then be found in a drawer that was part of every wax figure shrine: by this means the user would, as it were, have access to a complete three-dimensional anatomical treatise (ill. p. 21).

Around 1790 Fontana began another immensely ambitious project which was, however, never realised according to the technical difficulties encountered: the plan was to make a series of painted wooden anatomical models – either life-sized or somewhat larger – which could be taken apart for teaching purposes in order to demonstrate how the organs connect with each other. After a few initial attempts of which only a few examples have survived – one now in the museum, one in Paris and another enormous bust – Fontana had to abandon his plans. Not only was the wood harder to work but it also had a tendency to warp, which meant that all the craftsmen's efforts to make the individual parts fit into each other were rendered useless with time. The only wax model in the museum which has the desired qualities is the so-called 'Venere medicea': a recumbent female figure which can be dismantled from above, layer by layer, eventually revealing a uterus with a small foetus in it. (ill. p. 39)

By this time the museum had increased its personnel to include a graphic artist (Claudio Valvani) and a wood-carver (Luigi Gelati) as well as various anatomists for carrying out the dissections. Amongst the most famous of these was Paolo Mascagni (1755–1815) who left behind extremely beautiful anatomical plates. He specialised in the study of the lymphatic system and it is no coincidence that many of the recumbent statues and the smaller works display this part of the human body in minute detail.

In 1805 Felice Fontana died (ill. pp. 17, 76) but the wax workshop continued with its work, and even after the death of Clemente Susini in 1814 other wax sculptors, such as the Calenzuolis (Francesco Calenzuoli and his son Carlo) and Luigi Calamai (1800–1851) took over. Under Calamai the work was concentrated on comparative anatomical models and botany as well as on pathological models that were made for the Santa Maria Nuova hospital, and which can still be seen today in the Department of Pathological Anatomy in the University. After Calamai's death, Egisto Tortori (1829–1893) took over. Besides making figures of animals and models for the Department of Pathological Anatomy, Tortori also made a bust of Clemente Susini. When Tortori died, however, no-one was appointed to replace him and the workshop was closed down for ever.

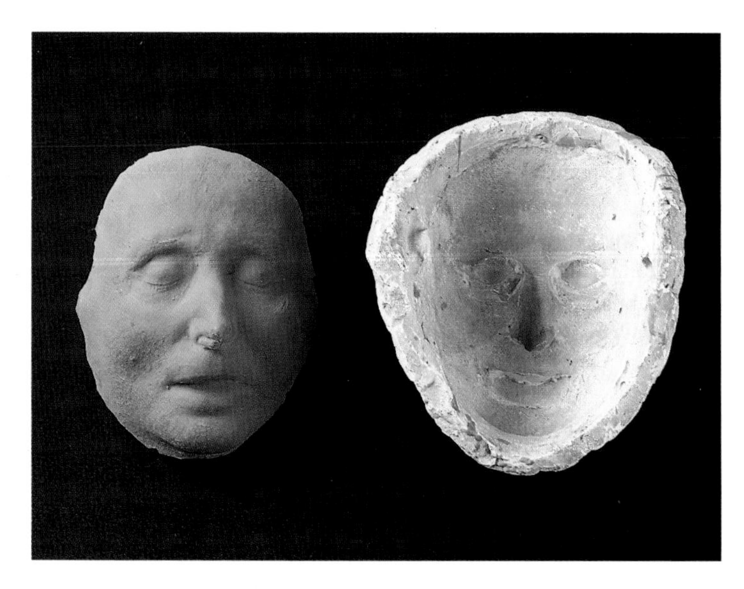

Wax death-mask of Felice Fontana and plaster relief

Totenmaske aus Wachs von Felice Fontana und Gipsabdruck

Masque mortuaire en cire de Felice Fontana et moulage en plâtre

Of the wax figures now stored in urns in La Specola, there are 513 of the human anatomy and 65 comparative figures; 5 made by G. G. Zumbo. There are 26 whole figures (including the half-finished figure of a young boy): 13 standing and 13 recumbent, 18 of which are life-sized (6 standing and 12 recumbent) and 8 approximately 60 cm in height (7 standing and 1 recumbent). Almost all are on display: only 14 of the human figures and parts are kept in storage. There are more than 800 framed drawings and almost 900 explanatory notes, and yet it is clear that there must have been more that have gone missing with time. In recent years both the original drawings and the written explanations have been removed from display in order to prevent them suffering further damage through light and humidity. The original frames now contain photocopies so that the collection is as useful as it ever was for teaching purposes and also still constitutes a worthwhile display.

The scientific perfection of these works could hardly be surpassed and although this is mainly appreciated by experts in the field, most visitors are equally impressed by their artistic quality. For however scientifically accurate these exhibits are and whatever their didactic purpose was, they also artworks in their own right. In addition their aesthetic qualities are heightened still further by the beauty of the wooden shrines in which they are kept, despite the fact that many show signs of

Page · Seite · Page 18:

G. G. Zumbo: Specimen of a head (cf. pp. 13, 345, 348)

Präparat eines Kopfes (vgl. S. 13, 345, 348)

Préparation anatomique d'une tête (cf. p. 13, 345, 348)

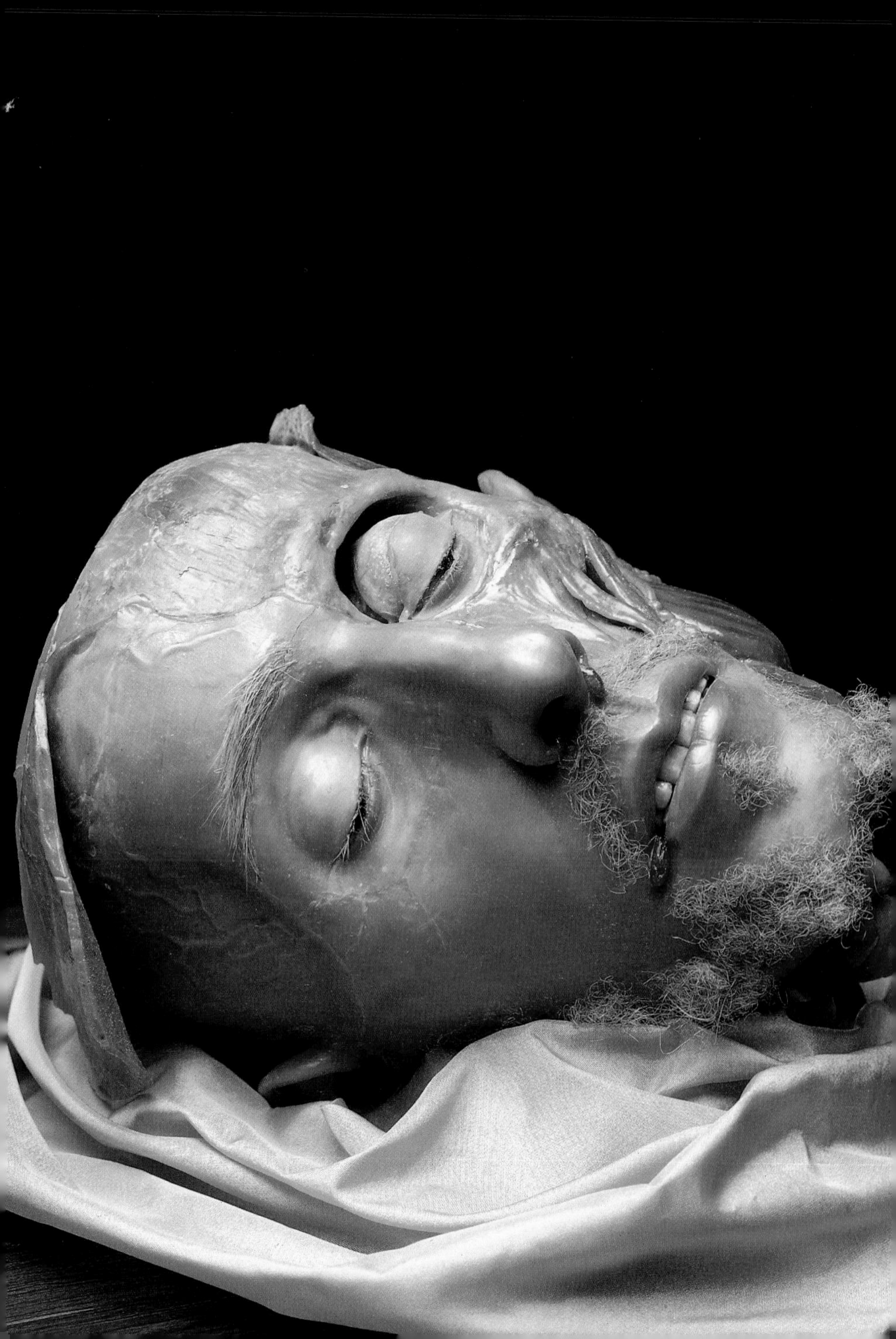

wear and tear and are badly in need of restoration. The care that originally went into the presentation of this collection is also evident from the equally fragile 18th-century silk drapes and veils – now faded but once bright green.

The importance that was placed on aesthetic appearances may be seen again in the collections of comparative anatomy, of human pathology and above all of botany: small ceramic vases were specially made in the workshop of Ginori von Sesto Fiorentino for the plant models.

We have no detailed knowledge of the technical processes involved in the production of wax figures, and all that is known has had to be deduced from various letters and documents in the state archives and in the archives of the Museum for the History of Science in Florence. Besides this, each wax modeller also had his own technique, and – like any other craftsmen or artists – they were not keen to have their methods spread abroad. However, the following is relatively certain: the dissected item would first of all be copied exactly in chalk or low-grade wax. A plaster cast would be taken of this, which could either be very large or made up of a number of sections. It was sometimes possible to make a cast directly from the pieces in question, such as bones for instance. These moulds, which are still stored in the museum (ill. p. 60), subsequently served as a kind of matrix which could also be used repeatedly for making additional casts of the same item. The hardest, most problematical part of the process was, however, the construction of the final model. This required great precision and knowledge of the substances that had to be added to the wax in order to achieve the desired colour and consistency – it was a matter of both skill and experience. The most crucial part of the process was the slow melting of the wax in a waterbath at exactly the right temperature so that it did not discolour. Most commonly the modellers used white wax from Smyrna, Chinese wax or Venetian wax; in order to make it more elastic, turpentine was added as well as paints (thinned with turpentine) and other substances that are recorded in the lists of purchases "needed for the wax models". Before the wax mixture was poured into the mould the latter would be moistened with luke-warm water and rubbed with soft soap to facilitate the removal of the model. Apart from a few which were solid, most of the wax models were hollow and in order to stabilise them they would be stuffed with rags, hemp waste or wood chippings. Figures which were constructed from a number of separate parts generally had a supporting metal frame. After the moulds had been removed, each item would be cleaned and finished, that is to say, equipped with the relevant organs, vessels, nerves and so on until it was complete; a final coating of clear varnish lent the whole a suitably glossy appearance. To ensure that the models were completely accurate each stage had to be supervised by the anatomists who were also responsible for deciding on how to place certain organs so that they might be seen to best advantage.

As has already been mentioned, each wax modeller developed his own technique and equipment, which would become increasingly refined, through the invention of new instruments such as a "drawing iron to turn the wax cylinders" which they used from 1786. Previously this process, which was essential for producing blood vessels, had to be carried out by hand. Towards the close of the century, the wax modellers kept notebooks in which they recorded what they had done each day and which tell us that even in those days it was frequently necessary to carry out restoration work, repairs, improvements and reworkings. In 1793 Susini made a note "about anatomical wax models in need of improvement as a result of flaws that should be treated and which are described in the following by me, Clemente Susini, modeller to the Royal Museum". For instance, on "examination of models of the pharyngeal artery and the jugular vein, leading off from the windpipe one sees two foreign muscles, that is to say, muscles which do not belong in the human form". Having noted further anomalies Susini ends with the comment that "these errors are too grave to be left uncorrected and, for the sake of brevity, I should rather not speak of the many others". Almost all the models have survived in good condition, and only a few have become at all discoloured, although here and there a foetus has darkened in colour and we know that some now greenish veins were once a clear "blue-purple".

The Wax Figures of Gaetano Giulio Zumbo

Gaetano Giulio Zumbo's wax figures deserve a chapter of their own, partly on account of their unusual conception and production but above all because they pre-empted the use of wax for anatomical models by a good century and laid the foundations for everything that was to follow.

Zumbo was born into a noble family in 1656 in Syracuse; the family name was probably Zummo, but this name has now disappeared without trace, and since the debased version 'Zumbo' is much better known and widely used it seems best to use it here. We have little information on his life – and what we have is only fragmentary. It is known that he was educated in the Jesuit Institute in his home town but had to leave following some – otherwise unspecified – "irksome incident". He may well have received his earliest inspiration as an artist from the Classical relics of his homeland, followed later by the pictorial world of Mannerism and the early Baroque. From 1687 to 1691 he lived in Naples and it may well have been there that he created two of the groups still preserved in La Specola: the 'teatrini' with the title *Il Trionfo del Tempo* (The Triumph of Time) and *La Peste* (The Plague). According to scholarly experts in the field who have studied Zumbo's work, this may be concluded from the painterly background and the particular arrangement of the figures, both of which point unmistakably to southern Italian precursors. From 1691 to 1694 he worked for Cosimo III de' Medici in Florence, where he created another two 'teatrini': *Il Sepolcro* (or *La Vanità della Gloria Umana*) and *Il Morbo Gallico* (or *Sifilide*) which display many Renaissance features. In 1695 he moved to Genoa where it seems he completed *The Anatomy of a Head* on show in the museum there, and he also made the acquaintance of the French surgeon Guillaume Desnoues. They later worked together on a number of anatomical wax models, including the life-sized figure of a woman giving birth, a smaller model of a woman who has died during child-birth, and the burial of Christ – unfortunately none of which has survived. In 1700 Zumbo and the French physician fell out and Zumbo moved to Marseilles

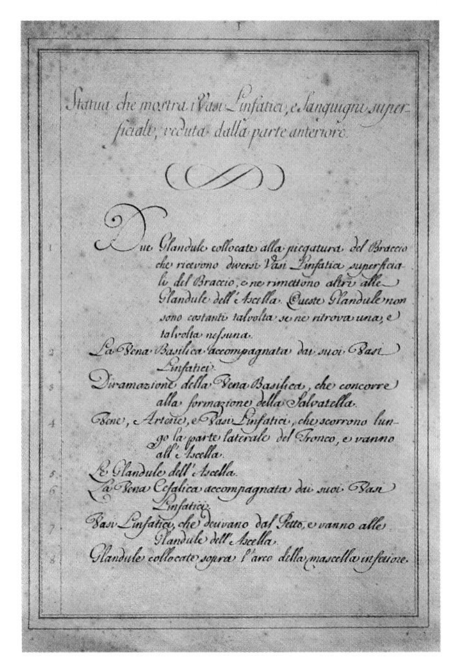

Statua che mostra i Vasi Linfatici, e Sanguigni superficiali, veduta dalla parte anteriore.

1 Due Glandule collocate alla piegatura del Braccio che ricevono diversi Vasi Linfatici superficiali del Braccio, e ne rimettono altri alle Glandule dell'Ascella. Queste Glandule non sono costanti talvolta se ne ritrova una, e talvolta nessuna.
2 La Vena Basilica accompagnata dai suoi Vasi Linfatici.
3 Diramazione della Vena Basilica, che concorre alla formazione della Salvatella.
4 Vene, Arterie, e Vasi Linfatici, che scorrono lungo la parte laterale del Tronco, e vanno all'Ascella.
5 Le Glandule dell'Ascella.
6 La Vena Cefalica accompagnata dai suoi Vasi Linfatici.
7 Vasi Linfatici, che derivano dal Petto, e vanno alle Glandule dell'Ascella.
8 Glandule collocate sopra l'arco della mascella inferiore.

Explanations of the drawings for *lo spellato* (Room XXVIII, display cabinet no. 740)

Erläuterungen zu den Zeichnungen für *lo spellato* (Saal XXVIII, Vitrine Nr. 740)

Explications des dessins réalisés pour *lo spellato* (salle XXVIII, vitrine n° 740)

where he created a wax head. His last move was to Paris where he entered the service of Louis XIV, and where he died from a brain haemorrhage in 1701. On his death his estate included a beautiful wax head, most probably the very one that is now in the Museum National d'Histoire naturelle. His grave in the L'Eglise de Saint Sulpice was destroyed during the Revolution.

The bulk of this artist's work, largely ignored until fifty years ago, is now in La Specola. The three 'teatrini' which are also known as the *Cera della peste* ('plague figures') and the anatomy of a head were in fact in the Royal gallery (that is to say in the Uffizi) when the Medici were succeeded by the Lotharingians, and were then passed to the Museo de Fisica e Storia Naturale, but never put on display there – perhaps because of the shocking nature of their subject matter. When Peter Leopold set off for Vienna in 1790 he gave the 'plague figures' to the court physician Giovanni Giorgio Asenöhrl, better known as Lagusius. The latter did not even take possession of them but commissioned first Fontana and then Agostino Renzi to sell them for 150 sequins. Renzi was well aware of the importance of these groups and offered them to the new Grand Duke Ferdinand III for the Royal Academy of Fine Arts. When the pieces were rejected as neither "useful nor suitable" he offered them instead to the Duke for the Royal Museum. This proposal was accepted in principle by Giovanni Fabbroni, the deputy director of the museum, and the three works were valued by various experts including Clemente Susini. Since the valuations turned out higher than the price set by Lagusius, the Grand Duke authorised their purchase. Thus Zumbo's works remained in the museum until 1878 when these three dramatic scenes were given for safe-keeping to the museum Nazionale del Bargello. From there they went to the Museo di Storia della Scienza. There they were badly damaged during the torrential rainstorms of 1966 but, on the basis of photographic records, were meticulously and miraculously restored over a period of eighteen months by Guglielmo Galli from the Pietre Dure workshop in Florence. *The Anatomy of a Head*, on the other hand, never left the Specola since it was rightly regarded as an essential component in the scientific collection of the Institute.

The technique used by Zumbo was considerably different from that used later in Florence. *The Anatomy of a Head* was made using a real human skull, which X-ray examination has shown must have belonged to a man of around 25 years old. By contrast, the head now in Paris and

Page · Seite · Page 23:

Drawing of a head: front view, 2nd half of the 18th century (Room XXVIII, display cabinet no. 707)

Zeichnung eines Kopfes: Frontalansicht, 2. Hälfte des 18. Jahrhunderts (Saal XXVIII, Vitrine Nr. 707)

Dessin d'une tête vue de face, 2e moitié du XVIIIe siècle (salle XXVIII, vitrine n° 707)

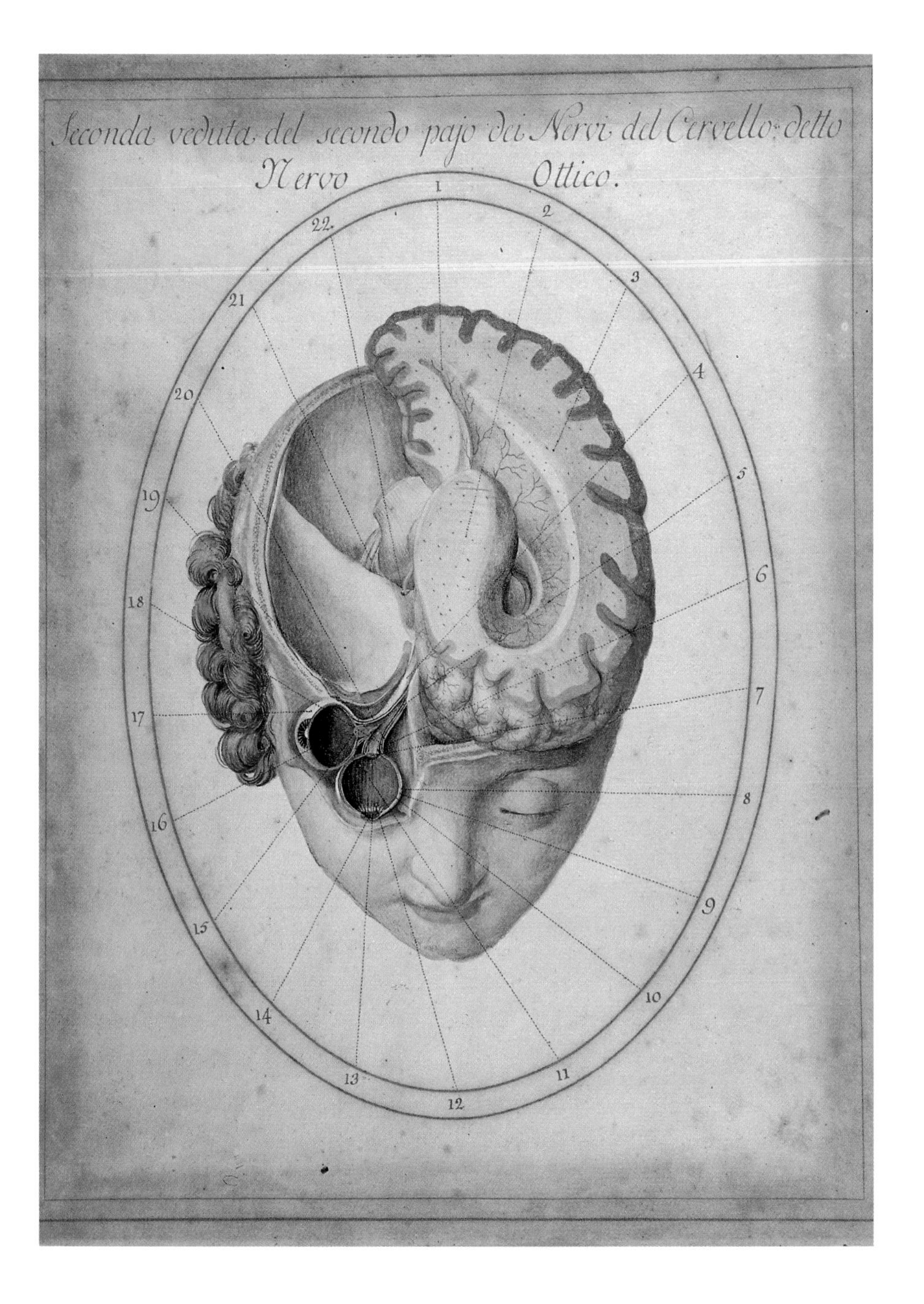
Seconda veduta del secondo pajo dei Nervi del Cervello: detto
Nervo Ottico.
1
2
3
4
5
6
7
8
9
10
11
12
13
14
15
16
17
18
19
20
21
22.

another head (now lost) were made completely from wax, which would seem to indicate advances in the master's technique, perhaps as a result of his collaboration with the French physician Desnoues. The figures in Zumbo's dramatic scenes, executed with the greatest artistry, were almost certainly produced using plaster moulds which in turn had been made from meticulously sculpted clay models. The more or less liquid wax (a mixture of bees' wax, resin, turpentine and pigments) was applied in relatively thin layers in order to achieve the desired consistency and colouration. The scant knowledge we have of Zumbo's technique we owe to the restoration work carried out by Guglielmo Galli after the floods of 1966. This work also revealed evidence of older repairs, for during his work Galli identified pigments that could not have existed in the late 17th century. It also seems that the small self-portrait (ill. p. 55) in the group *Trionfo del Tempo* cannot with any certainty be credited to Zumbo and could well be the work of a contemporary or a later work of Susini's. The *Morbo Gallico*, originally a similar composition, of which now only some fragments remain, was a present from Cosimo III to Filippo Corsini. Unfortunately, this was kept in the cellars of the Palazzo Corsini by the River Arno and was almost completely destroyed during the floods. A few of the figures were found in the gardens, but since there were no

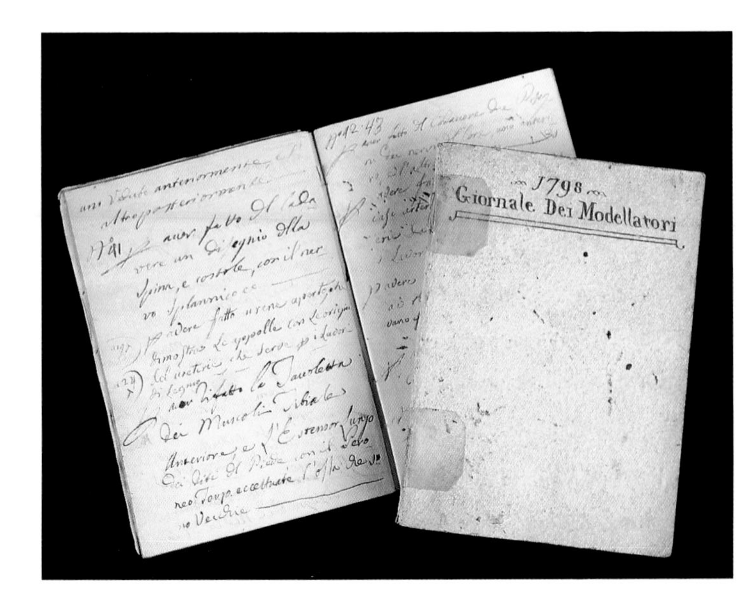

Diary kept by the wax modellers

Tagebuch der Wachsbildner

Cahiers des modeleurs

photographs of the group it was impossible to reconstruct it. On the occasion of the 200th anniversary of the opening of the museum, which was celebrated with a conference on wax modelling, the Corsini family donated the remains of the work to the museum.

For many years Zumbo's masterpieces were misread as the diseased caprices of an artist revelling in macabre and repellent details. It was not until the mid-20th century that they began to be seen in a more appropriate historical perspective: as realistic documentations of an era when death was omnipresent in the shape of war, famine and major epidemics. This vision of the destruction of human life, of its fragility, of the relentless passage of time, and the constant, specific reminders of death that face us in Zumbo's works are simply typical of the culture and thinking of the 17th century, and put him in the company of other great artists of the time such as Luca Giordano and Mattia Preti.

Pages · Seiten · Pages 26–27:

G. G. Zumbo: *Il Morbo Gallico* (or *Sifilide*)

Syphillis

Die Syphilis

Le Morbo Gallico

Die Wachsfigurensammlung des Museums La Specola in Florenz

Marta Poggesi

Die Geschichte

Als Peter Leopold von Habsburg-Lothringen (1747–1792), von 1765 bis 1790 Großherzog von Toskana (Abb. S. 7), 1771 beschloß, alle „wissenschaftlichen" Sammlungen der großherzoglichen Galerien in einem Museum zu vereinigen, führte er eine Neuerung ein, die in Europa und in der Welt ihresgleichen suchte. Viel früher als anderen Herrschern war dem aufgeklärten Landesfürsten, selbst ein begeisterter Student der Naturwissenschaften, deren Bedeutung für die kulturelle Entwicklung der Gesellschaft bewußt geworden. Als erster sann er über Möglichkeiten nach, die Errungenschaften der Naturwissenschaften allen Interessierten zugänglich zu machen.

Und in der Tat war das Imperial Regio Museo di Fisica e Storia Naturale (das Kaiserlich-Königliche Museum für Physik und Naturkunde, das später allgemein „La Specola" – „Sternwarte" im Italienischen – genannt wurde) das weltweit erste seiner Art, das von seiner Eröffnung am 21. Februar 1775 an für das allgemeine Publikum freigegeben wurde. Zwar gab es getrennte Besuchszeiten für Gebildete und für das gemeine Volk; letzteres hatte – „reinliche Bekleidung vorausgesetzt" – von acht bis zehn Uhr morgens Zutritt, um genügend Zeit verstreichen zu lassen bis zum Einlaß „der intelligenten und gelehrten Leute [...] um ein Uhr nachmittags". Auch wenn diese Unterscheidung heute diskriminierend erscheint, kann man doch ermessen, wie innovativ die Öffnung der Bestände für die breite Öffentlichkeit damals gewesen sein muß.

Die Einrichtung der Sammlungen ging auf die Medicis zurück, die sich auch um die Förderung der Wissenschaft verdient gemacht hatten, wie das Beispiel der Accademia del Cimento (1657–1667) zeigt, an der zur Zeit Ferdinando II. de' Medici so bedeutende Wissenschaftler wie Redi, Magalotti und Galileis Lieblingsschüler Viviani lehrten. 1737, nach dem Tod Giangastones, des letzten Abkömmling der Familie, fiel das Großherzogtum Toskana aufgrund der Bestimmungen des Wiener Friedens von 1735 an den späteren Kaiser Franz I. von Habsburg-Lothringen, der eine Aufstellung aller in seiner Residenz enthaltenen Schätze vornehmen ließ. Von 1763 bis 1764 inventarisierte der Arzt und Naturforscher Giovanni Targioni-Tozzetti (1712–1783) die Bestände.

Als Peter Leopold 1765 seinen Vater, der zum Kaiser von Österreich ernannt worden war, als Großherzog der Toskana ablöste, fand er eine gute Arbeitsgrundlage zur Reorganisation der wissenschaftlichen Sammlungen vor. Diese Aufgabe sollte dem Abt Felice Fontana (1730–1805) aus Rovereto (Abb. S. 29) zufallen, seines Zeichens Dozent für Logik an der Universität von Pisa, aber auch Anatom, Physiker, Chemiker und v. a. ein international renommierter Physiologe, der sich der geplanten Neustrukturierung mit solcher Leidenschaft widmete, daß bereits Ende 1771 der erste Teil der Bestände in die neuen Räumlichkeiten verlegt werden konnte. Bereits 1771 hatte Großherzog Peter Leopold den Palast der Torrigiani (zuvor im Besitz der Bini) in der Via Romana in unmittelbarer Nähe des Palazzo Pitti sowie einige angrenzende Häuser erworben und Fontana mit dem Entwurf des Umbaus der Gebäude beauftragt, um in ihnen die wissenschaftlichen Sammlungen unterzubringen.

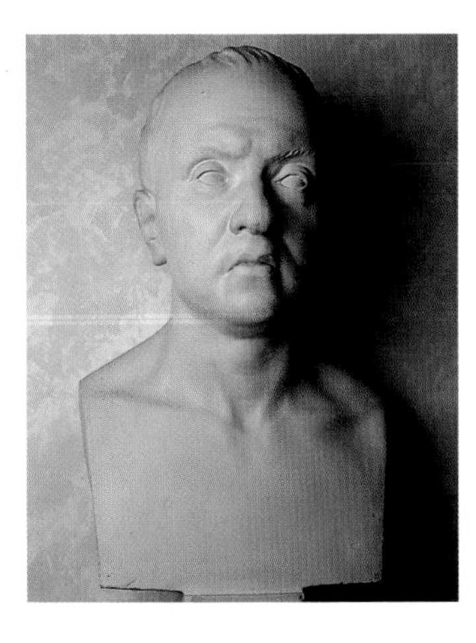

Bust of Felice Fontana

Büste von Felice Fontana

Buste de Felice Fontana

Als Direktor des neuen Museums bereiste Fontana in den ersten Jahren ganz Europa, um Bücher und Sammlungen zu erwerben und Kontakte zu Gelehrten in vielen Ländern herzustellen. So wurde das Florentiner Museum zu einem der bedeutendsten seiner Zeit, zumal es auch über eine reichhaltige wissenschaftliche Bibliothek verfügte. Fontana leitete das Museum bis zu seinem Tod 1805, wobei ihm Giovanni Fabbroni erst als Assistent und ab 1784 als Vize-Direktor (und permanenter Widersacher) zur Seite stand und ihn auf zahlreichen Reisen begleitete.

Das Geld zur Verwirklichung des Museums für Physik und Naturkunde (und auch vieler anderer Projekte Peter Leopolds) stammte aus dem Verkauf wertvoller Gegenstände aus dem Besitz der Medici; und dies trotz des Testaments der palatinischen Kurfürstin Anna Maria Luisa (1668–1743), der Schwester Giangastones, in dem das Vermögen der Medici unwiderruflich an die Stadt Florenz gefallen war. Zum Grundbestand des Museums, der vornehmlich aus den Uffizien stammte, gehörten Sammlungen (etwa von Mineralien und Muscheln), naturkundliche Kuriositäten aus der Zeit der Medici, die Instrumente Galileis und die Gerätschaften der Accademia del Cimento sowie vier Wachsfiguren des sizilianischen Bildhauers G. G. Zumbo.

Um den Bestand des Museums um die Gebiete Meteorologie und Astronomie zu erweitern, beauftragte Peter Leopold 1780 den Architekten Gaspare Paoletti (der bereits an der Umstrukturierung des Palastes beteiligt gewesen war) mit dem Bau des Osservatorio Astronomico (dem sog. „Türmchen"), der dem ganzen Komplex später den Namen „Specola" (Sternwarte) geben sollte. Das aufwendige Bauvorhaben, das – trotz der abweichenden Meinung vieler Experten, die einen Neubau auf den Hügeln von Acetri befürwortet hatten – zu einer beträchtlichen Erweiterung des Gebäudes führte, wurde 1789 abgeschlossen. In dasselbe Jahr fällt die Verlegung eines Teils des Boboli-Parks in den Botanischen Garten des Museums (Abb. S. 8).

Nach dem Tod seines Bruders Josef wurde Peter Leopold 1790 Kaiser von Österreich und überließ die Toskana seinem Zweitgeborenen Ferdinand III. (1769–1824), dem die Weitsicht seines Vaters ebenso abging wie dessen Geschick bei den Regierungsgeschäften. Erschwerend kam hinzu, daß die napoleonische Expansion die Lothringer zur Aufgabe der Toskana zwang. 1801 fiel die Toskana schließlich als Königreich Etrurien an Bourbon-Parma. Während dieser Periode richtete das Museum Lehrveranstaltungen in wissenschaftlichen Fächern ein, die auch nach der Restauration der Lothringer 1814 fortgeführt wurden.

Unter Ferdinand III. durchlebte das Museum unruhige Zeiten und verlor seinen Ruf als eines der bedeutendsten Zentren europäischen Wissens, was ihm auch von illustren Besuchern aus dem Ausland wie Goethe und Bernouilli attestiert worden war. Auch Ferdinand III. ließ wichtige Umbauten an dem Palast vornehmen: 1820 wurde unter der Leitung des Architekten Pasquale Poccianti der bereits bestehende Vasari-Korridor, der vom Palazzo Vecchio über die Uffizien zum Palazzo Pitti führt, um den Pocciantiani-Korridor bis zur Specola verlängert und so die Specola mit dem Flügel der Meridiana, des Palazzo Pitti, verbunden.

Auf Ferdinand III. folgte 1824 sein Sohn Leopold II. (1797–1870), von den Florentinern seiner blonden Haare wegen wohlwollend „Canapone" genannt. Ihm gebührt das Verdienst, den wissenschaftlichen Studien, insbesondere den anwendungsbezogenen wie etwa der Landwirtschaft, neue Impulse gegeben zu haben. In seine Regentschaft fällt die Realisierung der Tribuna di Galileo, eine Hommage an den großen Forscher, die 1841 anläßlich des dritten Kongresses italienischer Wissenschaftler in Florenz eingeweiht wurde. Bei der Tribuna handelt es sich um einen großen Saal im ersten Stock des Gebäudes, der von dem Architekten Giuseppe Marelli z. T. neu gebaut wurde. Die 1830 begonnenen Arbeiten sahen ursprünglich lediglich eine zusätzliche Apsis in einem bereits existierenden Saal vor. Um jedoch dem Vorhaben, dem berühmten Gelehrten ein würdiges Monument zu errichten, größeren Nachdruck zu verleihen, sollte ihm auf Wunsch des Großherzogs der gesamte Saal gewidmet werden und neben einer Statue alle Andenken Galileis sowie die Instrumente aus der Accademia del Cimento beherbergen. Der Bau der Tribuna, eines der seltenen Beispiele des späten Neoklassizismus in Florenz, führte zu bedeutsamen Änderungen der Palastarchitektur; insbesondere wurden die darunterliegenden Stockwerke umgestaltet und ein Teil des Hofes überdacht. Der Saal wurde mit toskanischem Marmor verkleidet, und für die Ausstattung zeichneten ausschließlich Künstler aus der Toskana – Bildhauer, Maler – verantwortlich.

Nach dem Fall Leopolds II., des letzten Großherzogs der Toskana, wurde 1859 das Istituto di Studi Superiori e di Perfezionamento (Institut für höhere Studien und Weiterbildung) gegründet, aus dem 1923 die königliche Universität von Florenz hervorgehen sollte. Die Gebäude des Museums wurden als Institut für Physik und Naturwissenschaften Teil der Universität.

Trotz der Bedeutung der Einrichtung dieses Instituts, markierte es andererseits für das Museum den Beginn seiner Zergliederung in die verschiedenen Disziplinen, die mit der stetig wachsenden Zahl der Studenten mitsamt der dazugehörigen Sammlungen und Bibliotheken ausgelagert werden mußten. Diesen Prozeß konnte selbst der Widerstand vieler Naturwissenschaftler jener Epoche nicht aufhalten, so daß im historischen Gebäude in der Via Romana letztlich nur die zoologischen Sammlungen und der Großteil der anatomischen Zeroplastiken verblieben.

Die Anfänge der Wachsbildnerei in der Anatomie

Die Verwendung von Wachs zur Modellierung von Figuren geht bis auf die Zeit der Römer zurück. Vor allem in der Kunst war das Material seit jeher sehr beliebt, sowohl aus ästhetischen Gründen als auch aus technischen. Wachs läßt sich einfach verarbeiten und eignet sich hervorragend für die Herstellung von Gußformen, wie etwa für den Bronzeguß. Insbesondere im europäischen Mittelalter wurden Weihgaben oder Heiligenstatuen, Krippen und auch Votivtafeln aus Wachs in großer Stückzahl gefertigt. In der Renaissance und während des gesamten 17. Jahrhunderts konnte man gewaltige Mengen dieser Wachsmodelle in Kirchen, besonders in Orsammichele und in Santissima Annunziata, bewundern, darunter Gliedmaßen, Organe oder Teile davon ebenso wie Porträts, gegenständliche Darstellungen und Statuen, nicht selten sogar lebensgroß. Im 17. Jahrhundert kamen dann in Wachs nachgebildete anatomische Objekte zu wissenschaftlichen Zwecken auf.

Obwohl der Staat und mehr noch die Kirche das Studium der menschlichen Anatomie am Leichnam nach Kräften behinderten und sanktionierten, tauchten Mitte des 15. Jahrhunderts erstmals anatomische Zeichnungen und Traktate auf, in der Regel aus der Feder so berühmter Maler und Bildhauer wie Leonardo da Vinci, Michelangelo, Raffael und Tizian, mit denen bedeutende Anatomie-Forscher wie Falloppio, Cesalpino oder Vesalius ihre Traktate illustrierten.

Gaetano Giulio Zumbo war gegen Ende des 17. Jahrhunderts der erste, der in der berühmten Anatomie-Schule von Bologna für die Modellierung anatomischer Präparate farbige Wachse verwendete. Der aus Syrakus stammende Künstler, dessen Werke im folgenden abgebildet sind, gestaltete zwei Männerköpfe (die sich heute in Florenz bzw. in Paris befinden), eine (leider verlorengegangene) Frauenfigur sowie einige kleinformatige Kompositionen. Ebenfalls in Bologna formierte sich

Pigments and other materials used in the production of wax models in their original containers

Farbstoffe und andere bei der Wachsfigurenherstellung erforderliche Materialien in ihren ursprünglichen Behältern

Colorants et autres matières nécessaires à la création de sculptures en cire dans leurs récipients d'origine

die erste Schule der Zeroplastik, aus der in der zweiten Hälfte des 18. Jahrhunderts auch die Wachsbildner des Florentiner Museums hervorgingen. Als Begründer der Bologneser Schule gilt der Maler, Bildhauer und Architekt Ercole Lelli (1702–1766), der – wie schon zuvor Zumbo – die Modellierung wächserner Muskeln und Eingeweide auf der Basis echter Skelette vornahm. Als Kenner und Liebhaber der Naturwissenschaften förderte Kardinal Prospero Lambertini, der 1740 zum Papst Benedikt XIV. ernannt wurde, die anatomischen Studien der Bologneser Schule und kaufte Lellis sämtliche Modelle, die man noch heute im Istituto di Anatomia Umana Normale der Universität Bologna bewundern kann. Von besonderer Bedeutung war auch seine Unterstützung für die Beschaffung der für diese Arbeit notwendigen Leichname. Zu den Assistenten Lellis gehörte außerdem Giovanni Manzolini (1700–1755), der mit Hilfe seiner Frau Anna Morandi (1716–1774) zahlreiche Modelle ausführte, die ebenfalls in der Bologneser Universität aufbewahrt werden.

In jener Zeit wurde das Studium der Anatomie zwar in ganz Europa vorangetrieben, doch die zeroplastische Schule von Florenz stammt direkt von derjenigen Bolognas ab. Die Vermittlung besorgte der Chirurg und Geburtshelfer Giuseppe Galletti, der unter dem Eindruck

Copper vat with modelling tools

Kupferwanne und Modellierbesteck

Bassine en cuivre et instruments à modeler

der Werke von Lelli und Manzolini zusammen mit dem Wachsbildner Giuseppe Ferrini eine Reihe von Modellen aus Wachs und Terracotta anfertigte (die sich zur Zeit im Museo di Storia della Scienza in Florenz befinden), um verschiedene Geburtsvorgänge – normale und dystokische – zu demonstrieren.

Die anatomischen Wachsfiguren der Specola: Schule, Werke, Techniken

Die zeroplastische Werkstatt in Florenz bestand etwa ein Jahrhundert lang – von 1771 bis in die zweite Hälfte des 19. Jahrhunderts hinein –, aber der überwiegende Teil der im Museum aufbewahrten Wachsbildnisse entstand ebenso wie die Auftragsarbeiten für andere Florentiner Institutionen und Auftraggeber in den ersten fünfzig bis sechzig Jahren ihres Bestehens.

Die Gründung der Specola geht auf Felice Fontana zurück, der sich mit voller Unterstützung Großherzog Peter Leopolds deren Ausbau widmete, zumal er sich als Anatom und Pathologe persönlich an der Erstellung der Modelle beteiligte. Anfangs arbeitete lediglich der Wachsbildner Giuseppe Ferrini unter Fontanas Anleitung in der Werkstatt. In der Folge wurden dann der Anatom Antonio Matteucci und der damals noch sehr junge Clemente Susini eingestellt, der aufgrund seines Geschicks und seiner Produktivität zum bedeutendsten und berühmtesten Zeroplastiker der Florentiner Schule avancierte (Abb. S. 47).

Die Wachse wurden vermutlich in den Räumen des Erdgeschosses verarbeitet, deren Fenster zur Via Romana im Südflügel des Palastes mit seinen vielen kleinen Höfen hinausgingen. Von den einstigen Gerätschaften ist kaum etwas erhalten, aber aus Archivunterlagen gehen folgende Ankäufe hervor: Kupferwannen verschiedener Größe, in denen das Wachs geschmolzen wurde; Modellierbesteck (Abb. S. 33); Eisendraht in unterschiedlichen Stärken; Marmorplatten zum Pressen des Wachs; Waagen; Dreifüße zum Erhitzen des Materials; Schiefertafeln für Anmerkungen und Zeichnungen während der Sektionen; Kisten mit Handgriffen zum Transport der Leichen, Holzkisten mit Stangen zum Transport der Wachsfiguren; Behälter, Vasen und Flaschen aus Keramik oder Glas für Farbmittel und andere Substanzen, die dem Wachs beigegeben wurden. Von letzteren wurden in den alten Magazinen des Museums noch etliche mitsamt Inhalt gefunden (Abb. S. 32).

Aus den Archivdokumenten geht ferner hervor, wie viele Leichname oder Leichenteile für die Ausführung eines Modells benötigt wurden:

über zweihundert für eine einzige Figur! Die erstaunlich hohe Zahl erklärt sich aus dem Umstand, daß es keine Konservierungsmöglichkeiten gab, so daß für genaue anatomische Sektionen ständig neue Leichname gebraucht wurden. Um alle Leichen oder Leichenteile, die vom (etwa zwei Kilometer entfernten) Krankenhaus Santa Maria Nuova herbeigeschafft wurden, numerisch genau zu erfassen, war an der Eingangstür eigens ein Register angebracht, auf dem alle Eingänge, aber auch alle für den Friedhof bestimmten Abgänge notiert wurden: „[...] bezüglich der Leichname aus Santa Maria Nuova, von denen man nicht weiß, wie viele anfallen, und weil man die Leute nicht passend für die Wachssammlungen sterben lassen kann: Darum wurde an der Tür des Museums ein Register angebracht, in dem die Leichname, die daselbst gebraucht werden und deren Zahl beträchtlich erscheint, festzuhalten sind und ebenso zu vermerken, wie oft und wie viele zum Friedhof geschickt werden." (aus einem Dokument von 1793).

Ungeachtet der geringen Zahl der Präparatoren gingen die Arbeiten recht schnell voran, wenn man bedenkt, daß um 1790, also zwanzig Jahre nach Beginn, die Modelle bereits acht Säle füllten, nicht gerechnet all jene, die für das Florentiner Krankenhaus und für andere Institutionen in Italien und im Ausland angefertigt worden waren (vgl. Abb. S. 48). Unter den über ganz Europa verstreuten Wachsplastiken aus der Specola-Werkstatt ragen die vom österreichischen Kaiser Josef, dem älteren Bruder von Peter Leopold, für die militärische Medizinschule Wien (das nach ihm benannte Josephinum) 1781 in Auftrag gegebenen hervor. Die Sammlung umfaßt 1 200 Stücke, die 1786 fertiggestellt und in zwei Lieferungen auf dem Rücken von Maultieren nach Wien gebracht wurden. Weitere Plastiken waren für Pavia, Cagliari, Bologna, Budapest, Paris (heute in Montpellier), Uppsala, London, Leiden und andere Städte bestimmt. Diese externen Aufträge warfen gelegentlich Organisationsprobleme auf, da dem Großherzog daran gelegen war, die Herstellung der für das Museum bestimmten Modelle nicht zu verzögern. Darum wurde die Durchführung dieser Aufträge an die Bedingung geknüpft, daß die Plastiker und Anatomen – im Fall der Wiener Sammlung Fontana selbst – zwar museumseigene Abdrücke und Utensilien benutzen durften, das Material aber selbst besorgen und mit Hilfe auswärtiger Handwerker verarbeiten mußten.

Fontana hatte den Ehrgeiz, so viele anatomische Wachsmodelle wie nur irgend möglich zu schaffen, um einen Fundus für didaktische

Zwecke einzurichten, der die direkte Exhumierung von Leichen für das Studium der Anatomie entbehrlich machen sollte. Hierfür stattete er seine Sammlungen mit einer Reihe von Tempera-Zeichnungen aus (Abb. S. 10, 11), auf denen die einzelnen Teile dargestellt werden; von einem Nummernkranz um die gezeichnete Figur ziehen sich dünne gestrichelte Linien zu den verschiedenen Organen, Muskeln, Knochen usw., die anhand der Ziffern identifiziert werden können. Die Blätter mit den entsprechenden Erläuterungen finden sich in einer Schublade, die an jeder Vitrine angebracht wurde: So erhielt man zu jedem Präparat einen regelrechten Anatomie-Traktat (Abb. S. 21).

Um 1790 nahm Fontana ein weiteres ambitioniertes Projekt in Angriff: Es ging um eine Serie anatomischer Teile aus bemaltem Holz, die – im Orignalmaßstab oder in Vergrößerungen – zu didaktischen Zwecken auseinandergenommen werden konnten, um die Beziehungen zwischen den Organen veranschaulichen zu können. Nach einigen Versuchen, von denen nur einige wenige Stücke erhalten sind (Abb. 26), darunter zwei ganze Plastiken (eine im Museum, die andere in Paris) und eine enorme Büste, mußte Fontana das Vorhaben aufgeben, weil das Holz sich nicht nur schwerer bearbeiten läßt, sondern darüber hinaus als organisches Material nicht formbeständig ist, wodurch mit der Zeit die einzelnen Teile ihre Passung zueinander verlieren. Die einzige Wachsplastik des Museums, die die gewünschten Merkmale aufweist, ist die sogenannte „Venere medicea": eine liegende weibliche Figur, von der mehrere Schichten entfernt werden können, bis der Uterus mit einem kleinen Fötus zum Vorschein kommt (Abb. S. 39).

Das Museum hatte mittlerweile seine Mannschaft um einen Zeichner (Claudio Valvani) und einen Holzschnitzer (Luigi Gelati) erweitert und beschäftigte diverse Anatomen für die Sezierung der Leichname. Zu den berühmtesten zählte Paolo Mascagni (1755–1815), der wunderschöne anatomische Tafeln hinterlassen hat und sich auf das Studium des lymphatischen Systems spezialisiert hatte: Nicht zufällig stellen viele liegende Statuen, aber auch kleinere Präparate diesen Apparat des Organismus minutiös dar.

Auch nach dem Tod Felice Fontanas 1805 (Abb. S. 17) setzte die zeroplastische Werkstatt ihre Arbeit fort, und nach dem Tod seines Nachfolgers Clemente Susinis 1814 übenahmen andere Wachsbildner wie die beiden Calenzuoli (Vater Francesco und Sohn Carlo) oder Luigi Calamai (1800–1851) den Betrieb. Unter Calamai konzentrierte sich die

Aktivität auf Modelle der vergleichenden Anatomie und der Botanik sowie auf jene der pathologischen Anatomie, die für das Krankenhaus von Santa Maria Nuova bestimmt waren und heute im Museum für pathologische Anatomie der Universität zu sehen sind. Auf Calamais Stelle rückte nach dessen Tod Egisto Tortori (1829–1893) nach, der neben Tierpräparaten und solchen der anatomischen Pathologie auch eine Büste Clemente Susinis modellierte. Nach Tortoris Tod 1893 löste sich die Werkstatt endgültig auf.

Von den Wachsfiguren, die das Museum La Specola gegenwärtig in Urnen verwahrt, befassen sich 513 mit menschlicher und 65 mit vergleichender Anatomie; 5 Exponate stammen von G. G. Zumbo. Es gibt 26 Figuren (einschließlich eines halbfertigen Jünglings): 13 stehende und 13 liegende Exemplare, davon 18 in Originalgröße (6 stehende und 12 liegende), während acht Figuren etwa 60 cm messen (7 stehende und 1 liegende). Mit Ausnahme von 14 Zeroplastiken zur menschlichen Anatomie, die sich im Magazin befinden, sind alle öffentlich ausgestellt. Hinzu kommen mehr als 800 gerahmte Zeichnungen und annähernd 900 erläuternde Blätter, doch es müssen mehr gewesen sein, da im Laufe der Zeit einiges verlorengegangen ist. In den letzten Jahren wurden die Zeichnungen und die Blätter mit den Erläuterungen entfernt, um sie vor weiteren, durch Licht und Feuchtigkeit bedingte Schäden zu bewahren, und durch Farbkopien ersetzt, so daß die Struktur der Sammlung unter didaktischen Gesichtspunkten wie auch unter solchen einer ausstellungsgerechten Präsentation erhalten blieb (Abb. S. 18).

Während Experten der Materie vor allen Dingen die wissenschaftliche Perfektion der Exponate

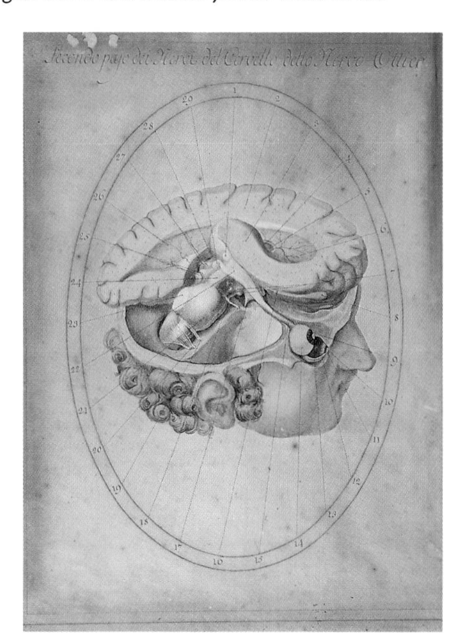

Drawing of a head: lateral view, 2nd half of the 18th century (Room XXVIII, display cabinet no. 707)

Zeichnung eines Kopfes: Seitenansicht, 2. Hälfte des 18. Jahrhunderts (Saal XXVIII, Vitrine Nr. 707)

Dessin d'une tête vue latérale, 2e moitié du XVIIIe siècle (salle XXVIII, vitrine n° 707)

beeindrucken werden, steht für die meisten Besucher in erster Linie die künstlerische Leistung im Vordergrund, denn bei aller wissenschaftlichen Strenge und didaktischem Anspruch sind diese Exponate regelrechte Kunstwerke. Ihre ästhetischen Qualitäten unterstreichen überdies die kunstvoll gearbeiteten Holzvitrinen, in denen sie aufbewahrt werden, auch wenn viele zeitbedingte Abnutzungserscheinungen aufweisen und dringend restauriert werden müßten. Welche Sorgfalt bei der Präsentation dieser Sammlung aufgewendet wurde, bezeugen auch die mittlerweile verblaßten, ursprünglich hellgrünen seidenen Draperien und Schleier aus dem 18. Jahrhundert, die sich ebenfalls in einem äußerst prekären Zustand befinden. Welche Aufmerksamkeit der Ästhetik gewidmet wurde, belegen die Sammlungen der vergleichenden Anatomie, der Humanpathologie und v. a. der Botanik, wo für die kleinen Pflanzenmodelle eigens Keramikvasen aus der Werkstatt Ginori von Sesto Fiorentino angefertigt wurden.

Die technischen Verfahren zur Herstellung der Wachsfiguren können nur aus verschiedenen Briefen und Dokumenten aus dem Staatsarchiv und dem des Museums für Wissenschaftsgeschichte in Florenz erschlossen werden. Außerdem hatte jeder Wachsbildner seine eigene Technik, die er – wie alle Handwerker und Künstler – geheimhielt. Von der in der anatomischen Sektion vorbereiteten Vorlage wurde zuerst eine exakte Kopie aus Kreide oder minderwertigem Wachs hergestellt, von der ein Gipsabdruck genommen wurde, der, wenn er sehr groß war, aus verschiedenen Einsatzstücken bestand. Diese Abdrücke, die in einem Depot des Museums aufbewahrt werden, stellten eine Art Matrix dar, die mehrfach für die Reproduktion desselben Modells verwendet werden konnte. Der schwierigste und heikelste Teil war die Konstruktion des definitiven Modells, die große Präzision, Kenntnis der Substanzen, die dem Wachs beigemischt werden mußten, um die gewünschte Farbe und Konsistenz zu erhalten, sowie große Erfahrung und Geschick verlangten. Das Wachs mußte bei der richtigen Temperatur langsam im Wasserbad schmelzen, damit es sich nicht verfärbte. Am häufigsten wurden weißes Wachs aus Smyrna, chinesisches Wachs sowie venezianisches Wachs verwendet. Um das Wachs formbarer zu machen, fügte man Terpentin hinzu, ferner – ebenfalls mit Terpentin verdünnte – Farbstoffe und andere Substanzen, die in den Einkaufslisten als „für die Wachsarbeiten benötigte Sachen“ nachgewiesen sind. Damit sich das Modell später leicht aus der Form lösen ließ, wurde die Gußform mit

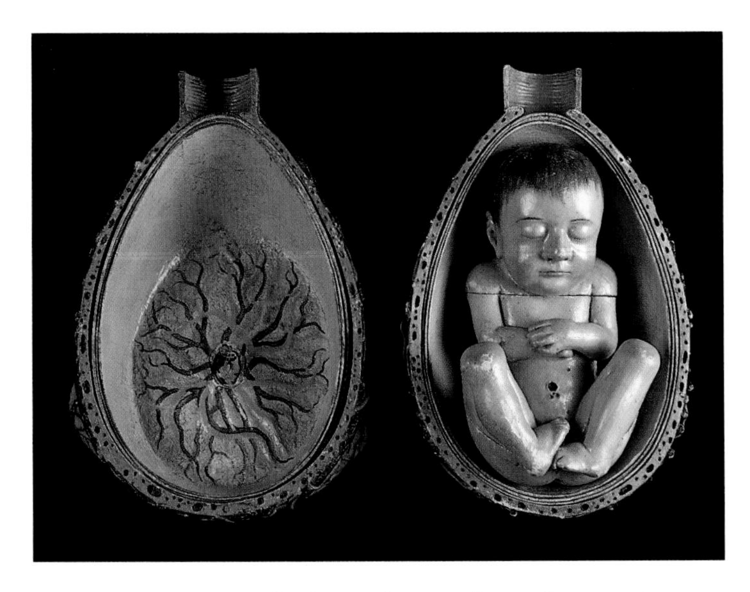

Uterus with foetus; wood, made of individual sections which can be taken apart

Uterus mit Fötus; Holz, in Einzelteile zerlegbar

Utérus et fœtus ; bois, démontable

lauwarmem Wasser angefeuchtet und mit weicher Seife eingerieben, bevor man das Wachsgemisch in den Abdruck goß. Abgesehen von einigen Vollwachspräparaten waren die meisten innen hohl und wurden zur Stabilisierung mit Lappen, Werg oder Holzstücken ausgestopft. Figuren, die aus mehreren Teilstücken bestehen, wurden in der Regel durch eine Stützarmatur aus Metall verstärkt. Nach Entfernung des Abdrucks wurde jede Komponente gesäubert und mit entsprechenden Geräten ausgekehlt sowie zur Vervollständigung mit Organen, Gefäßen, Nervensträngen usw. versehen. Eine letzte Schicht durchsichtigen Lacks verlieh dem Ganzen den nötigen Glanz. Um die Originaltreue des Modells sicherzustellen, mußten all diese Arbeitsgänge ständig von den Anatomen überwacht werden, denen auch die Entscheidung über die jeweils günstigste Position zur Hervorhebung bestimmter Organe oblag.

Jeder Wachsbildner verfügte über eine eigene Technik, die mit der Zeit immer mehr verfeinert wurde, u. a. dank neuer Instrumente wie z. B. einem „Zieheisen zum Drehen der Wachszylinder", das ab 1786 den Plastikern zur Verfügung stand. Früher mußte dieser Arbeitsgang, der für die Gestaltung der Blutgefäße unumgänglich war, von Hand bewältigt werden. Um 1790 führten die Wachsbildner einige Jahre lang „Hefte" (Abb. S. 24), in denen sie täglich die ausgeführten Arbeiten eintrugen, und aus denen hervorgeht, daß bereits damals des häufigeren Reparaturen, Verbesserungen und Überarbeitungen anfielen. 1793 legte Susini eine „Notiz [an] über zu korrigierende anatomische Wachspräparate aufgrund im folgenden beschriebener, von mir, Clemente Susini, Modellierer des Königlichen Museums, erkannter Fehler, die es verdienen, behoben zu werden." Zum Beispiel: „Überprüfung des Präparats der Schlundader und der Halsschlagader: Man sieht von der Luftröhre zwei fremde Muskeln abgehen, das heißt solche, die nicht zum menschlichen Körper gehören, und einen, der am Schilddrüsenfortsatz ansetzt, der aber weder der Zungen- noch der Schlund- und schon gar nicht der Schilddrüsengriffel zu sein scheint." Susinis Notiz endet mit der Feststellung: „Diese Fehler sind zu gravierend, um nicht berichtigt zu werden und von den vielen anderen schweige ich lieber, um mich kurz zu fassen." Fast alle Modelle sind gut erhalten, lediglich bei einigen sind Verfärbungen aufgetreten: mancher Fötus ist nachgedunkelt, und von den heute grün schimmernden Venen wissen wir, daß sie einst „blauviolett" waren.

Die Wachsfiguren des Gaetano Giulio Zumbo

Ein besonderes Kapitel gebührt den Wachsfiguren Gaetano Giulio Zumbos, zum einen wegen ihrer ungewöhnlichen Konzeption und Herstellungsmethode, v. a. aber, weil sie der allgemeinen Entwicklung bei der Verwendung von Wachs für anatomische Darstellungen ihrer Zeit um ein Jahrhundert voraus waren.

Zumbo wurde 1656 in Syrakus als Sohn einer alten Adelsfamilie geboren, deren Nachname vermutlich Zummo war und von der es heute keinerlei Spuren mehr gibt; da die entstellende Version „Zumbo" sich allgemein eingebürgert hat, scheint es angebracht, dabei zu bleiben. Von seinem Leben ist wenig überliefert, und auch das nur bruchstückhaft; man weiß, daß er im Jesuiten-Internat seiner Geburtsstadt erzogen wurde und wegen eines nicht näher benannten „lästigen Unfalls" die Schule aufgeben mußte. Die klassischen Zeugnisse seiner Heimat dürften ihm erste künstlerische Anregungen gegeben haben. Später wurde sein künstlerischer Werdegang vom Manierismus und der Frühphase des Barocks beeinflußt. Von 1687 bis 1691 lebte er in Neapel und schuf vermutlich hier zwei der in der Specola aufbewahrten Gruppenszenen, der „teatrini", nämlich *Il Trionfo del Tempo* (Der Triumph der Zeit) und *La Peste*. Seine Urheberschaft schließen die Gelehrten, die sich mit Zumbos Werk befaßt haben, aus dem malerischen Hintergrund und der Anordnung der Figuren, die deutlich auf süditalienische Vorbilder verweist.

Von 1691 bis 1694 arbeitete Zumbo für Cosimo III. de' Medici in Florenz, wo er zwei weitere „teatrini" mit Bezügen zur Renaissancekunst fertigstellte: *Il Sepolcro* (oder *La Vanità della Gloria Umana*) und *Il Morbo Gallico* (oder *Sifilide*). 1695 unternahm er einige Reisen nach Bologna, wo er mit großem Interesse die gerade im Aufschwung begriffenen anatomischen Studien verfolgte. Ende 1695 zog er nach Genua, wo er vermutlich die im Museum ausgestellte *Anatomie des Kopfes* vollendete. Hier lernte er den französischen Chirurgen Guillaume Desnoues kennen; auf ihre Zusammenarbeit gehen diverse anatomische Wachsmodelle zurück, darunter die lebensgroße Figur einer Gebärenden, eine bei der Geburt gestorbene Frau in kleinerem Format sowie Geburt und Grablegung Christi – Werke, die leider nicht mehr erhalten sind. 1700 entzweite sich Zumbo mit Desnoues und siedelte zuerst nach Marseille über, wo er einen weiteren Wachskopf anfertigte, und schließlich nach Paris, wo er in die Dienste Ludwigs XIV. trat, bis er 1701 an einer Gehirnblutung verstarb. Unter seinen Sachen fand man einen wunderschönen Wachskopf, der höchstwahrscheinlich mit demjenigen identisch ist, der im Museum National d'Historie Naturelle aufbewahrt wird. Sein Grab, das sich in der Kirche von Saint Sulpice befand, wurde während der Revolution zerstört.

Ein Großteil der Werke dieses bis noch vor einem halben Jahrhundert weitgehend verkannten Künstlers befindet sich im Museum La Specola. Die drei „teatrini", die auch als *Cera della peste* („Pestfiguren") bekannt sind, und die *Anatomie des Kopfes* befanden sich, als die Medici von den Lothringern abgelöst wurden, in der Königlichen Galerie (also in den Uffizien) und wurden anschließend dem Museo di Fisica e Storia Naturale überantwortet, dort aber, vielleicht wegen der Drastik der dargestellten Szenen, nie ausgestellt. Als Peter Leopold 1790 nach Wien aufbrach, schenkte er die „Pestfiguren" dem Hofarzt Giovanni Giorgio Asenöhrl, besser bekannt als Lagusius. Dieser beauftragte erst Fontana und dann Agostino Renzi, Superintendent der Königlichen Apotheke, sie für 150 Zechinen zu verkaufen. Letzterer war sich der Bedeutung der Ensembles durchaus bewußt und bot sie dem neuen Großherzog Ferdinand III. an, erst für die Königliche Akademie der Schönen Künste und daraufhin, da diesem die Stücke weder „nützlich noch angemessen" erschienen, für das Königliche Museum. Dieser Vorschlag wurde vom Vizedirektor des Museums Giovanni Fabbroni befürwortet, und die drei Werke wurden von verschiedenen Experten, darunter Clemente Susini, geschätzt. Da der Schätzwert höher war als der von Lagusius geforderte

View of Room XXVII

Gesamtansicht des Saals XXVII

Vue d'ensemble de la salle XXVII

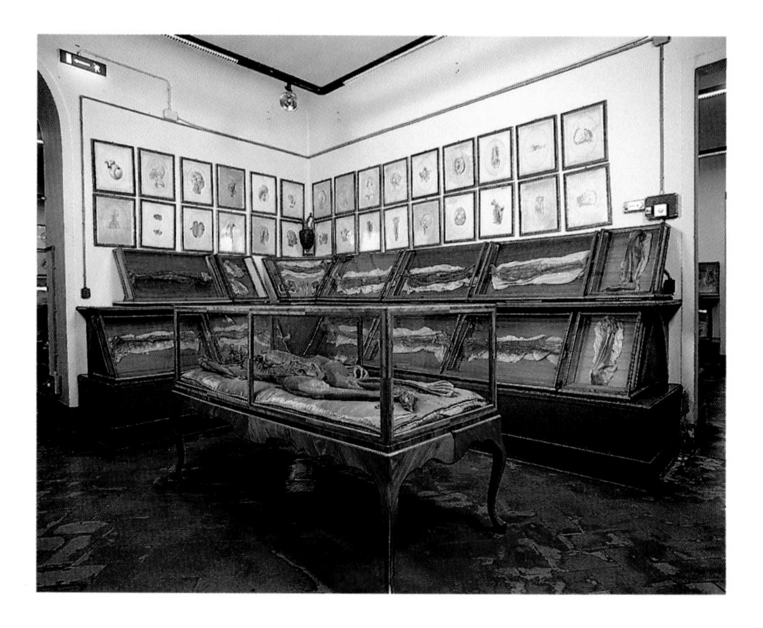

Preis, autorisierte der Großherzog den Ankauf, und so blieben Zumbos Werke im Museum, bis die drei Theaterszenen 1878 dem Museum Nazionale in Bargello zur Aufbewahrung übergeben wurden; von diesem kamen sie zum Museo di Storia della Scienza, um 1974 schließlich wieder zur Specola zurückzukehren. Archivdokumente belegen, daß die Wachsfiguren verschiedentlich restauriert wurden, u. a. von Susini und Tortori. Die Pestdarstellungen, die sich im Museo di Storia della Scienza befanden, wurden während der sintflutartigen Regenfälle im Jahre 1966 schwer beschädigt. In über 18monatiger Kleinarbeit gelang es Guglielmo Galli aus der Werkstatt der Pietre Dure in Florenz, sie nach vorhandenen Fotografien wiederherzustellen. Die *Anatomie des Kopfes* hingegen hat die Specola nie verlassen, da sie zu Recht als einschlägig für die wissenschaftliche Sammlung des Instituts betrachtet wurde.

Die von Zumbo angewandte Technik unterscheidet sich stark von jener der späteren Florentiner Schule. Die *Anatomie des Kopfes* wurde an einem echten Schädel ausgeführt, der nach dem Befund der Röntgenaufnahmen einem etwa 25jährigen Mann gehört haben muß. Der in Paris befindliche Kopf und ein weiteres, verschollenes Exemplar sind hingegen ganz aus Wachs modelliert, was auf eine Fortentwicklung seiner Verarbeitungstechnik schließen läßt, die möglicherweise auf die

Zusammenarbeit Zumbos mit Desnoues zurückgeht. Die Figuren der „teatrini", deren Gestaltung bis in die kleinsten Details von großer Kunstfertigkeit zeugt, wurden mit an Sicherheit grenzender Wahrscheinlichkeit von Gipsabdrücken hergestellt, denen ihrerseits sorgfältig geformte Tonmodelle zugrunde lagen. Das mehr oder weniger flüssige Wachs (ein Gemisch aus Bienenwachs, Kolophonium, Terpentin und Farbstoffen) wurde in dünnen Schichten aufgetragen, um die jeweils gewünschte Konsistenz und Farbigkeit zu erhalten. Die wenigen überlieferten Notizen über Zumbos Technik verdanken wir den von Guglielmo Galli nach den Wasserschäden von 1966 durchgeführten Restaurierungen, die zudem auch Spuren älterer Reparaturen anhand des Nachweises von Farbstoffen, die es Ende des 17. Jahrhunderts noch nicht gab, zutage förderten. Auch das kleine Selbstporträt (Abb. S. 55) aus dem Ensemble *Il Trionfo del Tempo* kann nicht mit Bestimmtheit Zumbo zugeschrieben werden, es könnte das Werk eines Zeitgenossen oder des späten Susini sein. Der *Morbo Gallico*, von der Anlage her eine vergleichbare Komposition, von der nur noch wenige Bruchstücke existieren, war ein Geschenk Cosimos III. an Filippo Corsini, das in den Kellern des Palazzo Corsini am Arno aufbewahrt wurde und von der Überschwemmung praktisch vollständig zerstört wurde; man fand ein paar Figuren im Garten wieder, aber da es keine Fotografien des Ensembles gab, war eine Rekonstruktion nicht möglich. Anläßlich des 200jährigen Jubiläums der Museumseröffnung, das mit einem Kongreß über Wachsplastik zelebriert wurde, übergab die Familie Corsini die Reste dieses Werks dem Museum.

Lange Zeit hindurch wurden Zumbos Meisterwerke als krankhafte Kaprizen eines sich an der Darstellung makabrer und abscheulicher Details ergötzenden Künstlers erachtet. Erst Mitte dieses Jahrhunderts ging man dazu über, sie in ihrer historischen Perspektive wahrzunehmen: als realistische Dokumente einer Epoche, in der der Tod mit Kriegen, Hungersnöten und großen Epidemien allgegenwärtig war. Diese Vision der Zerstörung, der Hinfälligkeit menschlichen Lebens, des unerbittlichen Vergehens der Zeit, dieses beständige Memento mori, mit dem uns Zumbos Werke in einer Fülle von Details konfrontieren, ist ein typisches Beispiel für die Kultur des 17. Jahrhunderts, wie sie uns auch von anderen großen Künstlern jener Zeit, etwa von Luca Giordano oder Mattia Preti, überliefert wird.

Pages · Seiten · Pages 44–45:

G. G. Zumbo: *Il Sepolcro*

The burial

Das Begräbnis

L'Enterrement

La collection de figures de cire du musée La Specola à Florence

Marta Poggesi

Le musée et son histoire

Lorsqu'en 1771 Pierre Léopold de Habsbourg-Lorraine (1747–1792), grand-duc de Toscane de 1765 à 1790 (ill. p. 7), décida de réunir en un seul musée toutes les collections « scientifiques » des galeries grand-ducales, il entreprenait quelque chose d'absolument inédit en Europe et dans le monde entier. Homme éclairé, bien en avance sur les autres souverains de son époque, le prince était passionné de sciences naturelles, conscient de leur importance pour le développement culturel de la société. C'est lui qui songea le premier à rendre accessibles à tous les découvertes réalisées dans ce domaine.

Appelé plus tard « La Specola », observatoire en italien, le « Imperial Regio Museo di Fisica e Storia Naturale » (Musée impérial royal de physique et d'histoire naturelle) ouvrit ses portes le 21 février 1775, au grand public, ce qui le rendit unique en son genre dans le monde entier. Certes, les heures d'ouverture étaient différentes pour les hommes cultivés et le commun des mortels : ces derniers dont on exigeait une « tenue propre » ne pouvaient visiter le musée que de huit heures à dix heures du matin afin de laisser suffisamment de temps entre leur passage et l'arrivée « des personnes intelligentes et savantes [...] à une heure de l'après-midi ». Mais même si cette distinction nous paraît discriminatoire de nos jours, on peut se figurer que l'accès des collections au grand public représentait jadis une incroyable innovation.

La création de ces collections remontait aux Médicis, grands mécènes et amateurs d'art, qui encouragèrent également les sciences avec la fondation, par exemple, de l'Accademia del Cimento (1657–1667) où enseignaient, à l'époque de Ferdinand II de Médicis, d'illustres savants comme Redi, Magalotti et Viviani, l'élève favori de Galilée. Après la mort de Jean Gaston, le dernier rejeton de la famille en 1737, le grand-duché de Toscane passa, conformément au traité de Vienne, aux mains du futur empereur François Ier de Habsbourg-Lorraine (1745–1765) qui fit dresser la liste des trésors de sa nouvelle résidence. En un peu moins d'un an, de 1763 à 1764, le médecin et naturaliste Giovanni Targioni-Tozzetti (1712–1783) parvint à faire l'inventaire des collections.

Lorsqu'en 1765 Pierre Léopold succéda à son père devenu empereur d'Autriche, il s'aperçut qu'en raison du travail déjà effectué il pouvait aisément entreprendre la réorganisation des collections scientifiques. Cette tâche devait revenir à l'abbé Felice Fontana (1730–1805), originaire de Rovereto (ill. p. 29). Professeur de logique à l'université de Pise, il était également anatomiste, physicien, chimiste et surtout physiologiste de réputation internationale. L'abbé se lança dans cette entreprise avec tant d'ardeur qu'à la fin de 1771, on pouvait déjà transporter la première partie des collections dans les nouvelles salles. La même année, le grand-duc Pierre Léopold avait fait l'acquisition du palais des Torrigiani (qui appartenait auparavant à la famille Bini), situé dans la via Romana à deux pas du palais Pitti, ainsi que de quelques maisons avoisinantes. Il chargea ensuite Fontana de transformer les bâtiments afin qu'ils puissent abriter les collections scientifiques. Nommé par ailleurs directeur du nouveau musée, l'abbé sillonna l'Europe durant les premières années pour acheter des livres et des collections et nouer des contacts avec les savants d'autres pays. C'est ainsi que le musée florentin devint l'un des plus importants de son époque d'autant plus qu'il possédait une vaste bibliothèque scientifique.

Egisto Tortori : Coloured plaster relief of Clemente Susini

Gipsbüste von Clemente Susini, farbig gefaßt

Buste en plâtre polychrome de Clemente Susini

Fontana dirigea le musée jusqu'à sa mort en 1805. Il fut secondé dans son travail par Giovanni Fabbroni qui l'accompagna dans ses nombreux voyages et devint son adversaire permanent en accédant au poste de directeur adjoint en 1784.

L'argent pour la réalisation du Musée de physique et d'histoire naturelle (et de beaucoup d'autres projets de Pierre Léopold) fut obtenu en vendant des objets précieux appartenant aux Médicis et ce, malgré le testament de la princesse palatine Anna Maria Luisa (1668–1743), la sœur de Jean Gaston, qui avait légué d'une façon irrévocable toute la fortune des Médicis à la Ville de Florence.

Les fonds du musée, provenant essentiellement des Offices, comprenaient des collections diverses (minéraux, coquillages ...), des curiosités de sciences naturelles de l'époque des Médicis, les instruments de Galilée, les appareils de l'Accademia del Cimento et quatre figures de cire du sculpteur Giulio Gaetano Zumbo.

Voulant élargir le musée aux domaines de l'astronomie et de la météorologie, Pierre Léopold demanda en 1780 à l'architecte Gaspare

Paoletti (qui avait déjà participé à la transformation du palais) de construire l'Osservatorio Astronomico (la « petite tour ») qui devait donner plus tard à l'ensemble du complexe le nom de « Specola » (observatoire). Il s'agissait d'un projet ambitieux qui agrandissait considérablement le bâtiment et fut achevé en 1789, malgré les avis divergents de nombreux experts, lesquels auraient préféré bâtir une nouvelle construction sur les collines de l'Acetri. La même année, une partie du parc Boboli fut intégrée au jardin botanique du musée (ill. p. 8).

A la mort de son frère Joseph, Pierre Léopold lui succéda à la tête de l'Empire en 1790 et céda la Toscane à son fils cadet Ferdinand III (1769–1824) qui était bien loin de posséder la largeur d'esprit de son père et son habileté à gouverner. Pour couronner le tout, la Toscane fut prise par les Français en 1799. Après maintes complications, elle devint en 1801 le royaume d'Etrurie et fut remise aux Bourbon-Parme. Durant ces années, le musée organisa des cours dans les disciplines scientifiques, qui se poursuivirent après la restauration des Lorrains en 1814.

Le musée vécut une période mouvementée sous Ferdinand III. Il perdit son importance scientifique qui l'avait caractérisé à ses débuts, époque où il était considéré comme l'un des plus grands centres du savoir européen, ce qu'attestaient d'ailleurs des visiteurs étrangers aussi illustres que Goethe et Bernouilli. A l'instar de son père, Ferdinand III entreprit des transformations importantes dans le palais : en 1820, un

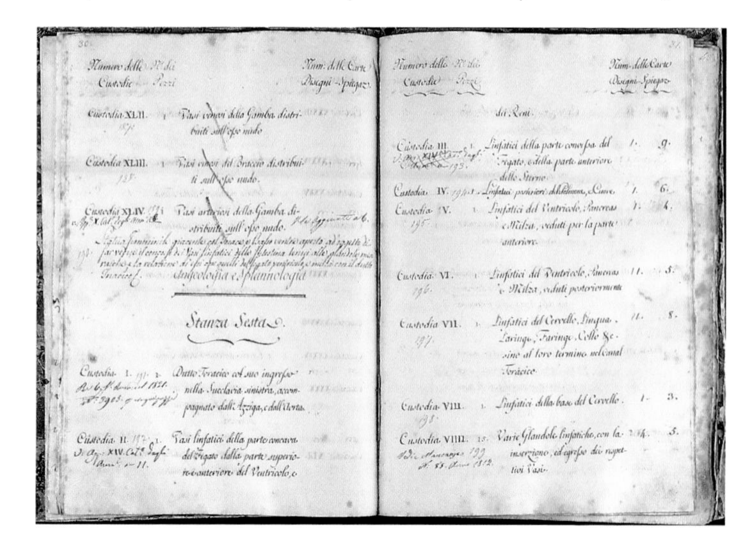
Numero delle Custodie – N. di Pezzi – Num. delle Carte Disegni – Spiegaz.
Custodia XLII. 1 Vasi venosi della Gamba distribuiti sull'osso nudo
Custodia XLIII. 1 Vasi venosi del Braccio distribuiti sull'osso nudo.
Custodia XLIV. 1 Vasi arteriosi della Gamba distribuiti sull'osso nudo.
Angeologia e Splannologia
Stanza Sesta.
Custodia I. 1 Dutto Toracico col suo ingresso nella Succlavia sinistra, accompagnato dall'Azziga, e dall'Aorta.
Custodia II. 1 Vasi linfatici della parte concava del Fegato dalla parte superiore e anteriore del Ventricolo, e
Numero delle Custodie – N. di Pezzi – Num. delle Carte Disegni – Spiegaz.
dei Reni.
Custodia III. 1 Linfatici della parte convessa del Fegato, e della parte anteriore dello Sterno 1. 9.
Custodia IV. 1 Linfatici posteriori del Polmone, e Cuore 1. 6.
Custodia V. 1 Linfatici del Ventricolo, Pancreas e Milza, veduti per la parte anteriore. 1. 4.
Custodia VI. 1 Linfatici del Ventricolo, Pancreas e Milza, veduti posteriormente 11. 5.
Custodia VII. 1 Linfatici del Cervello, Lingua, Laringe, Faringe, Collo &c. sino al loro termine nel Canal Toracico. 11. 8.
Custodia VIII. 1 Linfatici della base del Cervello. 1. 3.
Custodia VIIII. 15 Varie Glandole linfatiche, con la inserzione, ed egresso dei respettivi Vasi. 5.

1793 catalogue of the wax models

Katalog der Wachspräparate von 1793

Catalogue des préparations en cire de 1793

corridor reliant la Specola à l'aile de la Méridienne, un pavillon du palais Pitti, fut construit sous la direction de l'architecte Pasquale Poccianti. Ce nouveau passage prolongeait jusqu'à la Specola le corridor Vasari existant qui conduisait du palais Vecchio au palais Pitti en passant par les Offices.

A la mort de Ferdinand III en 1824, le grand-duché passa aux mains de son fils Léopold II (1797–1870), surnommé gentiment « Canapone » par les Florentins en raison de ses cheveux blonds. C'est à lui que revient le mérite d'avoir donné un nouvel élan aux études scientifiques, en particulier à celles que l'on pouvait appliquer concrètement comme dans l'agriculture par exemple. C'est également sous sa régence que fut construite la Tribuna di Galileo, hommage au grand chercheur, qui fut inaugurée à Florence, en 1841, à l'occasion du troisième congrès des scientifiques italiens. La Tribuna est une grande salle au premier étage du bâtiment en partie reconstruit par l'architecte Giuseppe Marelli. Les travaux commencés en 1830 ne prévoyaient à l'origine qu'une abside supplémentaire dans une salle déjà existante, mais les plans furent modifiés à la demande du grand-duc. Voulant honorer avec grandeur la mémoire de l'illustre savant, il décida de lui consacrer la salle entière qui abriterait, outre une statue de Galilée, tous les objets que l'on avait conservés de lui ainsi que ses instruments se trouvant à l'Accademia del Cimento. La construction de la Tribuna, l'un des rares témoignages du néoclassicis-

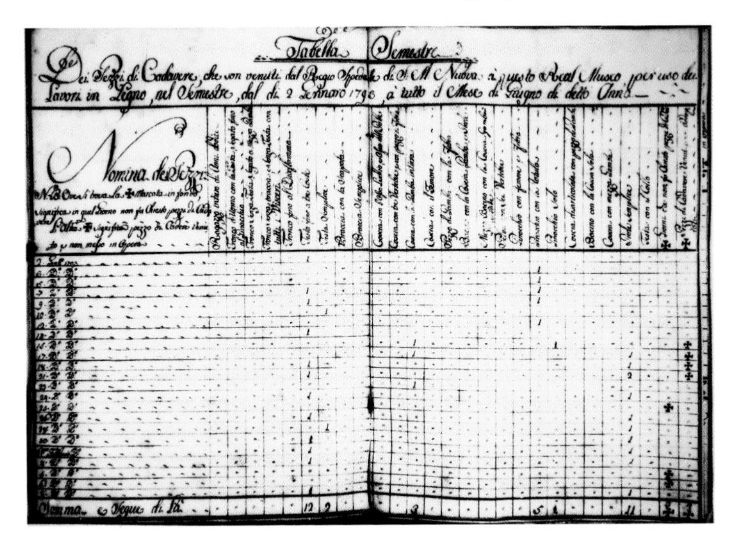

A page of the register recording receipts for corpses from the hospital

Registerseite mit Vermerk der Eingänge der Leichname aus dem Krankenhaus

Page du registre avec les indications des arrivées des cadavres provenant des hospices

me à Florence, entraîna d'importantes modifications dans l'architecture du palais : les étages inférieurs furent transformés et la cour fut en partie couverte. On fit venir des peintres et des sculpteurs de la région pour décorer la salle recouverte entièrement de marbre de Toscane.

L'Istituto di Studi Superiori e di Perfezionamento (institut d'études supérieures et de perfectionnement), transformé en 1923 en université royale, fut fondé en 1859 après la chute de Léopold II, le dernier grand-duc de Toscane. Evénement déjà important en soi, la fondation de l'Institut a marqué par ailleurs pour le musée le début de son démembrement en différents départements qui, en raison du nombre toujours croissant des étudiants, durent s'installer en d'autres lieux avec leurs collections et leurs bibliothèques. Seul le département de physique et de sciences naturelles du musée demeura au siège de l'Institut. Malgré l'opposition de nombreux scientifiques de l'époque, le bâtiment historique de la via Romana n'abrita plus finalement que les collections zoologiques et la plus grande partie de la céroplastie anatomique.

Les débuts de la céroplastie dans l'anatomie

L'utilisation de la cire pour modeler des figures remonte à l'époque des Romains. De tout temps, la cire fut un matériau très apprécié, dans l'art en particulier, pour des raisons esthétiques et techniques. La cire se laisse aisément travailler, elle est idéale comme moule pour couler le bronze par exemple. C'est surtout dans l'Europe du Moyen Age que l'on fabriqua en cire un grand nombre d'offrandes, de statues de saints, de crèches et d'ex-voto. A la Renaissance et durant tout le XVIIe siècle, on pouvait admirer d'innombrables modèles en cire dans les églises italiennes, en particulier dans l'Orsammichele et la Santissima Annunziata. Ces modèles étaient des membres et des organes humains, mais aussi des portraits et des statues bien souvent grandeur nature. Au XVIIe siècle apparurent ensuite des modèles anatomiques reproduits pour les besoins de la science.

Bien que l'Etat et encore plus l'Eglise eussent interdit et puni l'étude du corps humain sur les cadavres, les premiers dessins anatomiques firent leur apparition au milieu du XVe siècle. Ils furent réalisés en général par des grands peintres et sculpteurs comme Léonard de Vinci, Michel-Ange, Raphaël et Titien, pour ne citer que les plus célèbres. D'illustres chercheurs, tels Falloppio, Cesalpino ou Vesalius, utilisèrent les dessins de ces artistes pour illustrer leurs traités d'anatomie.

A la fin du XVIIe siècle, Gaetano Giulio Zumbo fut le premier à se servir dans la célèbre école d'anatomie de Bologne de cire colorée pour modeler des préparations en cire. Originaire de Syracuse, l'artiste dont les œuvres sont reproduites plus loin, réalisa deux têtes d'homme (l'une conservée à Florence, l'autre à Paris), une figure féminine (malheureusement disparue) et quelques petites compositions. C'est à Bologne également que fut fondée la première école de céroplastie dont sortirent, durant la seconde moitié du XVIIIe siècle, les sculpteurs du musée florentin. Considéré comme le fondateur de l'école de Bologne, le peintre, sculpteur et architecte Ercole Lelli (1702–1766) modela, comme Zumbo avant lui, des muscles et des viscères en cire sur de vrais squelettes. Amateur de sciences naturelles, le cardinal Prospero Lambertini, élu pape en 1740 sous le nom de Benoît XIV, encouragea les études anatomiques de l'école de Bologne et acheta tous les modèles de Lelli, de magnifiques exemplaires que l'on peut admirer aujourd'hui encore à l'Istituto di Anatomia Umana Normale de l'université de Bologne. Enfin, et ceci est d'une grande importance, il apporta son soutien aux chercheurs devant se procurer des cadavres pour leurs travaux. Parmi les assistants de Lelli, signalons Giovanni Manzolini (1700–1755) qui, avec l'aide de sa femme Anna Morandi (1716–1774), réalisa de nombreux modèles conservés également à l'université de Bologne.

A cette époque, l'étude de l'anatomie était en plein essor dans toute l'Europe. L'école céroplastique de Florence fut fondée sur le modèle de l'école de Bologne à l'instigation de Giuseppe Galletti. Chirurgien et accoucheur, il s'inspira des œuvres de Lelli et de Manzolini pour créer avec le sculpteur Giuseppe Ferrini toute une série de modèles en cire et en terre cuite (elles se trovent actuellement au Museo di Storia della Scienza de Florence) devant illustrer différents types d'accouchements, normaux et dystociques.

Les figures en cire de la Specola : l'école, les œuvres, les techniques

L'atelier de céroplastie à Florence fut actif pendant un siècle environ, de 1771 à la seconde moitié du XIXe siècle, mais la majeure partie des objets en cire conservés au musée, de même que les travaux réalisés pour les institutions florentines et d'autres commanditaires, furent effectués durant les soixante premières années de son existence (ill. p. 42).

La fondation de la Specola remonte à Felice Fontana qui, pleinement soutenu par le grand-duc Pierre Léopold, s'y consacra avec une énergie d'autant plus grande qu'il participait lui-même à la création des modèles en tant qu'anatomiste et pathologiste. Au début, le sculpteur Giuseppe Ferrini travailla seul à l'atelier sous la direction de Fontana. Plus tard, on engagea l'anatomiste Antonio Matteucci ainsi que Clemente Susini, un tout jeune homme à l'époque, qui se révéla si habile et si productif qu'il ne tarda pas à devenir le plus célèbre céroplasticien de l'école florentine (ill. p. 47).

On ignore aujourd'hui où on travaillait la cire, probablement dans les salles du rez-de-chaussée dont les fenêtres donnent sur la Via Romana. Ces salles sont situées dans l'aile sud du palais, qui est pourvue de nombreuses courettes. La plupart des appareils d'origine ont disparu, mais leur acquisition figure encore dans les archives : bassines en cuivre de plusieurs tailles dans lesquelles on faisait fondre la cire ; instruments à modeler (ill. p. 33) ; fils de fer de grosseur différente ; plaques de marbre servant à presser la cire en fines couches ; balances ; trépieds pour chauffer le matériau ; ardoises pour les notes et les dessins réalisés pendant les dissections ; caisses munies de poignées pour le transport des cadavres, caisses en bois munies de tiges pour le transport des figures de cire ; récipients, vases et bouteilles en céramique ou en verre pour les couleurs et autres substances ajoutées à la cire. On retrouva dans les anciens entrepôts du musée un grand nombre de ces récipients avec leur contenu (ill. p. 32).

Les archives nous indiquent par ailleurs qu'il fallait réunir plus de deux cents cadavres ou parties de cadavres pour fabriquer une seule figure. Ce nombre incroyablement élevé s'explique pour la bonne raison que le formol et la réfrigération étant inconnus, il fallait constamment recourir à des cadavres frais pour les dissections. Afin de recenser exactement les cadavres provenant des hospices de Santa Maria Nuova à deux kilomètres de là, on tenait un registre dans lequel étaient notées toutes les arrivées, mais aussi tous les départs pour le cimetière : « [...] à propos des cadavres de Santa Maria Nuova dont on ignore le nombre et parce qu'on ne peut pas faire mourir les gens pour les besoins des collections de cire, il a été décidé de placer un registre à la porte du musée afin de consigner les cadavres utilisés en ce lieu, dont le nombre semble considérable, et afin de noter la fréquence des envois au cimetière. » (extrait d'un document de 1793).

Malgré le petit nombre de préparateurs, les travaux allaient bon train si l'on considère qu'aux environ de 1790, soit vingt ans après leur commencement, les modèles remplissaient déjà huit salles, sans compter ceux que l'on avait fabriqués pour les hospices de Florence et d'autres institutions en Italie et à l'étranger (cf. ill. p. 48). De toutes les sculptures en cire qui sortirent de l'atelier de la Specola pour être envoyées aux quatre coins de l'Europe, les plus remarquables sont celles effectuées pour l'école militaire de médecine à Vienne. Commandée en 1781 par l'empereur d'Autriche Joseph II, le frère aîné de Pierre Léopold, la collection fut achevée en 1786. Elle comprenait 1200 pièces qui furent acheminées à dos d'âne jusqu'à Vienne et nécessitèrent deux voyages. D'autres sculptures prirent la route de Pavie, de Cagliari, de Bologne, de Budapest, de Paris (elles se trouvent aujourd'hui à Montpellier), d'Uppsala, de Londres, de Leyde et de bien d'autres villes encore. Ces commandes de l'extérieur soulevaient parfois des problèmes d'organisation car le grand-duc ne tolérait aucun retard dans la fabrication de certains modèles pour le musée. On décida donc que pour ces commandes, les sculpteurs et les anatomistes – Fontana lui-même dans le cas de la collection viennoise – pouvaient certes utiliser les moulages et les instruments du musée, mais qu'ils devaient se procurer eux-mêmes le matériel et se faire assister d'artisans étrangers au musée.

Si Fontana tenait absolument à créer le plus grand nombre possible de modèles anatomiques en cire, c'est parce qu'il voulait constituer une réserve qui rendrait superflue l'exhumation de cadavres pour les cours d'anatomie. C'est également dans ce but qu'il réalisa toute une série de dessins (ill. p. 10, 11) illustrant les différentes parties du corps : les organes, les muscles, les os, etc., étaient reliés par des lignes discontinues à des chiffres qui permettaient leur identification. Les dessins et les explications étaient conservés dans un tiroir du coffre où se trouvait la figure de cire correspondante. Chaque préparation était ainsi accompagnée d'un véritable traité d'anatomie (ill. p. 21).

Vers 1790, Fontana se lança dans un projet encore plus ambitieux qu'il ne put mener à bien à cause des difficultés qu'il rencontra. Il s'agissait d'une série de pièces d'anatomie en bois peint – de taille réelle ou agrandies – que l'on aurait pu démonter afin de montrer les liens entre les différents organes. Après quelques tentatives dont on a gardé seulement un petit nombre d'exemplaires, à savoir deux sculptures entières (une au musée, l'autre à Paris) et un buste gigantesque, Fontana dut abandonner son projet. Non seulement le bois se laissait difficilement travailler, mais étant un matériau organique, il se déformait avec le temps et les différentes pièces ne s'adaptaient plus. La seule sculpture du musée présentant les caractéristiques souhaitées fut fabriquée en cire. Il s'agit de la « Venere medicea », une figure féminine allongée dont on peut retirer les éléments superposés jusqu'à l'apparition de l'utérus et d'un petit fœtus (ill. p. 39).

Le musée avait entre-temps agrandi son équipe. Il avait engagé le dessinateur Claudio Valvani, le sculpteur sur bois Luigi Galati ainsi que plusieurs anatomistes pour la dissection des cadavres. L'un des plus célèbre fut Paolo Mascagni (1755–1815) qui laissa à la postérité de magnifiques planches anatomiques. Il s'était spécialisé dans l'étude du système lymphatique, ce n'est donc pas un hasard si à la Specola cette partie de l'organisme est représentée avec une grande minutie sur de nombreuses statues, mais aussi sur des préparations plus petites.

La mort de Felice Fontana (ill. p. 17) en 1805 n'empêcha pas l'atelier de poursuivre son travail, et à la mort de son successeur Clemente Susini, en 1814, ce furent d'autres sculpteurs, comme les Calenzuoli (le père Francesco et le fils Carlo) ou Luigi Calamai (1800–1851), qui se chargèrent de la direction. Sous Calamai, les activités se concentrèrent sur des modèles d'anatomie comparative et de botanique ainsi que sur les modèles d'anatomie pathologique qui étaient destinés aux hospices de Santa Maria Nuova et que l'on peut voir aujourd'hui au musée d'anatomie pathologique de l'université. Après sa mort, Calamai fut remplacé par Egisto Tortori (1829–1893) qui, à côté de préparations d'animaux et d'anatomie pathologique, modela également un buste de Clemente Susini. A la mort de Tortori, en 1893, l'atelier cessa définitivement ses activités.

De nos jours, le musée La Specola abrite 513 modèles en cire d'anatomie humaine, soixante-cinq d'anatomie comparative et cinq réalisés par G. G. Zumbo. Il possède aussi vingt-six figures (dont un tronc de jeune garçon) : treize debout et treize couchées, dont dix-huit grandeur nature (six debout et douze couchées) et huit mesurant soixante centimètres environ (sept debout et une allongée). A l'exception de quatorze sculptures sur l'anatomie humaine se trouvant dans les entrepôts, toutes sont exposées au public. Il faut également ajouter plus de huit cents dessins encadrés et près de neuf cents feuillets explicatifs dont le nombre était jadis certainement supérieur, mais qui ont été égarés au cours des siècles. Ces dernières années, on a retiré des salles d'exposition les dessins et les feuilles explicatives pour les préserver des dommages causés par la lumière et l'humidité. Ils ont été remplacés par des photocopies en couleur de sorte que la structure de la collection a pu être conservée aussi bien du point de vue de la présentation que du point de vue didactique.

Tandis que les experts en la matière sont impressionnés par la perfection scientifique des objets exposés, la plupart des visiteurs s'intéressent surtout à leur côté artistique, car ils sont malgré toute leur rigueur scientifique et leur vocation didactique, de véritables œuvres d'art. La beauté de ces objets est mise en valeur par les coffres en bois très travaillés, dans lesquels ils sont conservés. Malheureusement beaucoup de ces coffres se sont abîmés avec le temps et auraient grand besoin d'être restaurés. Datant du XVIIe siècle, les draperies et voiles en soie, à l'origine vert clair, et se trouvant eux aussi dans un bien triste état, montrent quel soin fut toujours apporté à la présentation de la collection.

G. G. Zumbo: Self-portrait (detail from *The Triumph of Time*)

Selbstporträt (Ausschnitt aus *Der Triumph der Zeit*)

Autoportrait (détail de *Le Triomphe du Temps*)

Ce souci d'esthétique se constate par ailleurs dans les collections d'anatomie comparative, d'anatomie pathologique et en particulier dans les collections de botanique où des vases en céramique furent réalisés spécialement pour les petits modèles de plantes par l'atelier Gionori de Sesto Fiorentino.

Comme les procédés techniques pour la fabrication des figures en cire ne nous ont pas été transmis de façon explicite, nous ne pouvons les reconstituer qu'à partir des diverses lettres et documents des archives nationales et du Musée des sciences de Florence. Pour compliquer le tout, chaque sculpteur avait sa propre technique qu'il se gardait bien de divulguer. Une chose est sûre cependant : une fois les parties du corps disséquées, on en faisait une copie exacte en craie ou en cire de qualité inférieure dont on tirait un moulage en plâtre qui, s'il s'avérait de grande taille, pouvait être constitué de plusieurs pièces. Pour certains organes, comme les os, on pouvait faire directement un moulage en plâtre (ill. p. 60). Ces moulages conservés aujourd'hui encore dans les entrepôts du musée constituaient une sorte de matrice que l'on pouvait utiliser plusieurs fois pour reproduire le même modèle. L'étape la plus complexe et la plus délicate était la construction du modèle définitif. Elle exigeait une grande précision, une connaissance approfondie des substances qu'il fallait mélanger à la cire pour obtenir la couleur et la consistance souhaitées, ainsi que beaucoup d'expérience et de savoir-faire. La cire devait fondre lentement au bain-marie pour que sa couleur ne soit pas altérée. On utilisait le plus souvent de la cire blanche de Smyrne, de la cire chinoise et de la cire de Venise. Pour la rendre plus malléable, on y ajoutait de la térébenthine ainsi que des colorants, dilués eux aussi dans de la térébenthine, et d'autres substances apparaissant sur les listes d'achat sous la mention « matières nécessaires pour les travaux en cire ». Afin que plus tard, le modèle se détache facilement du moule, celui-ci était humidifié à l'eau tiède et frotté avec un savon doux avant que l'on y coule le mélange de cire. Mis à part quelques préparations complètement en cire, la plupart étaient creuses à l'intérieur et remplies de chiffons, d'étoupe ou de morceaux de bois pour une plus grande stabilité. Les statues constituées de plusieurs pièces assemblées les unes aux autres étaient en général renforcées par une armature métallique. Une fois le moule ôté, la figure était soigneusement nettoyée, puis cannelée avec les outils appropriés et, finalement, dotée de ses organes, vaisseaux, nerfs, etc. Une dernière couche de vernis transparent lui conférait le brillant nécessaire. Afin d'être sûr que le modèle soit conforme à l'original, toutes les étapes étaient constamment surveillées par les anatomistes qui décidaient également de la meilleure position pour mettre en valeur tel ou tel organe.

Comme nous l'avons déjà dit, chaque sculpteur possédait sa propre technique et ses méthodes particulières qu'il perfectionnait avec le temps en utilisant de nouveaux instruments, comme par exemple cette « filière à tirer les cylindres en cire » mise à leur disposition dès 1786. Avant, cette opération permettant de modeler les veines et les artères devait être faite à la main. Autour de 1790, les sculpteurs tinrent pendant quelques années des « cahiers » (ill. p. 43) dans lesquels ils notaient quotidiennement les travaux effectués, et qui nous permettent de constater qu'à l'époque déjà ils devaient entreprendre des restaurations, des réparations, des améliorations et des corrections. En 1793, Susini rédigea une « [...] note sur les préparations anatomiques en cire à corriger en raison des erreurs constatées par moi, Clemente Susini, modeleur du musée royal, erreurs qui méritent d'être réparées. » Par exemple : « Vérification de la préparation de l'artère pharyngée et de la carotide : on peut voir partant de la trachée-artère deux muscles étrangers n'appartenant pas au corps humain et un muscle rattaché au prolongement de la thyroïde, mais qui ne semble être ni le muscle de la langue, ni du pharynx et encore moins de la thyroïde. » La note de Susini se termine avec la constatation suivante : « Ces erreurs sont trop graves pour ne pas être corrigées. Je passe toutes les autres sous silence car je désire être bref. » Presque tous les modèles se trouvent en bon état de conservation, quelques-uns seulement ont changé de couleur : certains fœtus sont devenus foncés et les veines qui présentent aujourd'hui une teinte verte étaient jadis ‹ bleu violet ›.

Les figures en cire de Gaetano Giulio Zumbo

Les sculptures en cire de Gaetano Giulio Zumbo méritent qu'on leur consacre un chapitre à part, non seulement en raison de leur conception et de leur méthode de fabrication inhabituelles, mais surtout parce qu'étant en avance d'un siècle sur les autres représentations anatomiques en cire, elles constituent la base de toutes celles qui suivront.

Né en 1656 à Syracuse, Zumbo est issu d'une famille de la vieille noblesse qui s'appelait probablement Zummo et dont il ne reste plus aucune trace aujourd'hui. Comme le nom de « Zumbo » est plus connu et est passé dans l'usage, il nous semble opportun de le garder. Nous savons peu de choses sur sa vie, quelques fragments tout au plus. Elevé chez les Jésuites dans sa ville natale, il dut quitter le collège à cause d'un « incident ennuyeux » qui n'est pas précisé. Les œuvres classiques qu'il put voir dans sa patrie ont certainement eu au début une influence sur sa formation artistique et culturelle. Plus tard, ce furent les thèmes et les sujets du maniérisme et du baroque qui exercèrent un ascendant sur lui. Il vécut à Naples de 1687 à 1691 où il réalisa probablement deux des groupes allégoriques appartenant à la Specola, les « teatrini », à savoir *Il Trionfo del Tempo* (Le Triomphe du Temps) et *La Peste*. Les spécialistes qui se sont penchés sur l'œuvre de Zumbo affirment que le fonds pittoresque et l'agencement des figures trahissant l'influence des maîtres d'Italie du Sud sont le signe de sa paternité. De 1691 à 1694, Zumbo travailla à Florence pour Cosme III de Médicis et acheva d'autres « teatrini » : *Il Sepolcro* (ou *La Vanità della Gloria Umana*) et *Il Morbo Gallico* (ou *Sifilide*) qui se réfèrent à l'art de la Renaissance. En 1695, il entreprit quelques voyages pour Bologne où il manifesta beaucoup d'intérêt pour les études anatomiques alors en plein essor. A la fin de 1695, il s'installa à Gênes. C'est là probablement qu'il acheva *L'anatomie de la tête* exposée à la Specola. C'est là aussi qu'il rencontra le chirurgien français

Guillaume Desnoues avec lequel il collabora pour des modèles anatomiques en cire comprenant, entre autres, une parturiente grandeur nature, une femme morte en couches de format plus petit ainsi que la naissance et la mise au tombeau du Christ. Malheureusement, ces œuvres n'ont pas été conservées. En 1700, Zumbo se brouilla avec le médecin français et partit pour Marseille où il effectua une tête en cire, puis pour Paris où il entra au service de Louis XIV. Il demeura dans cette ville jusqu'à sa mort, en 1701, emporté par une hémorragie cérébrale. Parmi ses affaires, on trouva une magnifique tête en cire qui est, selon toute probabilité, celle conservée de nos jours au Musée national d'histoire naturelle. Enterré en l'église de Saint-Sulpice, sa tombe fut détruite pendant la Révolution.

Une grande partie des œuvres de cet artiste encore largement méconnu il y a un demi-siècle est conservée au musée La Specola. Les trois « teatrini », connus également sous le nom de *Cera della peste* (figures de la peste), et l'anatomie de la tête se trouvaient à la Galerie royale (c'est-à-dire aux Offices) lorsque les Lorrains succédèrent aux Médicis. Ils furent ensuite envoyés au Museo di Fisica e Storia Naturale où ils ne furent jamais exposés, peut-être à cause du réalisme brutal des scènes. Lorsque Pierre Léopold partit pour Vienne en 1790, il offrit les « figures de la peste » au médecin de la cour, Giovanni Giorgio Asenöhrl, plus connu sous le nom de Lagusius. Celui-ci ne les garda pas mais chargea Fontana et, plus tard, Agostino Renzi, surintendant de la Pharmacie royale, de les vendre pour 150 sequins. Conscient de leur valeur, le surintendant les proposa au nouveau grand-duc Ferdinand III, d'abord pour l'Académie royale des beaux-arts où l'on jugea ces figures « ni utiles ni appropriées », puis pour le Musée royal. Cette proposition fut appuyée par Giovanni Fabbroni, directeur adjoint du musée, et les trois œuvres furent estimées par différents experts dont Clemente Susini. Comme les estimations dépassaient le prix réclamé par Lagusius, le grand-duc consentit à leur acquisition. Les « teatrini » restèrent au musée jusqu'en 1978, date à laquelle ils furent remis au Museo Nazionale de Bargello. Après avoir été cédés au Museo di Storia della Scienza, ils revinrent finalement à la Specola en 1974. Comme le prouvent les archives, les sculptures en cire furent plusieurs fois restaurées dans le passé, entre autres par Susini et Tortori. Se trouvant au Museo di Storia della Scienza, celles « de la peste » furent gravement endommagées par les pluies diluviennes de 1966. Très méticuleusement, Guglielmo Galli

Page · Seite · Page 59:

G. G. Zumbo:
Il Trionfo del Tempo

The Triumph of Time (detail)

Der Triumph der Zeit (Detail)

Le Triomphe du Temps (détail)

Plaster mould of a heart

Gipsabdruck eines Herzens

Moulage d'un cœur

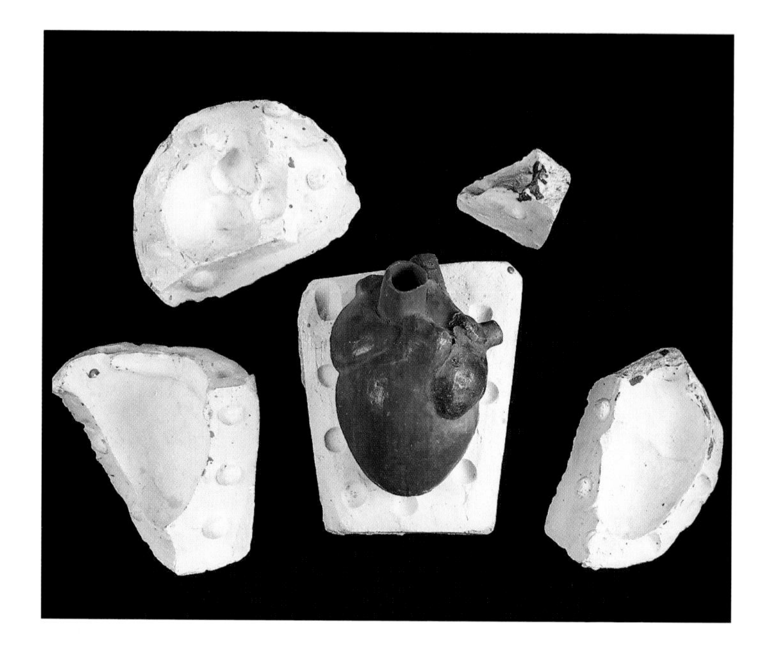

de l'atelier de Pietre Dure à Florence parvint au bout de dix-huit mois à les remettre en état en s'aidant de photographies. *L'anatomie de la tête* en revanche n'a jamais quitté la Specola puisqu'elle était à juste titre considérée comme présentant intérêt particulier pour la collection scientifique de l'institut.

La technique employée par Zumbo se distingue fortement de celle de l'école florentine à venir. *L'anatomie de la tête* fut exécutée sur un vrai crâne ayant appartenu, si l'on en croit les radios, à un homme âgé de vingt-cinq ans. En revanche, la tête conservée à Paris et une autre disparue aujourd'hui ont été fabriquées entièrement en cire, ce qui permet de conclure à une évolution de sa technique, due probablement à la coopération du céroplasticien sicilien avec Desnoues. Les figures des « teatrini » exécutées avec une grande habileté dans les moindres détails, ont été très probablement obtenues à partir de moulages de pièces qui, à leur tour, se basaient très soigneusement sur des modèles en argile. La cire plus ou moins liquide (un mélange de cire d'abeille, de colophane, de térébenthine et de colorants) était appliquée en couches de diverses épaisseurs suivant la consistance et la coloration souhaitées. Si l'on possède aujourd'hui quelques précisions sur la technique de

Zumbo, c'est grâce aux travaux de restauration de Guglielmo Galli entrepris après les inondations de 1966, qui révélèrent par ailleurs les traces de réparations plus anciennes du fait de la présence de colorants n'existant pas encore à la fin du XVIIe siècle. Le petit autoportrait (ill. p. 55) du groupe *Il Trionfo del Tempo* ne peut pas être attribué à Zumbo avec certitude, il pourrait être l'œuvre d'un contemporain ou de Susini plus tard. Celui intitulé *Il Morbo Gallico*, à l'origine une composition du même type et dont il n'existe plus que quelques fragments, fut offert à Filippo Corsini par Cosme III. Conservé en permanence dans les caves du palais Corsini sur l'Arno, il fut pratiquement détruit par les inondations. On retrouva quelques figures dans les jardins mais, n'ayant pas de photographie de l'ensemble, une reconstruction s'avéra impossible. A l'occasion du bicentenaire de l'ouverture du musée, qui fut célébrée avec un congrès sur la céroplastie, la famille Corsini fit don au musée des restes de cette œuvre.

Pendant longtemps, les chefs-d'œuvre de Zumbo furent considérés comme les caprices malsains d'un artiste se plaisant à reproduire des détails macabres et repoussants. Ce n'est qu'au milieu de notre siècle qu'on les considéra dans leur véritable perspective historique, comme des documents réalistes d'une époque où les guerres, les famines et les grandes épidémies étaient le lot quotidien des hommes. Cette vision de la destruction, de la précarité de l'existence humaine, de la fuite inexorable du temps, ce « memento mori » continuel auquel les œuvres de Zumbo nous confrontent avec une foule de détails, est un exemple typique de la culture du XVIIe siècle, telle qu'elle nous a été également transmise par d'autres grands artistes de l'époque, comme Luca Giordano ou Mattia Preti.

Pages · Seiten · Pages 62–63:

G. G. Zumbo: *La Peste*

The plague

Die Pest

La peste

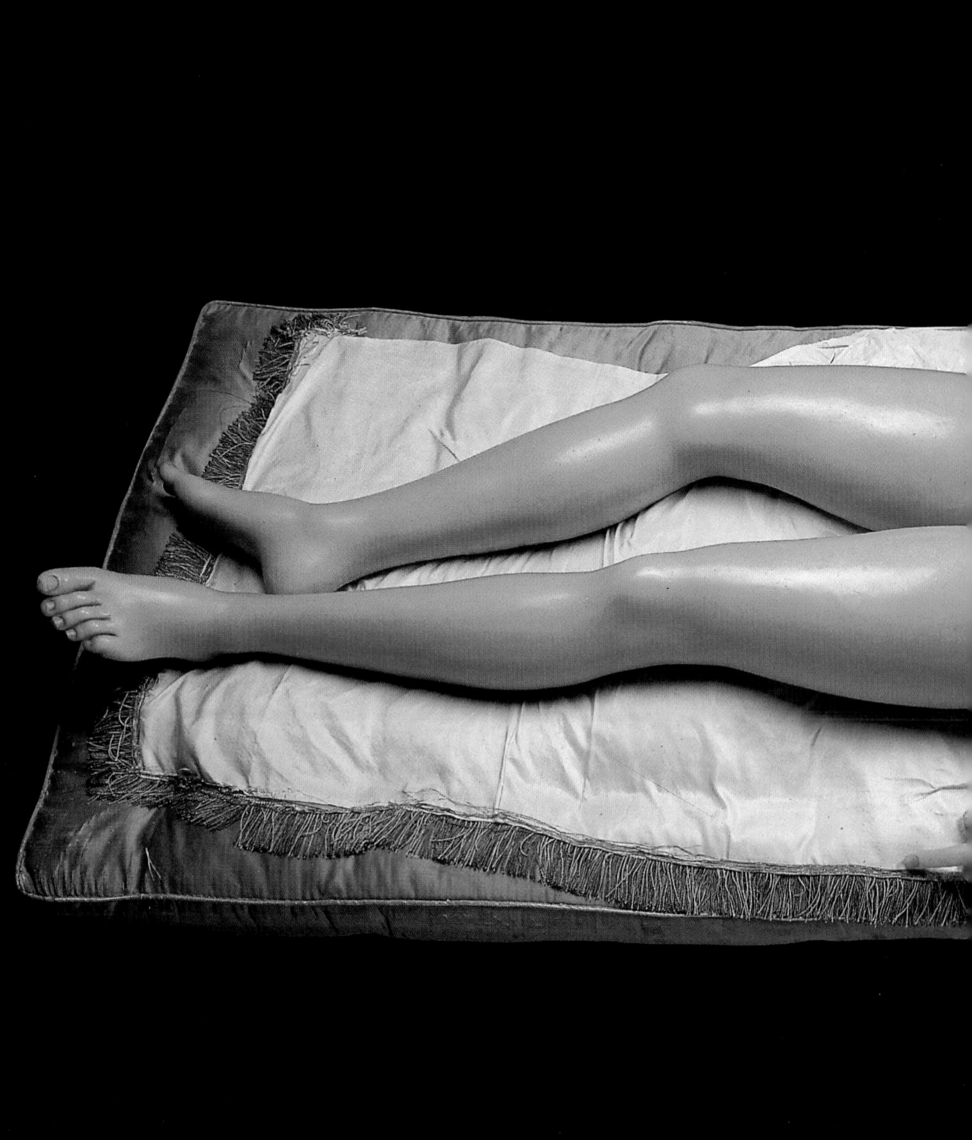

Human Anatomy
Die Anatomie des Menschen
L'anatomie humaine

The Anatomy of the Human Body – A unique collection of the late 18th century

Monika v. Düring

This unique collection of anatomical wax models from the museum 'La Specola', which is embedded in the tradition of European thought, gives us great insight into the knowledge and understanding of the anatomy of the human body as it existed at the end of the 18th century. If one compares the wax models with the representation of anatomical sections in woodcuts and copperplate engravings, it is clear that progress had been made since the traditional anatomical sections and illustrations of the 16th and 17th centuries, and that the production of these specimens was based upon the earlier examples.

These wax models in their three-dimensional form represent for the first time the original specimen much more accurately than a flat drawing could possibly do. The fascination of these specimens lies in the precision with which the details of the anatomical structure are reproduced. What is more, the artistic presentation is so perfected that in the model the whole body is felt to be alive — an impression which results from the representation of the human body in typical poses and gestures and complete refusal to depict a dead body. The form and shape of the human body thus appears as a living unit, with the greatest possible economy of power. Those who observe pictures of the models as much as those who visit the museum are equally impressed by the esthetic quality of human anatomy.

These models do not only represent one of the earliest macroscopic collections, but also one of the most manyfold. It is an impressive record of the skeleton and locomotor apparatus, the internal organs, the cavities of thorax, abdomen and pelvis, the circulatory and lymphatic systems, the central and peripheral nervous systems – including the sense organs – together with a large number of models depicting the relationship of these various entities in the different regions and cavities of the body.

These many exemplary specimens take the visitor on a voyage of discovery through the inner mysteries of our bodies, right down into the deeper regions where organs, vessels and nerves lie hidden. While learning anatomy, the examination of such specimens as these provides the student with a wealth of pictorial detail, bringing before his or her inner eye the image of a "transparent man".

This encyclopaedic collection, however, contains not only models for the scientific teaching of anatomy, but also models which the viewer initially regarded with displeasure and consternation. In this context one should realise that the burghers of that time were both enlightened and interested. A dissection was regarded and expirienced as a special public occasion which one not only attended but paid to attend. The promoters for their part likewise strove to satisfy the curiosity and sensationalism of the visitor, for which spectacular displays were particularly suitable.

In this presentation, the Florentine collection is for the first time made available to a wider readership in the form of a complete edition of colored illustrations. Even today it is of practical use. It offers both to the interested layman and to the medical student a variety of interesting experiences. The latter can test his anatomical knowledge by naming the various structures depicted, and will further recognize misinterpretat-ions based upon the scientific knowledge of the time. But it will also become apparent how much detail was already known even in those days, and at the same time we become aware of the advances medicine has made with the development of new techniques of visual display.

Die anatomische Gestalt des Menschen
– Ein einmaliges Zeitdokument des ausgehenden 18. Jahrhunderts

Monika v. Düring

Diese einmalige Modellsammlung der anatomischen Wachse des Museums La Specola vermittelt, eingebettet in die Tradition der europäischen Geistesgeschichte, den Erkenntnis- und Wissensstand der Anatomie des menschlichen Körpers im ausgehenden 18. Jahrhundert. Vergleicht man die Wachspräparate mit den Darstellungen anatomischer Sektionen auf Holzschnitten und Kupferstichen, wird deutlich, daß die anatomische Sektions- und Abbildungstradition des 16. und 17. Jahrhunderts fortgeschrieben wurde und man sich bei der Herstellung der Präparate an diesen Vorbildern orientierte.

In Form des gestalteten Wachsmodells gewinnt das Präparat hier zum ersten Mal eine dreidimensionale Dimension und kommt somit dem Originalpräparat sehr viel näher als die zweidimensionale Buchabbildung. Das Faszinierende dieser Präparate liegt in der Genauigkeit, mit der die Details der anatomischen Strukturen wiedergegeben werden. Darüber hinaus ist die künstlerische Darstellung so perfektioniert, daß der tote Körper im Modell als lebendig empfunden wird – ein Eindruck, der durch die Darstellung des menschlichen Körpers in typischen Posen und Gesten und durch den vollständigen Verzicht auf die Darstellung eines vom Tode gezeichneten Körpers entsteht. Die Form und Gestalt des menschlichen Körpers wird auf diese Weise zu einer vitalen Einheit von höchster Ökonomie der Kräfte. Die Modelle vermitteln dem Betrachter der Abbildungen genau wie dem Besucher des Museums eine Humananatomie von besonderer Ästhetik.

Diese Modellsammlung ist nicht nur eine der ersten makroskopischen Sammlungen, sondern auch eine der vielfältigsten. Auf beeindruckende Weise dokumentieren dies die zahlreichen anatomischen Präparate des Skelett- und Bewegungsapparates, der inneren Organsysteme, des Brust-, Bauch- und Beckenraums, des Blut- und Lymphgefäßsystems, des zentralen und peripheren Nervensystems – einschließlich der Sinnesorgane – sowie eine große Anzahl von Modellen, die die Komplexität der räumlichen Beziehungen verschiedener Systeme in einer Körperregion oder Körperhöhle wiedergeben. Die Vielzahl der Detailpräparate führt den Betrachter auf eine Entdeckungsreise durch das Innere unseres Körpers, systematisch von der Oberflächenanatomie bis in die tieferen Regionen der Körperhöhlen mit Organen, Gefäßen und Nerven. Im anatomischen Unterricht schulte das Studium derartiger Präparate das räumliche Vorstellungsvermögen und half den Studenten, sich anhand des Modells den „gläsernen Menschen“ vor seinem geistigen Auge vorzustellen.

Die enzyklopädisch reiche Sammlung enthält allerdings nicht nur Demonstrationsobjekte für den wissenschaftlichen anatomischen Unterricht, sondern auch Modelle, die der Betrachter aufgrund einer drastischen Darstellung zunächst mit Befremden und Betroffenheit betrachtet. Hierzu sollte man wissen, daß der Bürger der damaligen Zeit aufgeklärt und interessiert war. Eine Sektion wurde als ein besonderes öffentliches Ereignis empfunden und erlebt, das man besuchte und für das Eintritt zu bezahlen war. Die Veranstalter waren andererseits auch bemüht, die Neugier und Sensationslust der Besucher zu befriedigen, wozu spektakuläre Darstellungen am besten geeignet waren.

In der hier vorgelegten Ausgabe wird die Florentiner Sammlung erstmalig einem breiten Leserkreis als komplette Farbtafelsammlung zugänglich gemacht. Auch heute noch ist das Werk von praktischem Nutzen. Es bietet sowohl dem interessierten Laien wie auch dem angehenden jungen Arzt vielfältige Anregungen, der Medizinstudent kann durch Benennung der einzelnen Strukturen sein anatomisches Wissen überprüfen und wird natürlich auch die historisch bedingten Fehlinterpretationen erkennen. Im Vergleich wird deutlich, über welch genauen Detailkenntnisse man bereits zur damaligen Zeit verfügte, und gleichzeitg werden wir der Fortschritte gewahr, die die Medizin mit der Entwicklung neuer bildgebender Verfahren gemacht hat.

L'Anatomie du corps humain – Une collection unique de la fin du XVIIIe siècle

Monika v. Düring

La collection, unique en son genre, des cires anatomiques de la Specola, replacée dans le contexte de l'histoire des sciences de l'Homme en Europe, renseigne sur l'état des connaissances en anatomie humaine à la fin du XVIIIe siècle. La comparaison des préparations en cire avec les représentations des dissections anatomiques gravées sur bois ou sur cuivre permet de constater que la tradition des dissections et de l'iconographie anatomique des XVIe et XVIIe siècles a non seulement été maintenue, mais a influencé l'élaboration des modèles en cire.

Sous la forme du modèle en cire, l'illustration anatomique devient pour la première fois tridimensionnelle, et par là se rapproche beaucoup plus de la préparation originale que de la représentation plane. Le caractère fascinant des cires de la Specola tient à la précision avec laquelle sont restitués les détails des structures anatomiques. La représentation artistique, poussée à la perfection, fait percevoir le modèle cadavérique comme un sujet vivant. Cette impression résulte de la volonté de l'artiste de faire abstraction des stigmates de la mort ; à leur place sont évoqués des positions et des gestes de la vie courante laissant transparaitre l'unicité des formes et des fonctions vitales. Ces modèles procurent à l'observateur le sentiment de cette esthétique particulière qu'éprouve le visiteur d'un musée d'anatomie.

Cette collection de modèles en cire est non seulement l'une des plus anciennes collections anatomiques macroscopiques, mais aussi l'une des plus grandes et des plus variées. Les nombreuses préparations anatomiques apportent des données remarquables sur le squelette et l'appareil locomoteur, les viscères du thorax, de l'abdomen et du petit bassin, les systèmes circulatoires sanguin et lymphatique, les systèmes nerveux central et périphérique, les organes des sens ... Un grand nombre de modèles restituent la complexité des rapports topographiques entre les différents éléments anatomiques d'une région donnée du corps ou de l'une des cavités naturelles. La multiplicité des préparations concernant des régions isolées permet à l'observateur une exploration systématique du corps depuis l'anatomie de surface jusqu'aux régions profondes où sont logés les organes, vaisseaux et nerfs. Au cours de l'enseignement de l'anatomie, l'étude de telles préparations permet l'acquisition de la notion de volume et la représentation mentale de « l'homme transparent ».

A côté des objets de démonstration pour l'enseignement de la science anatomique constituant cette collection encyclopédique, se trouvent encore des modèles qui peuvent déconcerter. voire troubler le spectateur d'aujourd'hui en raison de leur représentation radicale. Il faut savoir que l'« honnête homme » du XVIIIe siècle était instruit, et que la dissection était une manifestation publique suivie avec intérêt dont le spectacle était payant. Les organisateurs s'efforçaient en conséquence de satisfaire la curiosité et le goût du sensationnel des visiteurs par des représentations spectaculaires.

Dans le présent ouvrage, la collection de cires anatomiques de la Specola de Florence est rendue, pour la première fois, accessible à un large public sous la forme d'une série complète de photographies en couleur. A l'heure actuelle, un tel ouvrage présente encore un intérêt pratique, et ouvre de nombreuses perspectives tant au profane intéressé qu'au médecin. L'étudiant en médecine pourra notamment tester ses connaissances anatomiques en identifiant et nommant chaque structure, mais aussi en reconnaissant les fautes d'interprétation liées au contexte historique. Par comparaison avec les connaissances précises disponibles à l'époque de cette collection, il est possible de mesurer les progrès de la médecine réalisés grâce au développement de nouvelles techniques.

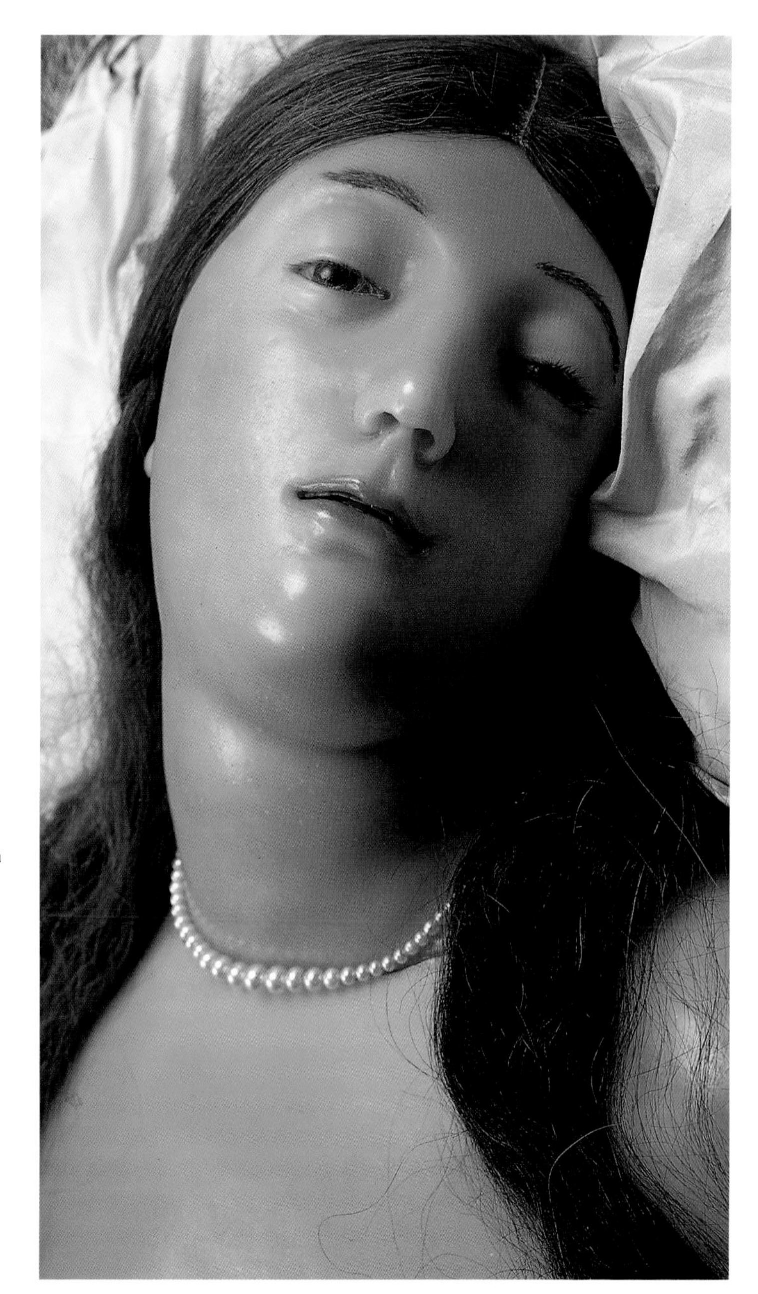

Head of whole body specimen of a pregnant woman. The model can be taken apart.

Kopfstudie des zerlegbaren Ganzkörperpräparates einer Schwangeren

Tête d'une préparation démontable du corps entier d'une femme enceinte

[OSTETRICIA, 968]

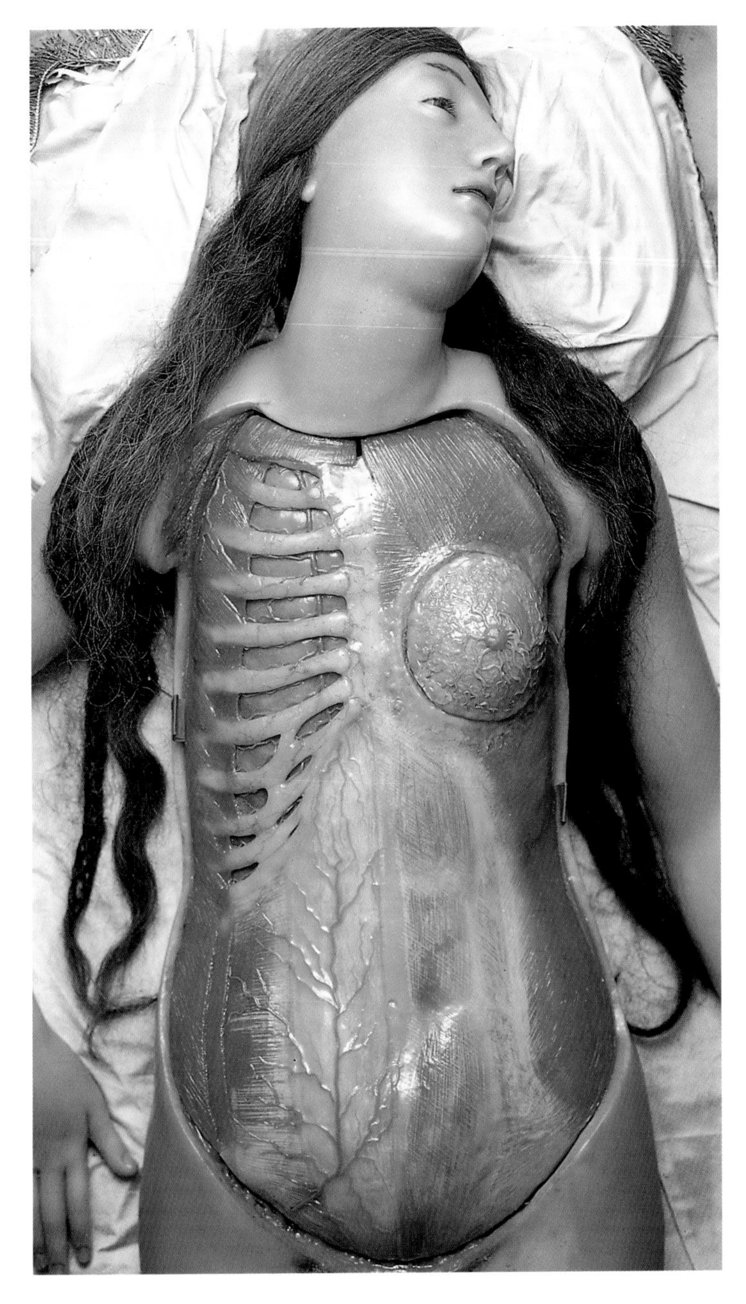

Specimen showing the thoracic wall, mammary gland and abdominal wall

Darstellung der Brustwand, der Brustdrüse und Bauchwand

Parois du thorax et de l'abdomen, sein

[OSTETRICIA, 968]

Specimen showing the thoracic and abdominal walls

Darstellung der Brust- und Bauchwand

Parois du thorax et de l'abdomen

[OSTETRICIA, 968]

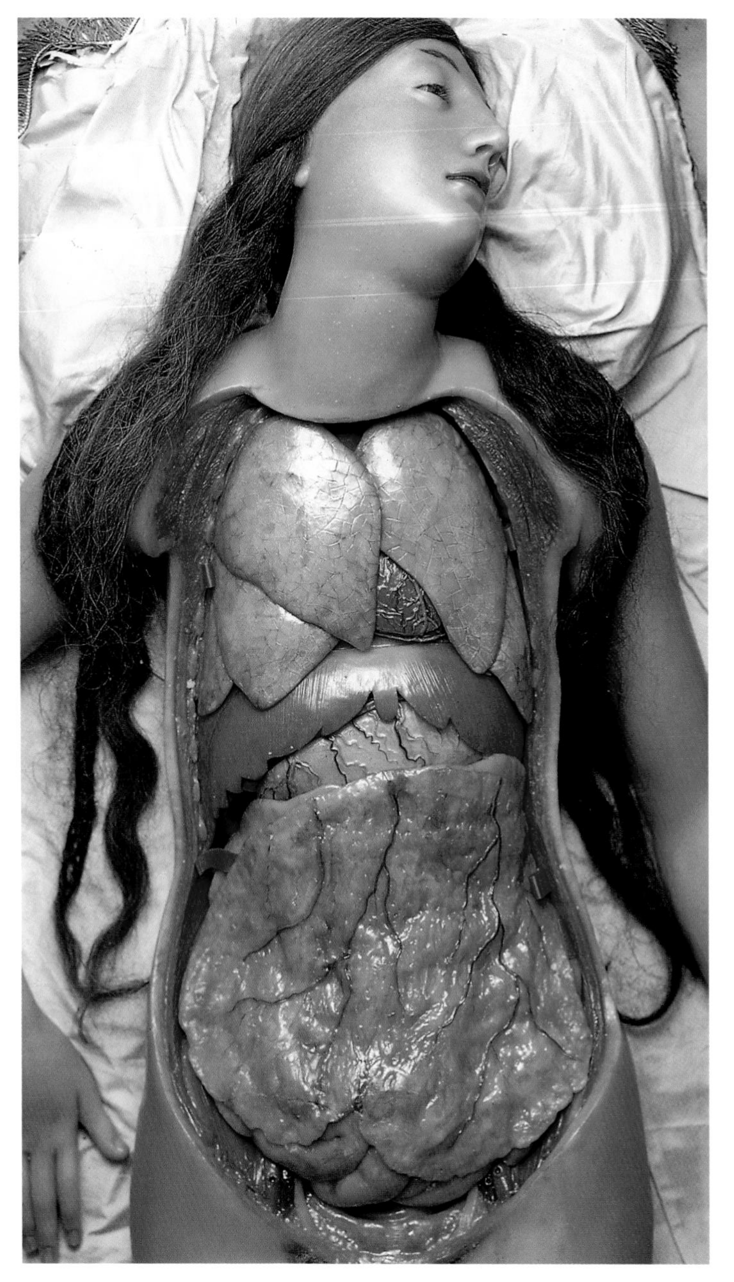

View into the opened thoracic and abdominal cavities. In the thorax the right and left lungs can be seen, and part of the anterior surface of the heart. The intestines are hidden behind the greater omentum.

Einblick in die eröffnete Brust- und Bauchhöhle. In der Brusthöhle sind rechter und linker Lungenflügel sowie ein Teil der Herzvorderfläche zu erkennen. Die Darmschlingen sind vom großen Netz bedeckt.

Viscères du thorax et de l'abdomen : cœur et poumons vus de l'avant, anses de l'intestin grêle recouvertes par le grand épiploon

[OSTETRICIA, 968]

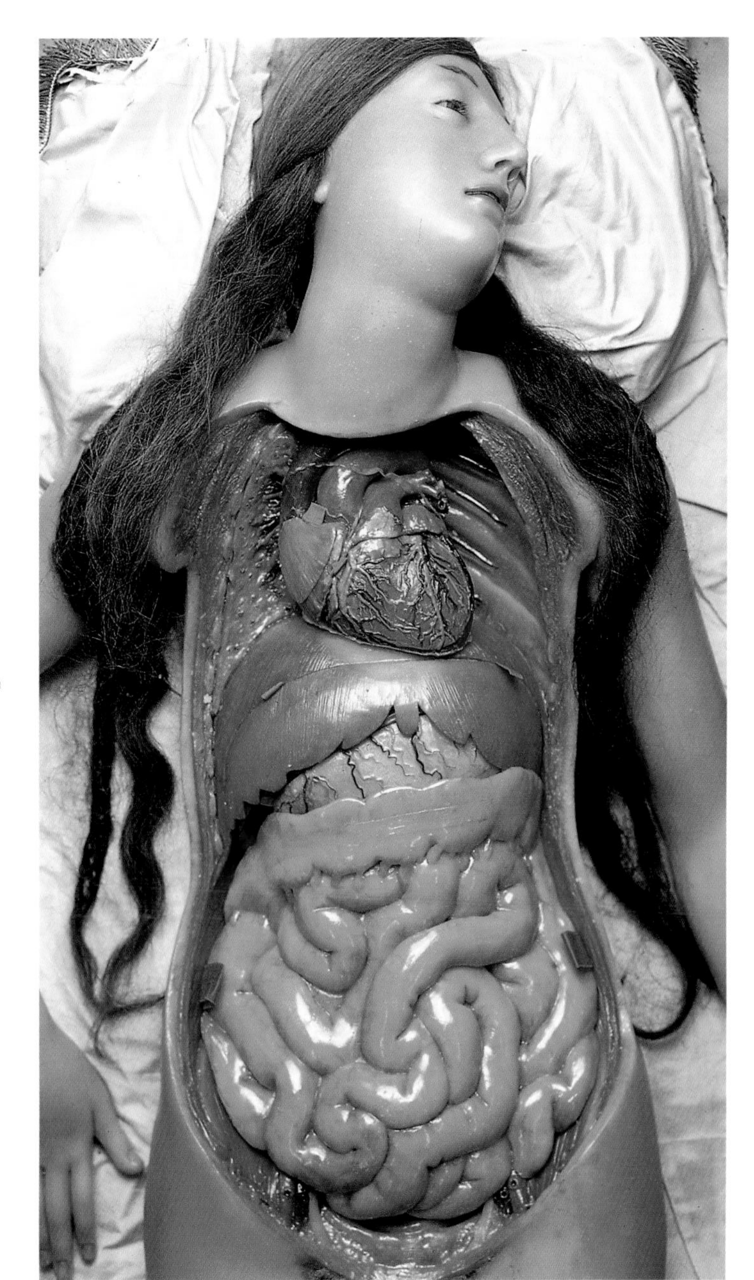

Specimen of the heart showing the coronary vessels and the great vessels that enter and leave the heart. In the abdominal cavity the greater omentum has been removed.

Präparation des Herzens mit Darstellung der Herzkranzgefäße und der abgehenden großen Gefäße aus dem Herzen. In der Bauchhöhle ist das große Netz entfernt.

Viscères du thorax et de l'abdomen : cœur, vaisseaux coronaires, gros vaisseaux de la base du cœur, ablation du grand épiploon

[OSTETRICIA, 968]

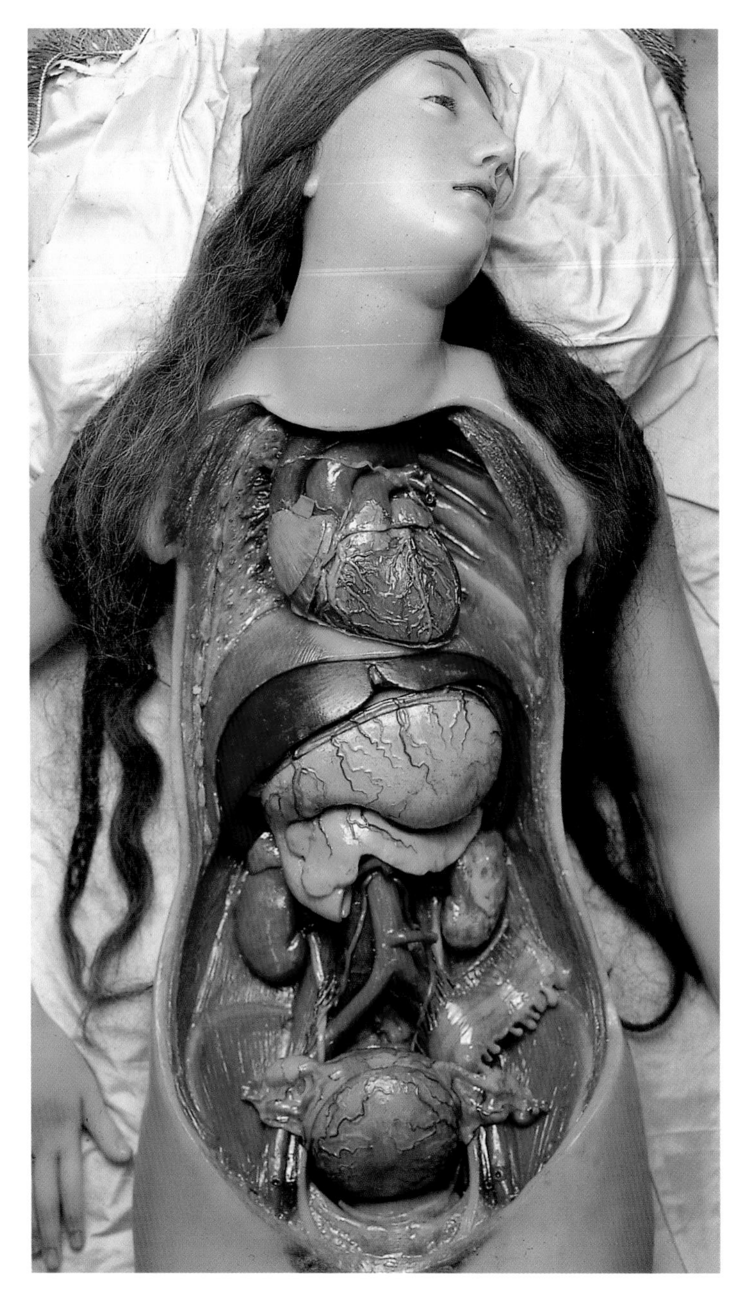

Display showing the organs in the upper part of the abdomen, such as the stomach, liver and duodenum, and also the right and left kidneys, the right adrenal gland, the abdominal aorta and the uterus

Blick auf die Oberbauchorgane wie Magen, Leber und Zwölffingerdarm sowie die rechte und linke Niere, die rechte Nebenniere, die Bauchschlagader und die Gebärmutter

Viscères de l'abdomen : estomac, foie, duodénum, reins, glande surrénale droite, utérus

[OSTETRICIA, 968]

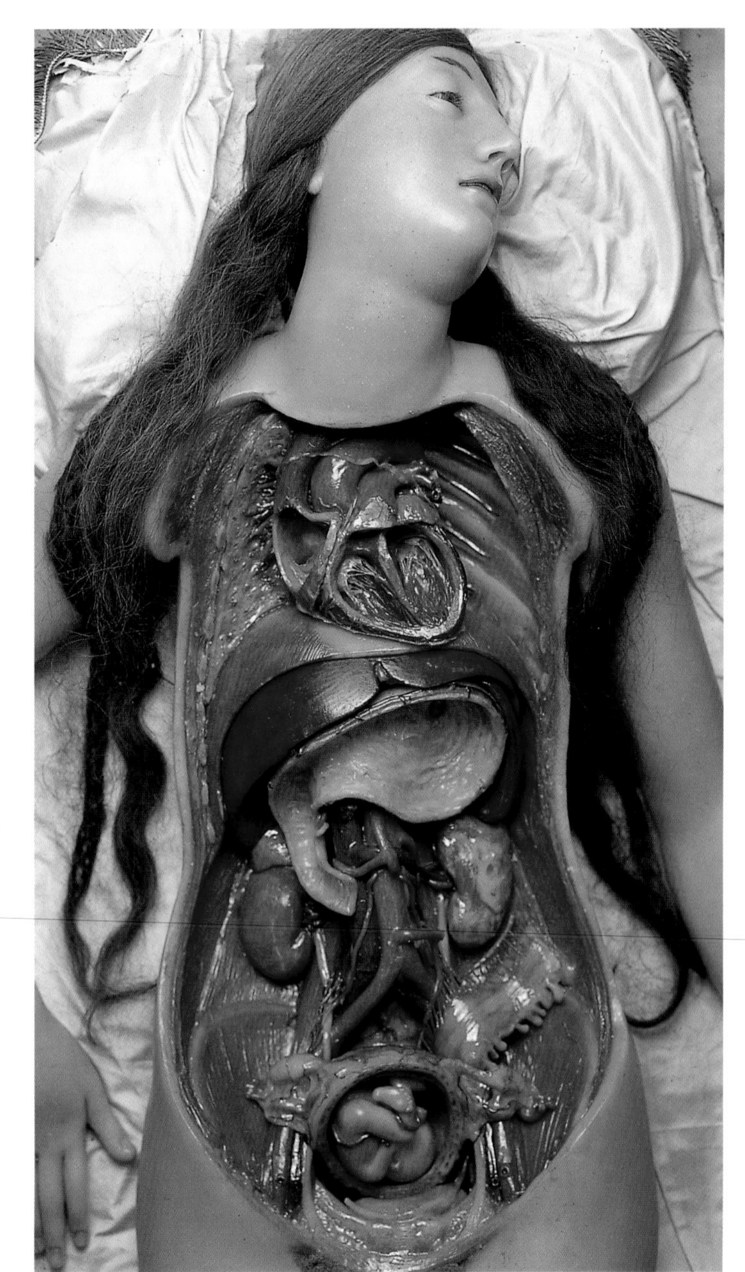

The right atrium, ventricles, stomach, duodenum and uterus have been opened

Eröffnung des rechten Herzvorhofes der Herzkammern, des Magens, des Zwölffingerdarmes und der Gebärmutter

Viscères du thorax et de l'abdomen : ouvertures de l'oreillette droite, des ventricules, de l'estomac, du duodénum et de l'utérus

[OSTETRICIA, 968]

Whole body specimen of a pregnant woman. The model can be taken apart.

Zerlegbares Ganzkörperpräparat einer Schwangeren

Préparation démontable du corps entier d'une femme enceinte

[OSTETRICIA, 968]

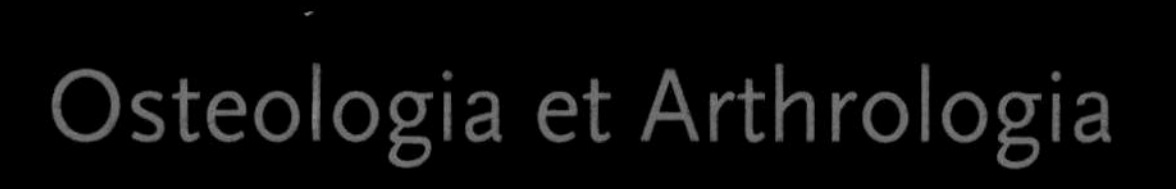

Osteologia et Arthrologia

Bones and Joints

Knochen und Gelenke

Squelette et articulations

Bones and Joints

The form of the human body, which is genetically determined, is a particular example of the general plan underlying that of all vertebrate animals. As the supporting element in this plan, the skeleton forms the bony framework of our bodies. The high proportion of crystalline minerals in bone is responsible for the skeleton being the organic system that remains intact long after the death of the individual.
Under certain circumstances, such as the exclusion of air, or even the various forms of embalming to which the body was subjected in some cultures before burial, the bones may even survive for thousands of years. This means that long after death, the age, sex, dietary habits and health or disease of a person can be determined. The preservation of the skeleton after death also explains why it has become a symbol of death and the transitoriness of human life in the religions and cultural history of Europe.

In the first table, examples are given of the structural characteristics of various bones. These consist of a compact outer shell (the cortex) and an inner layer of fine bony trabeculae (the spongiosa), the particular structure and architecture of which is the morphological correlate revealing the organization of the pressure and tension lines of the bone under normal loading. These patterns remain dynamic throughout life, and alter in response to the functional demands made upon the bone. The red bone marrow lies between the cancellous trabeculae of the spongiosa. It contains the stem cells which during life give rise to the newly developing cells of the blood.

In the following table the many different shapes and rich variation in form of the individual bones of our bodies make up an impressive picture. Differences in size, projections, ridges and crests of whole bones or sections through bones, symmetry or asymmetry, are further clues to the interpretation of the function of the individual components of the entire skeletal system.

Pages · Seiten · Pages 80–81:

Detail from pages 96–97

Detail der Seiten 96–97

Détail des pages 96–97

Considerable space is taken up by the different aspects of the bones which make up the skull, and which in their entirety form the cranium and facial skeleton. In addition to these, single vertebrae, sections of the vertebral column, the bones of the thoracic cage, the pelvis, and also the bones of the upper and lower limbs, including those of the hands and feet, are depicted.

A few specimens of joints displaying the ligaments which strengthen them and which stabilize and define their movements are also represented. Some of the joints have been opened to reveal the articular surfaces, or particular structures such as, for instance, the menisci of the clavicular and knee joints.

Knochen und Gelenke

Die Gestalt des menschlichen Körpers ist genetisch festgelegt und repräsentiert den allgemeinen Bauplan der Wirbeltiere. Als tragender Bestandteil dieses Bauplans bildet das Skelettsystem das knöcherne Gerüst unserer Körperformen. Der hohe Anteil an kristallisierten Mineralien im Knochen macht das Skelett zu jenem Organsystem, das über den Tod des Menschen hinaus erhalten bleibt. Unter bestimmten Voraussetzungen – wie etwa Luftabschluß, aber auch durch Konservierungsmaßnahmen, die in einigen Kulturen vor der Bestattung des Leichnams durchgeführt wurden – kann der Knochen sogar über Jahrtausende erhalten bleiben. So lassen sich am Skelett noch lange nach dem individuellen Tod, Alter, Geschlecht, Ernährungsgewohnheiten, körperliche Leistung und Erkrankungen nachweisen und ablesen. Die postmortale Erhaltung des Skeletts erklärt auch, warum es in der europäischen Religions- und Kulturgeschichte zum Symbol für den Tod an sich wie auch für die individuelle Vergänglichkeit geworden ist.

Die erste Tafel zeigt Beispiele des prinzipiellen Aufbaus verschiedener Knochen, bestehend aus einer kompakten äußeren Rindenschicht (Corticalis) und einer aus feinen Knochenbälkchen aufgebauten Innenschicht (Spongiosa). Die besondere Strukturierung und Architektur der Spongiosa aus unterschiedlich feinen Knochenbälkchen ist das morphologische Korrelat für die Ausrichtung der mechanischen Druck- und Zugspannungslinien des unter natürlicher Belastung stehenden Knochen. Diese Strukturen sind zeitlebens dynamisch und verändern sich entsprechend der jeweiligen funktionellen Belastungen des Knochens. Zwischen den Spongiosabälkchen befindet sich das rote Kochenmark mit den Blutstammzellen, die zeitlebens Ursprung der Neubildung unserer Blutzellen sind.

Auf den nachfolgenden Tafeln wird die vielfältige Gestalt sowie der Formenreichtum der einzelnen Knochen unseres Körpers eindrucksvoll dokumentiert. Größenunterschiede, Knochenvorsprünge und Knochenleisten an einzelnen Knochen oder Knochenabschnitten, Symmetrie oder auch Asymmetrie sind weitere Kriterien der funktionellen Interpretation einzelner Knochen im Verband des Gesamtsystems.

Einen breiten Raum nehmen die unterschiedlichen Ansichten der Schädelknochen ein, die in ihrer Gesamtheit den Hirn- und Gesichtsschädel bilden. Darüber hinaus werden die einzelnen Wirbel, ganze Wirbelsäulenabschnitte, die Knochen des Brustkorbes, des Beckens sowie der oberen und unteren Extremitäten, einschließlich der Hand- und Fußknochen, dargestellt. Einige Knochen- und Gelenkpräparate sind um die Darstellung ihrer Bänder, die ein Gelenk stärken, stabilisieren und durch ihren Verlauf die Gelenkbewegungen definieren, ergänzt und erweitert. Zum Teil sind die Gelenke eröffnet, um die Form der miteinander artikulierenden Gelenkflächen oder auch Sondereinrichtungen, wie beispielsweise die Menisken am Schlüsselbeingelenk und Kniegelenk zu demonstrieren.

Squelette et articulations

Le squelette constitue l'armature osseuse du corps, élément de charpente de l'architecture du corps humain, comme pour tous les vertébrés. La richesse en sels minéraux des os rend leur conservation possible après la mort. Des os peuvent ainsi être conservés pendant des milliers d'années, notamment grâce à l'utilisation de techniques de conservation pratiquées dans certaines cultures avant l'inhumation. Le squelette permet de révéler, longtemps après la mort de l'individu, son âge, son sexe, ses habitudes alimentaires, ses capacités physiques, ses maladies ... La conservation des os après la mort explique pourquoi le squelette est devenu, dans l'histoire des religions et des cultures européennes, le symbole de la mort et de l'existence éphémère de l'individu.

L'architecture générale des os est démontrée par la première planche qui en présente des exemples. Les os sont constitués d'une couche compacte périphérique, l'os cortical, entourant une zone profonde formée par de fines travées osseuses, l'os spongieux. La disposition et l'architecture spécifique des différentes travées de l'os spongieux sont la traduction morphologique des forces de pression et de traction exercées de façon physiologique. Ces structures évoluent de façon dynamique au cours de la vie et se transforment sous l'action des contraintes fonctionnelles. Dans l'interstice entre les travées d'os spongieux se trouve la mœlle osseuse rouge contenant les cellules souches des éléments figurés du sang permettant le renouvellement des globules sanguins durant toute la vie.

La grande variété et la richesse des formes des os du corps apparaissent de façon remarquable sur la deuxième planche. La variation de taille, la présence de tubercules et d'apophyses, la symétrie ou l'asymétrie constituent des critères pour l'interprétation fonctionnelle dans l'ensemble squelettique de pièces osseuses isolées.

Différentes vues des os du crâne, articulés pour former la voûte et la base du crâne ainsi que le squelette de la face occupent une place importante de la collection. Les os de la colonne vertébrale, les os de la cage thoracique, les os du bassin, et les os des membres supérieurs et inférieurs sont ensuite représentés.

Les préparations concernant le squelette et les articulations sont complétées par la représentation de ligaments ; ceux-ci renforcent et stabilisent chaque articulation et guident les mouvements. La capsule articulaire est ouverte dans certains cas pour démontrer la cavité articulaire, la forme des surfaces articulaires, ou la présence de dispositifs intra-articulaires particuliers tels que le disque sterno-claviculaire ou les ménisques du genou.

Skeleton humanum

Human Skeleton

Menschliches
Skelett

Squelette humain

[XXXI, 429]

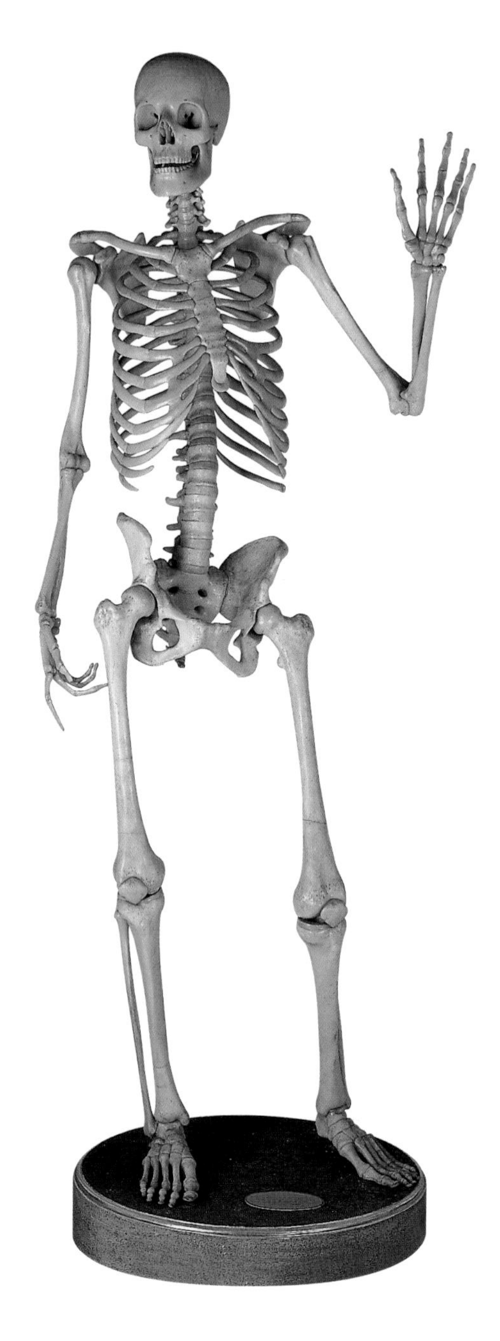

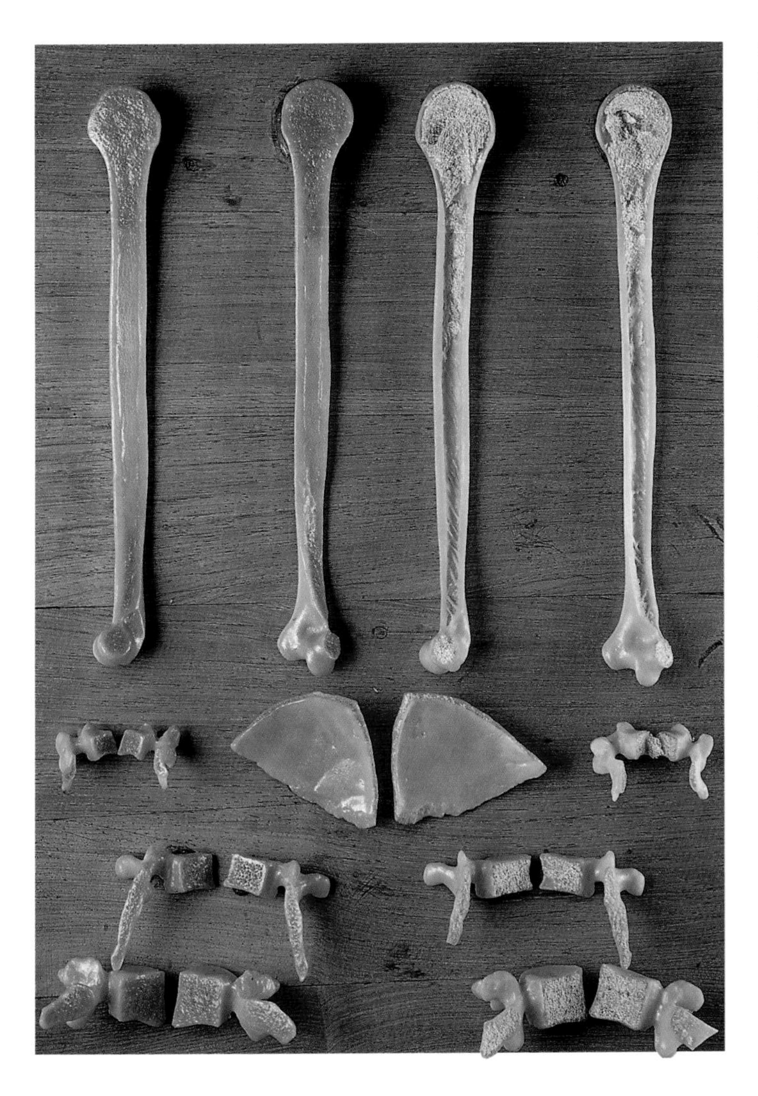

Internal structure of long bones, skull bones and vertebral bodies

Innere Struktur von Röhrenknochen, Schädelknochen und Wirbelkörpern

Structure interne d'os longs, d'os du crâne, et de vertèbres

[XXXI, 383]

View of the calvaria, facial skeleton and skull base

Ansicht des Hirn- und Gesichtsschädels sowie der Schädelbasis

Crâne vu du dessus (en haut), et crâne vu du dessous (en bas)

[XXXI, 381]

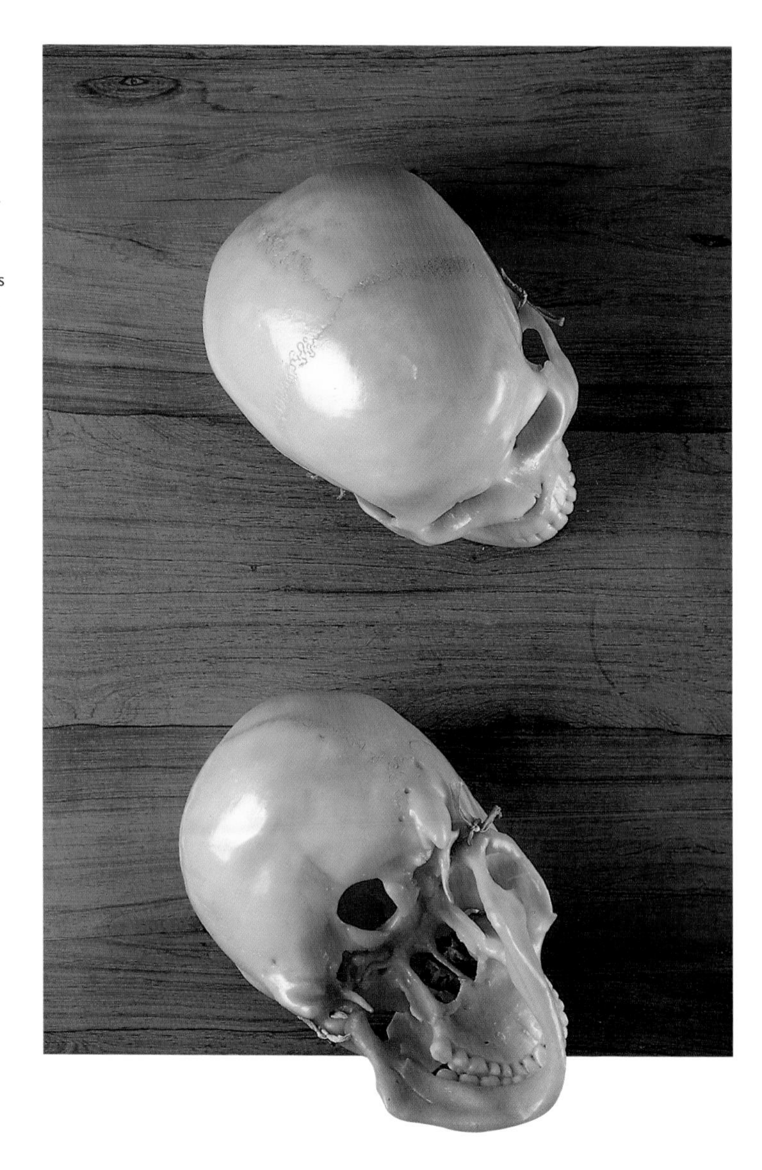

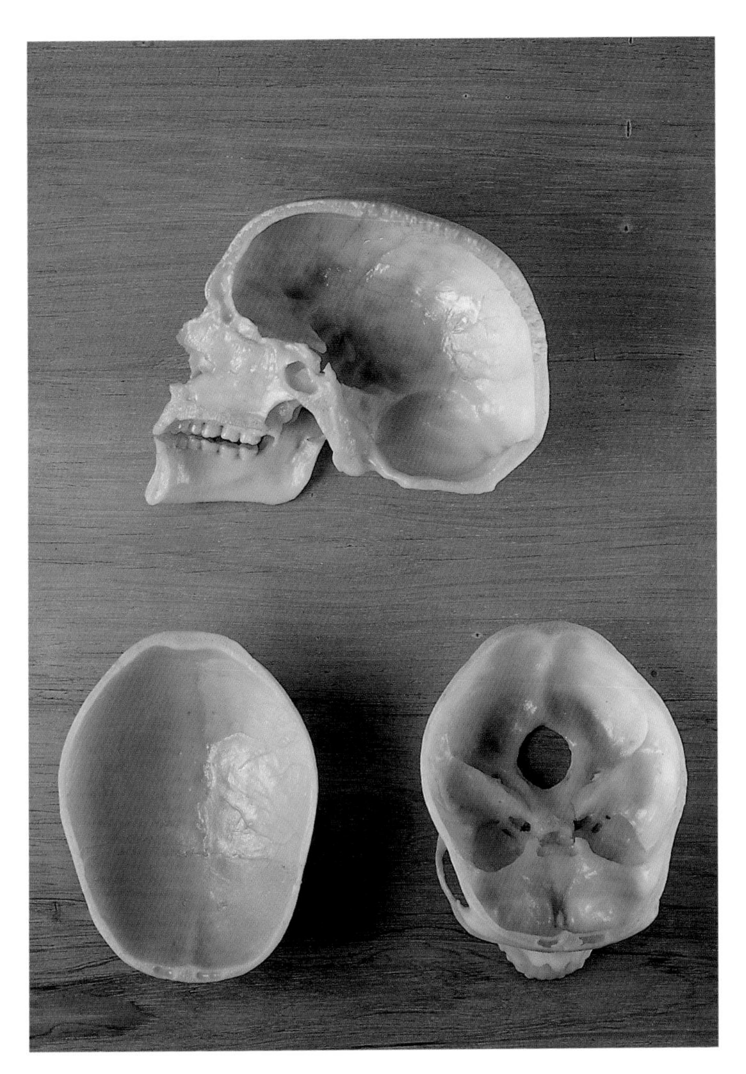

Above: median section through calvaria and facial skeleton; inner view of right half of the skull. Below: horizontal section with inner view of the skull roof and skull base

Oben: Medianschnitt durch Hirn- und Gesichtsschädel, Innenansicht der rechten Schädelhälfte. Unten: Horizontalschnitt mit Innenansicht von Schädeldach und Schädelbasis

En haut : coupe médiane du crâne : vue interne de la moitié droite.
En bas : coupe horizontale : vues endocrâniennes de la voûte et de la base du crâne

[XXXI, 380]

Frontal bone (upper and lower specimens) together with two views of the upper jaw bone (specimen at side). In the center is the ethmoid bone.

Stirnbein (oberes und unteres Präparat) sowie Oberkieferknochen (seitlich gelegene Präparate) in zwei Ansichten. In der Mitte ist das Siebbein zu sehen.

Os frontal (en haut et en bas) ; os maxillaire (à droite et à gauche) ; os ethmoïde (au centre)

[XXXI, 379]

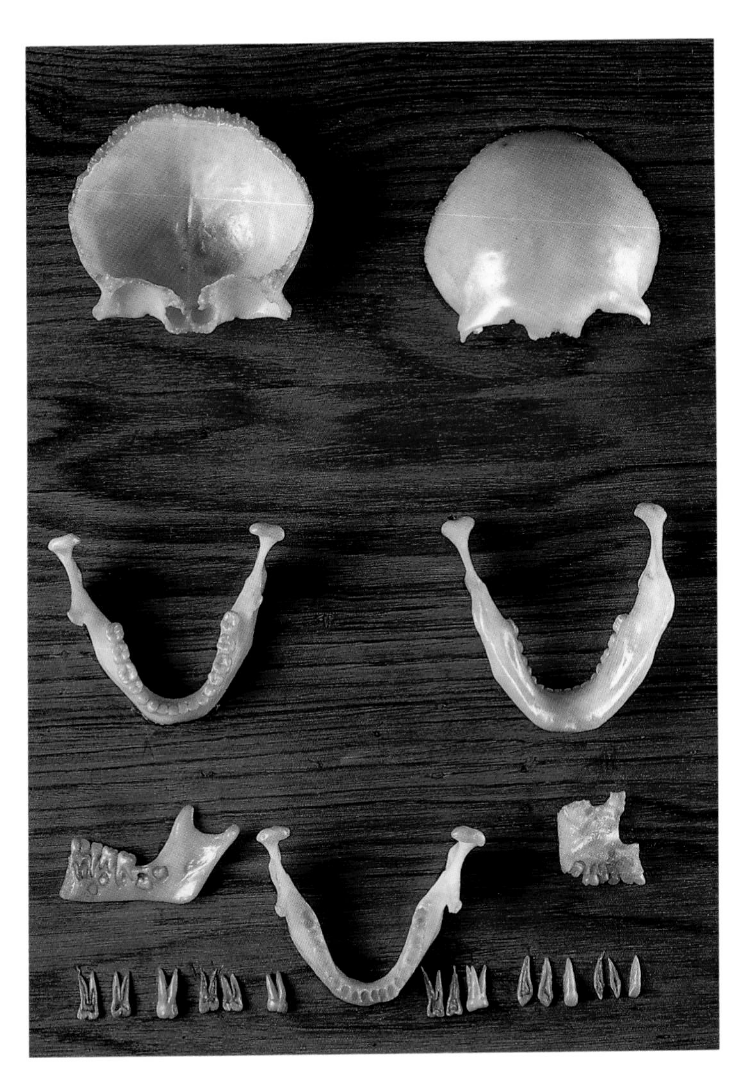

Upper specimen: two views of the frontal bone. Lower specimen: lower jaw bone with teeth. Left: milk teeth with tooth germs

Obere Präparate: Stirnbein in zwei Ansichten. Untere Präparate: Unterkieferknochen mit Zähnen; links: Milchgebiß mit Zahnanlagen

En haut : os frontal vu de l'avant et de l'arrière. En bas : mandibules d'adultes et d'enfant avec la dentition

[XXXI, 375]

Temporal bone, partly chiseled away to reveal the inner ear

Schläfenbein, zum Teil aufgemeißelt zur Darstellung des Innenohres

Série d'os temporaux partiellement trépanés pour démontrer l'oreille interne

[XXXI, 374]

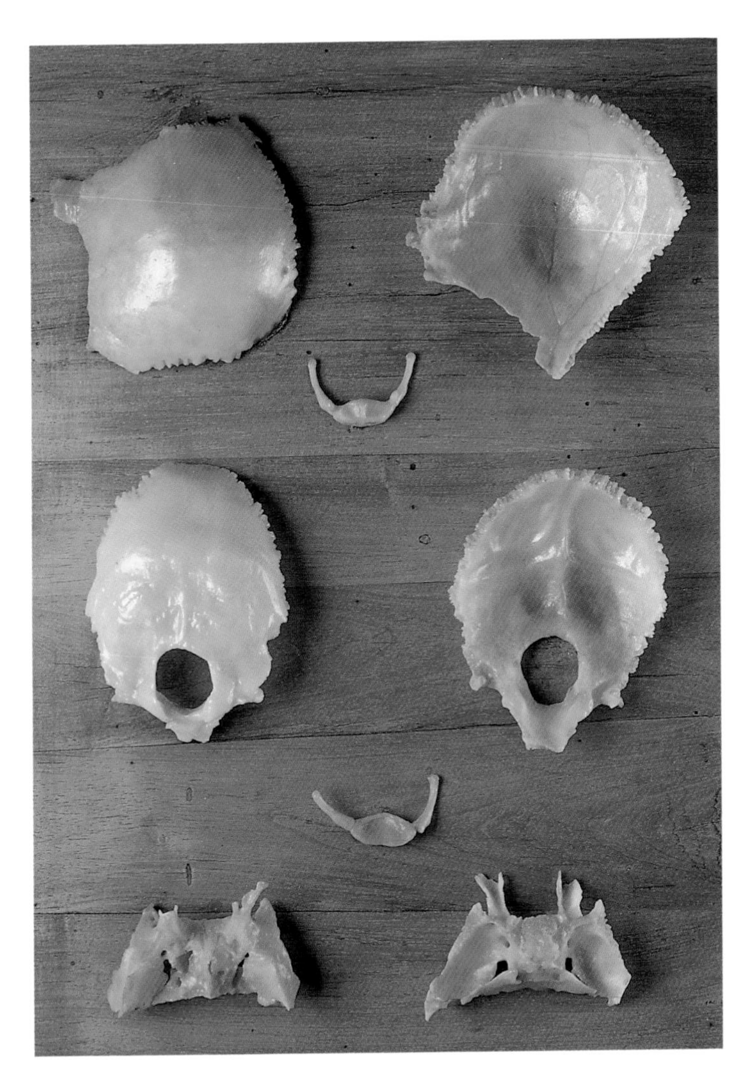

Parietal bone, occipital bone with foramen magnum, two views of the sphenoid. Middle specimen: hyoid bone

Scheitelbein, Hinterhauptsbein mit Hinterhauptsloch sowie Keilbein in zwei Ansichten, in der Mitte das Zungenbein

Os pariétal (en haut) ; os occipital avec trou occipital (au milieu) ; os sphénoïde (en bas) ; os hyoïde (au centre)

[XXXI, 914]

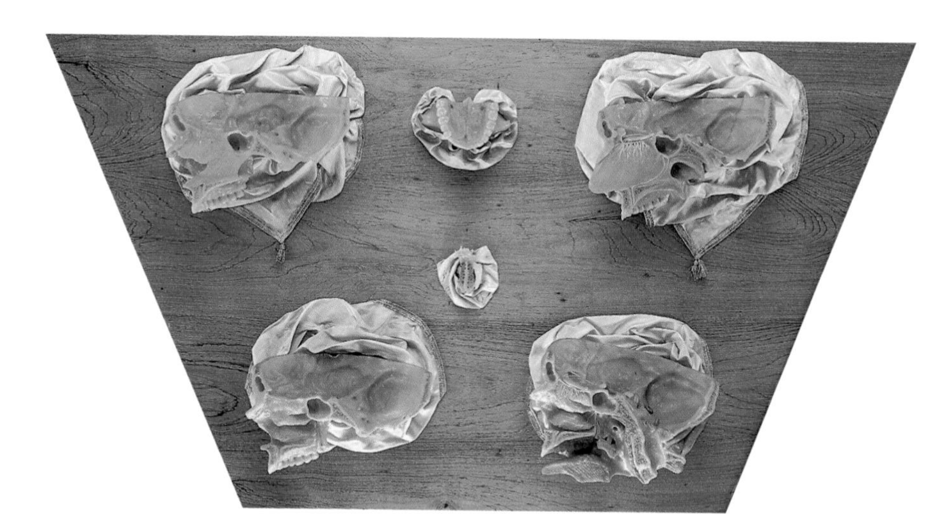

Median section through the skull showing nasal septum, side wall of nasal cavity and nasal conchae

Medianschnitt durch den Schädel mit Darstellungen der Nasenscheidewand sowie der seitlichen Nasenwand mit ihren Nasenmuscheln

Coupes médianes du crâne : représentations de la cloison nasale et de la paroi nasale latérale avec les cornets

[XXVIII, 733]

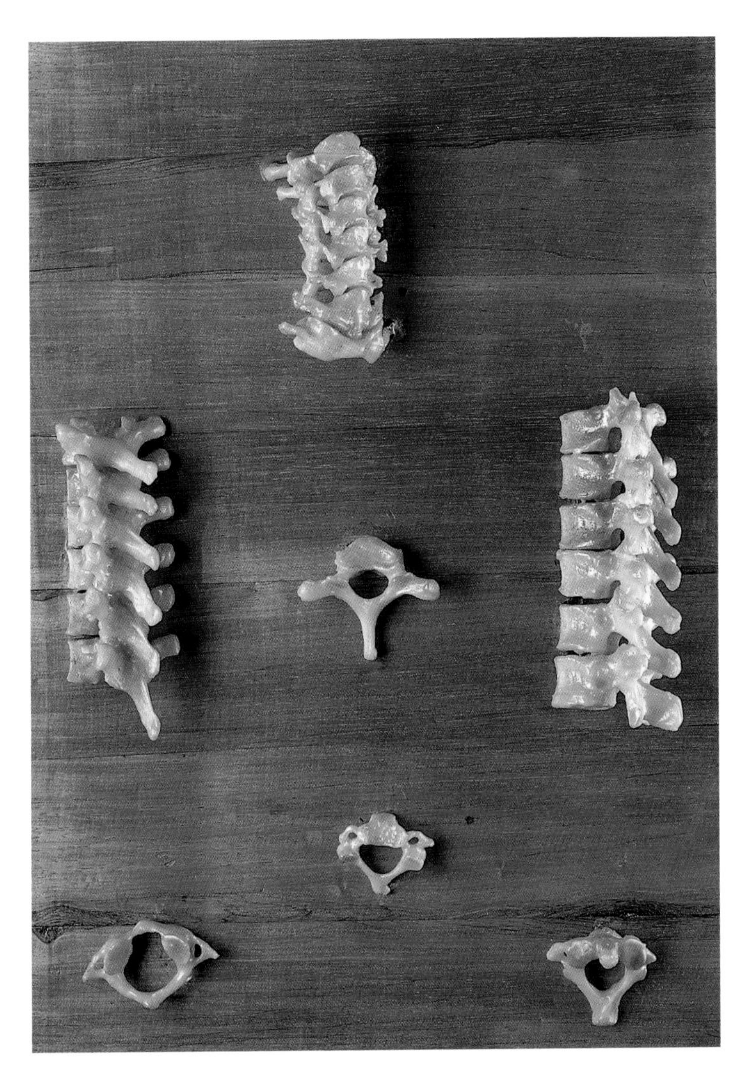

Cervical and thoracic vertebral column. Below left: the first vertebra (atlas), which supports the skull

Hals- und Brustwirbelsäule. Unten links : der den Schädel tragende erste Halswirbel, der Atlas

Segments de la colonne vertébrale cervicale et thoracique; vertèbres cervicales isolées : atlas portant le crâne (en bas à gauche)

[XXXI, 368]

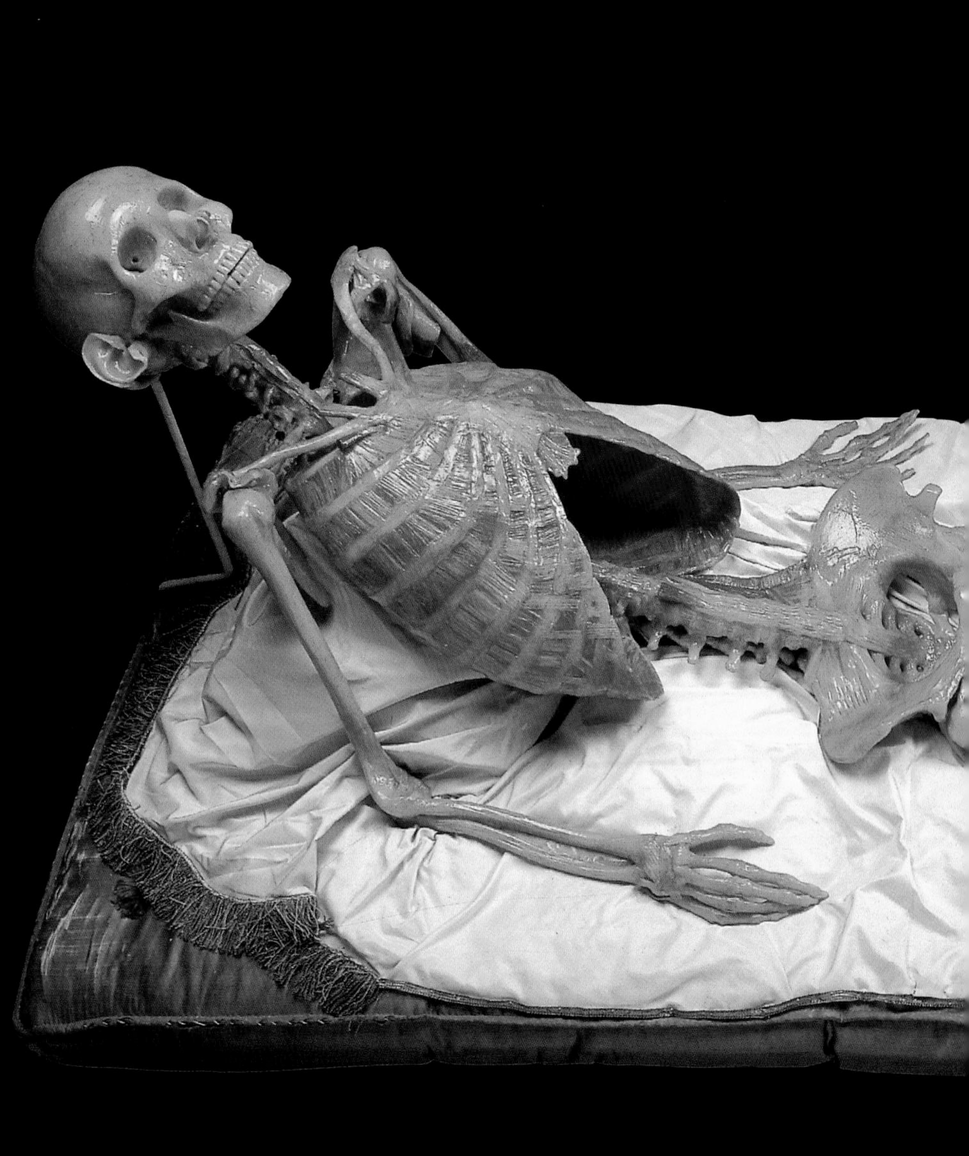

Articulationes, Ligamenta

Whole body specimen showing the ligaments, joints and some single muscles and tendons

Ganzkörperpräparat mit Darstellung von Bändern, Gelenken sowie einzelner Muskeln und Sehnen

Préparation du corps entier représentant les articulations, les ligaments et quelques muscles et tendons

[XXVI, 428]

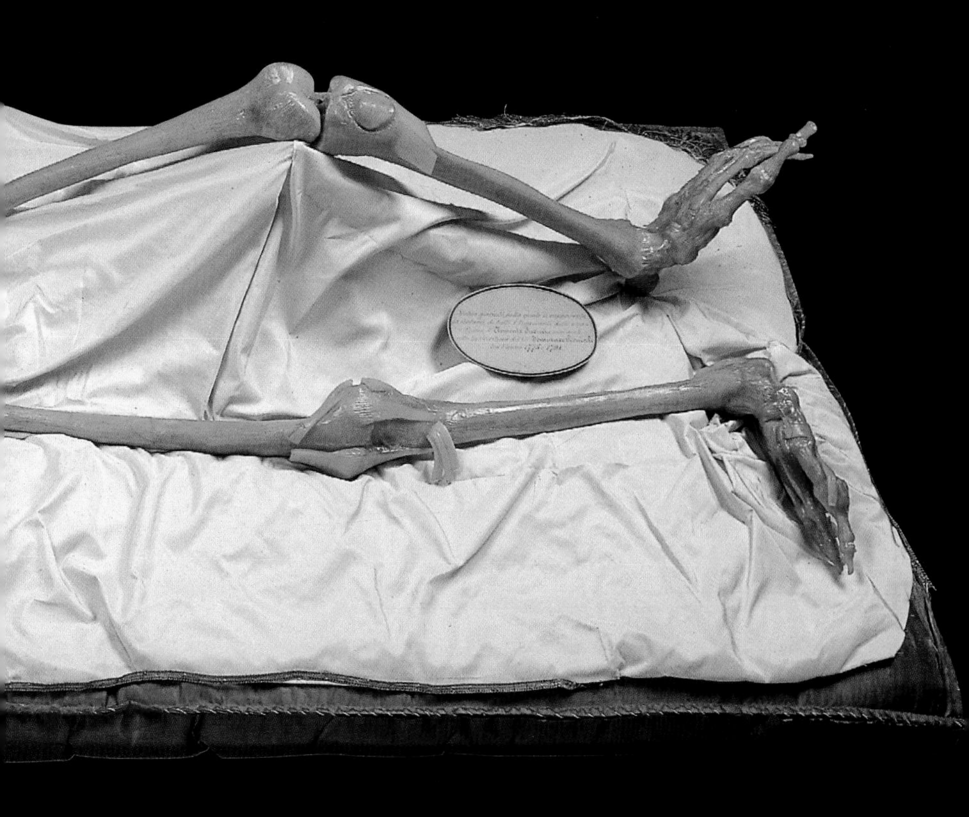

Lumbar vertebral column, sacrum and coccyx

Lendenwirbelsäule, Kreuzbein und Steißbein

Colonne vertébrale lombaire ; sacrum ; coccyx

[XXXI, 369]

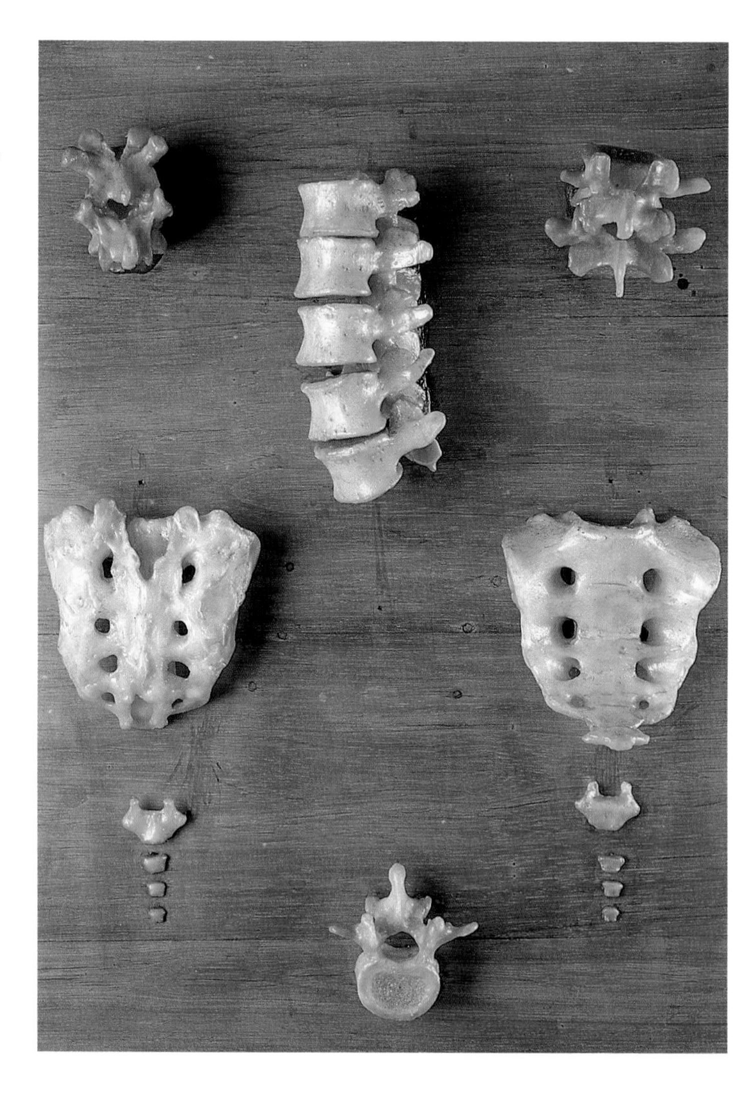

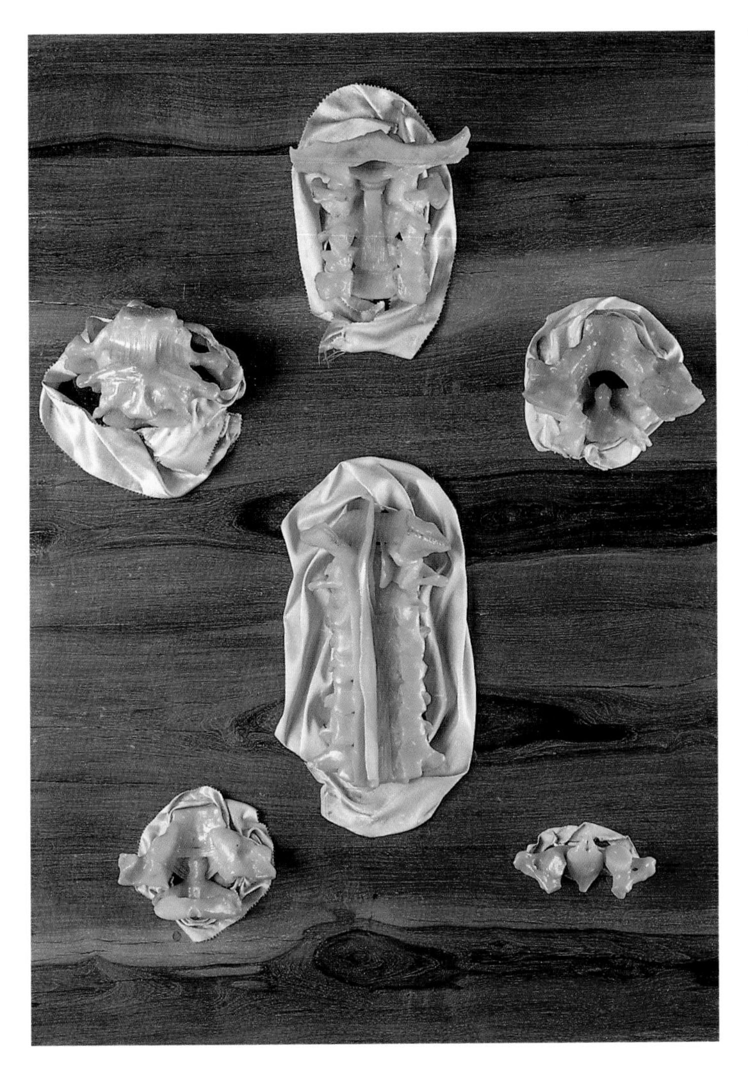

View from behind into the vertebral canal showing ligaments of the upper cervical vertebral column

Blick von hinten in den Wirbelkanal, dargestellt sind Anteile des Bandapparates der oberen Halswirbelsäule

Canal vertébral cervical vu de l'arrière ; ligaments de la jonction crânio-vertébrale

[XXXI, 373]

Ligaments of the upper cervical vertebral column seen from behind

Anteile des Bandapparates der oberen Halswirbelsäule in der Ansicht von hinten

Ligaments de différents segments de la colonne vertébrale vus de l'arrière

[XXXI, 372]

Ligaments of the thoracic vertebral column and of the costavertebral joints thoracic vertebrae and ribs

Bandverbindungen der Brustwirbelsäule sowie der Brustwirbel mit den Rippen

Ligaments des articulations de la colonne vertébrale thoracique et des articulations costovertébrales

[XXXI, 467]

Collarbone, breastbone, shoulder blade

Schlüsselbein, Brustbein und Schulterblatt

Clavicule ; sternum ; scapula (omoplate)

[XXXI, 362]

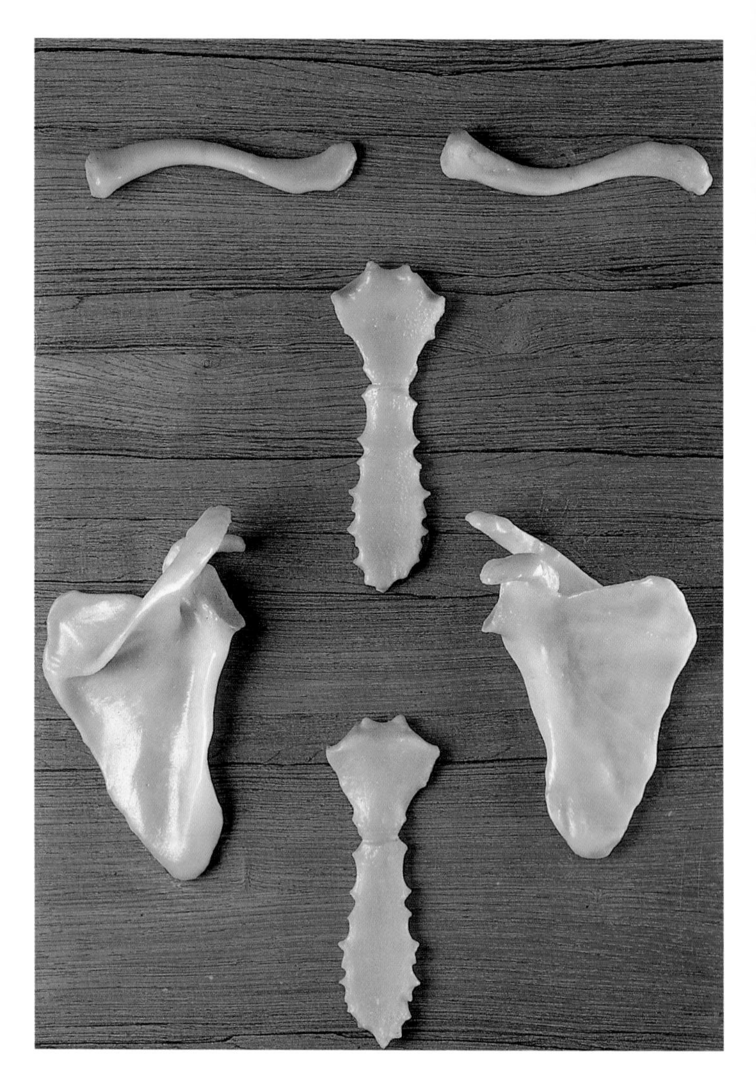

▶
Breastbone with rib attachments, ligaments and muscles, seen from in front

Brustbein mit Rippenanteilen, Bandverbindungen und Muskeln in der Ansicht von vorne

Plastron sternocostal vu de l'avant : os, ligaments, muscles intercostaux

[XXXI, 421]

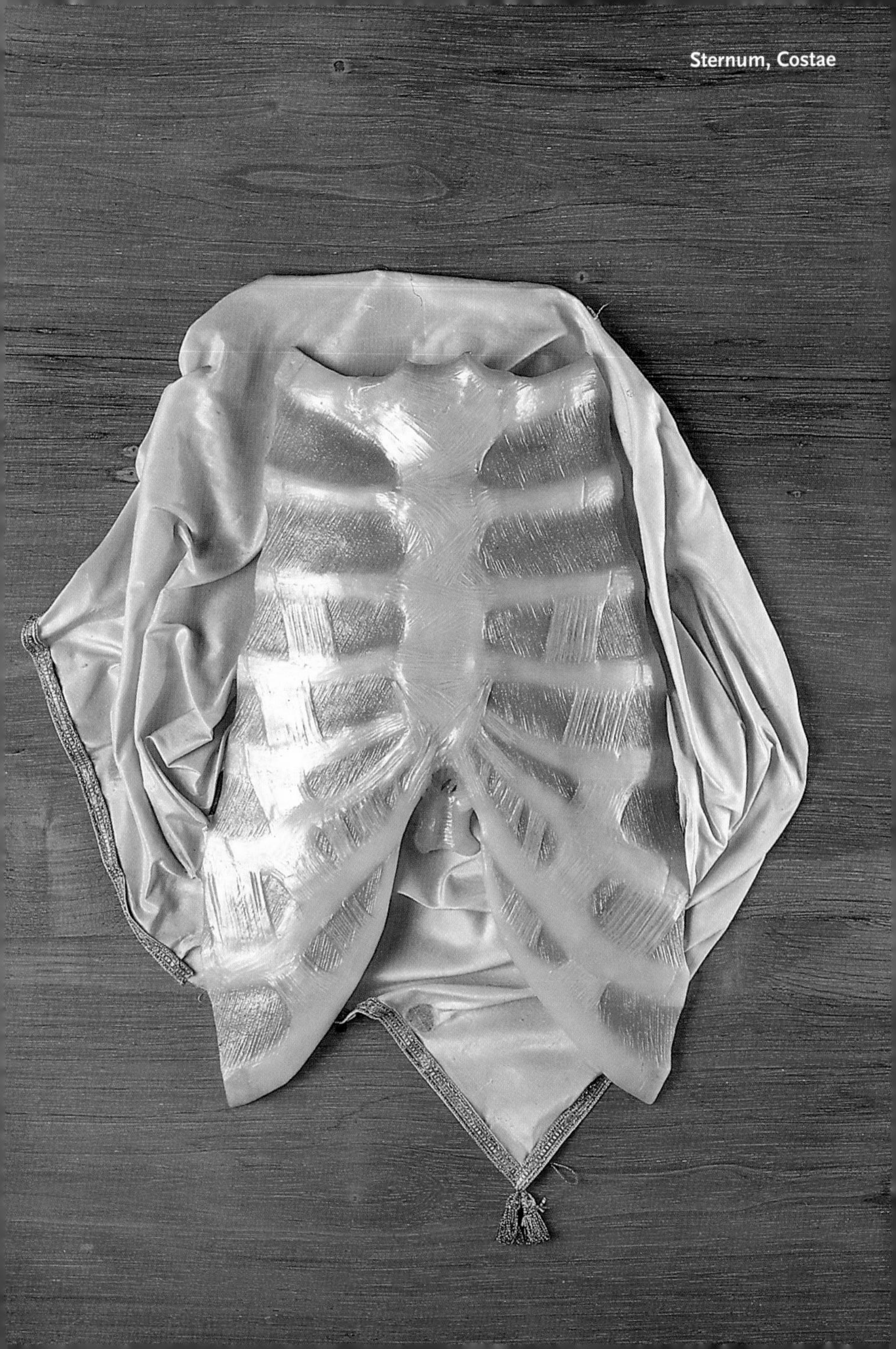

Upper arm and radius

Oberarmbein und Speiche

Humérus et radius

[XXXI, 384]

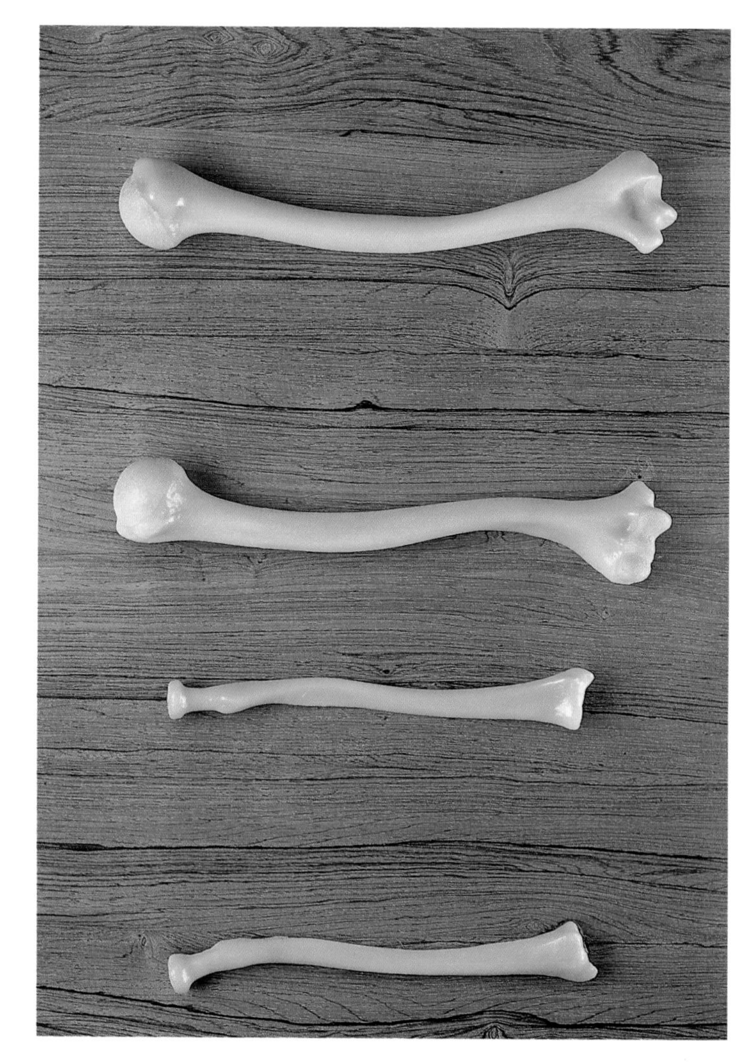

►
Two views of the skeleton of the right hand

Skelett der rechten Hand in zwei Ansichten

Squelette de la main droite, vues palmaire et dorsale ; os du poignet

[XXXI, 363]

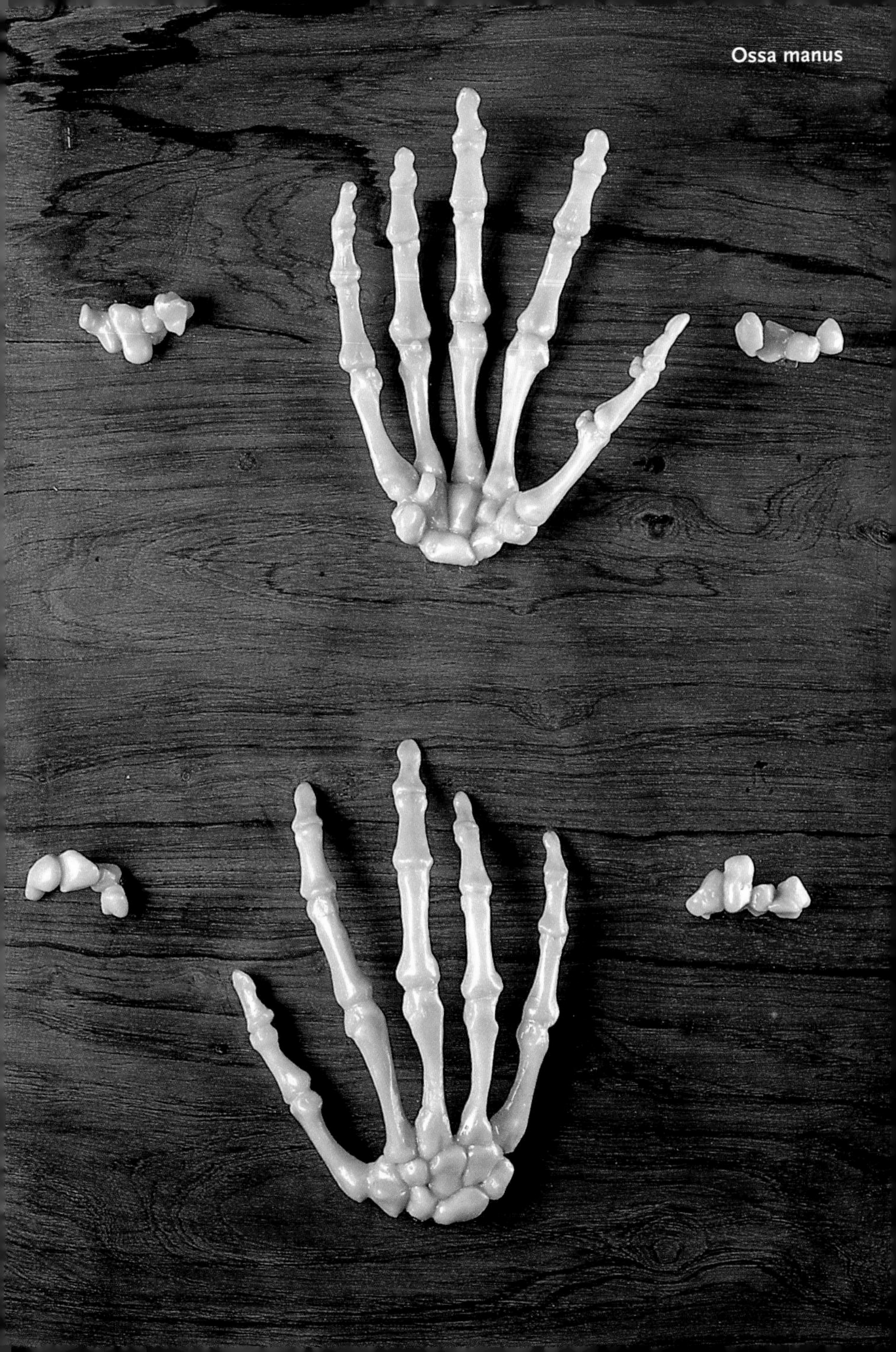
Ossa manus

Articulatio humeri

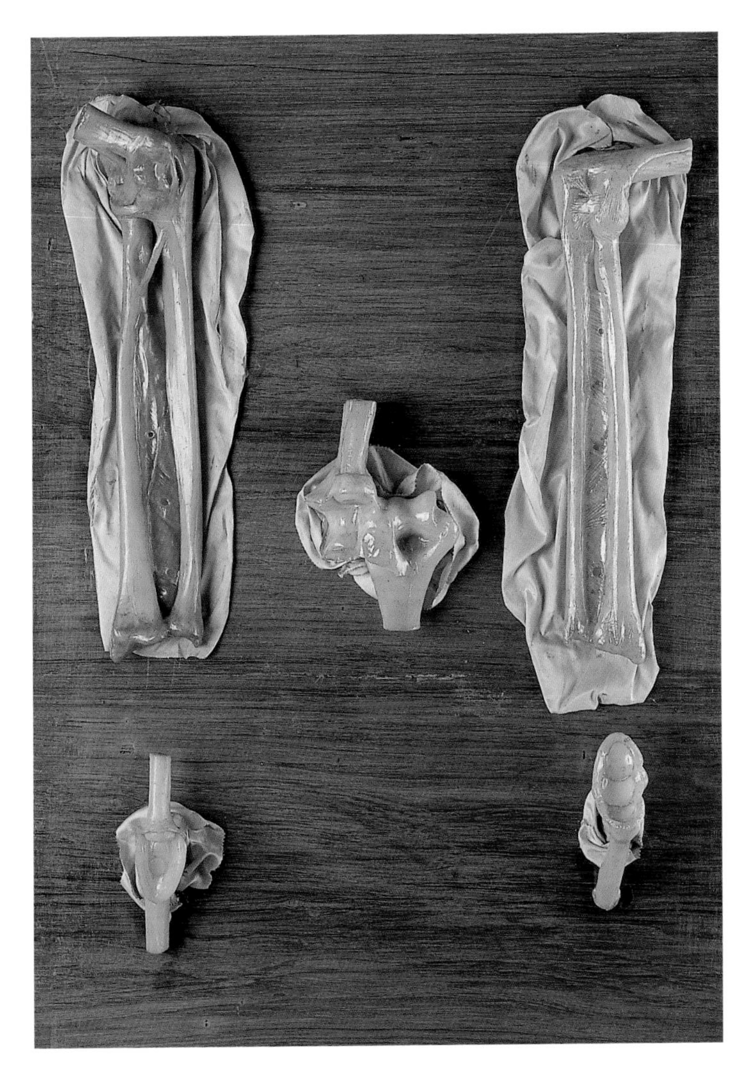

Display showing various aspects of the elbow joint

Verschiedene Darstellungen des Ellenbogengelenkes

Articulations du coude et des os de l'avant-bras

[XXXI, 409]

◀
The shoulder joint, different views

Schultergelenk in mehreren Ansichten

Articulations de l'épaule (différentes vues)

[XXXI, 411]

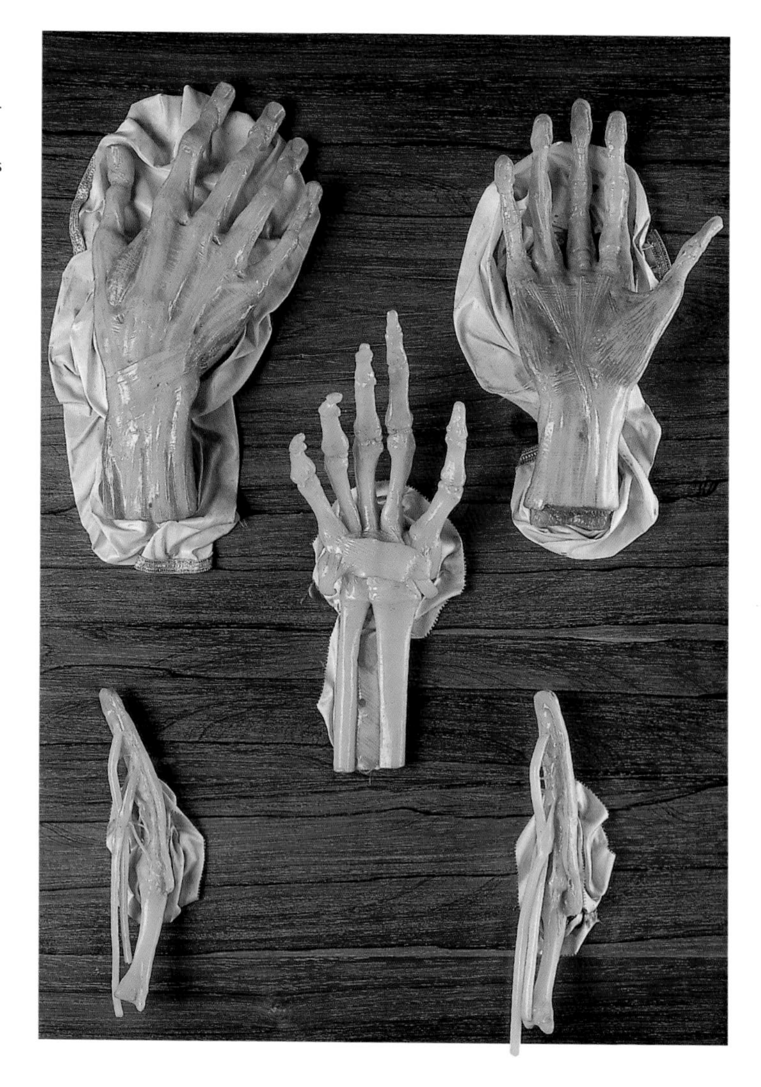

Tendons and tendon sheaths of the back of the hand. Tendinous sheet of palm, and the insertions of tendons of the flexor and extensor muscles of the fingers

Darstellungen der Sehnen und Sehnenscheiden des Handrückens sowie der Sehnenplatte der Handinnenfläche und der Ansätze der Sehnen der Fingerbeuge- und Fingerstreckmuskeln

Articulations de la main ; tendons et gaines tendineuses des muscles fléchisseurs et extenseurs des doigts

[XXXI, 405]

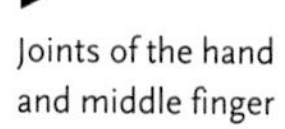

Joints of the hand and middle finger

Gelenke der Handwurzel und Mittelfinger

Articulations de la main et d'un doigt (majeur)

[XXXI, 403]

Articulatio carpometacarpea, Ligamenta carpometacarpea dorsalia et palmaria

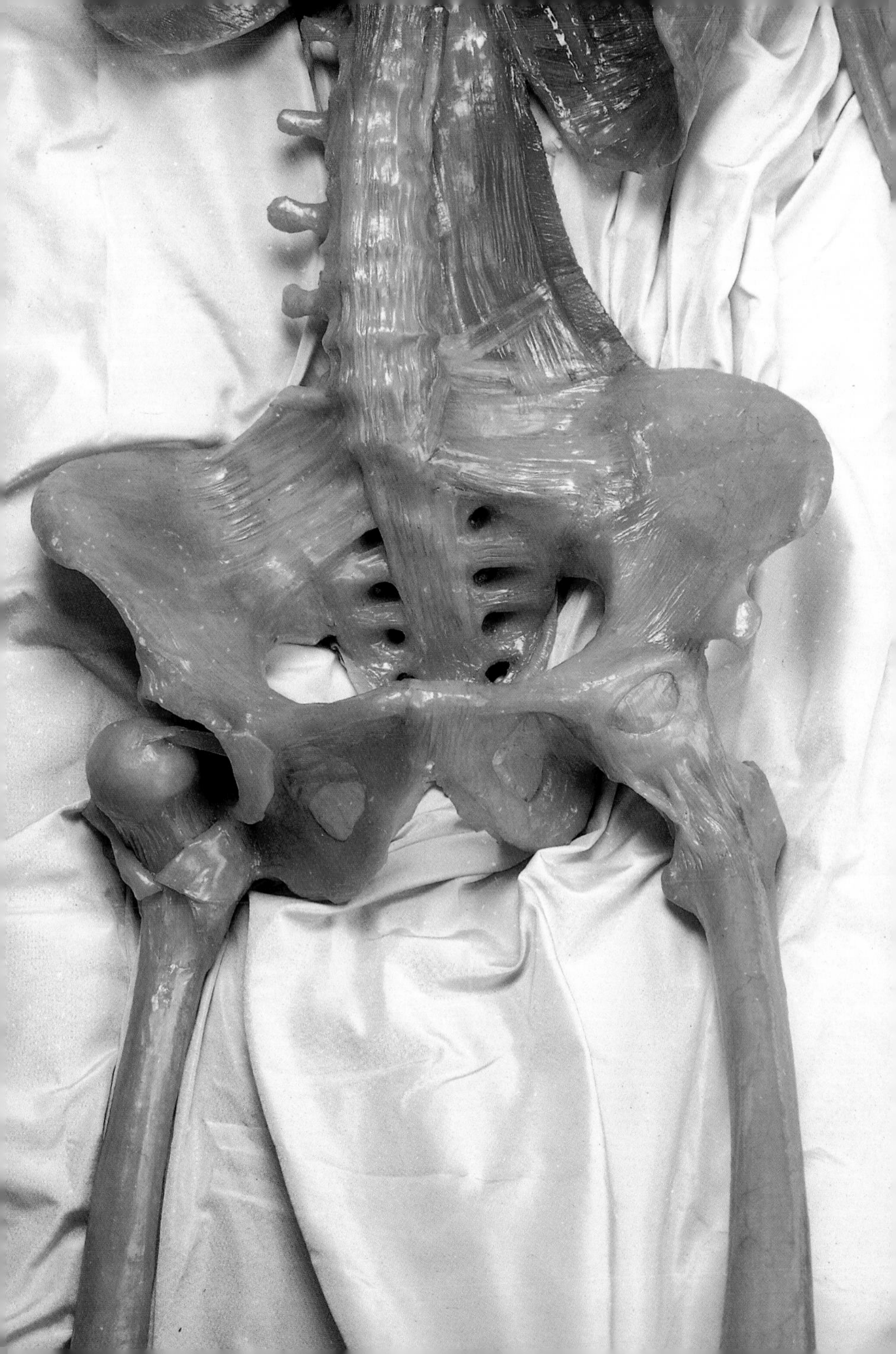

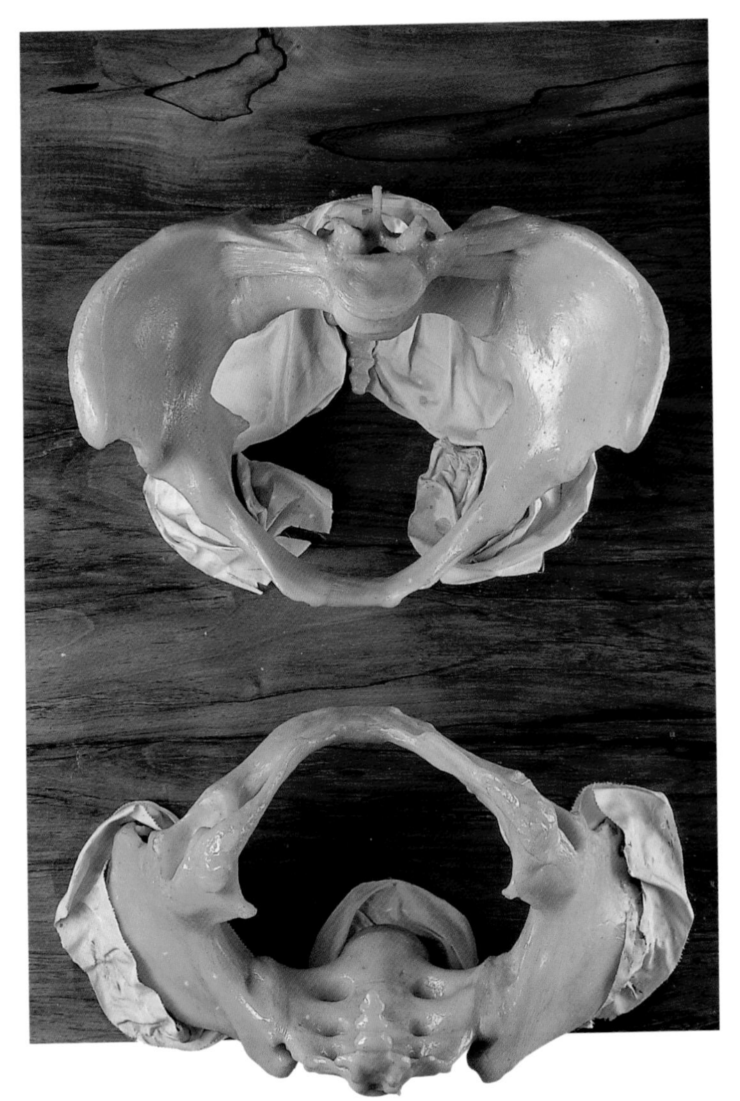

Pelvic girdle, consisting of the sacrum and two hip bones, seen from above in the upper specimen and from below in the lower specimen

Beckengürtel bestehend aus Kreuzbein und zwei Hüftbeinen, im oberen Präparat in der Ansicht von vorn oben, im unteren Präparat von unten

Articulations du bassin : vue supérieure (en haut), vue inférieure (en bas)

[XXXI, 356]

◄
detail from page | Detail von Seite | détail de la page 96-97

Pathologically deformed pelvis. Ligaments between sacrum, iliac and pubic bones have been divided.

Krankhaft verformtes Becken. Die Bandverbindungen zwischen Kreuzbein, Darmbein und Schambein sind durchtrennt.

Bassins pathologiques déformés : section des ligaments et de la symphyse pubienne

[XXXI, 360]

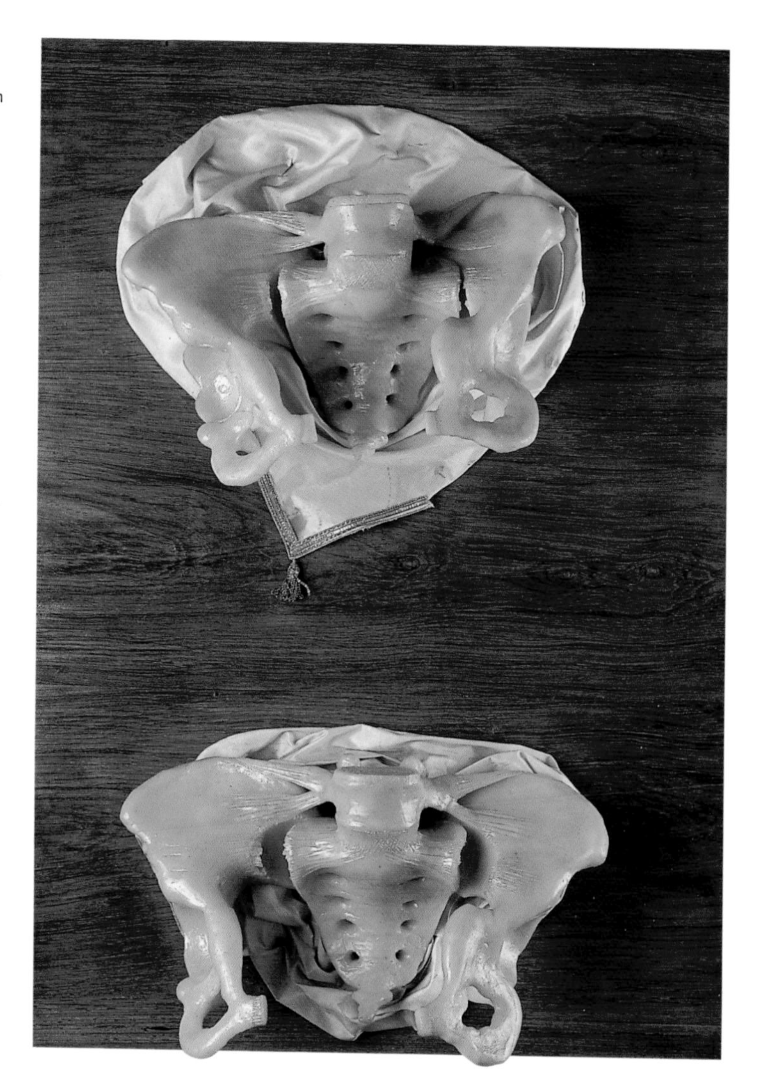

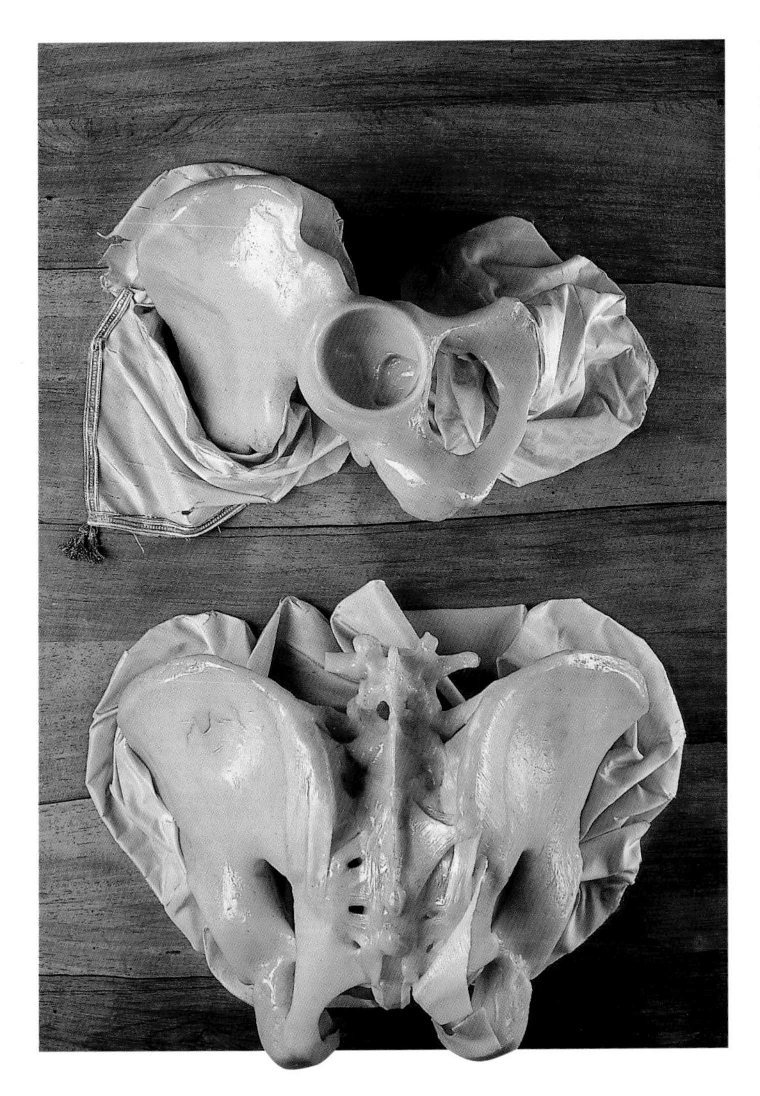

Upper specimen: hip bone with socket for the head of the femur. Lower specimen: pelvis with ligaments seen from behind

Oberes Präparat: Hüftbein mit Gelenkpfanne für den Oberschenkelkopf. Unteres Präparat: Becken von hinten mit Bändern

En haut : os coxal avec la cavité articulaire pour la tête du fémur. En bas : bassin vu de l'arrière avec les ligaments

[XXXI, 358]

Hip bone with
its ligaments

Darstellung des
Hüftbeines mit
seinen Bandver-
bindungen

Os coxal avec cer-
tains ligaments

[XXXI, 925]

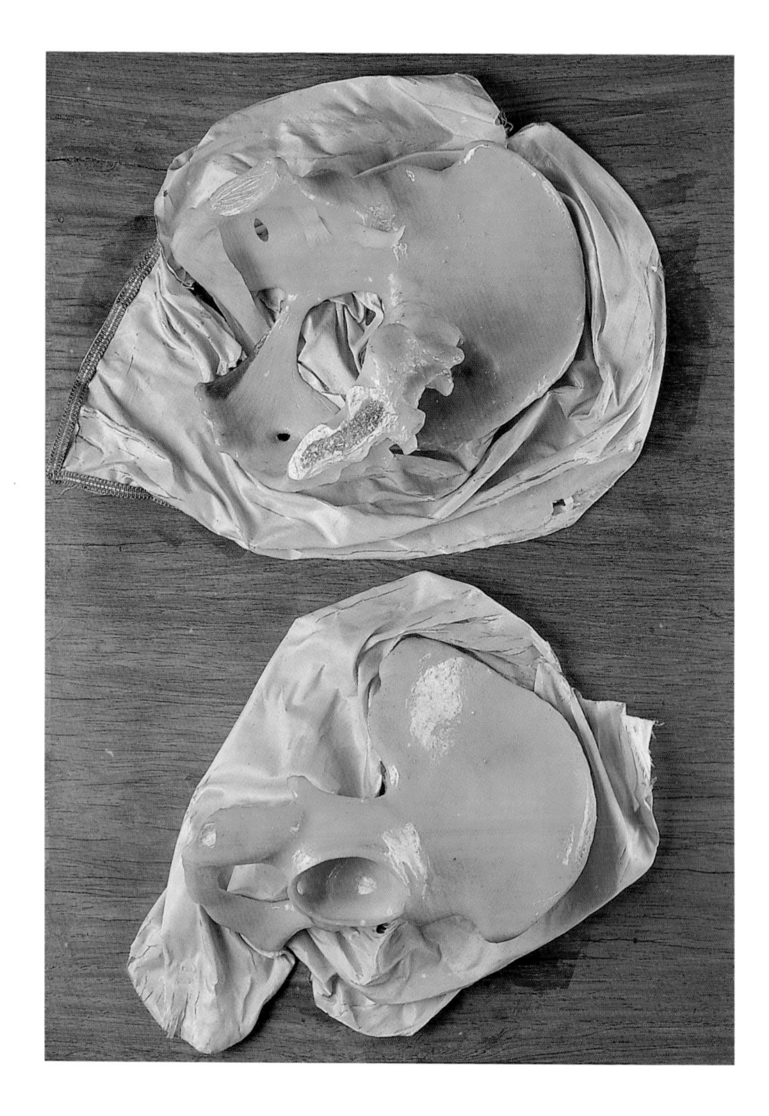

Upper specimen: right hip joint with ligaments. Lower specimen: the head of the femur has been disarticulated from the socket after dissection of the joint capsule

Oberes Präparat: rechtes Hüftgelenk mit Bandverbindungen. Unteres Präparat: der Gelenkkopf des Oberschenkelknochens ist nach Durchschneidung der Gelenkkapsel aus der Gelenkpfanne herausgedreht

En haut : articulation de la hanche droite et ligaments. En bas : tête du fémur luxée de la cavité articulaire après section de la capsule

[XXXI, 398]

Articulatio coxae, Articulatio genus

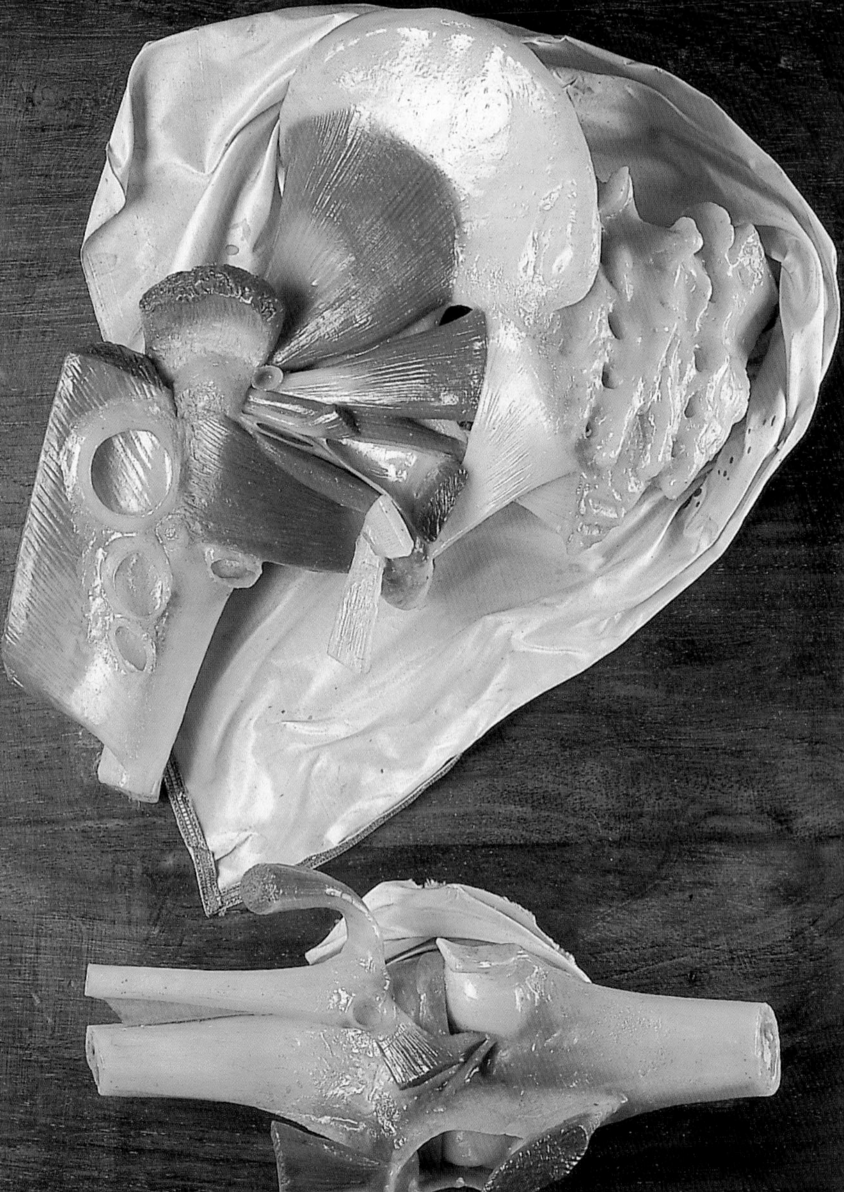

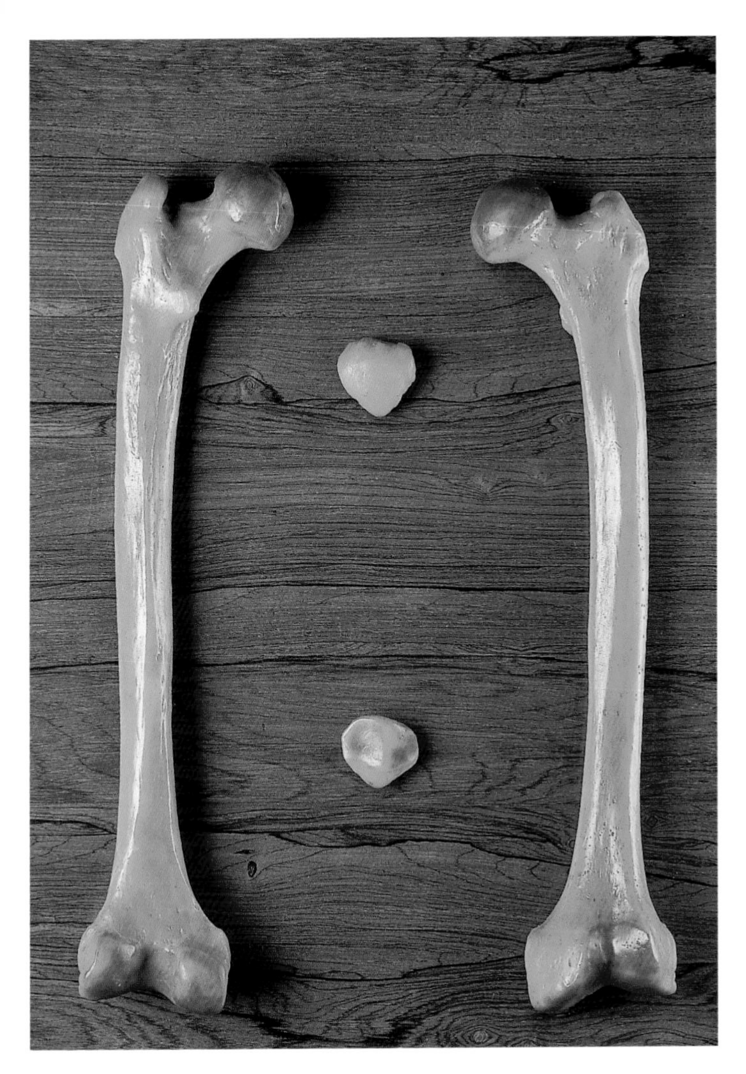

Left femur and kneecap seen from in front (left specimen) and from behind

Linker Oberschenkelknochen und Kniescheibe in der Ansicht von vorne (linkes Präparat) und hinten

Fémur et patella (rotule) gauches

[XXXI, 386]

◀

Left hip and knee joints with muscle insertions, tendons, ligaments and opened bursae seen from behind

Linkes Hüft- und linkes Kniegelenk mit Muskelansätzen, Sehnen und Bändern sowie eröffneten Schleimbeuteln in der Ansicht von hinten

Articulations de la hanche et du genou vues de l'arrière : rapports musculaires, tendons, ligaments, ouverture des bourses synoviales

[XXXI, 853]

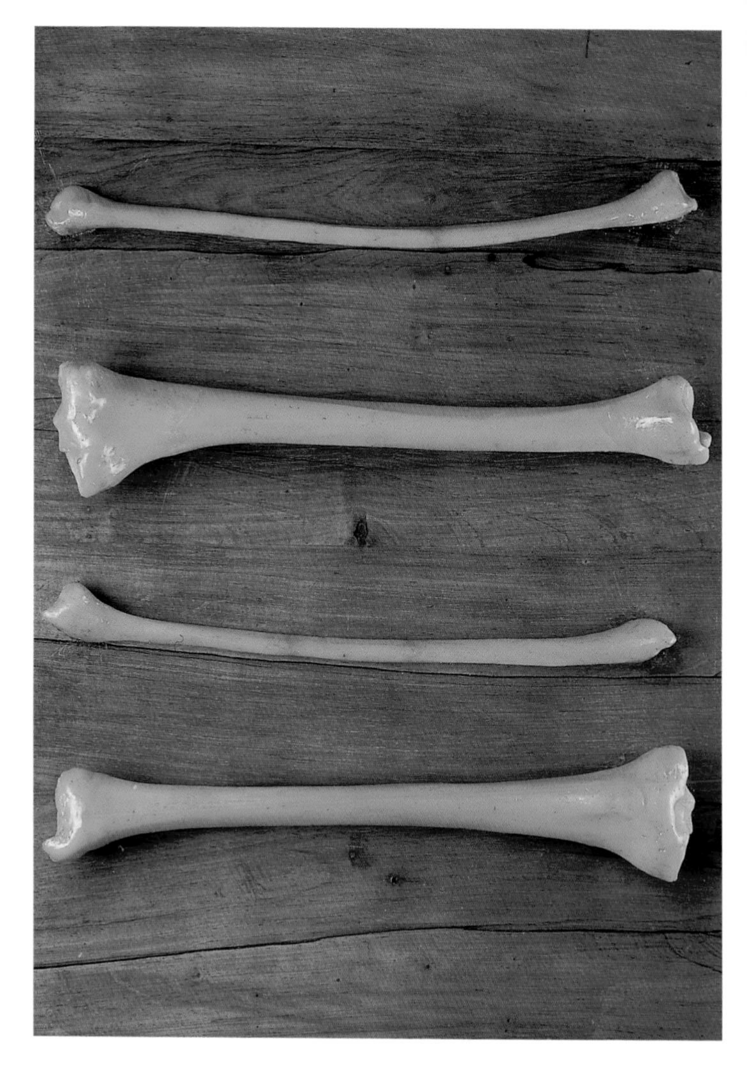

Various aspects of the left fibula and shin

Linkes Waden- und Schienbein in verschiedenen Ansichten

Os de la jambe gauche : tibia et fibula (péroné)

[XXXI, 950]

▶
Skeleton of the foot

Fußskelett

Squelette du pied et os isolés des orteils

[XXXI, 365]

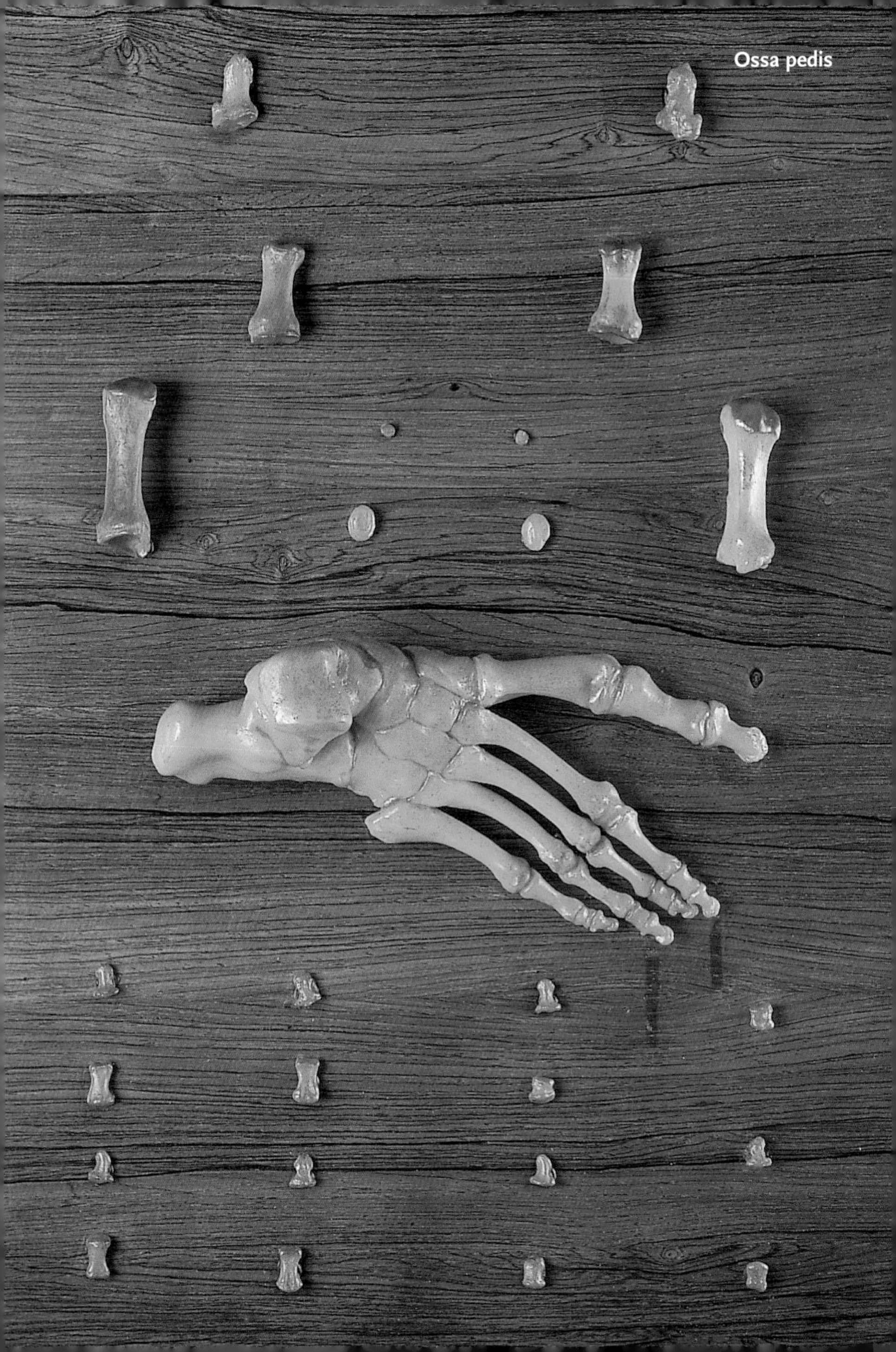

Ossa pedis

Heel-bone and bones of the foot

Fersenbein und Fußwurzelknochen

Os isolés du tarse

[XXXI, 905]

►

Bones of the middle part of the foot and of the toes

Mittelfußknochen und Zehenglieder

Os isolés du métatarse et des orteils

[XXXI, 366]

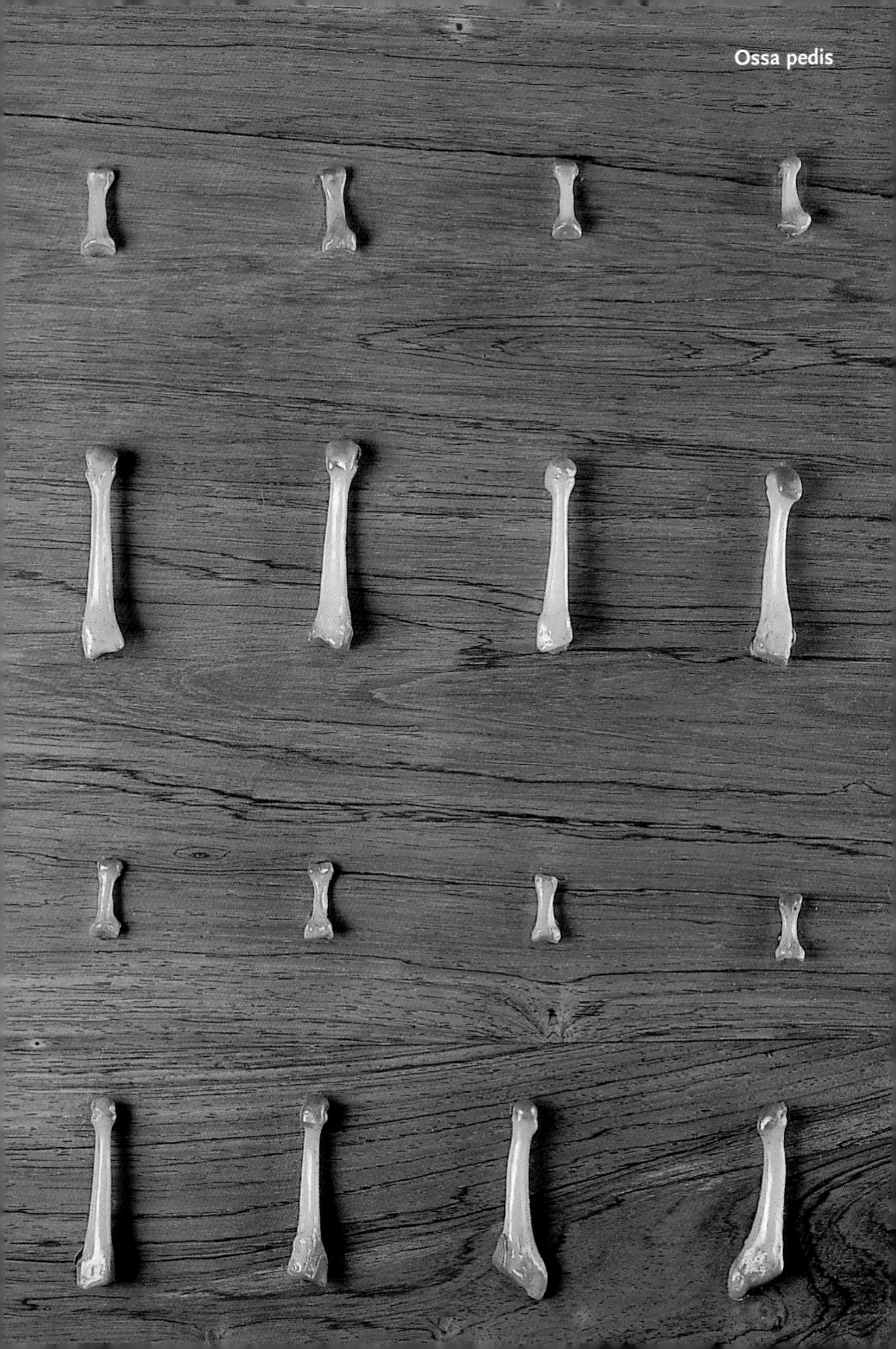
Ossa pedis

Upper specimen: knee joint from in front. Lower specimen: the capsule of the knee joint has been removed to show the cruciate ligaments and semilunar cartilages

Oberes Präparat: Kniegelenk in der Ansicht von vorne. Untere Präparate: das Gelenk ist zur Darstellung der Kreuzbänder und Menisken eröffnet

En haut : articulations du genou vues de l'avant. En bas : articulations ouvertes avec les ménisques et les ligaments croisés

[XXXI, 427]

Upper specimen: left knee joint with ligaments, tendons and muscle insertions seen from behind. Lower specimen: view of the semilunar cartilages and cruciate ligaments

Oberes Präparat: linkes Kniegelenk von hinten mit Bändern, Sehnen und Muskelansätzen. Unteres Präparat: Aufsicht auf die Menisken und Kreuzbänder

En haut : articulations du genou vues de l'arrière : ligaments, tendons et rapports musculaires. En bas : vue supérieure des ménisques et des ligaments croisés

[XXXI, 396]

Articulatio humeri, Articulatio genus

Shoulder and knee joints: the joint components, joint capsule and adipose tissue

Schultergelenk und Kniegelenk mit Darstellung von Gelenkkörpern, Gelenkkapsel und Fettgewebe

Articulation de l'épaule et articulation du genou : capsule, synoviale, corps adipeux

[XXXI, 410]

◄
Shoulder and knee joints with muscle insertions

Schulter- und Kniegelenk mit Darstellung der Muskelansätze

Articulation de l'épaule et articulation du genou avec tendons et rapports musculaires

[XXXI, 397]

The ligaments of the ankle joint and foot

Darstellungen der Bandverbindungen des Sprunggelenkes und des Fußes

Articulation de la cheville et articulations du pied avec les ligaments

[XXVI, 422]

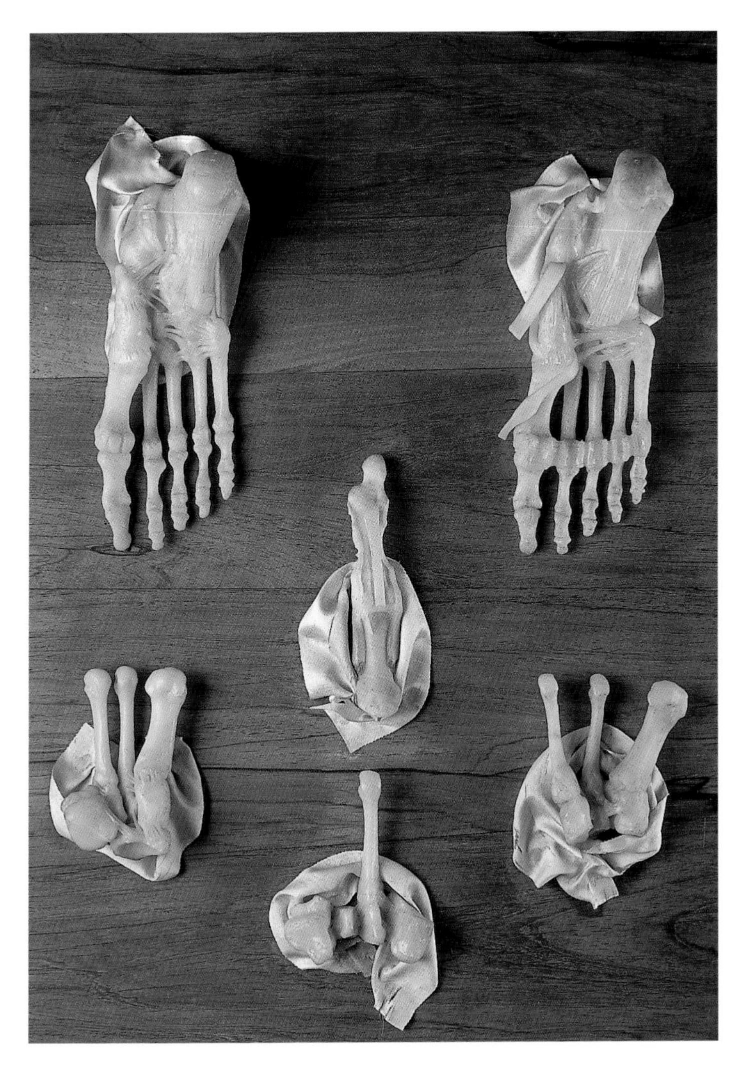

The ligaments of the ankle joint and foot

Darstellungen der Bandverbindungen der Sprunggelenke und des Fußes

Articulations du pied avec les ligaments (vue plantaire)

[XXVI, 424]

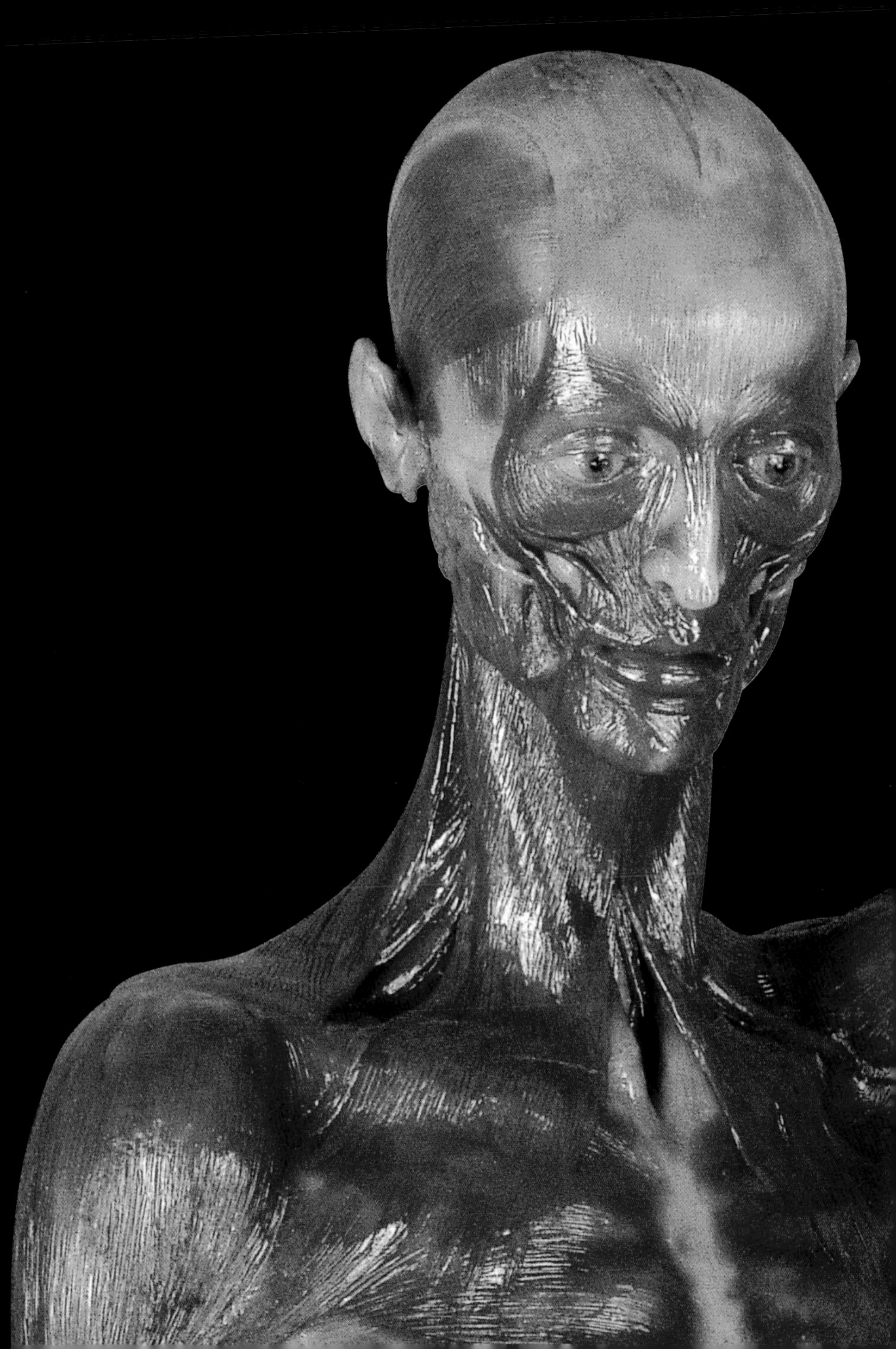

Myologia

Muscles
Muskeln
Muscles

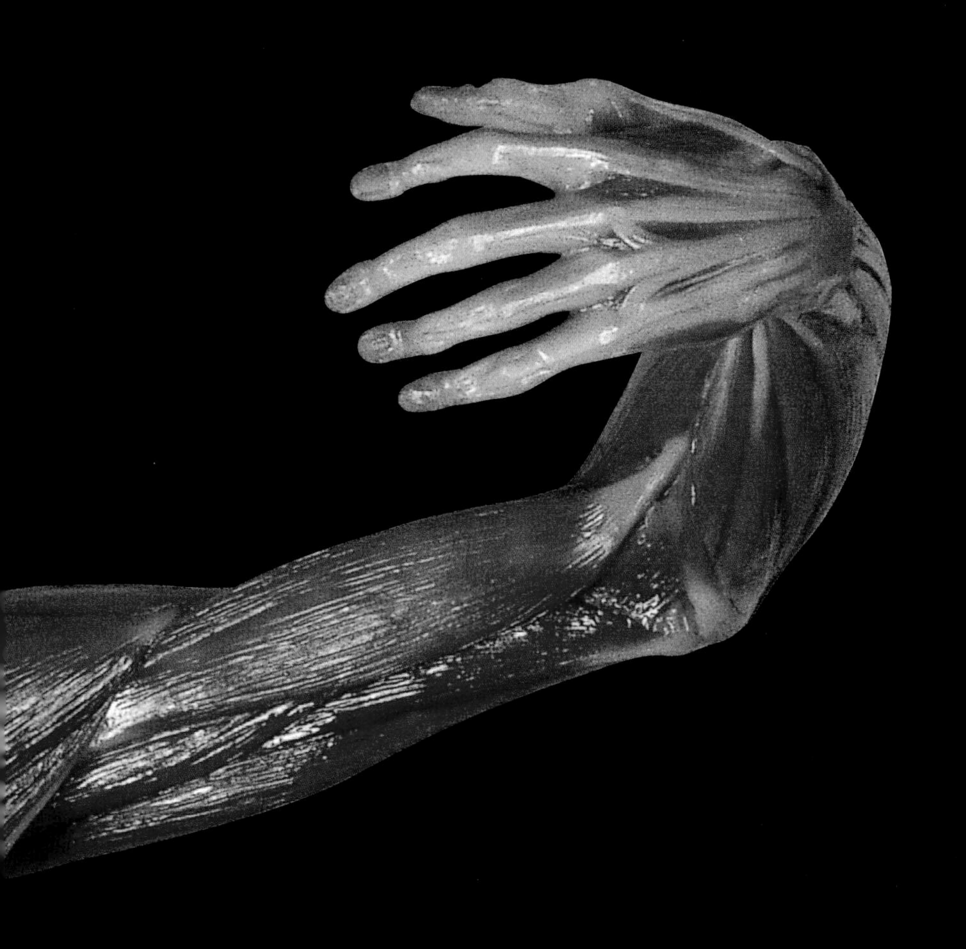

Muscles and Locomotor System

The ability of human beings to walk on two feet and the postures they take up when active or at rest necessitate highly sophisticated coordination between the different groups of muscles. Our central nervous system is responsible for the neuronal control of this complex conscious and unconscious activity, all of which is carried out by motor units standing in a particular functional relationship to one another and acting on the relevant joints as the individual muscles come into play together or in sequence. In order to produce controlled movement, those muscles which act together as agonists and antagonists must be finely tuned to work in accordance with each other, since only in this way does the overall achievement of harmonious movement become possible.

Posture, walking, grasping and eye movement are probably the kinds of movement most familiar to all people, and these can be extended to include the chewing of food, breathing and speech. The morphological basis of facial expression, which reaches the maximum variety only in our species, is provided by the numerous small muscles of the face. It is not, however, only these facial muscles but the entire musculature of the whole body, which contribute to the individual "body language" by which our conscious or unconscious non-verbal self-awareness and emotions are expressed.

This extensive and remarkable collection of exhibits certainly emphasizes the extent and variety of our locomotor system, but it also illustrates the particular relationship between muscles and bones, and thus provides a key to the understanding of both simple and complex activity. The models depicting the musculature of the hand, face, eyes and larynx are extraordinarily impressive, and their representation in this medium offers a clear insight into the finely graded and highly differentiated movement of which these organs are capable.

Pages · Seiten · Pages 128–129:

Detail from page 134

Detail der Seite 134

Détail de la page 134

Muskeln und Bewegungsapparat

Der aufrechte Gang des Menschen, seine Körperhaltung in Ruhe sowie seine Körperbewegungen erfordern ein überaus differenziertes Zusammenspiel verschiedener Muskelgruppen. Unser zentrales Nervensystem übernimmt dabei die neuronale Steuerung der komplexen bewußten und unbewußten Bewegungsabläufe. Alle Bewegungsabläufe werden von funktionell in Beziehung stehenden Muskelgruppen ausgeführt, die über mehrere Gelenke hinweg ziehen können und deren individuelle Muskeln im direkten Zusammenspiel bzw. nacheinander arbeiten. Für den harmonischen Bewegungsablauf sind die agonistisch und antagonistisch arbeitenden Muskeln in ihrer Aktivität immer aufeinander abgestimmt. Nur so entsteht der Eindruck eines harmonischen Bewegungsablaufs.

Haltungs-, Stütz-, Bewegungs-, Greif- und Augenmotorik sind für den Laien die wohl bekanntesten Bewegungsformen. Einzubeziehen sind ferner die Kaumotorik bei der Nahrungsaufnahme, die Atmungsmotorik sowie die Sprachmotorik. Morphologische Grundlage der Mimik sind die zahlreichen kleinen Gesichtsmuskeln, die in dieser Vielfalt ausschließlich im menschlichen Gesicht vorkommen. Aber nicht nur diese mimischen Muskeln, sondern alle Muskeln unseres Körpers haben ihren funktionellen Anteil an der individuellen Körpersprache, mit der wir bewußt oder unbewußt non-verbal unser Selbstwertgefühl und unsere Emotionen ausdrücken.

Die vielen außergewöhnlichen Exponate vermitteln nicht nur ein umfassendes Bild von der Formenvielfalt der Muskeln unseres Körpers, sondern geben auch Aufschluß über ihre spezielle Anordnung am Skelett und tragen so zu einem funktionellen Verständnis für einfache und komplexe Bewegungsabläufe bei. Besonders beeindruckend sind die Modelle der Muskeln von Hand, Gesicht, Augen und Kehlkopf, deren plastische Darstellung ein umfassendes Verständnis der feinabgestuften und differenzierten Bewegungen ermöglicht.

Les muscles et l'appareil locomoteur

Tous les mouvements du corps humain, et en particulier la marche bipède et l'attitude érigée au repos, nécessitent la mise en jeu coordonnée de plusieurs groupes musculaires. Le système nerveux central prend en charge la régulation des mouvements complexes volontaires et involontaires. Des groupes musculaires fonctionnellement interdépendants, pouvant croiser plusieurs articulations, sont mis en jeu lors de l'exécution des mouvements. Chaque muscle participe à l'ensemble du mouvement ou prend le relai d'un autre. Le déroulement de mouvements harmonieux nécessite une coordination permanente des muscles agonistes et antagonistes.

Les aspects les plus connus de la motricité sont pour le non spécialiste le maintien d'une position, le déplacement dans l'espace, la préhension, et les mouvements de l'œil ; la mastication, les mouvements respiratoires, ou la parole articulée sont plus rarement pris en considération. Les nombreux petits muscles de la face, particulièrement développés dans l'espèce humaine, sont les supports morphologiques de la mimique. Ces muscles de la mimique ne sont pas les seuls qui interviennent dans le langage corporel, mais tous les muscles de l'organisme participent à la manifestation consciente ou inconsciente de notre personnalité et de nos émotions.

Le nombre particulièrement important des pièces de cette série ne donne pas seulement un aperçu de la variété des formes musculaires, mais renseigne aussi sur l'agencement particulier des muscles par rapport au squelette ; ces rapports permettent de comprendre la fonction de chaque muscle dans les mouvements élémentaires ou complexes. Particulièrement remarquables sont les modèles concernant la main, la face, les yeux ou le larynx, dont la représentation plastique permet la compréhension globale de mouvements différenciés fins.

Page · Seite · Page 133:

Detail from page 320–321
Detail von Seite 320–321
Détail de page 320–321

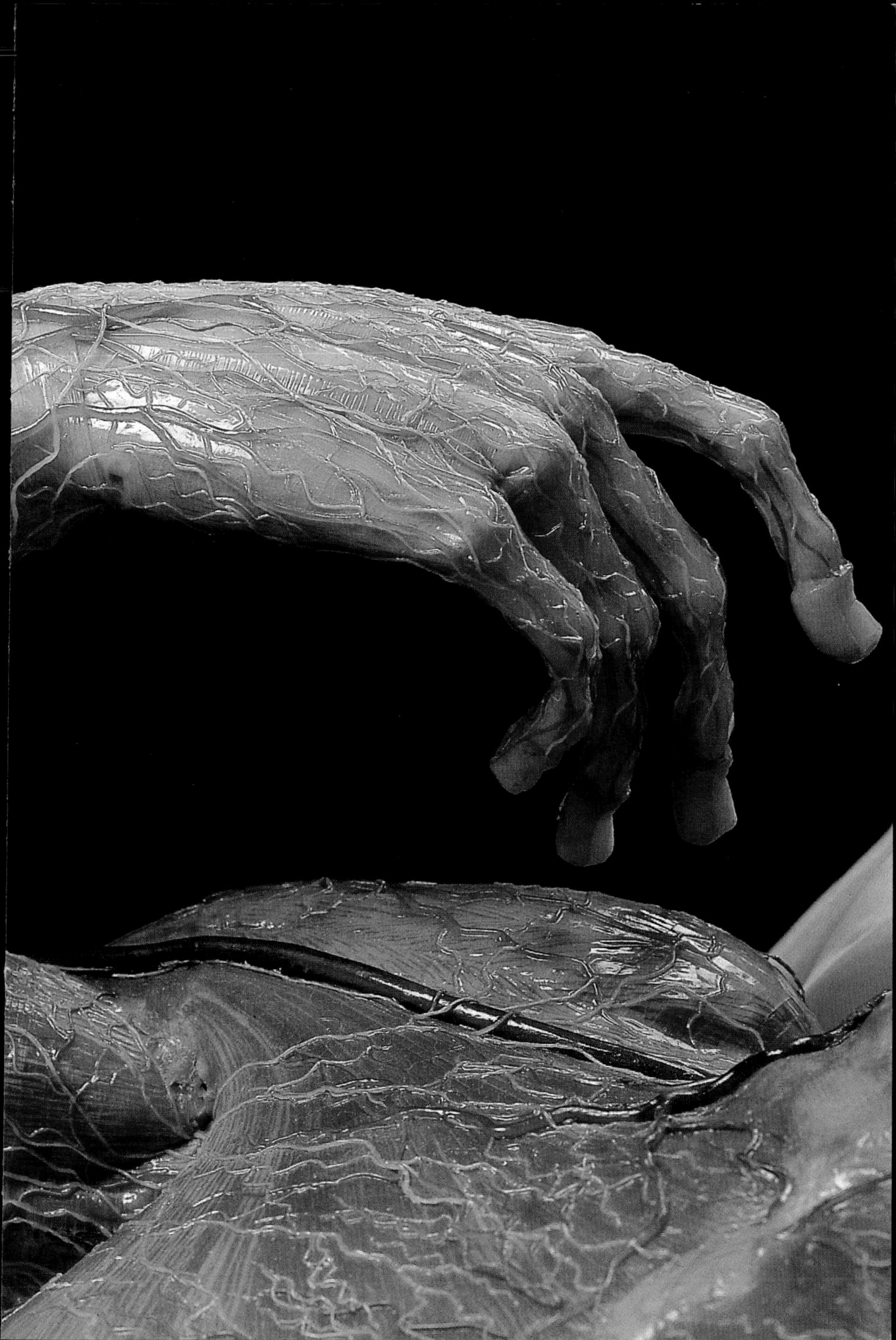

Musculi faciales, Musculi colli, Musculi thoracis, Musculi abdominis, Musculi membri superioris, Musculi membri inferioris

Whole body specimen displaying the superficial muscles

Ganzkörperpräparat zur Darstellung der oberflächlichen Muskeln

Préparation du corps entier représentant les muscles superficiels

[XXV, 444]

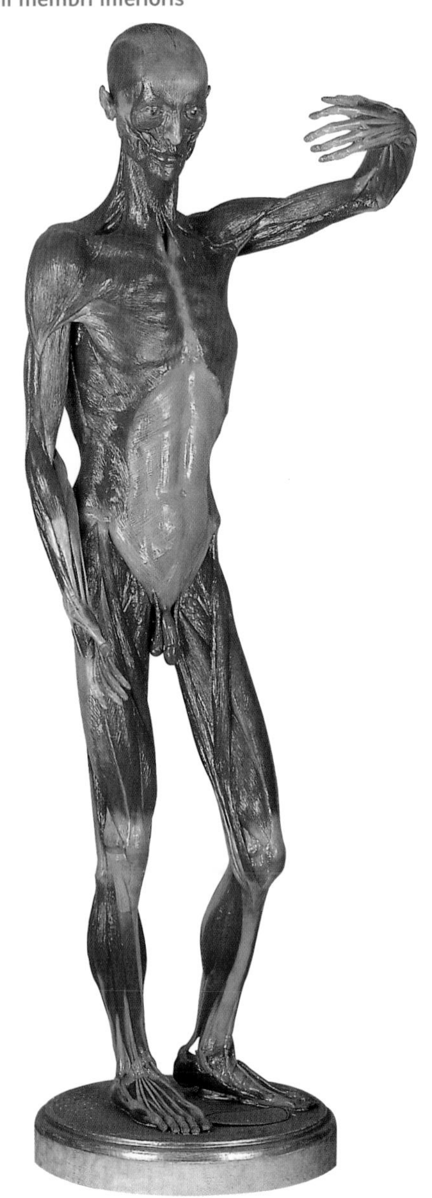

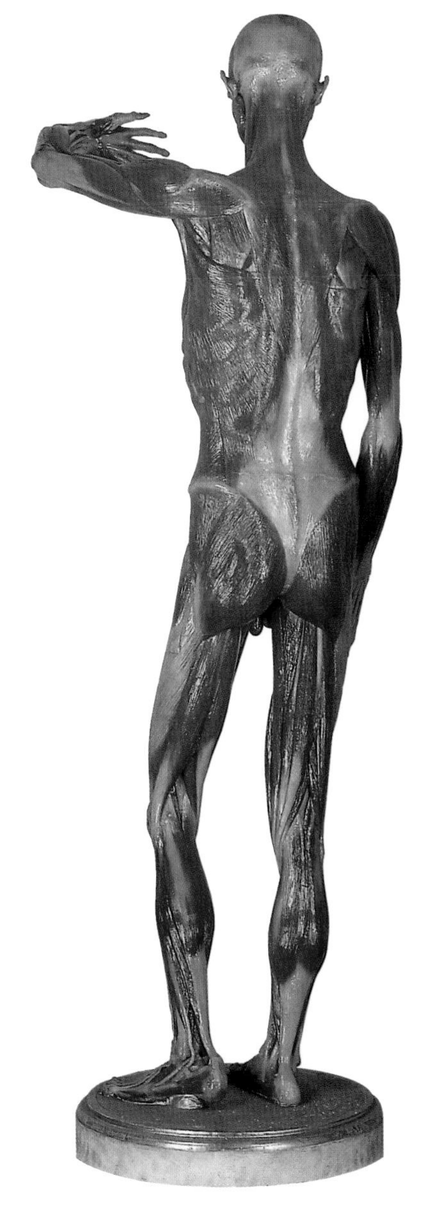

Whole body specimen displaying the superficial muscles

Ganzkörperpräparat zur Darstellung der oberflächlichen Muskeln

Préparation du corps entier représentant les muscles superficiels

[XXV, 444]

Musculi faciales, Musculi colli, Musculi thoracis, Musculi abdominis, Musculi membri superioris, Musculi membri inferioris, Musculi dorsi

Whole body specimen displaying the superficial and deep muscles

Ganzkörperpräparat zur Darstellung oberflächlicher und tiefer liegender Muskulatur

Préparation du corps entier représentant des muscles superficiels et profonds

[XXV, 442]

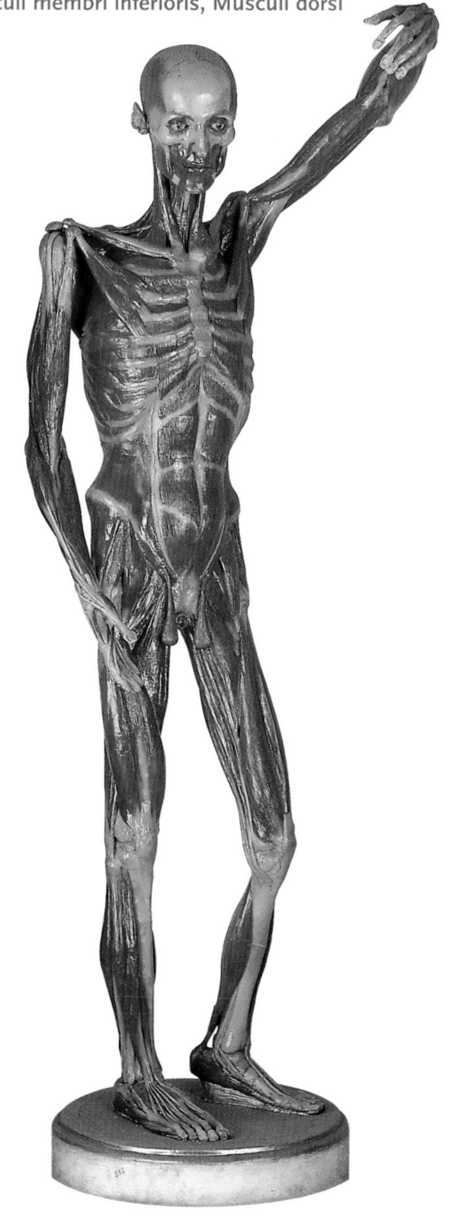

Whole body specimen displaying both the superficial and deep muscles

Ganzkörperpräparat zur Darstellung oberflächlicher und tiefer liegender Muskulatur

Préparation du corps entier représentant des muscles superficiels et profonds

[XXV, 442]

Muscles of facial expression, muscles of mastication

Mimische Muskeln und Kaumuskeln

Muscles peauciers de la mimique et muscle temporal

[XXV, 506]

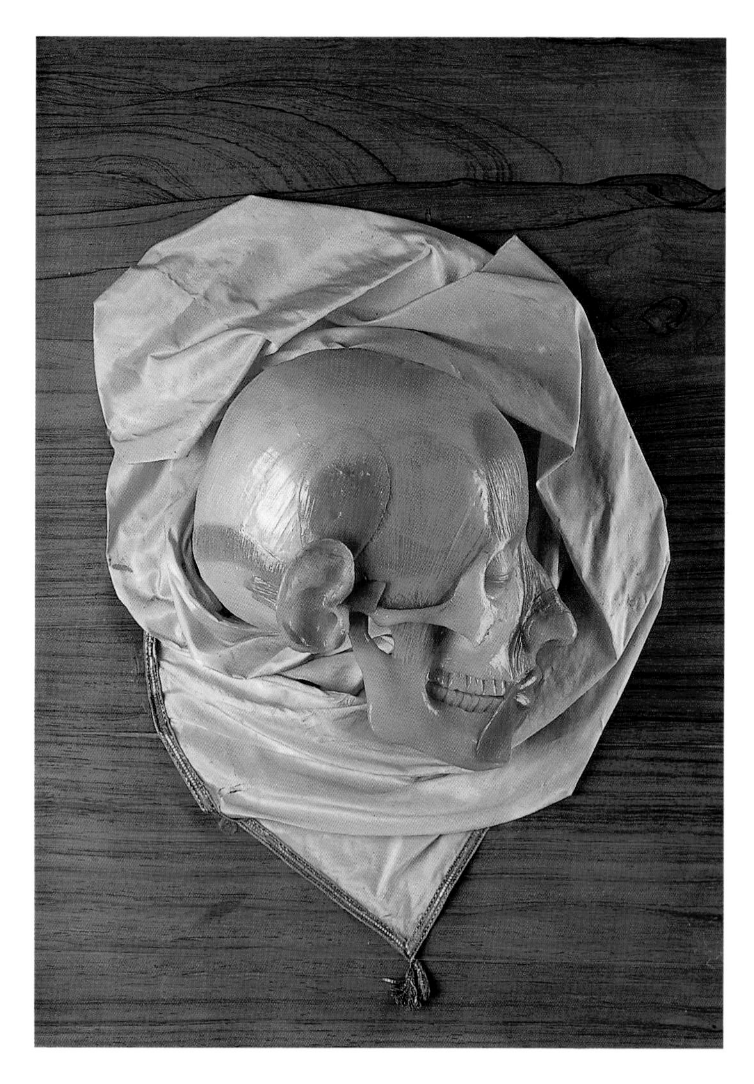

►
Views of the temporalis muscle. The lower specimen shows the attachment of this muscle of mastication to the lower jaw.

Darstellungen des Schläfenmuskels. Das untere Präparat zeigt den Ansatz dieses Kaumuskels am Unterkiefer.

Muscle temporal ; insertion du muscle temporal sur la mandibule (en bas)

[XXV, 495]

Musculus temporalis

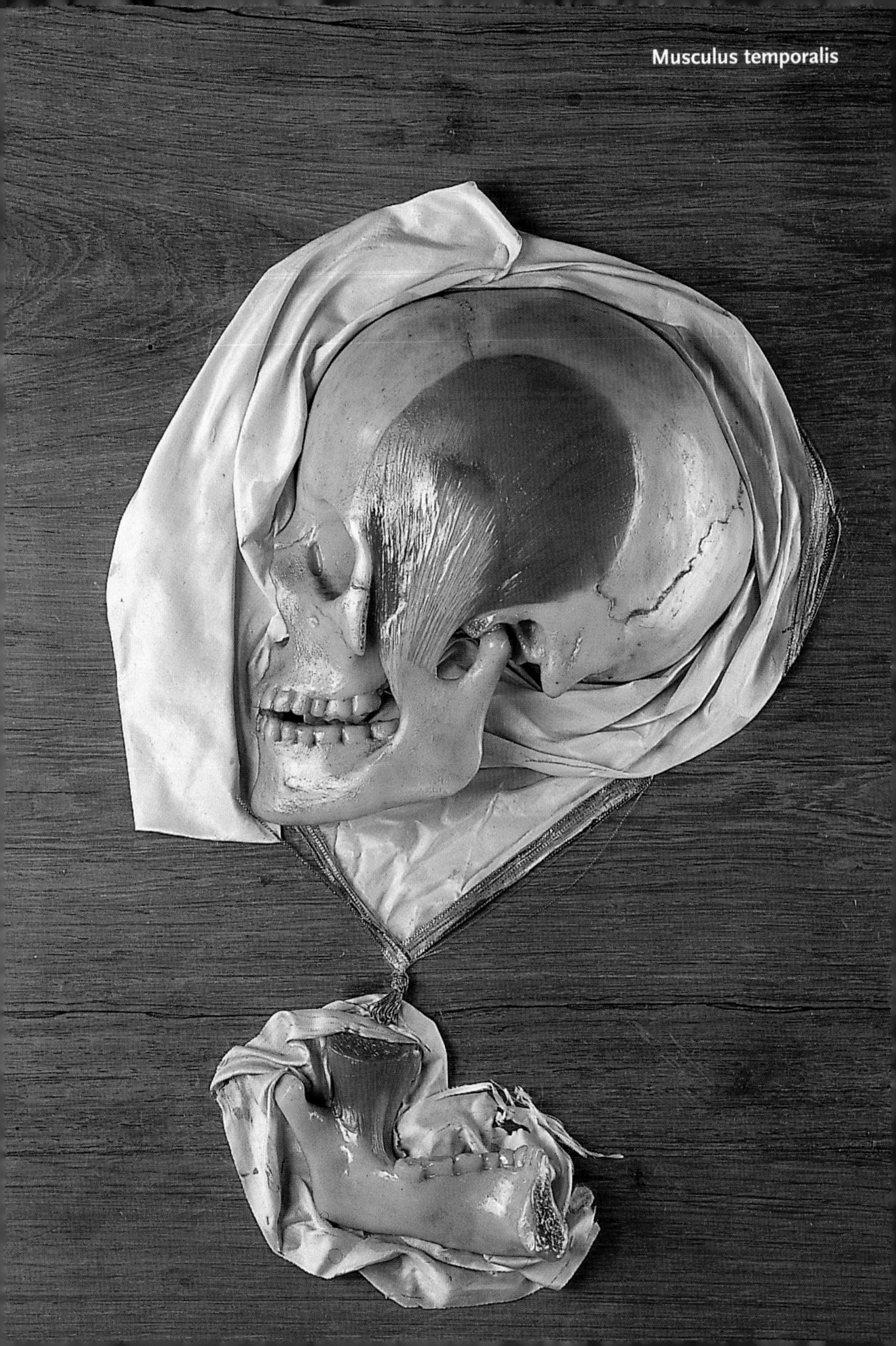

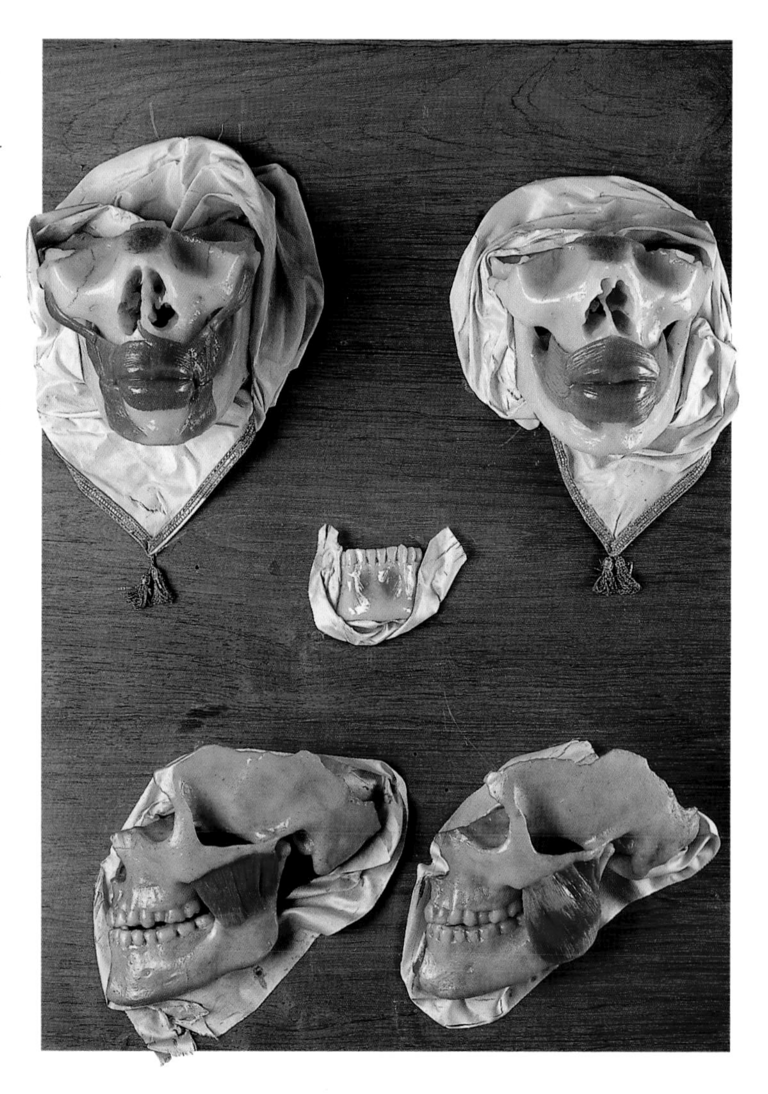

Muscles of facial expression (upper specimens) and muscle of mastication (lower specimens)

Darstellung von mimischer Muskulatur (obere Präparate) und Kaumuskeln (untere Präparate)

Muscles peauciers de la mimique (en haut) ; muscle masséter (en bas)

[XXV, 496]

▶
The more deeply placed muscles of mastication

Darstellung von Kaumuskeln aus einer tiefen Gesichtsregion

Muscles masticateurs des régions profondes de la face

[XXV, 505]

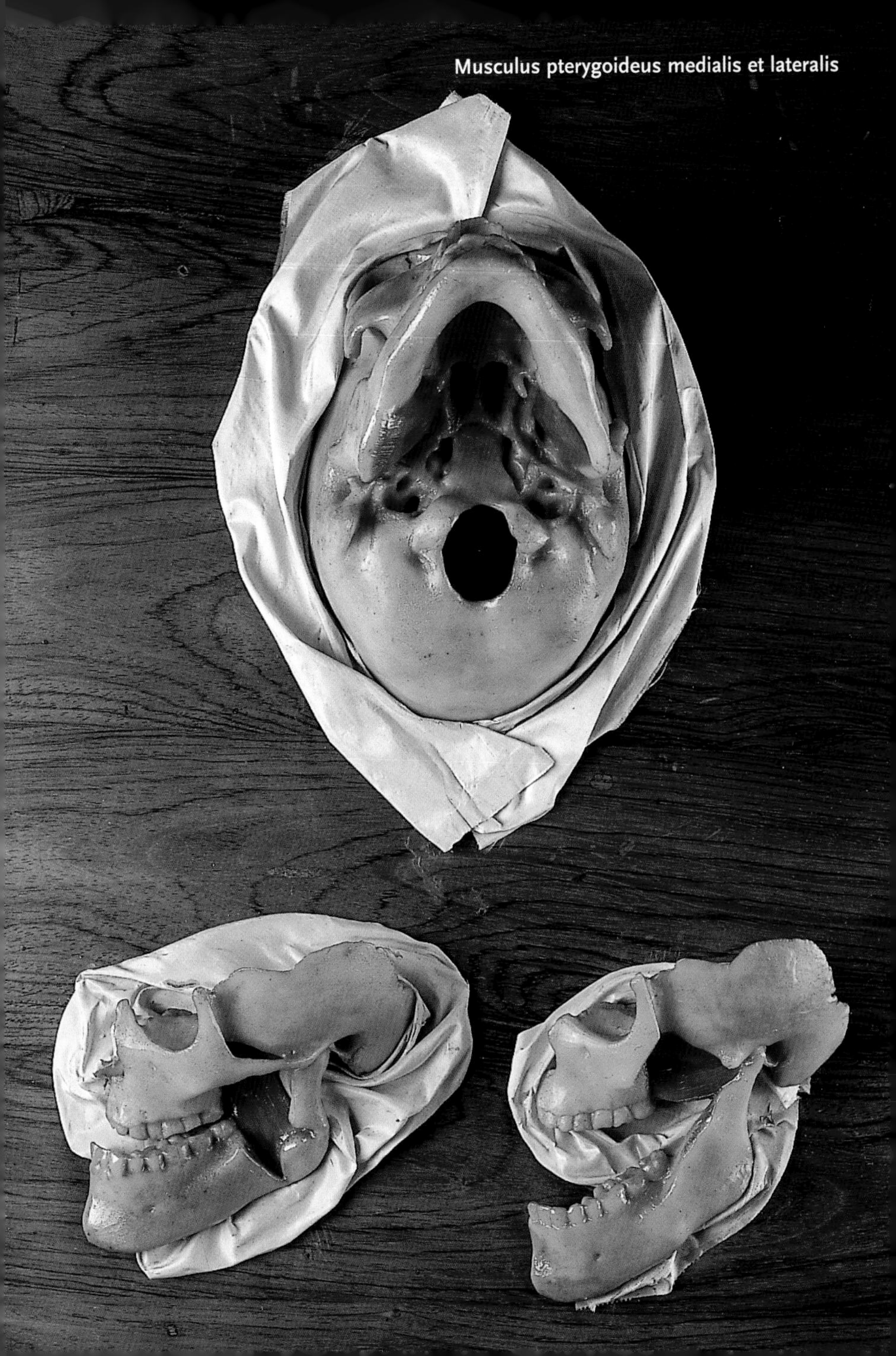

Musculus pterygoideus medialis et lateralis

Middle specimen: longitudinal section through the facial skeleton showing the oral cavity and throat. Above and to the side one can see the nasal conchae, orbital cavities and maxillary sinuses.

Mittleres Präparat: Längsschnitt durch den Gesichtsschädel mit Einblick in Mundhöhle und Rachen. Auf den oben und seitlich gelegenen Präparaten sind die Nasenmuscheln, die Augenhöhlen und die Kieferhöhlen zu erkennen.

Coupe médiane de la face : cavité nasale, cavité buccale, pharynx (au centre). Diverses préparations de muscles de la face

[XXVIII, 788]

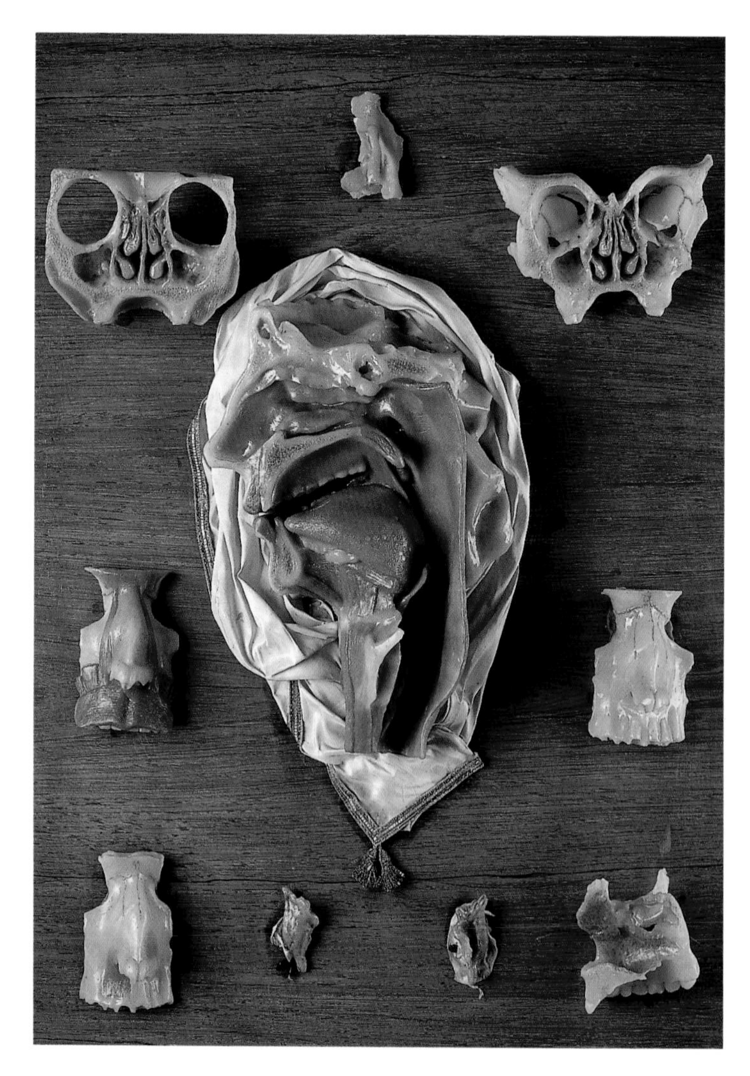

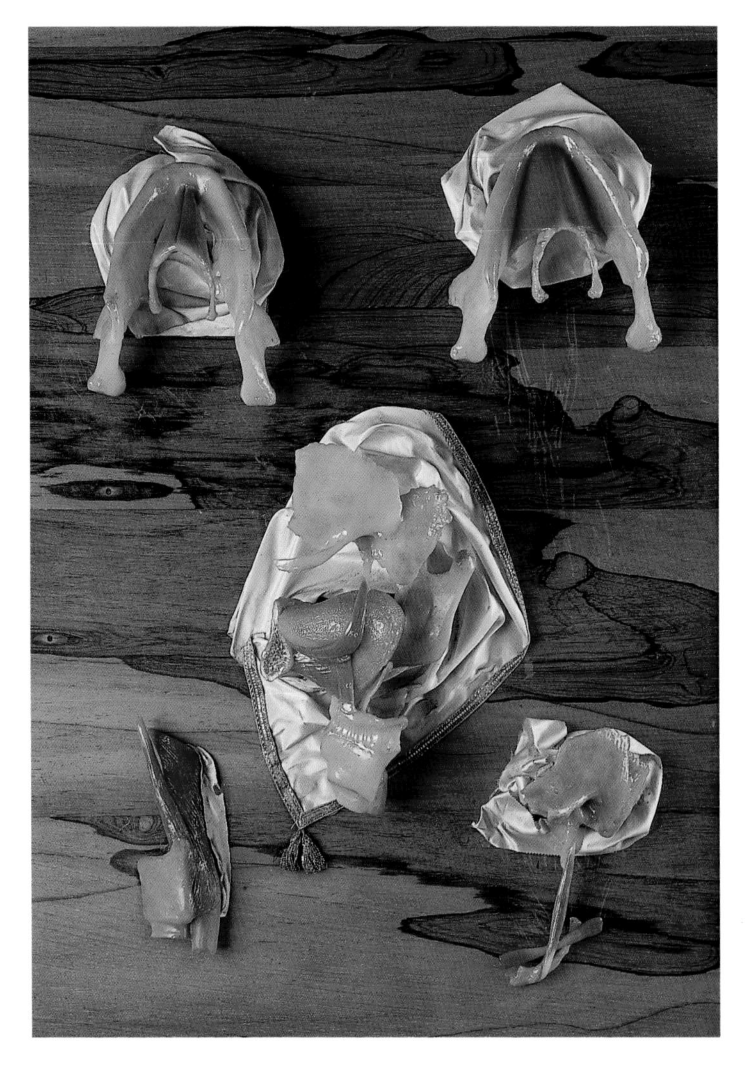

Muscles attached to the hyoid bone

Darstellungen von Zungenbeinmuskeln

Muscles insérés sur l'os hyoïde

[XXV, 504]

Muscles attached to the hyoid bone

Darstellungen der unteren Zungenbeinmuskeln

Muscles de la gorge (infra-hyoïdiens)

[XXV, 491]

Muscles of the throat seen from behind, including individual muscles of the floor of the mouth

Schlundmuskeln, von hinten gesehen, sowie einzelne Muskeln des Mundbodens

Muscles du pharynx (vus de l'arrière) ; muscles du plancher buccal

[XXV, 503]

Muscles of the throat seen from behind to show the arrangement of their various layers

Schlundmuskeln von hinten mit ihren unterschiedlich angeordneten Muskelschichten

Muscles du pharynx vus de l'arrière

[XXV, 479]

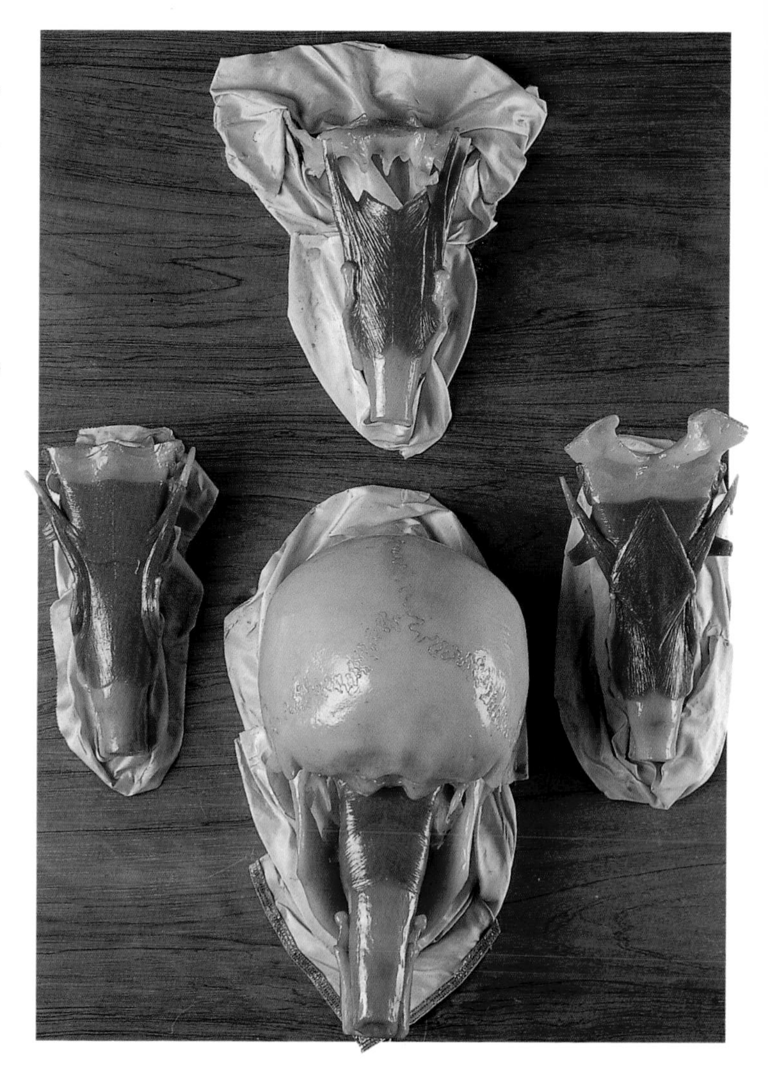

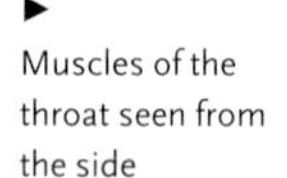

►
Muscles of the throat seen from the side

Schlundmuskulatur von der Seite

Muscles du pharynx vus du côté gauche

[XXV, 483]

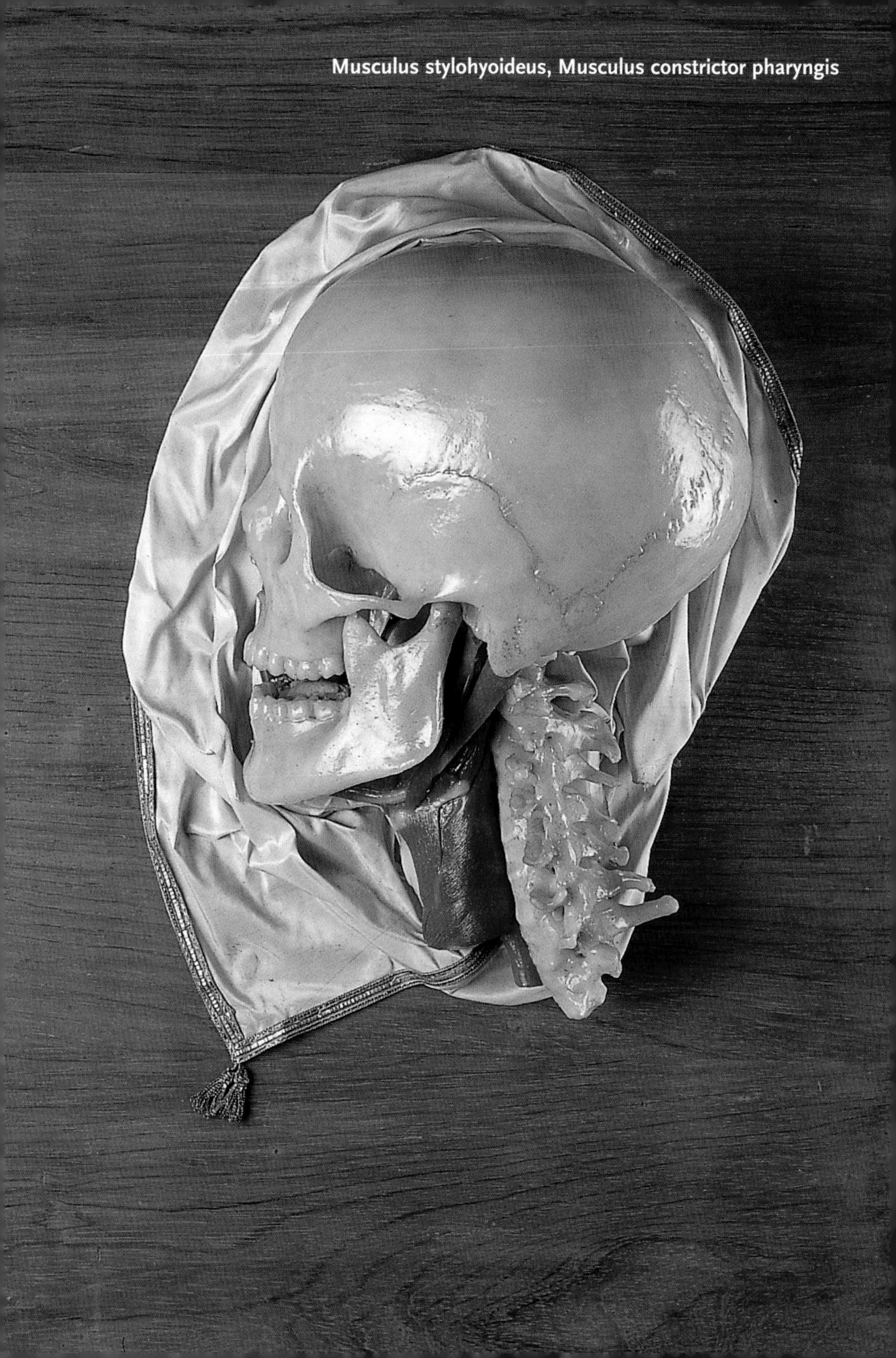

Musculus stylohyoideus, Musculus constrictor pharyngis

Musculus constrictor pharyngis, Musculus geniohyoideus, Musculus genioglossus, Musculus hyoglossus

Deeper layer of the muscles of the floor of the mouth, tongue and throat seen from the side

Tiefere Schicht der Mundboden-, Zungen- und Schlundmuskeln in seitlicher Ansicht

Muscles de la langue, du plancher buccal, et du pharynx, vus du côté gauche

[XXV, 484]

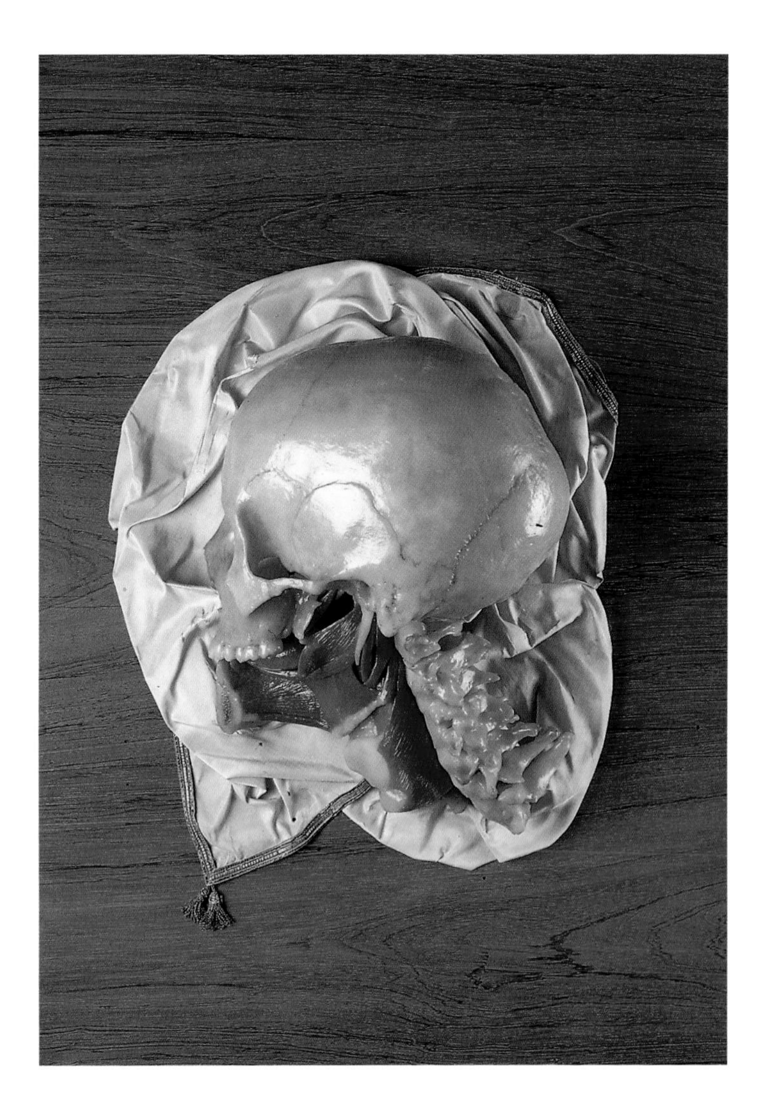

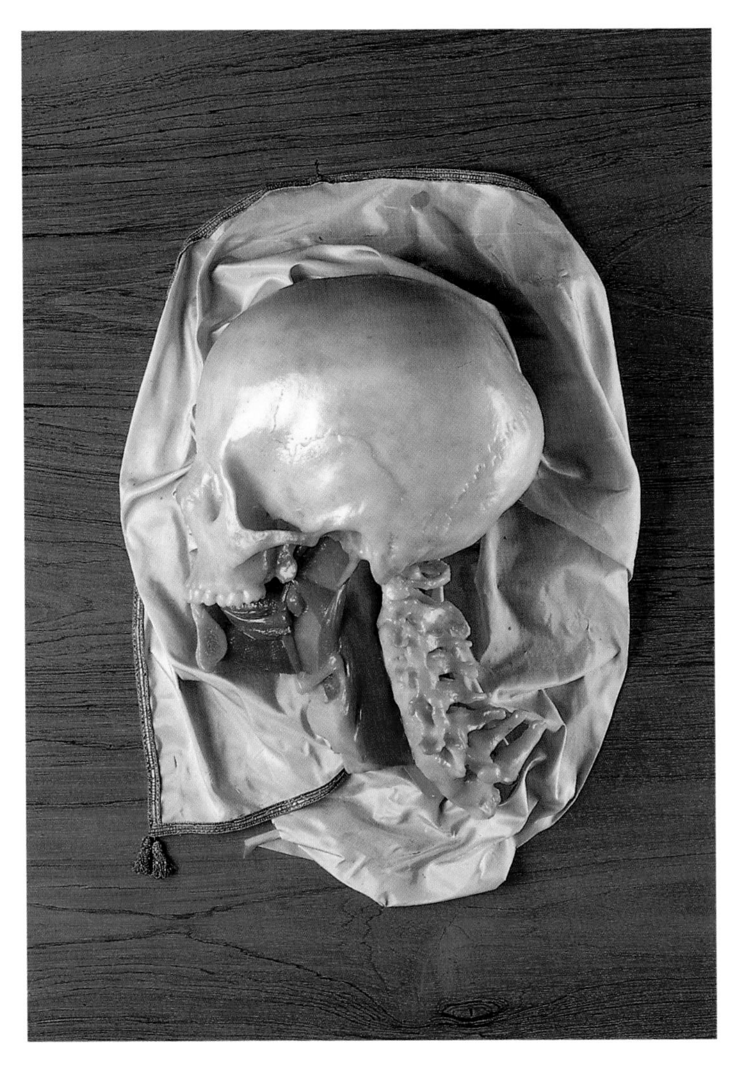

The deeper layer of muscles of the tongue, floor of the mouth and pharynx seen from the side after removal of part of the bone of the lower jaw

Zungen-, Mundboden- und Schlundmuskulatur von der Seite in einer tieferen Schicht nach Entfernung eines Teiles des Unterkieferknochens

Muscles de la langue, du plancher buccal, et du pharynx, vus du côté gauche

[XXV, 485]

View from the side into the opened pharynx. Parts of the cartilaginous skeleton of the larynx can be seen.

Blick von der Seite in den eröffneten Rachenraum. Anteile des knorpeligen Kehlkopfskeletts sind sichtbar.

Muscles de la langue et du pharynx (pharynx ouvert), cartilages du larynx, vus du côté gauche

[XXV, 492]

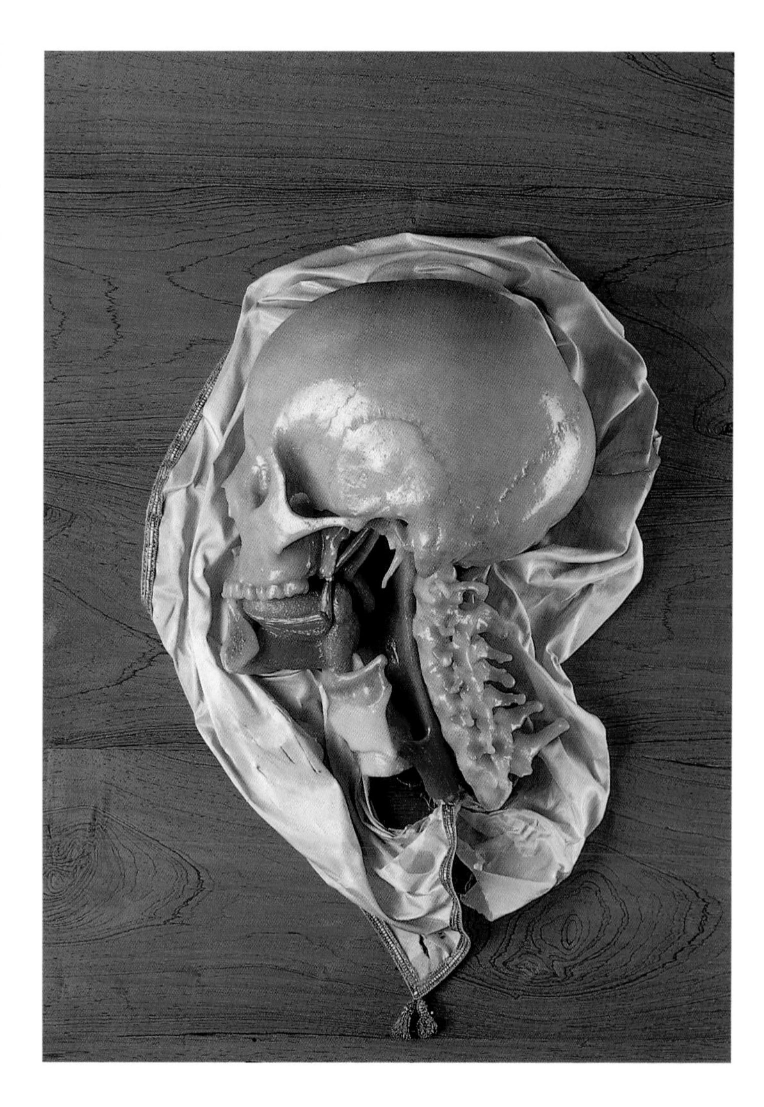

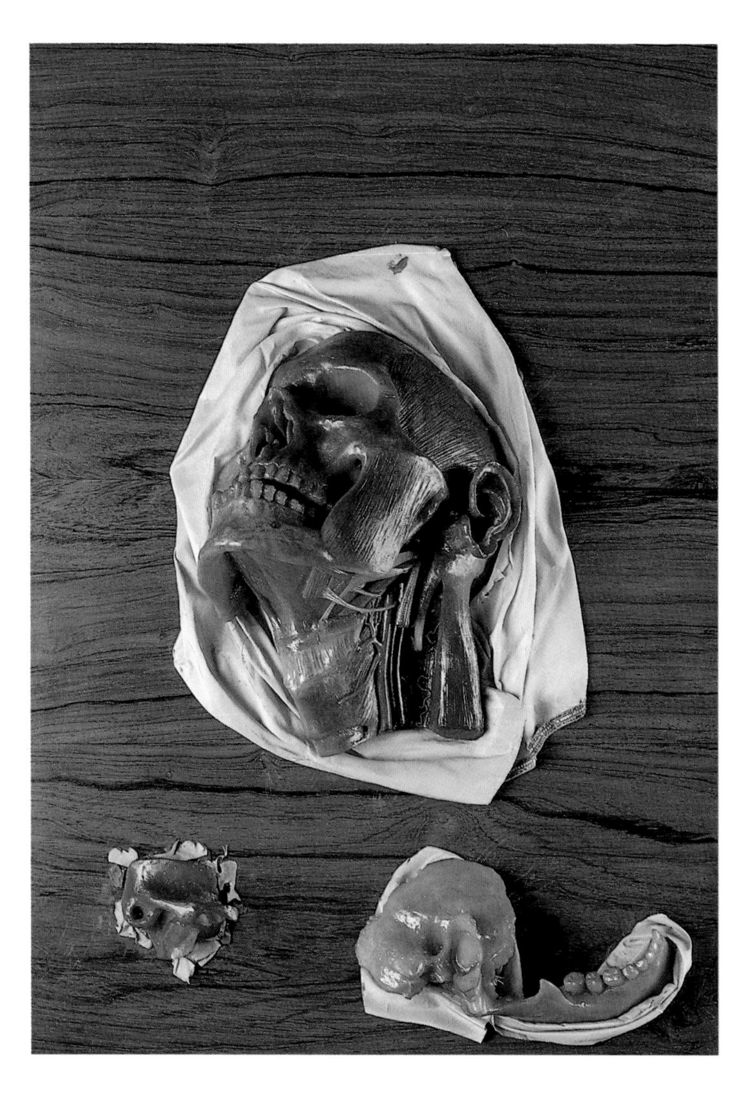

Display showing the floor of the mouth, muscles of mastication and jaw joint

Darstellung von Mundboden, Kaumuskulatur und Kiefergelenk

Muscles du plancher buccal et muscles masticateurs ; articulation temporo-mandibulaire

[XXV, 460]

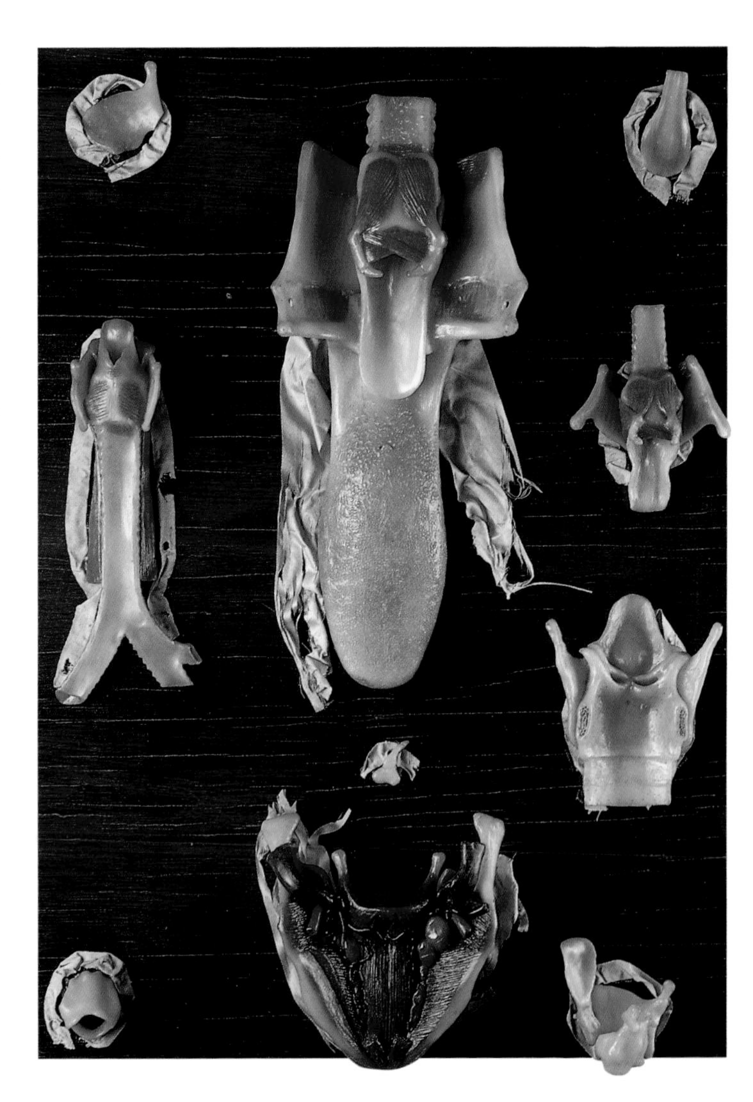

Display showing the larynx with ligaments and muscles, the trachea (left specimen) and the tongue (middle specimen)

Darstellungen des Kehlkopfes mit Bändern und Muskeln sowie der Zunge (mittleres Präparat) und der Luftröhre (Präparat links)

Muscles et ligaments du larynx : avec la langue (au centre), avec la trachée (à gauche)

[XXV, 786]

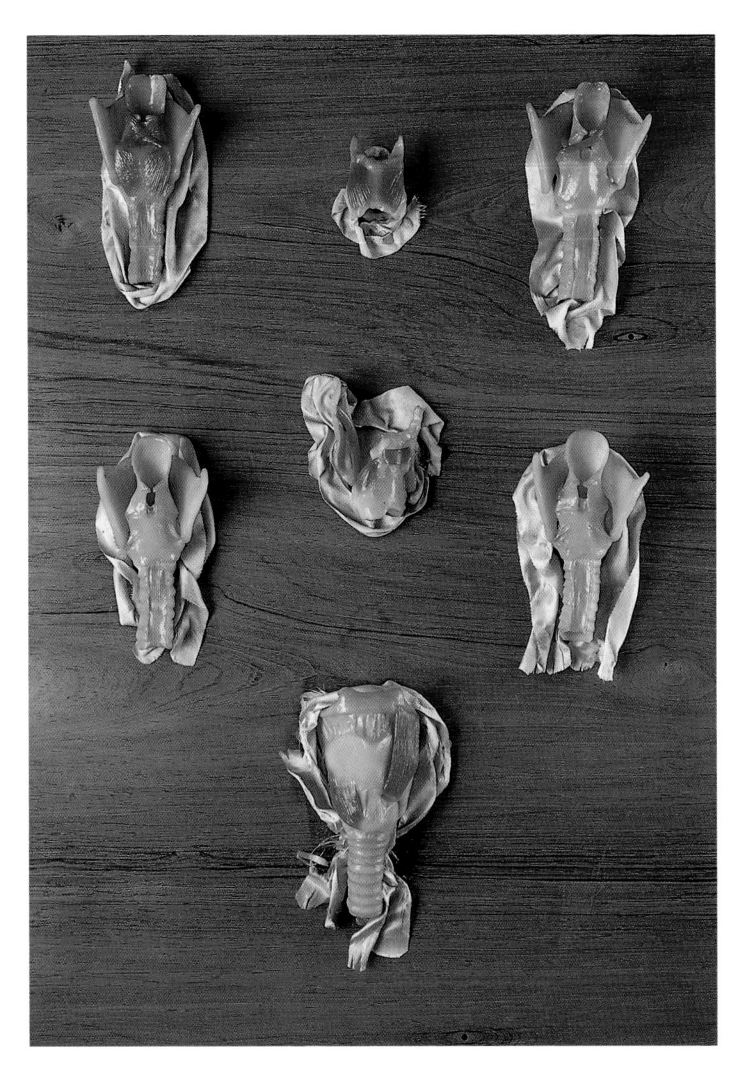

Display showing the internal and external muscles of the larynx

Darstellungen der inneren und äußeren Kehlkopfmuskeln

Muscles du larynx

[XXV, 493]

Display showing individual muscles of the larynx

Darstellungen einzelner Kehlkopfmuskeln

Muscles du larynx

[XXV, 494]

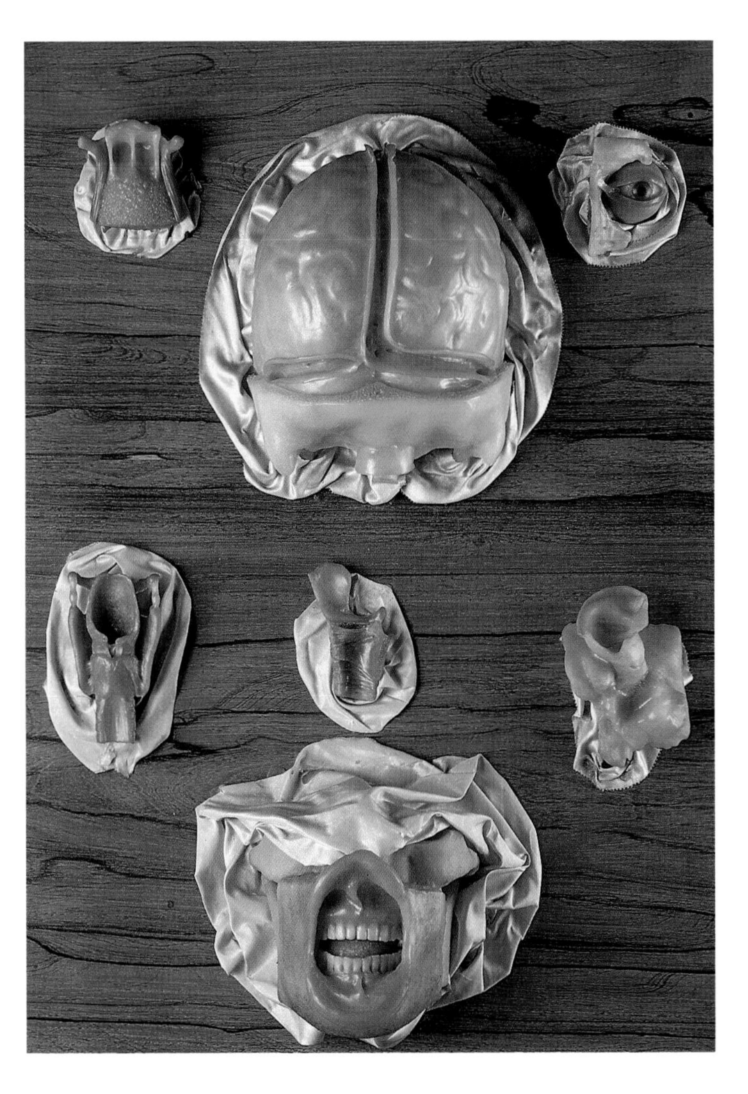

Selected specimens from the head and neck region

Ausgewählte Präparate aus dem Kopf- und Halsbereich

Différentes préparations de détails de la tête et du cou

[XXV, 472]

Display showing some muscles of the back

Darstellung eines Teiles der Rücken-muskulatur

Muscles érecteurs du rachis (spinaux)

[XXV, 514]

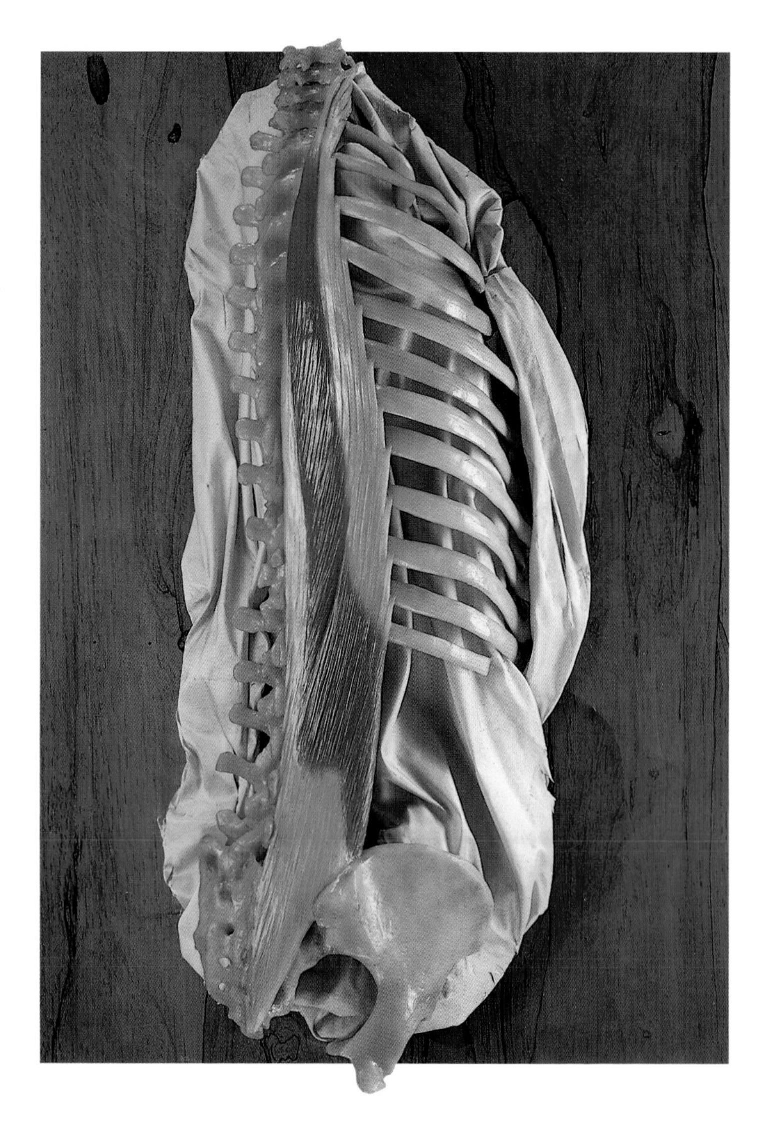

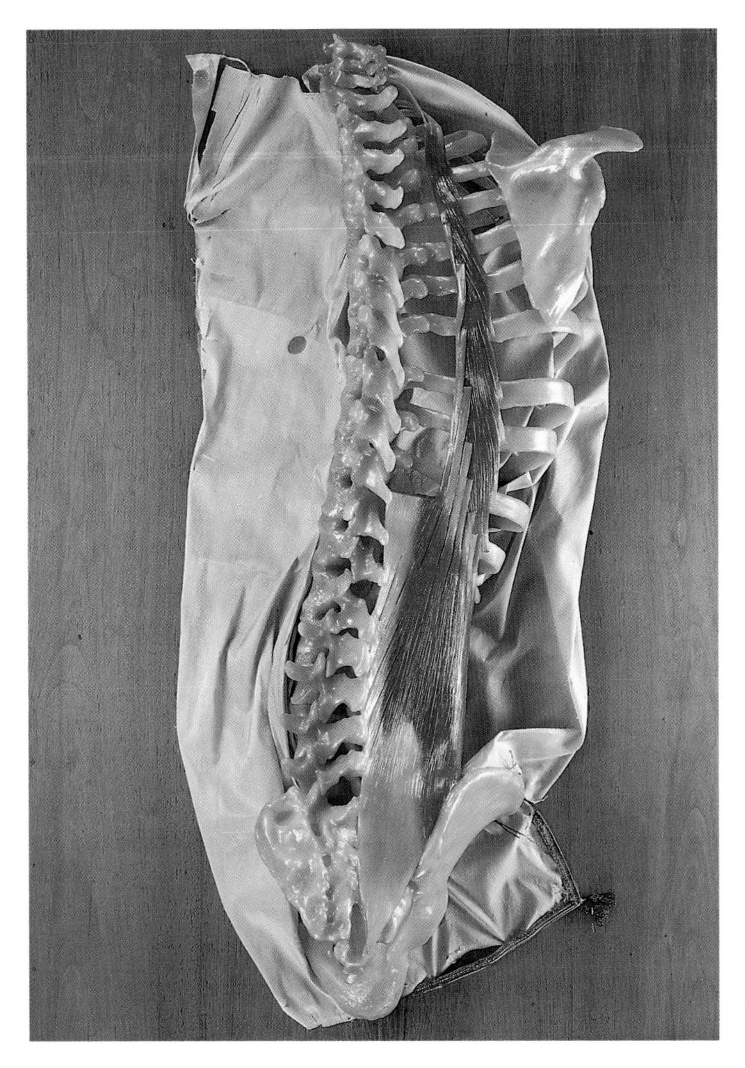

Display showing some muscles of the back

Darstellung eines Teiles der Rückenmuskulatur

Muscles érecteurs du rachis (spinaux)

[XXV, 511]

Musculus semispinalis capitis

Display showing a deep extensor muscle of the back

Darstellung eines tiefen Streckmuskels des Rückens

Muscles du plan profond de la nuque insérés sur la colonne vertébrale

[XXV, 489]

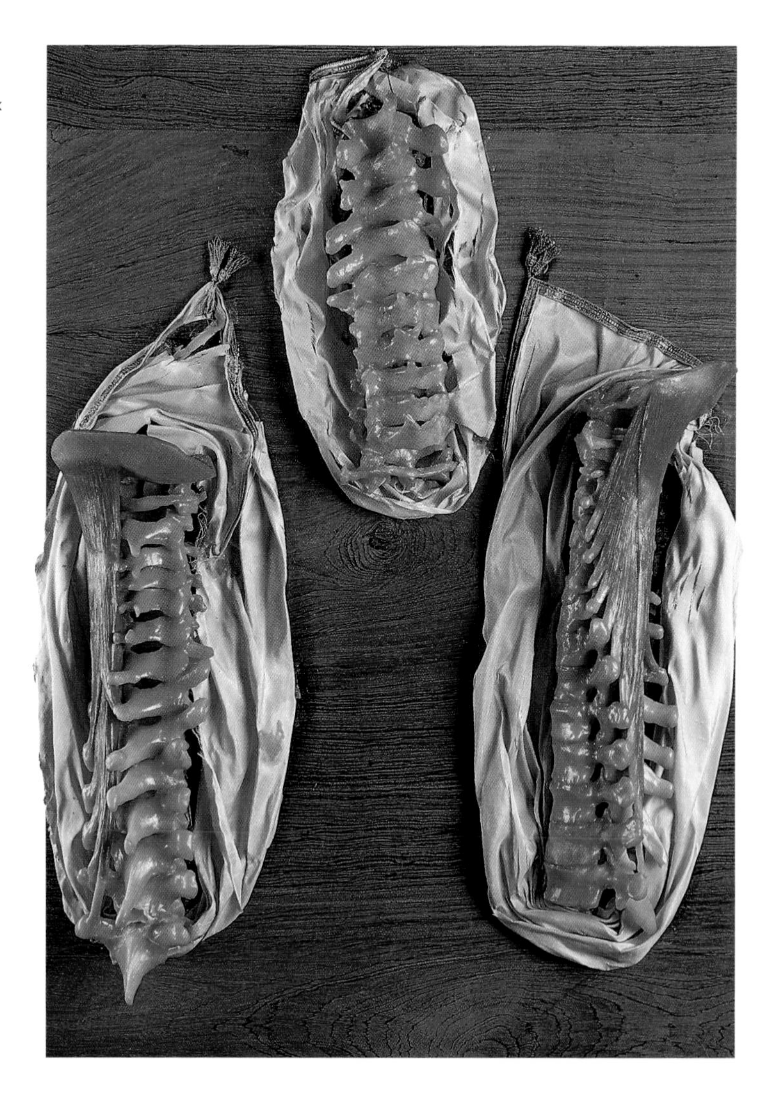

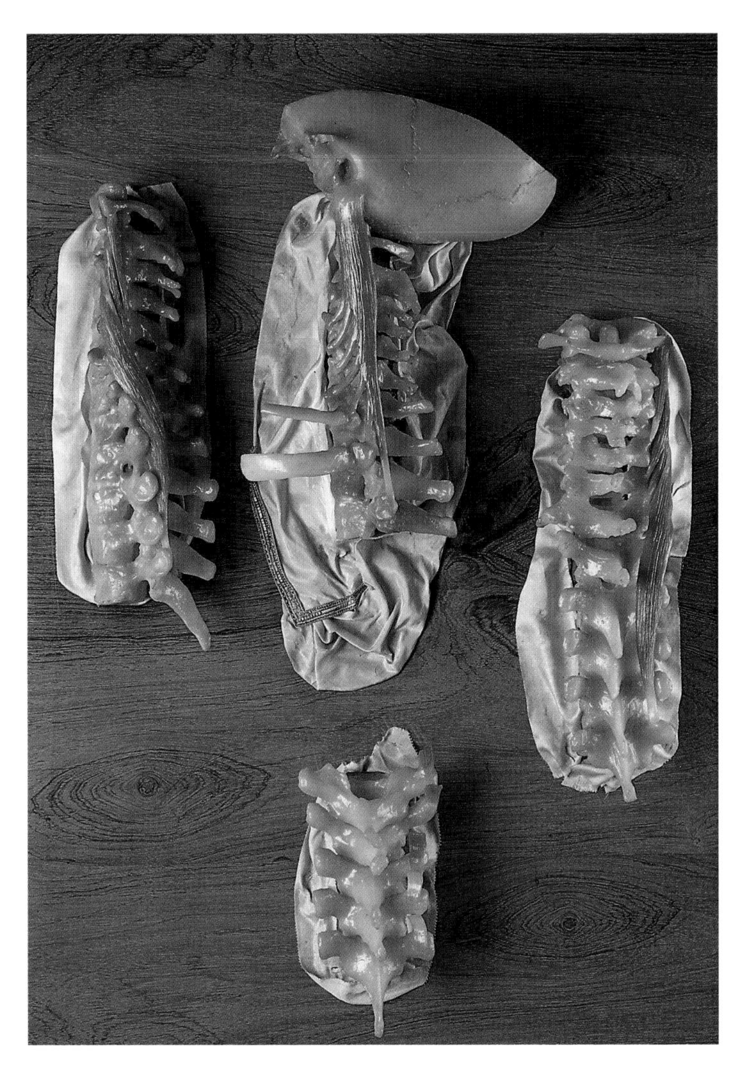

Display showing deep extensor muscles of the back

Darstellung von tiefen Streckmuskeln des Rückens

Muscles profonds de la nuque insérés sur la colonne vertébrale

[XXV, 481]

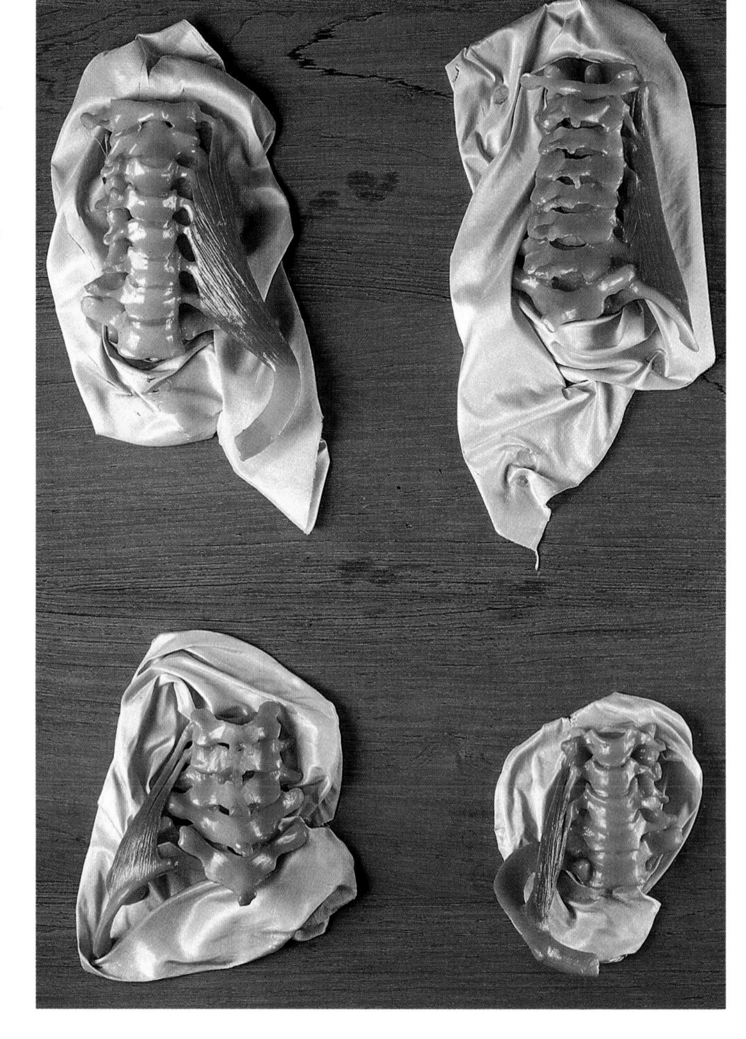

Display showing deep muscles at the side of the neck

Darstellung von tiefen seitlichen Halsmuskeln

Muscles profonds de la région latérale du cou (scalènes)

[XXV, 474]

►
A deep neck muscle in front of the cervical vertebral column

Tiefer Halsmuskel auf der Vorderseite der Halswirbelsäule

Muscles profonds du cou dans la région prévertébrale

[XXV, 488]

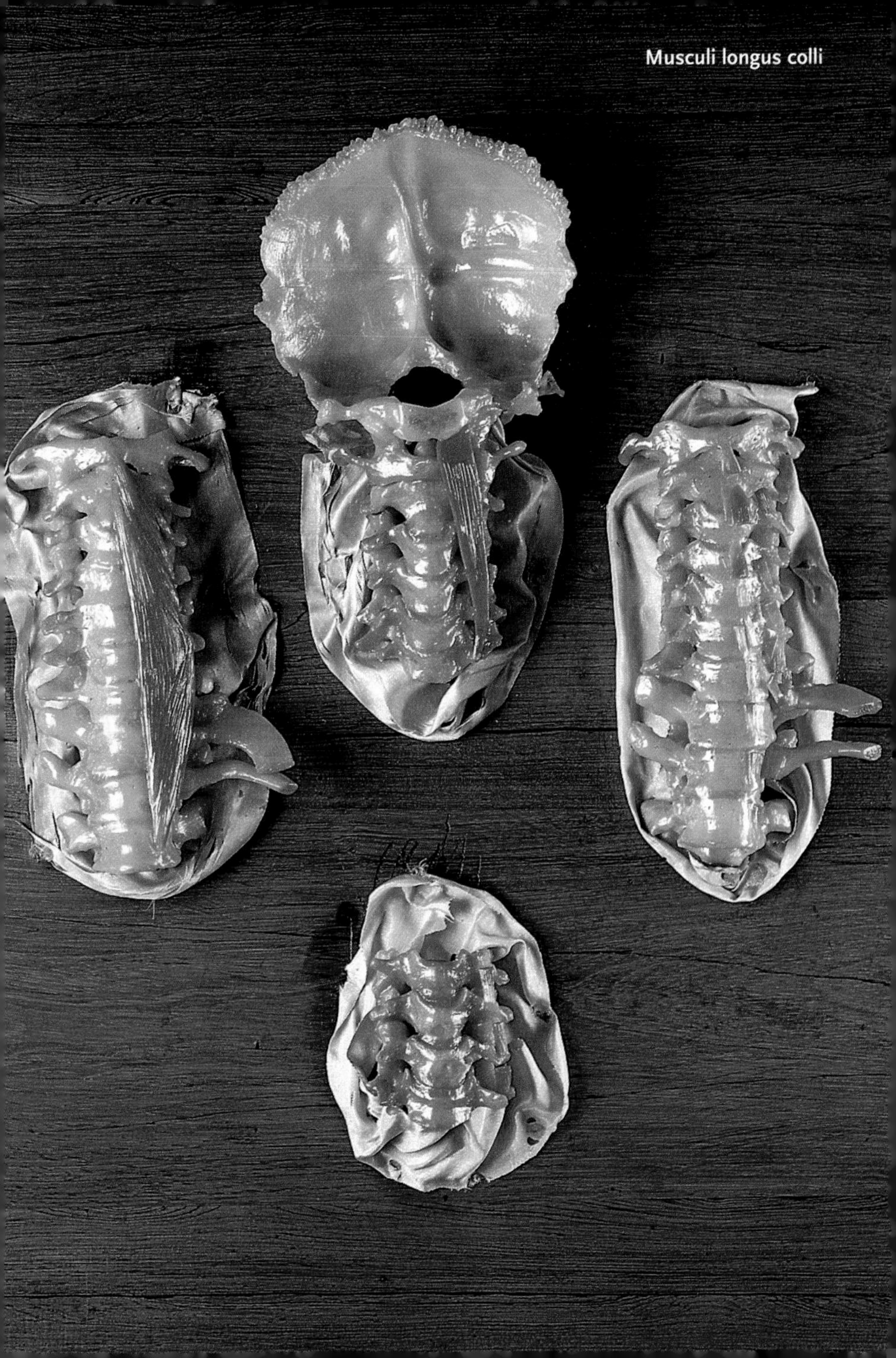

Musculi longus colli

Short muscles which connect the transverse processes of the vertebrae with one another

Kurze Muskeln, die die Querfortsätze der Wirbel miteinander verbinden

Muscles courts tendus entre les processus transverses des vertèbres cervicales

[XXV, 487]

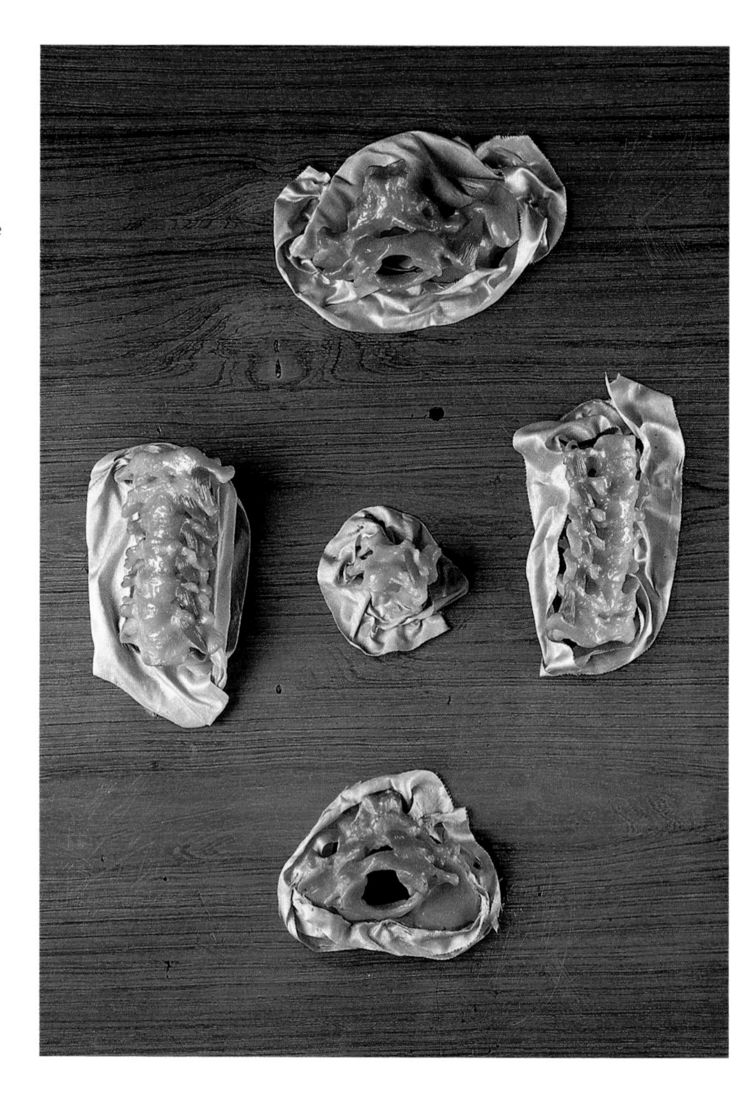

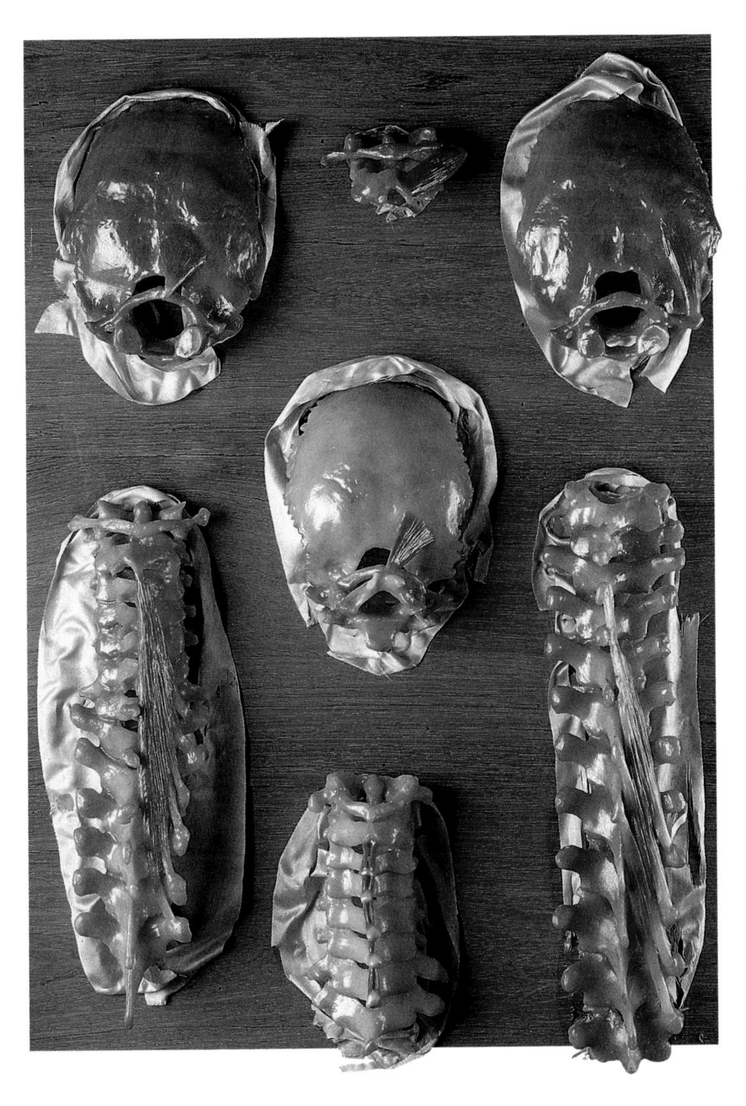

Display showing part of the deep musculature of the neck and back

Darstellung eines Teiles der tiefen Muskulatur des Nackens und des Rückens

Muscles profonds de la nuque et du dos insérés sur la colonne vertébrale

[XXV, 490]

Display showing part of the deep musculature of the neck and back

Darstellung eines Teiles der tiefen Nacken- und Rückenmuskulatur

Muscles profonds de la nuque et du dos insérés sur la colonne vertébrale

[XXV, 486]

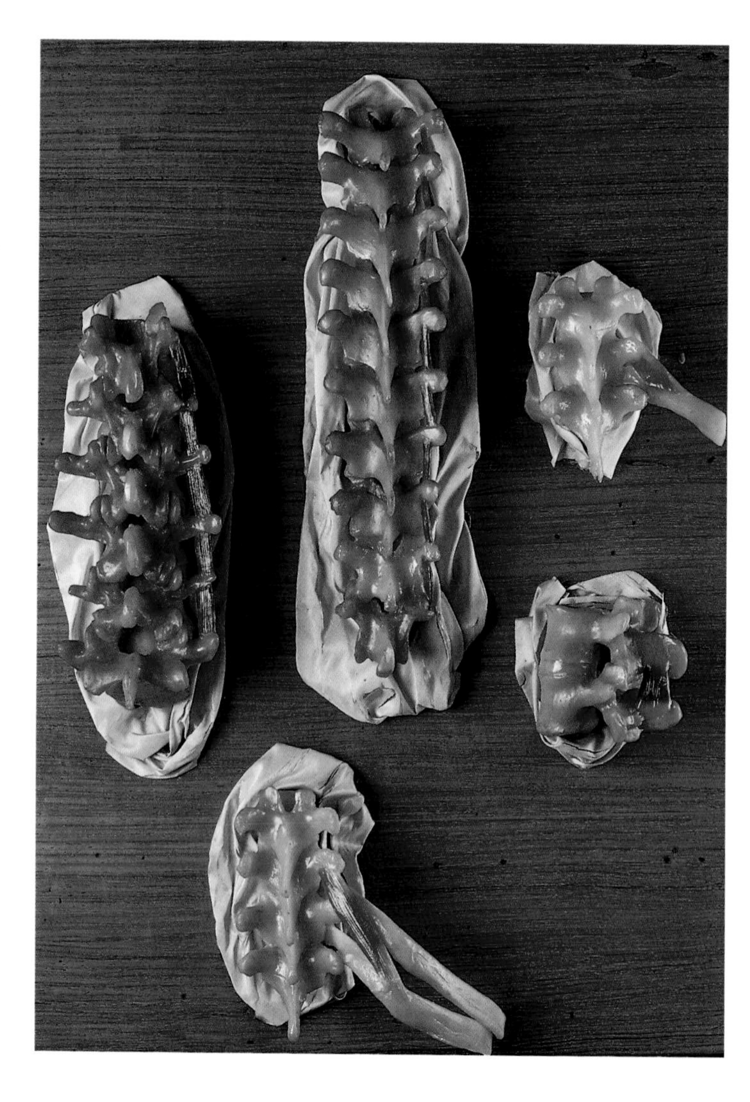

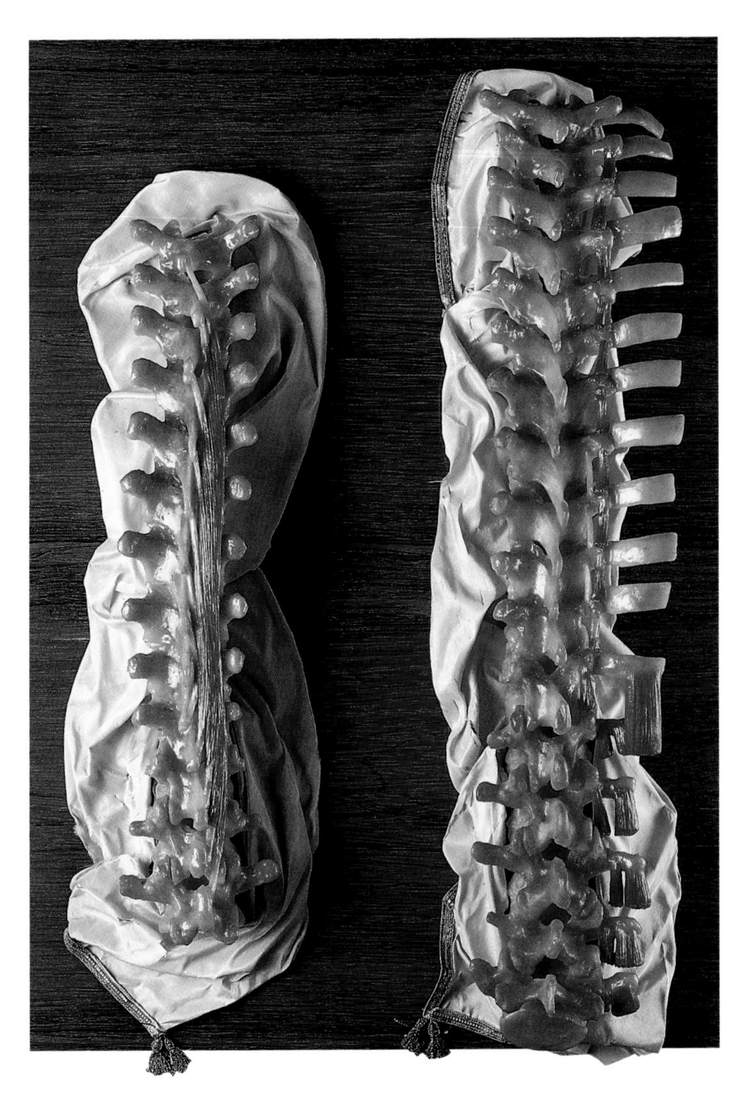

Display showing deep muscles of the back in the thoracic and lumbar regions of the vertebral column

Darstellung eines Teiles der tiefen Rückenmuskulatur im Bereich der Brust- und Lendenwirbelsäule

Muscles profonds du dos et des lombes insérés sur la colonne vertébrale

[XXV, 480]

Musculus multifidus

Deep muscles of the back in the lumbar region of the vertebral column

Tiefer Rückenmuskel im Bereich der Lendenwirbelsäule

Muscles profonds de la région lombaire (spinaux)

[XXV, 458]

A muscle which rotates the head (left specimen) and one which lies in the skin of the neck (right specimen)

Darstellung des Kopfwenders (linkes Präparat) und eines in der Haut des Halses gelegenen Muskels (rechtes Präparat)

Muscles du cou : sterno-cléido-mastoïdien (à gauche), et peaucier du cou (à droite)

[XXV, 500]

Musculus trapezius

View of the trapez-
ius muscle

Ansicht des Trapez-
muskels

Muscle superficiel
de la nuque et du
dos (trapèze)

[XXV, 471]

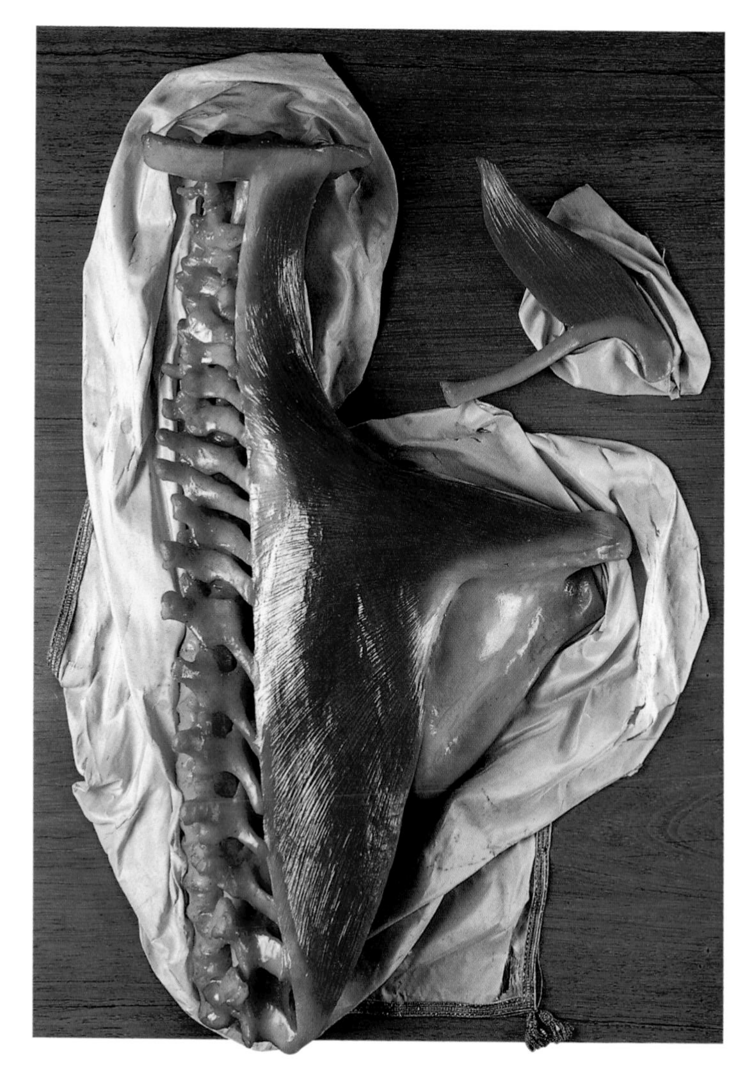

►
detail from page |
Detail von Seite |
détail de la page
252

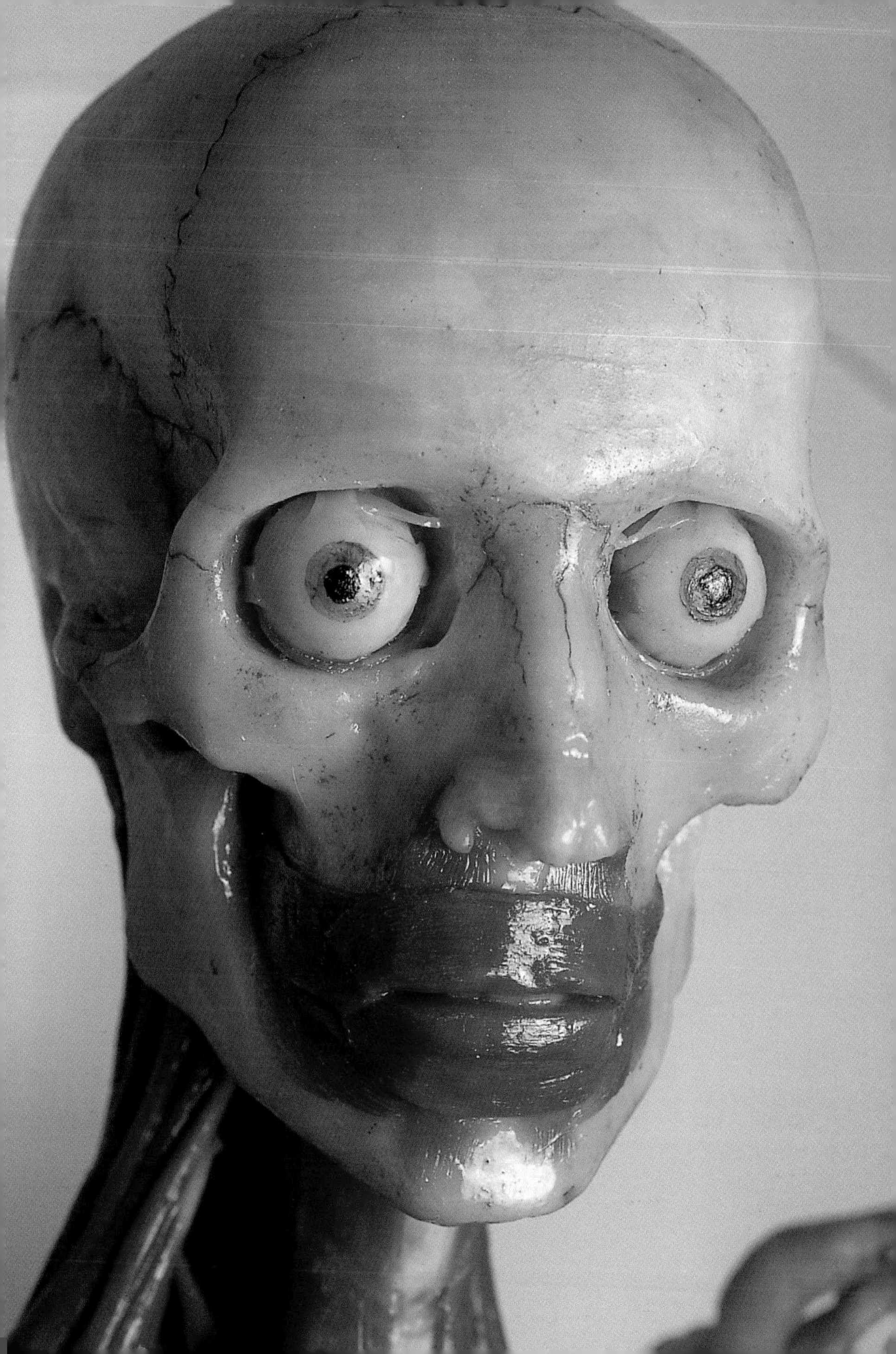

View of the flat muscles of the back

Anteile der platten Rückenmuskeln

Muscles du plan intermédiaire de la nuque

[XXV, 482]

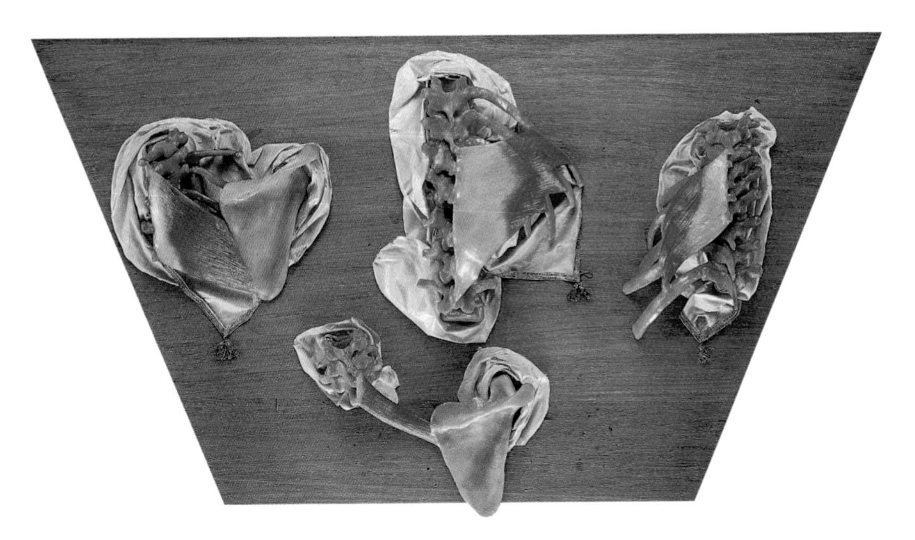

Some of the flat muscles of the back which connect the shoulder blade to the vertebral column

Anteile der platten Rückenmuskeln zwischen Schulterblatt und Wirbelsäule

Muscles fixateurs de l'omoplate à la colonne vertébrale

[XXVI, 579]

Display showing various muscles of the thorax and upper arm

Darstellungen verschiedener Muskeln von Brust und Oberarm

Muscles attachés à la clavicule et à l'omoplate

[XXV, 476]

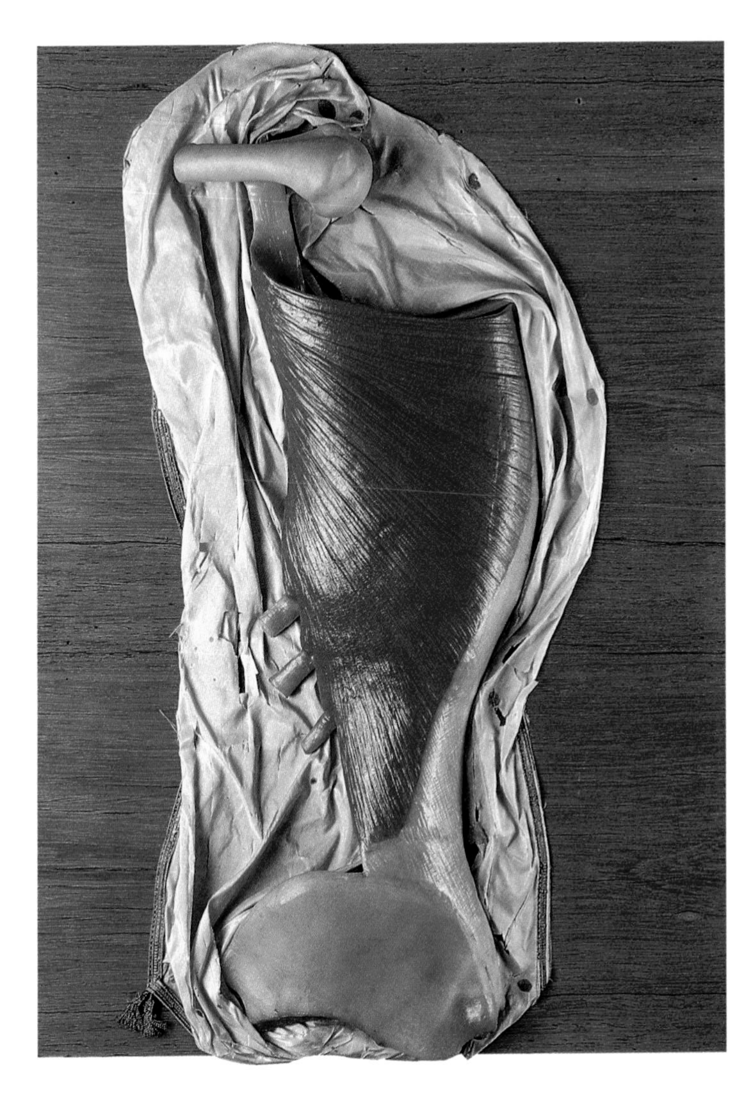

View of a superficial muscle of the back

Ansicht eines oberflächlichen Rückenmuskels

Muscle superficiel des lombes et du thorax (grand dorsal)

[XXV, 464]

Musculi faciales, Musculi colli, Musculi thoracis, Musculi abdominis, Musculi membri superioris, Musculi membri inferioris

Whole body specimen with the deep layer of muscles displayed

Ganzkörperpräparat zur Darstellung tiefer Schichten der Muskulatur

Préparation du corps entier représentant des muscles profonds

[XXX, 964]

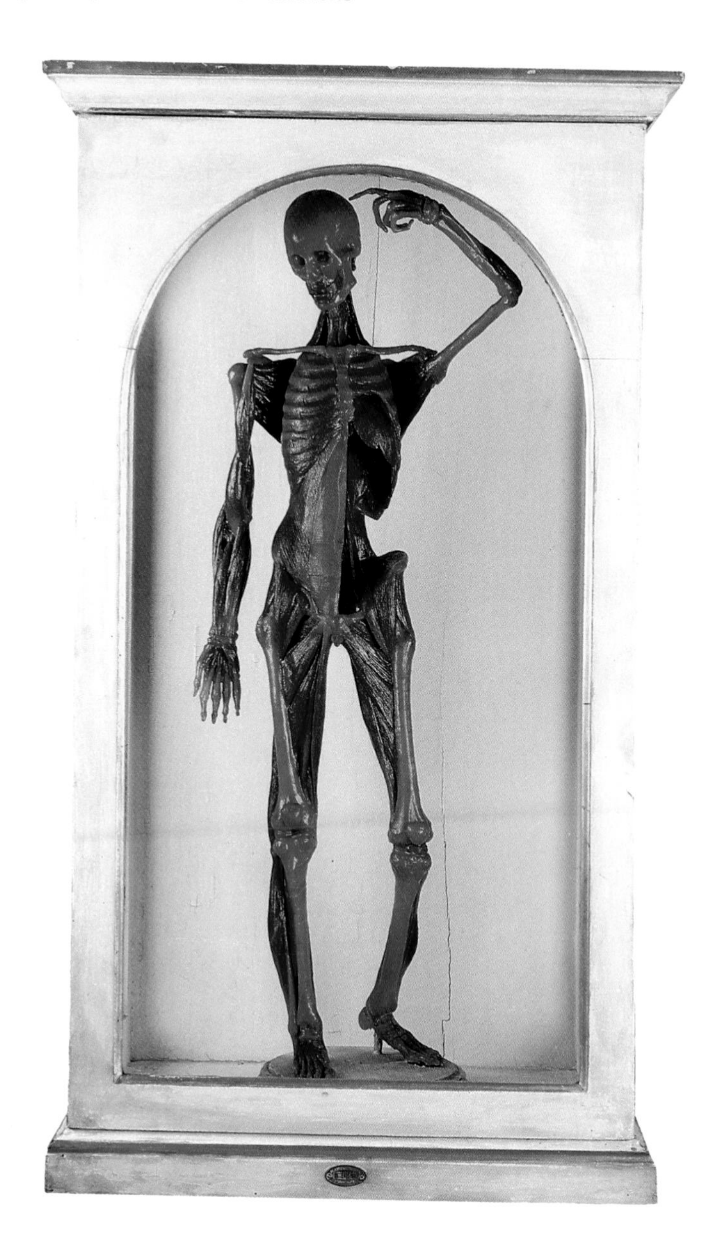

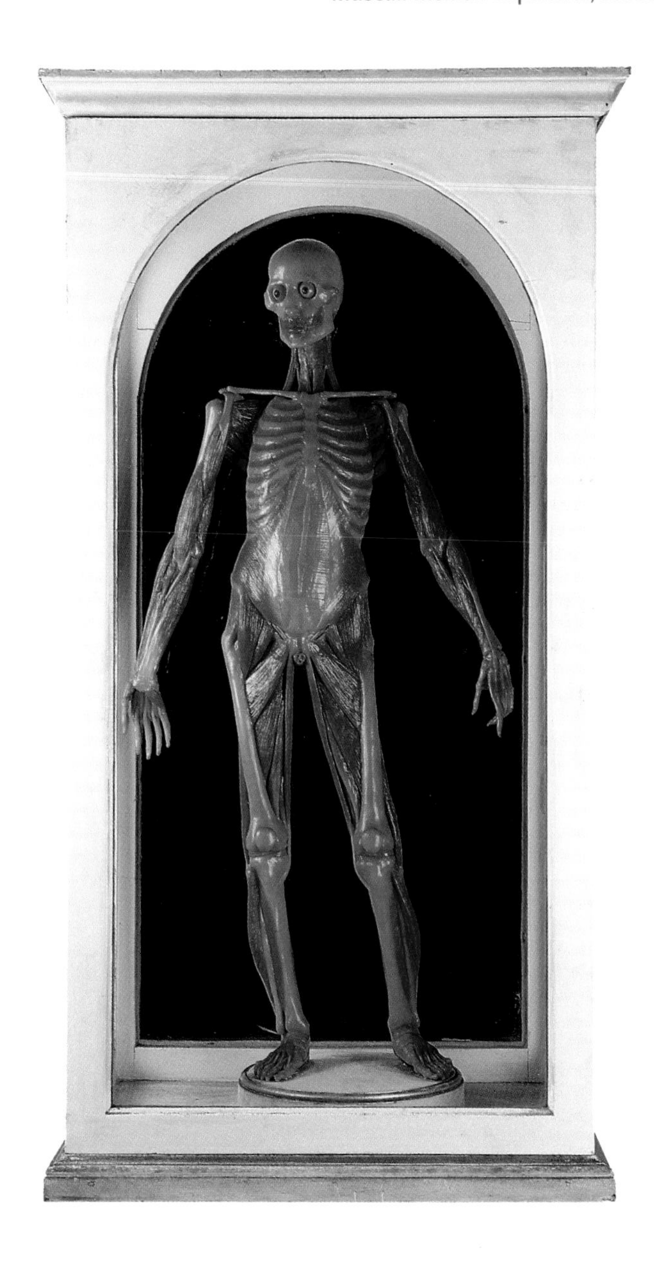

Whole body specimen with the deep layer of muscles displayed

Ganzkörperpräparat zur Darstellung tiefer Schichten der Muskulatur

Préparation du corps entier représentant des muscles profonds

[DEP. 26, 960]

Superficial and deep muscles of the anterior abdominal wall

Oberflächliche und tiefe Muskeln der vorderen Bauchwand

Muscles abdominaux superficiels et profonds

[XXV, 466]

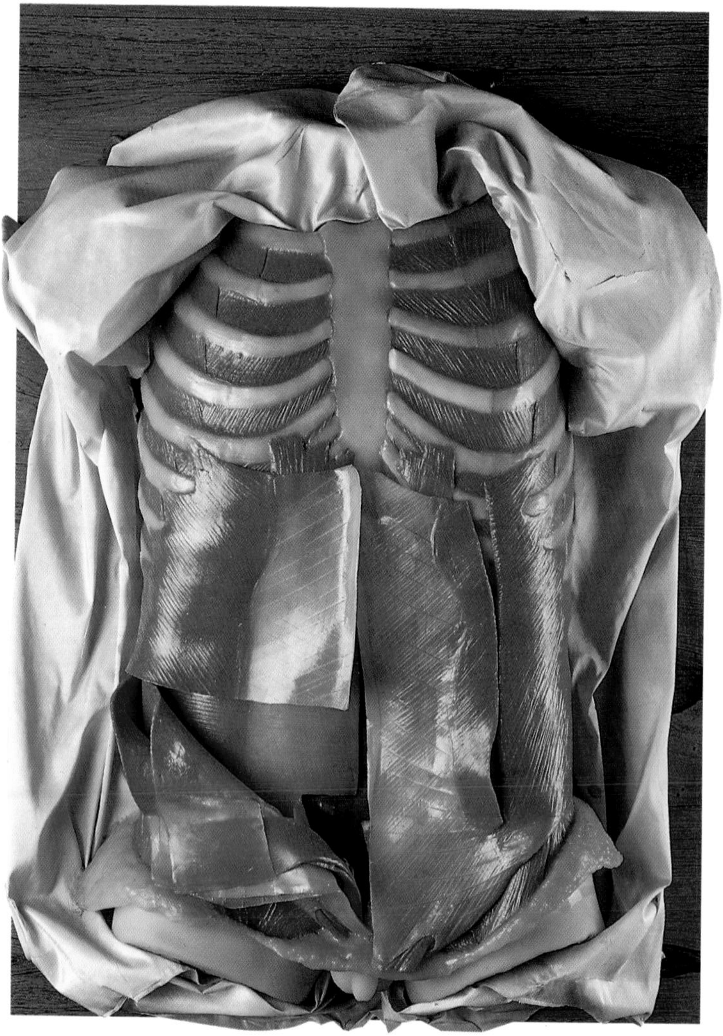

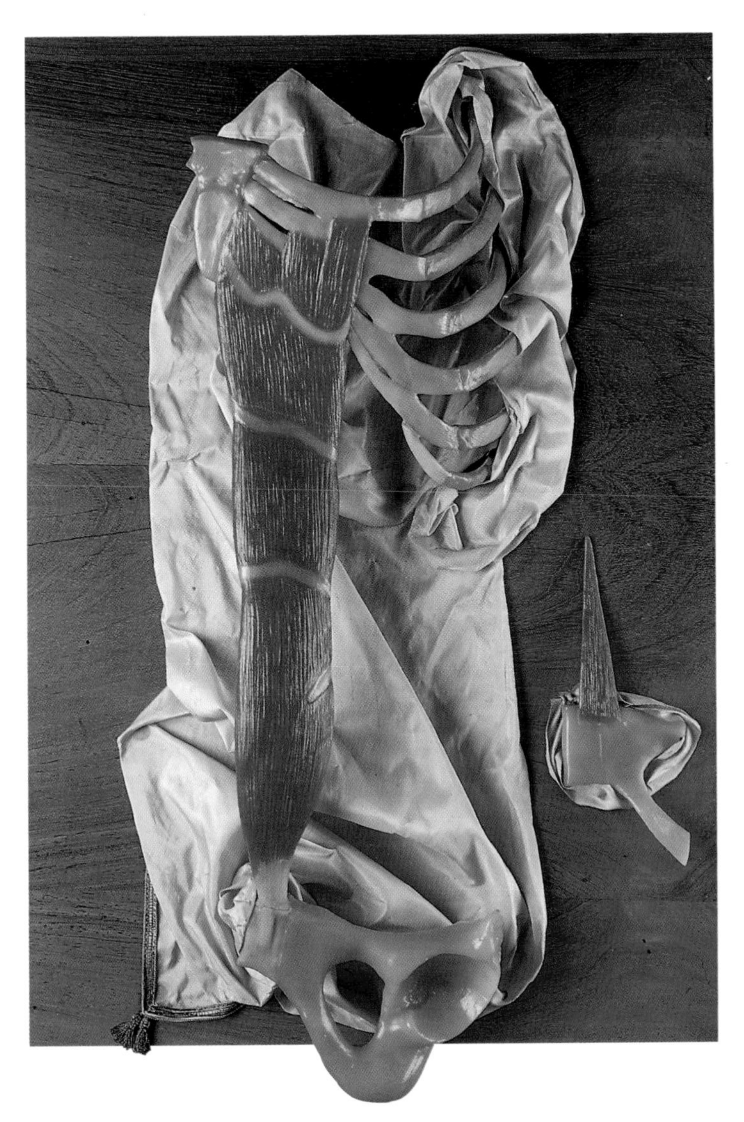

View of the straight abdominal muscle with its intermediate tendons

Ansicht des geraden Bauchmuskels mit seinen Zwischensehnen

Muscle droit de l'abdomen et ses intersections tendineuses

[XXV, 470]

The external oblique abdominal muscle together with its tendinous sheet

Darstellung des äußeren schrägen Bauchmuskels mit seiner Sehnenplatte

Muscle oblique externe de l'abdomen : corps musculaire et tendon

[XXV, 462]

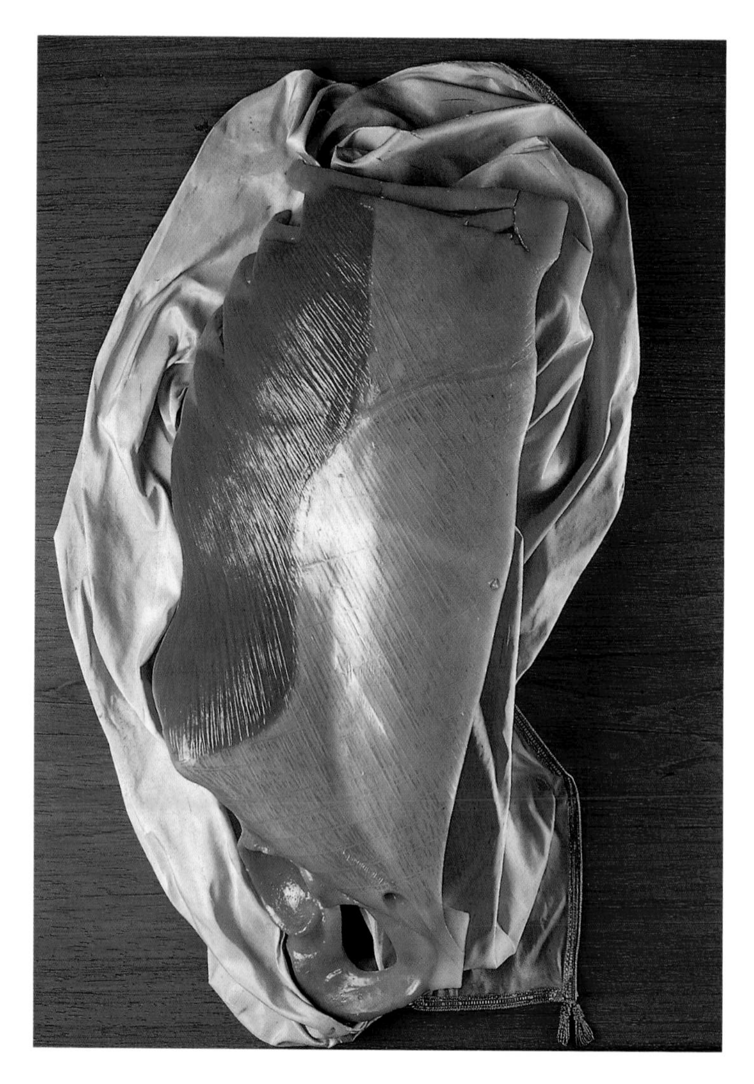

►
A deep oblique muscle of the anterior abdominal wall

Darstellung eines tiefen schrägen Muskels der vorderen Bauchwand

Muscle oblique interne de l'abdomen : corps musculaire et tendon ; muscle transverse du thorax

[XXV, 508]

Musculus obliquus abdominis internus

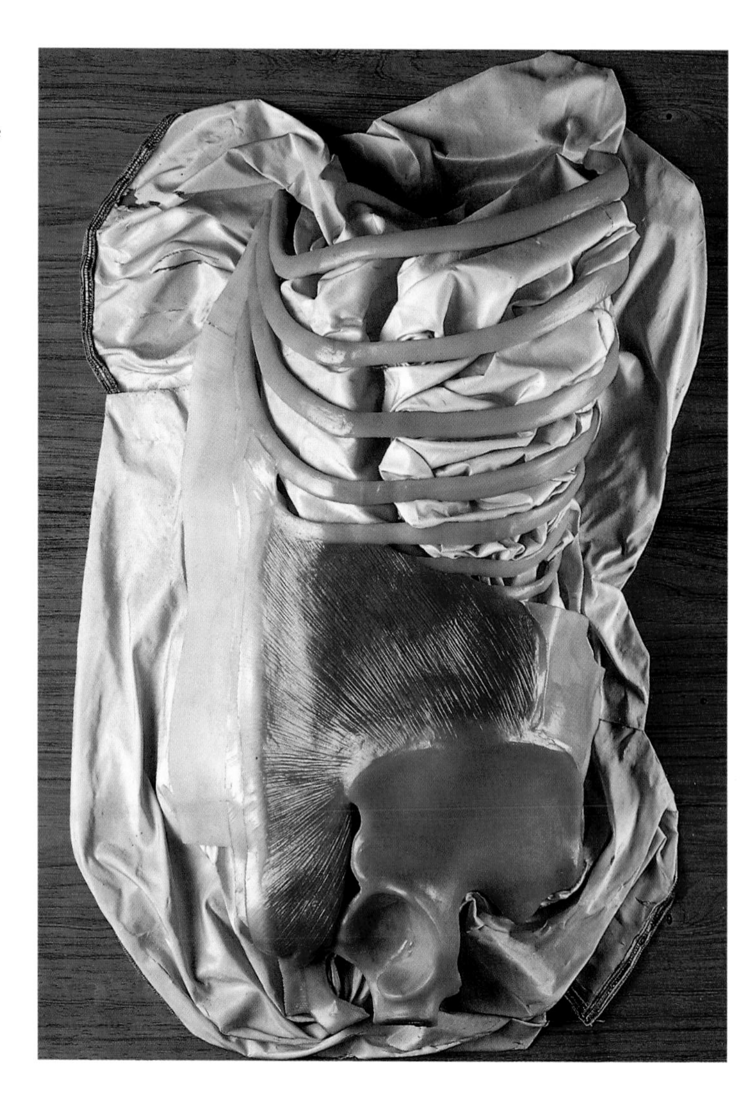

The internal oblique abdominal muscle seen from the side. Below: the hip bone with its joint socket

Innerer schräger Bauchmuskel von der Seite gesehen. Unten: das Hüftbein mit der Gelenkpfanne

Muscle oblique interne de l'abdomen (vu de la gauche)

[XXV, 468]

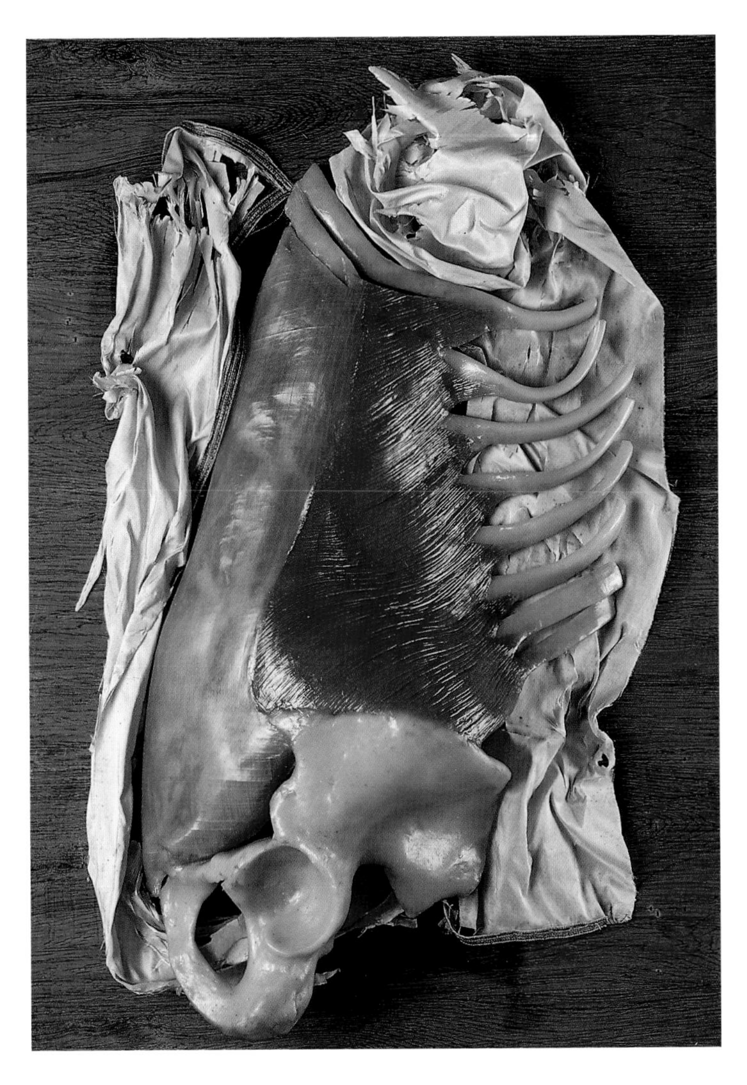

The external oblique abdominal muscle seen from the side

Äußerer schräger Bauchmuskel von der Seite gesehen

Muscle oblique externe de l'abdomen vu de la gauche

[XXV, 469]

The transverse abdominal muscle seen from behind

Ansicht des queren Bauchmuskels von hinten

Muscle transverse de l'abdomen vu de l'arrière

[XXV, 457]

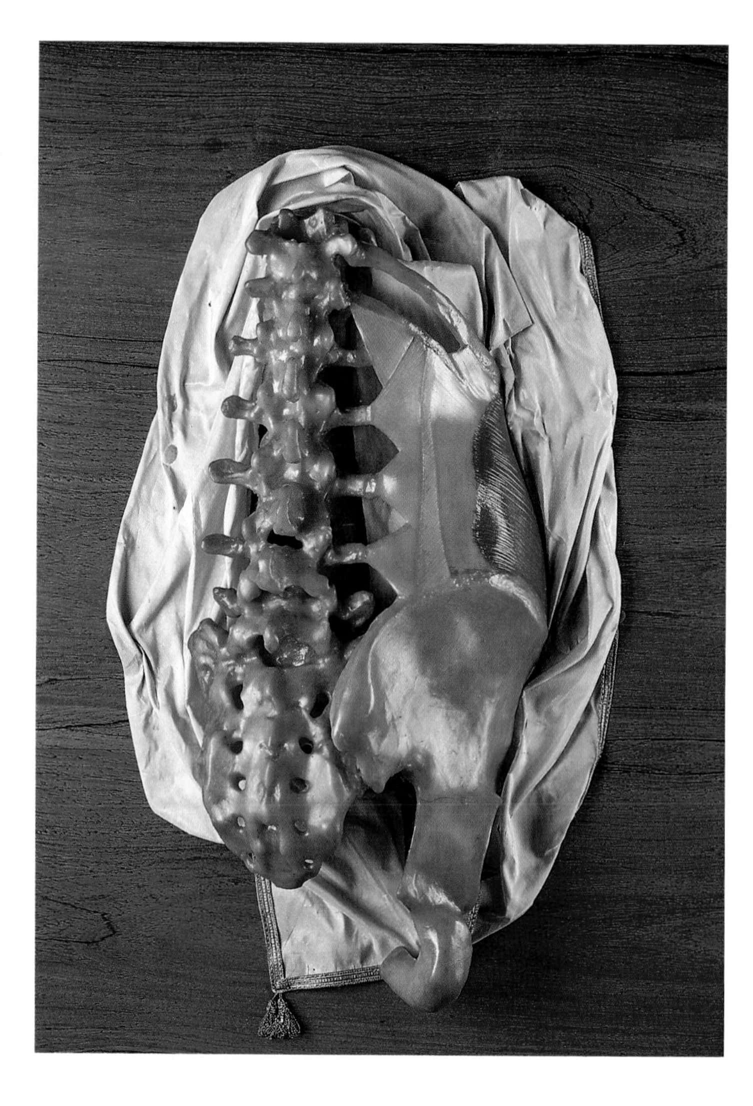

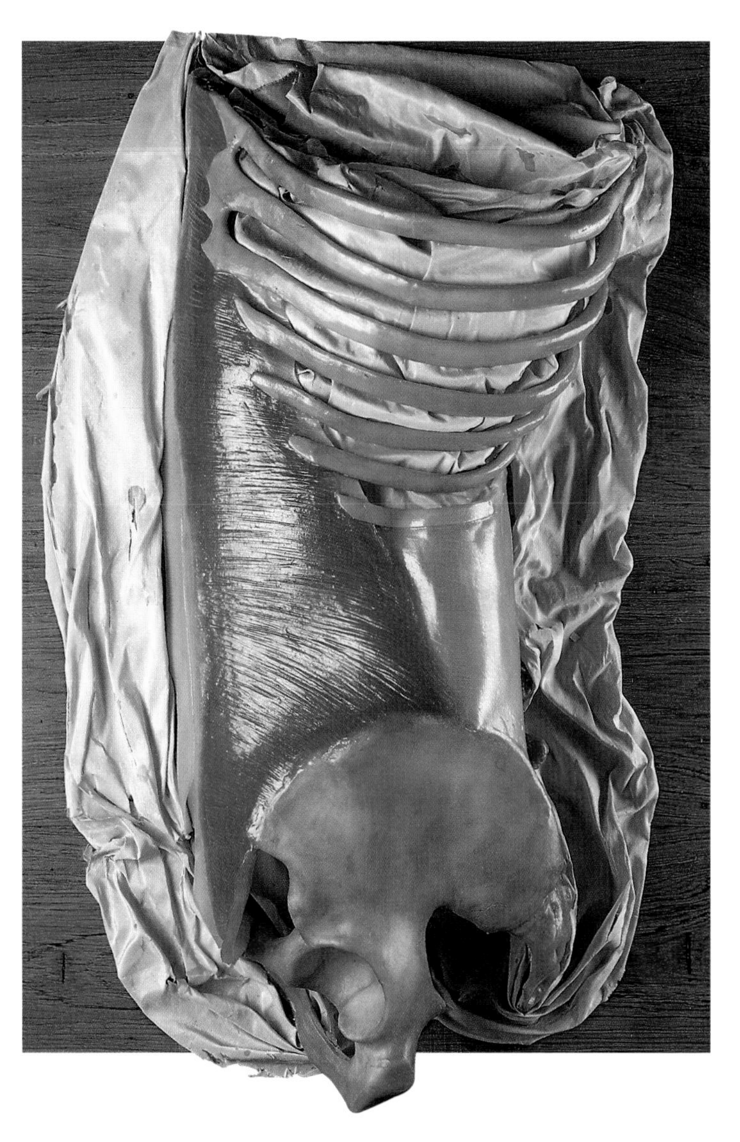

The transverse abdominal muscle seen from the side

Querer Bauchmuskel von der Seite gesehen

Muscle transverse de l'abdomen vu de la gauche

[XXV, 461]

The external and internal intercostal muscles

Äußere und innere Zwischenrippen-muskeln

Muscles intercos-taux interne et externe

[XXV, 473]

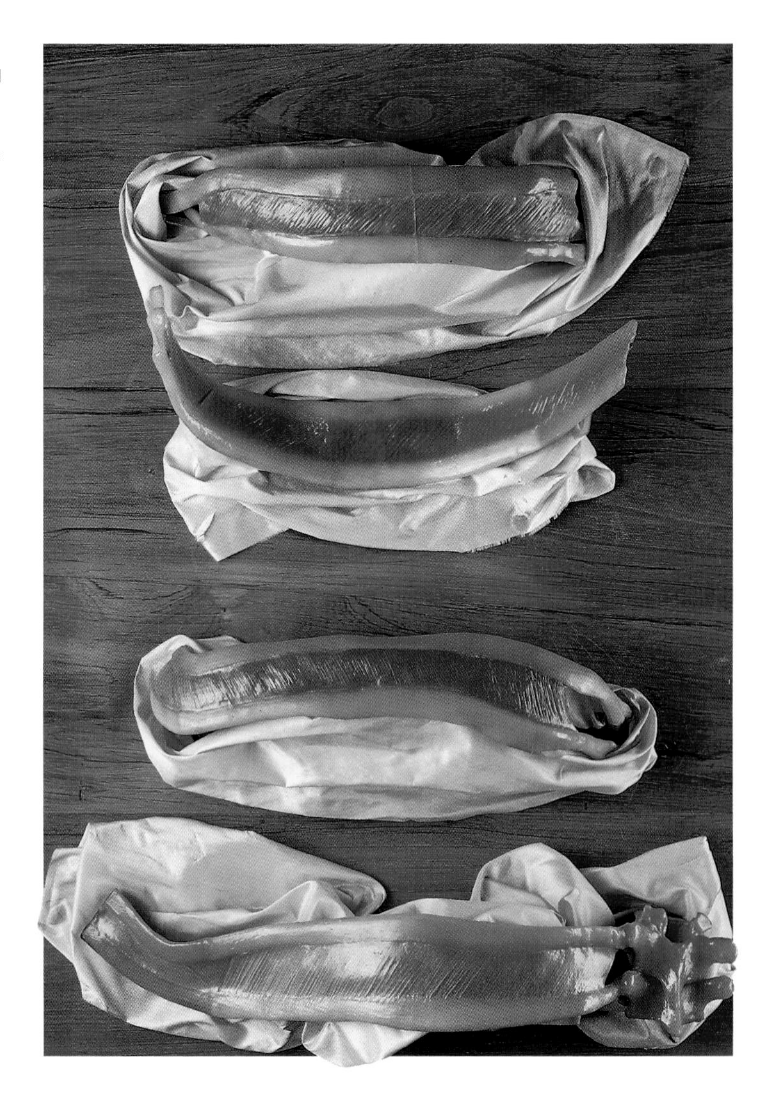

Display showing various intercostal muscles

Darstellung verschiedener Zwischenrippenmuskeln

Muscle transverse du thorax (en haut) ; muscles intercostaux

[XXV, 465]

The posterior thoracic and abdominal walls seen from in front

Blick von vorn auf die hintere Brust- und Bauchwand

Muscles intercostaux et carré des lombes vus de l'avant

[XXV, 509]

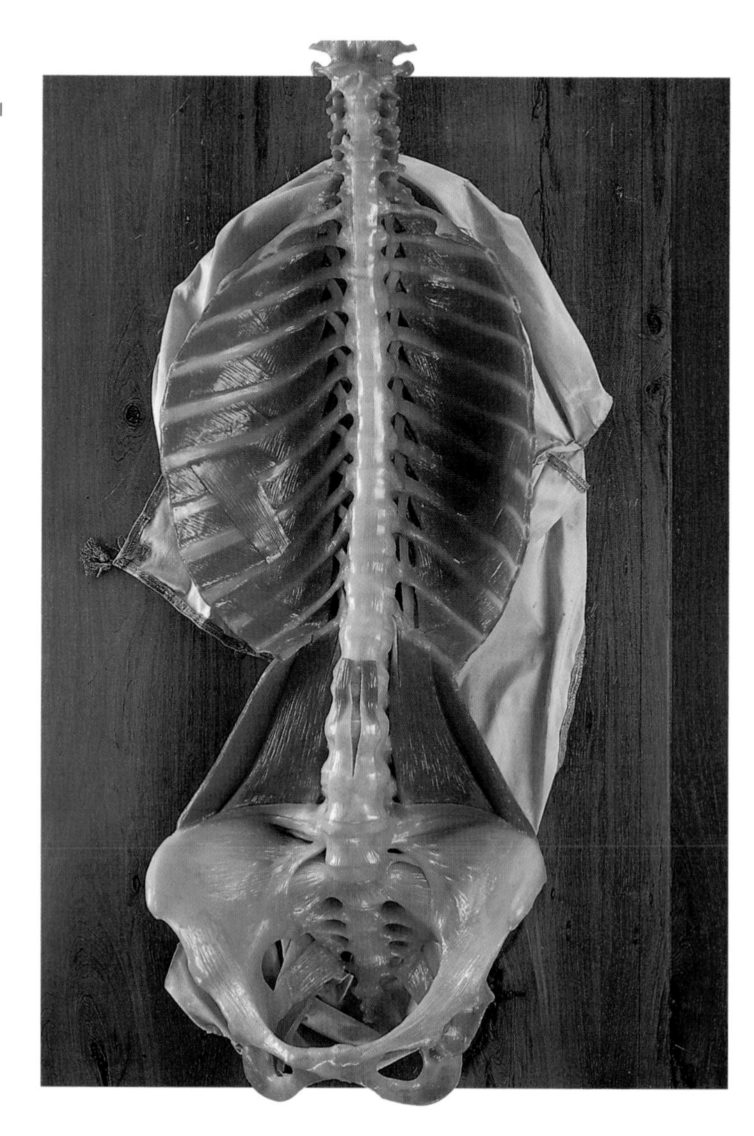

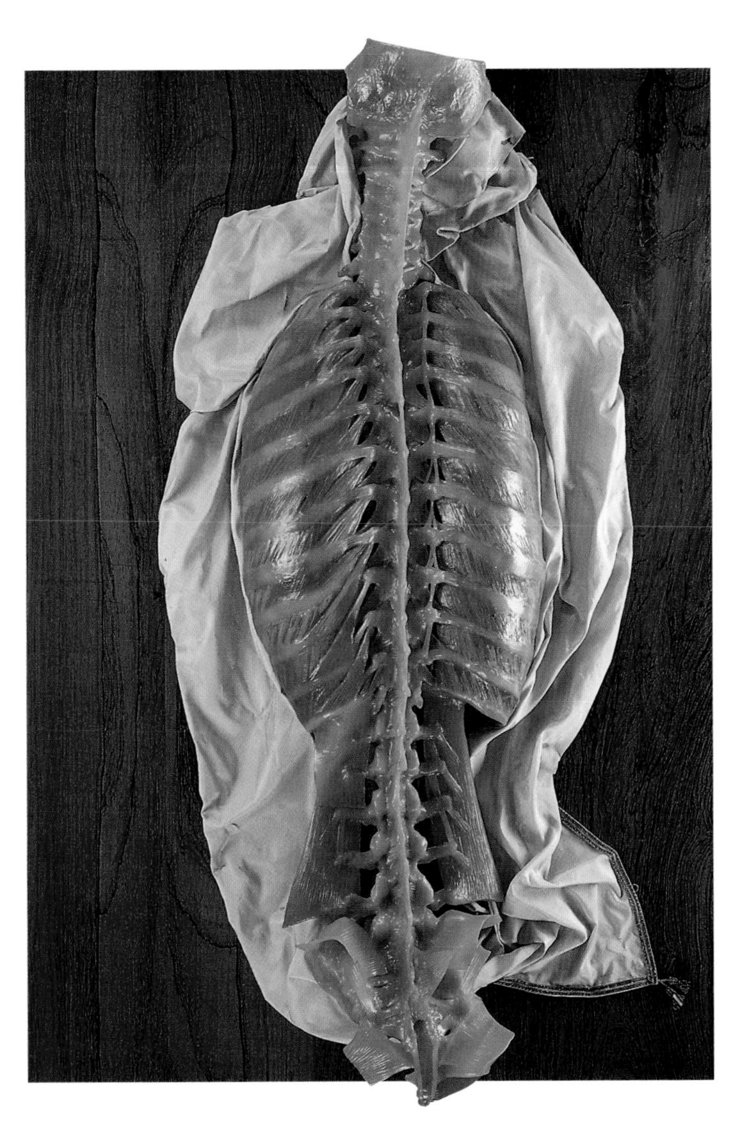

The intercostal muscles seen from behind

Darstellung der Zwischenrippen-muskeln in der Ansicht von hinten

Muscles inter-costaux et carré des lombes vus de l'arrière

[XXV, 510]

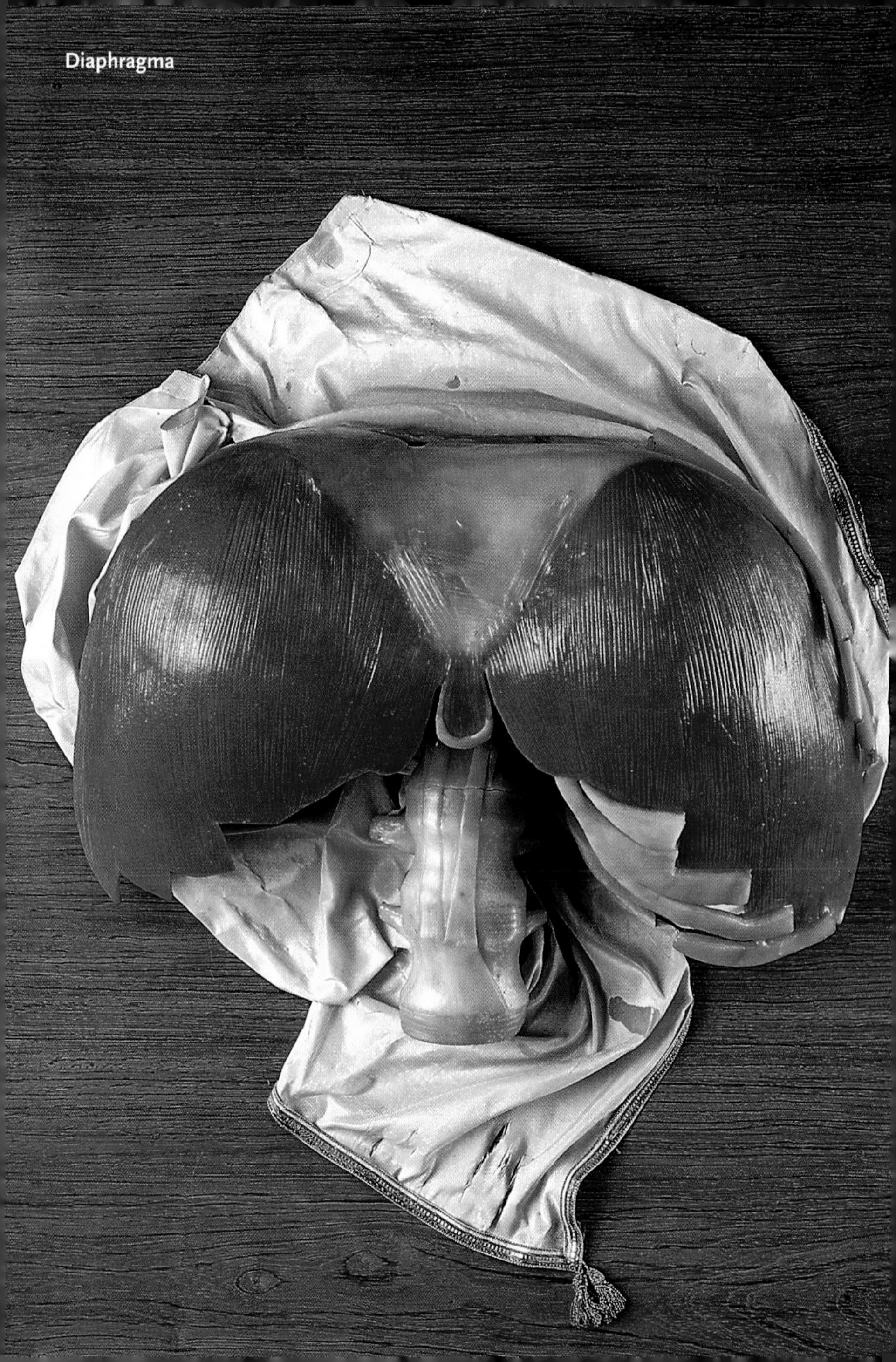
Diaphragma

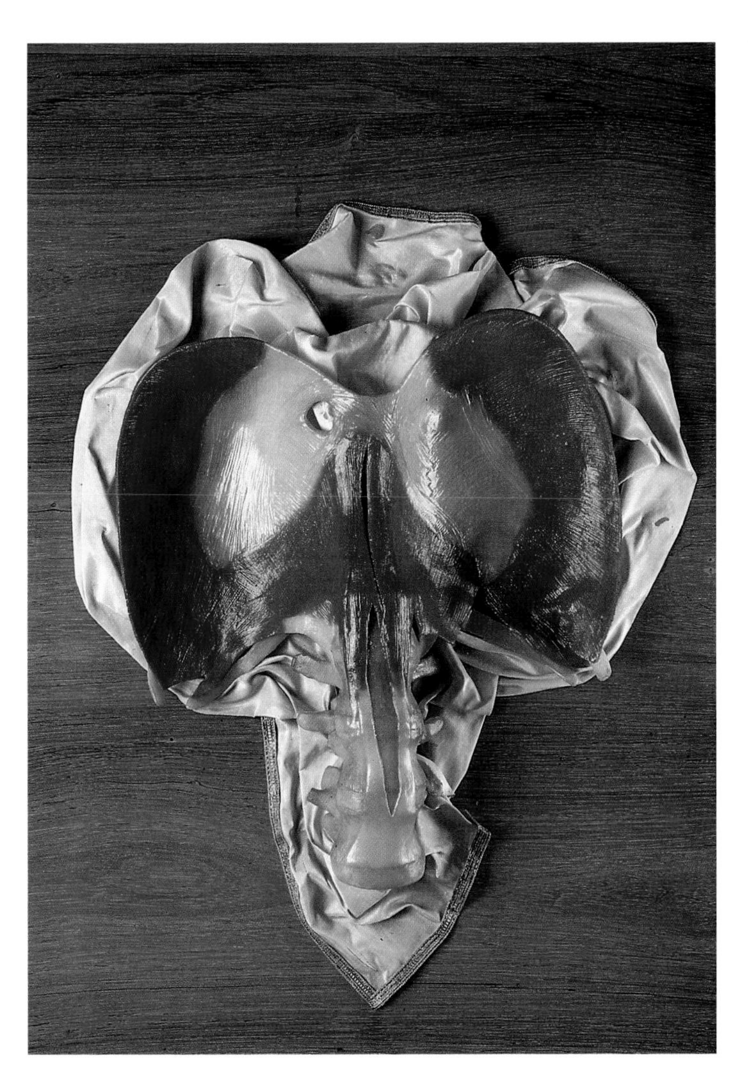

View of the abdominal surface of the diaphragm showing the openings for esophagus, aorta and inferior vena cava

Ansicht auf die dem Bauchraum zugewandte Seite des Zwerchfells mit den Durchtrittsstellen für Speiseröhre, Körperschlagader und untere Hohlvene

Diaphragme avec les orifices de passage de l'œsophage, de l'aorte, et de la veine cave inférieure (vue inférieure)

[XXV, 477]

◄
The diaphragm seen from in front

Ansicht von vorn auf das Zwerchfell

Diaphragme vu de l'avant

[XXV, 478]

Musculi faciales, Musculi colli, Musculi thoracis, Musculi abdominis, Musculi membri superioris, Musculi membri inferioris

Whole body specimen displaying the deeper layer of muscles

Ganzkörperpräparat zur Darstellung tieferer Schichten der Muskulatur

Préparation du corps entier représentant des muscles profonds

[XXIX, 957]

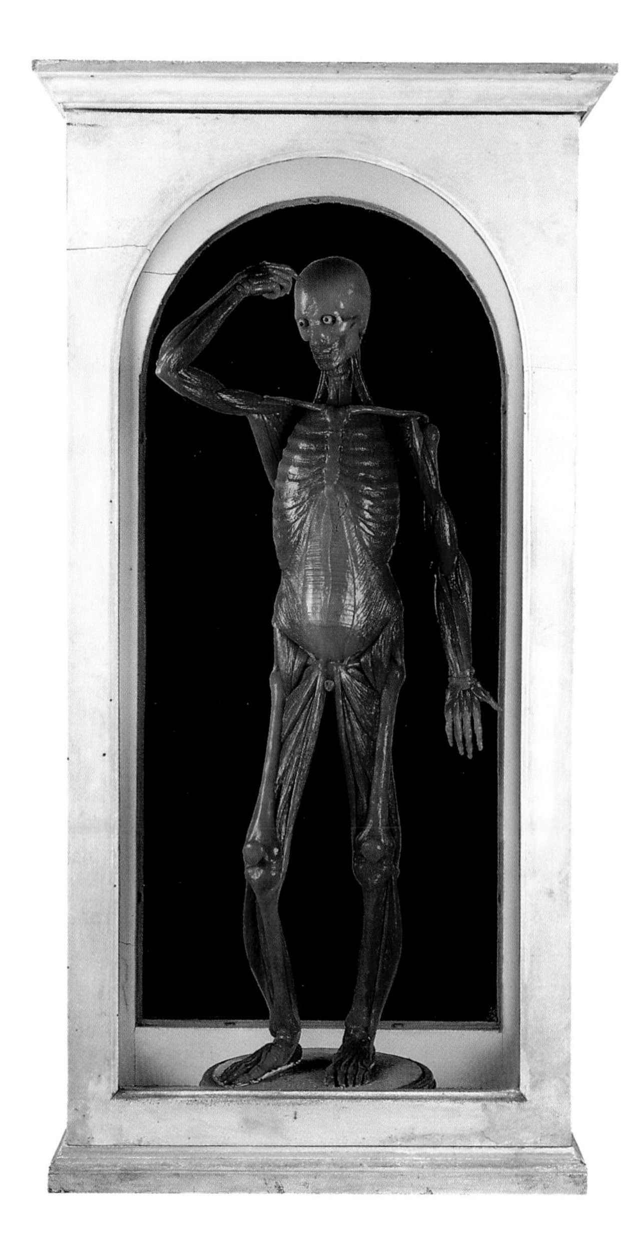

Whole body specimen displaying the superficial muscles

Ganzkörperpräparat zur Darstellung der oberflächlichen Muskulatur

Préparation du corps entier représentant des muscles superficiels

[DEP. 14, 956]

The large muscle of the chest seen from in front

Großer Brustmuskel in der Ansicht von vorne

Muscle grand pectoral vu de l'avant

[XXV, 453]

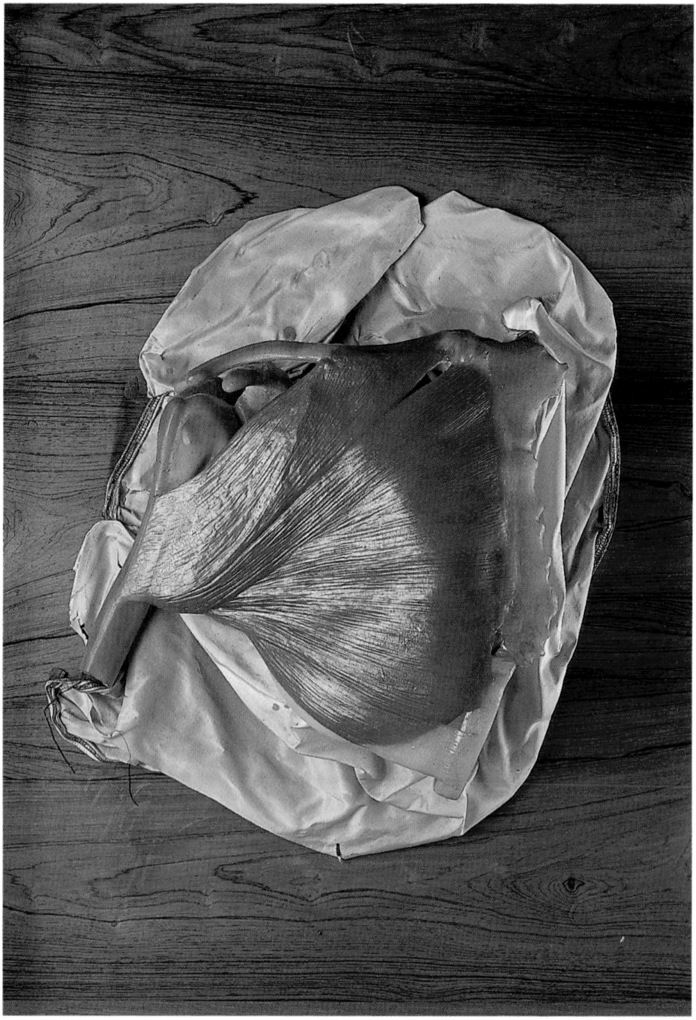

►
The deep surface of the large muscle of the chest, viewed from inside

Ansicht aus dem Brustraum auf die Innenfläche des großen Brustmuskels

Muscle grand pectoral vu à travers la paroi antérieure de la cage thoracique

[XXV, 463]

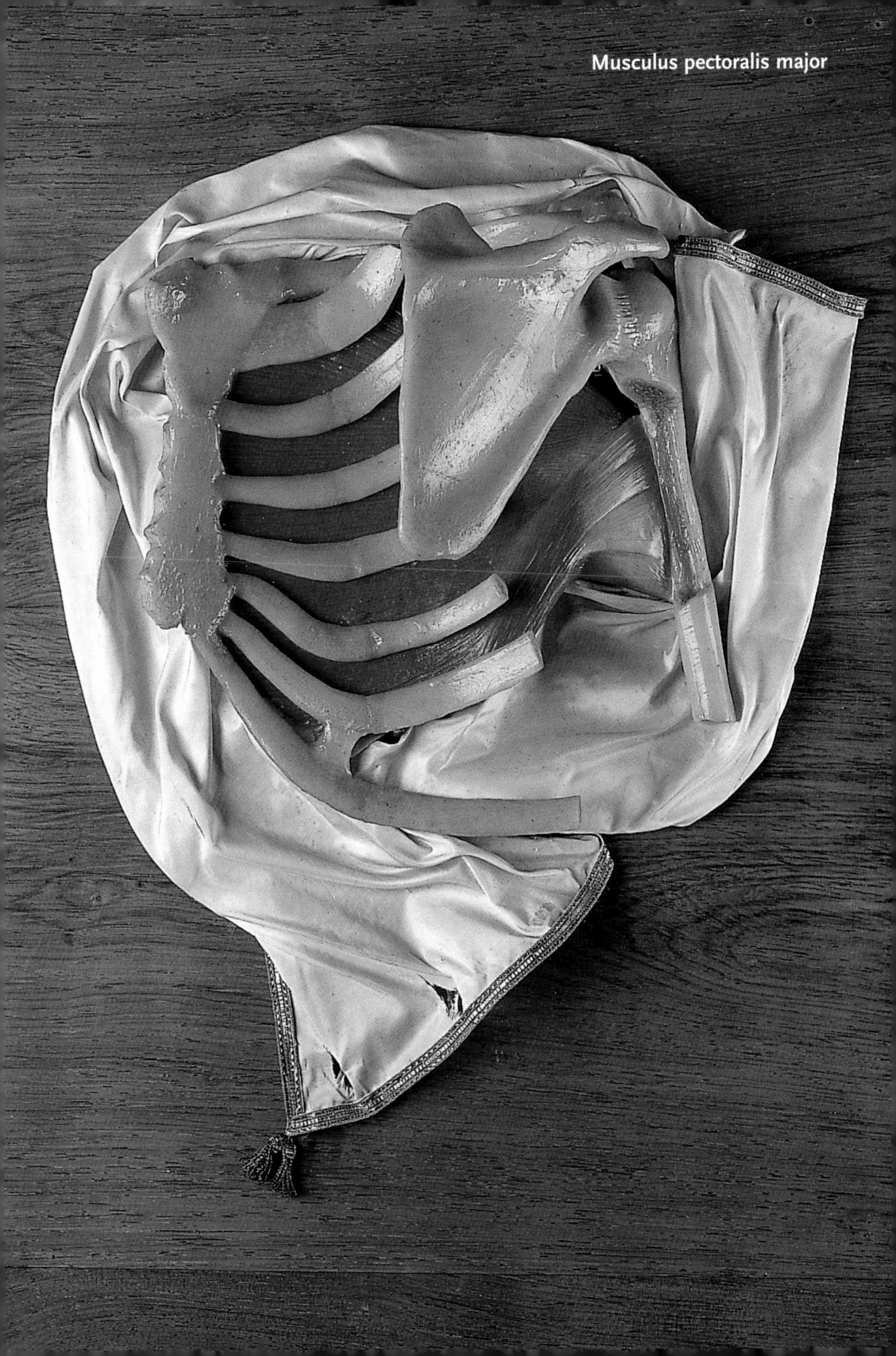
Musculus pectoralis major

Musculus deltoideus

The deltoid muscle

Ansicht des Delta-
muskels

Muscle deltoïde

[XXV, 451]

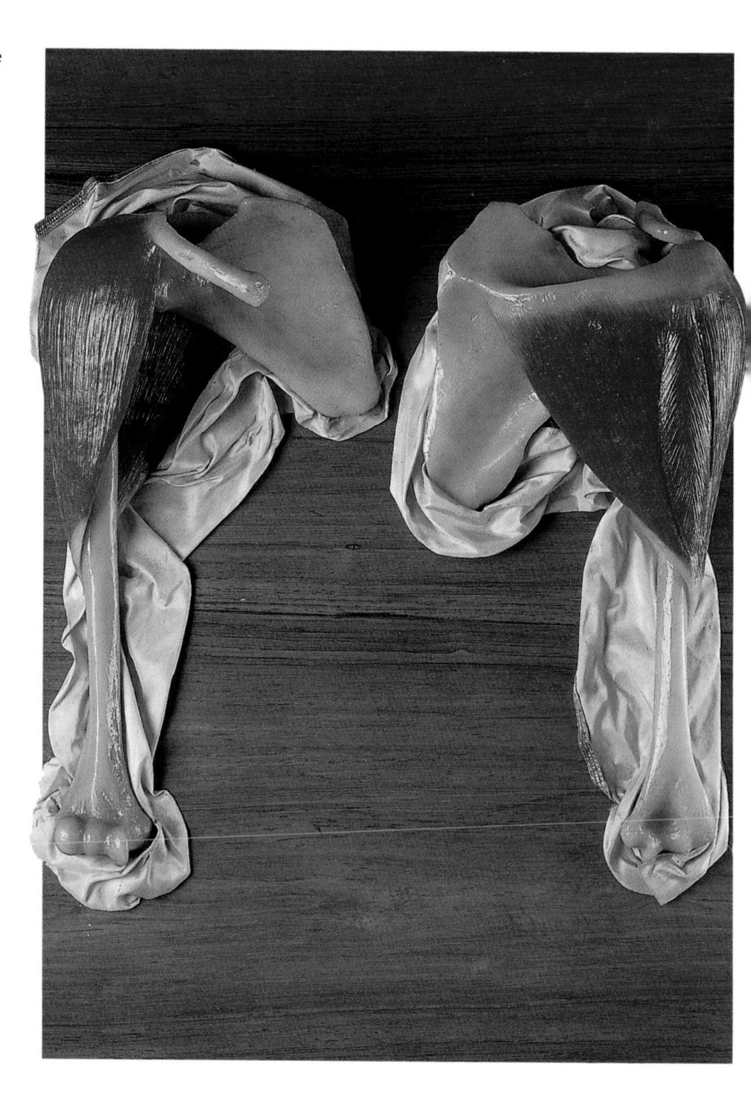

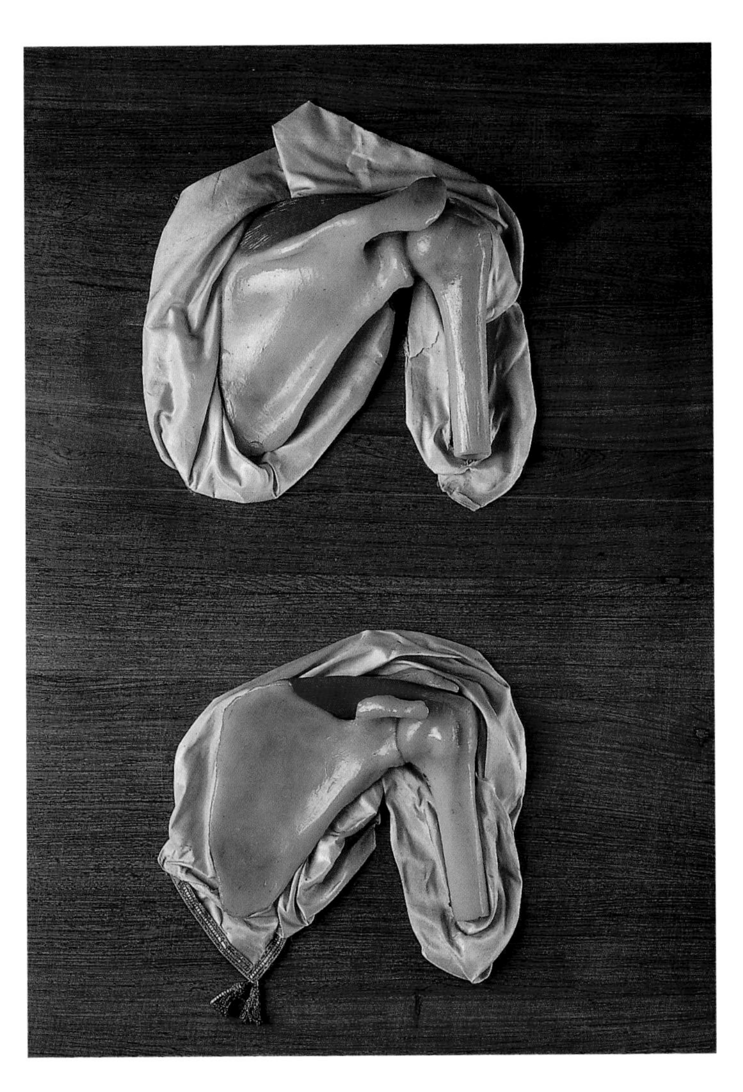

Display showing the shoulder joint with a muscle of its muscular cuff

Darstellung eines zur Muskelmanschette des Schultergelenkes gehörigen Muskels

Muscle de la coiffe des rotateurs de l'articulation de l'épaule (supra-épineux)

[XXVI, 573]

Musculus teres major

View of a muscle which connects the upper arm to the shoulder blade

Ansicht eines Muskels, der Schulterblatt und Oberarm verbindet

Muscle tendu entre la scapula et l'humérus (grand rond)

[XXVI, 575]

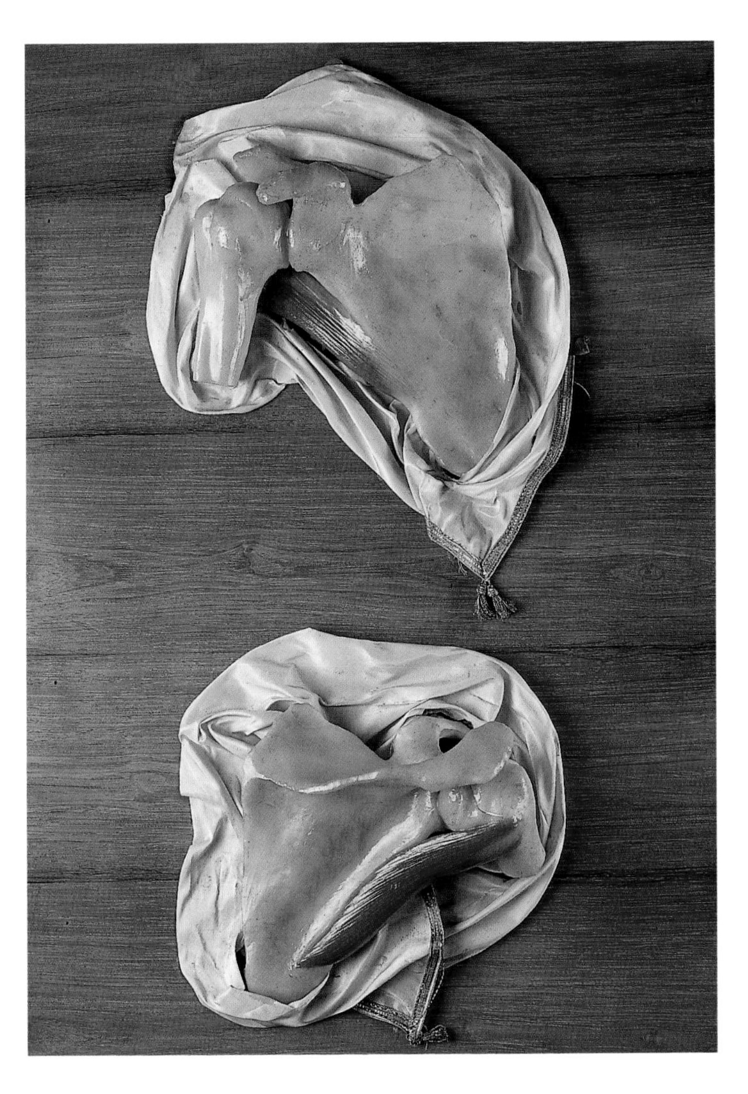

View of a muscle which connects the upper arm to the shoulder blade

Ansicht eines Muskels, der Schulterblatt und Oberarm verbindet

Muscle tendu entre la scapula et l'humérus (petit rond)

[XXVI, 577]

Musculus subscapularis, Musculus supraspinatus et infraspinatus

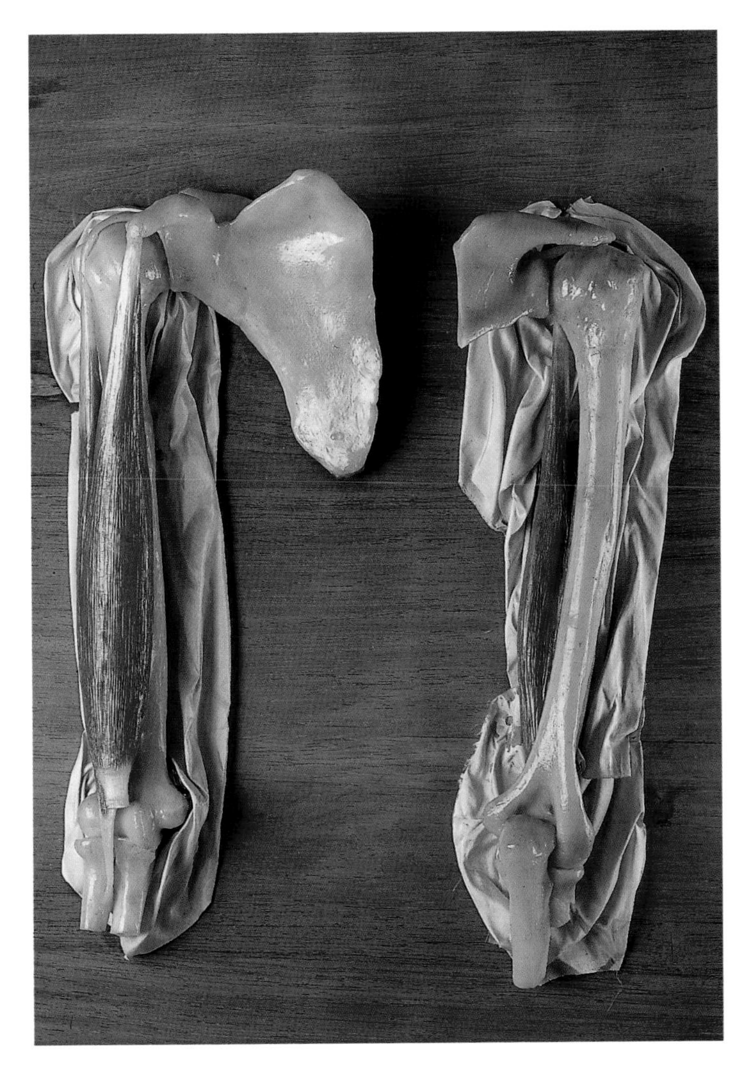

View of the biceps muscle from in front (left specimen) and from behind (right specimen)

Ansichten des Bizepsmuskels von vorn (linkes Präparat) und von hinten (rechtes Präparat)

Muscle superficiel de la région antérieure du bras (biceps brachial)

[XXVI, 551]

◄

Display showing shoulder muscles

Darstellung von Schultermuskeln

Muscles de la coiffe des rotateurs de l'articulation de l'épaule

[XXVI, 413]

Display showing the extensor muscle of the upper arm

Ansichten des Oberarmstreckmuskels

Muscle de la région postérieure du bras (triceps brachial)

[XXVI, 572]

The extensor muscles of the hand and thumb and their tendons

Streckmuskeln der Hand und des Daumens und ihre Sehnen

Muscles extenseurs du poignet et des doigts et leurs tendons

[XXIX, 750]

Musculus supinator

A muscle which contributes to the rotation of the forearm

Ansichten eines Muskels, der an der Unterarmdrehung beteiligt ist

Muscle profond de la loge latérale de l'avant-bras (supinateur)

[XXVI, 570]

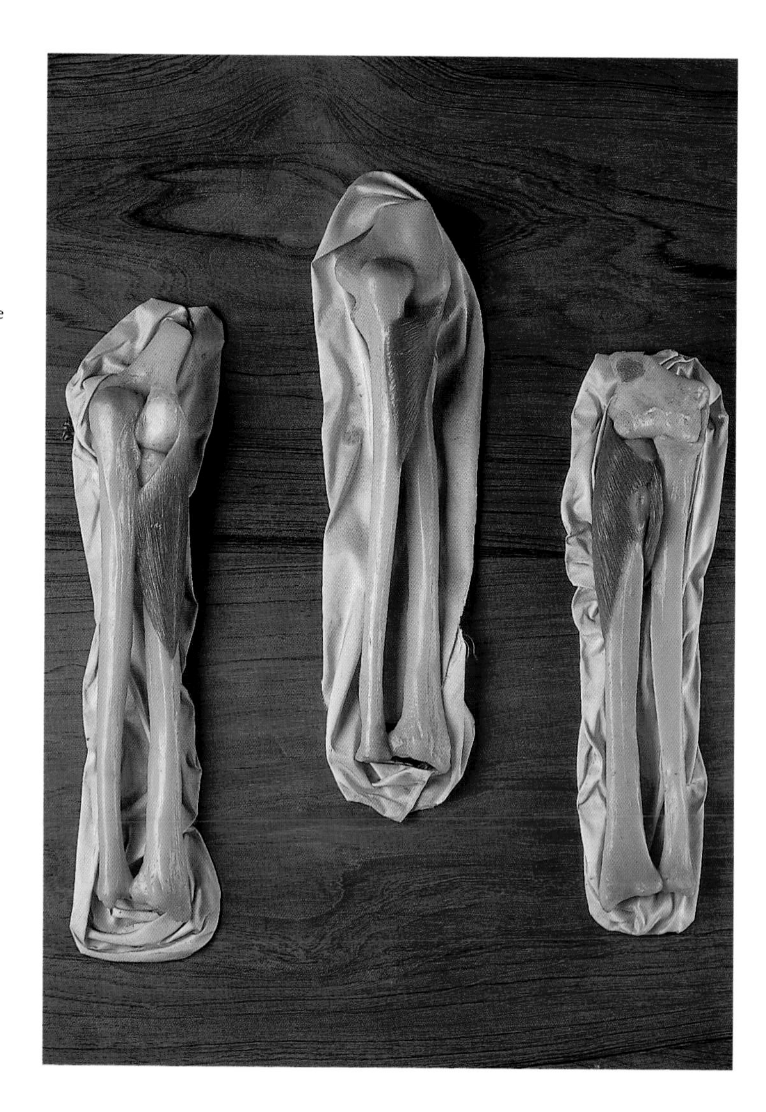

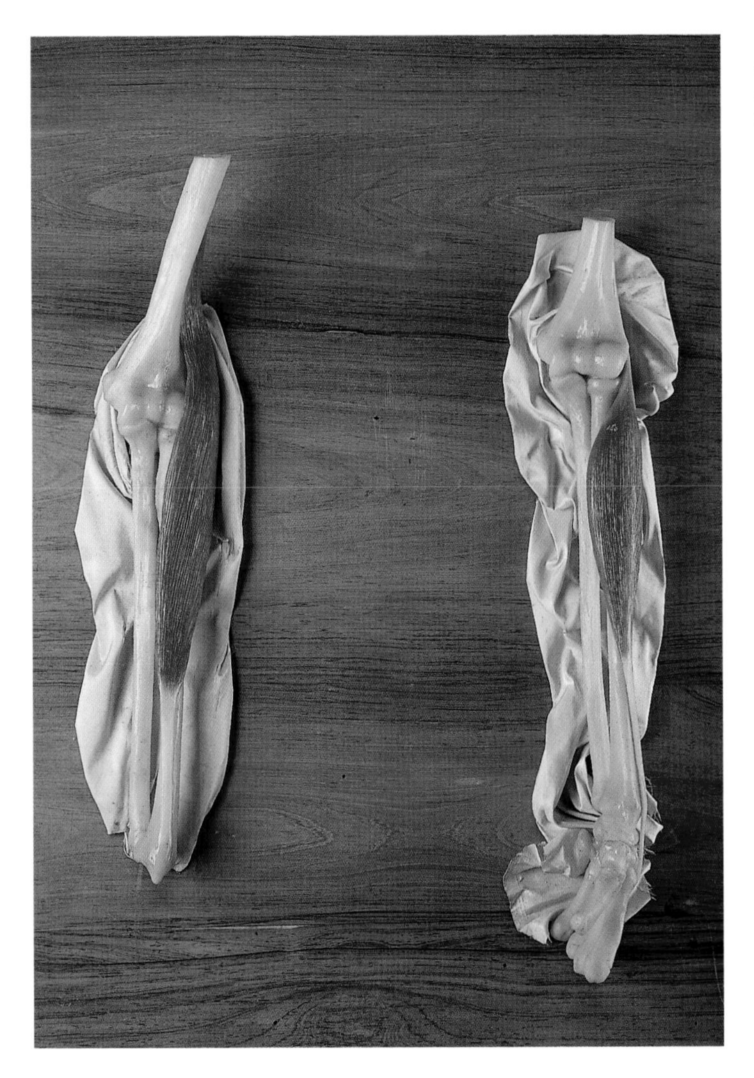

Display showing the extensor and rotator muscles of the forearm

Darstellungen von Unterarmstreck- und drehmuskeln

Muscles de la loge latérale de l'avant-bras (brachio-radial et long extenseur radial du carpe)

[XXVI, 552]

Musculus pronator teres, Musculus pronator quadratus

Display showing the long flexor muscles of the fingers with their tendons

Darstellung der langen Fingerbeugemuskeln mit ihren Endsehnen

Muscles long fléchisseur du pouce (à gauche), et fléchisseur profond des doigts (à droite) et leurs tendons

[XXVI, 553]

◄

Display showing muscles which contribute to rotation of the forearm

Darstellungen von Muskeln, die an Drehbewegungen des Unterarms beteiligt sind

Muscles de la loge antérieure de l'avant-bras (rond pronateur et carré pronateur)

[XXVI, 549]

Musculus extensor pollicis longus et brevis

The long extensor muscle of the thumb

Darstellung der langen Daumenstreckmuskeln

Muscles extenseurs du pouce

[XXVI, 569]

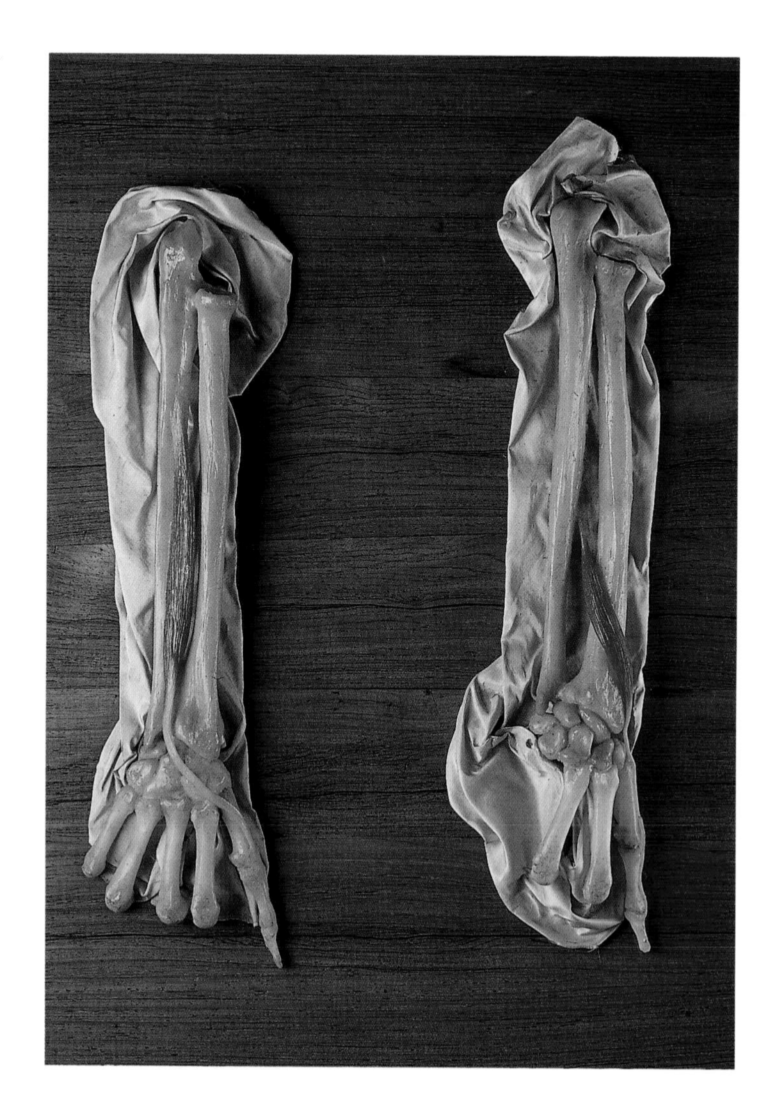

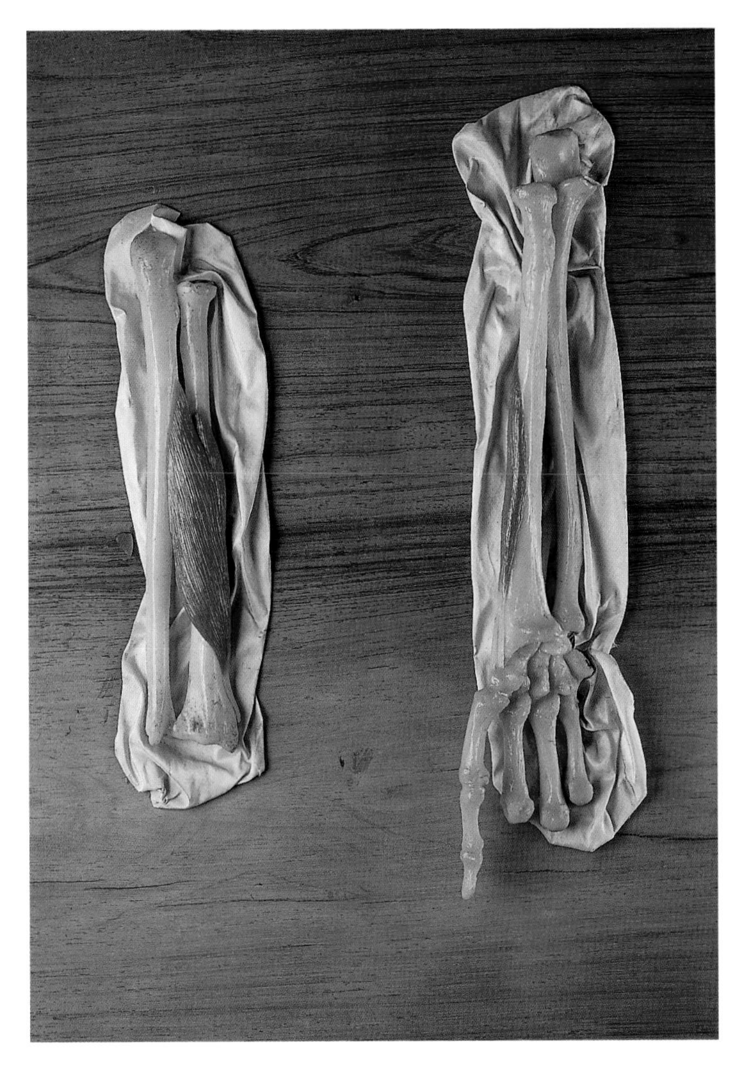

A long muscle acting on the thumb

Darstellung eines langen Daumenmuskels

Muscle long abducteur du pouce

[XXVI, 568]

Display showing forearm muscles which contribute to flexion and extension at the wrist joint

Darstellung von Unterarmmuskeln, die an der Beugung und Streckung im Handgelenk beteiligt sind

Muscles de l'avant-bras (fléchisseur ulnaire du carpe et court extenseur radial du carpe)

[XXVI, 548]

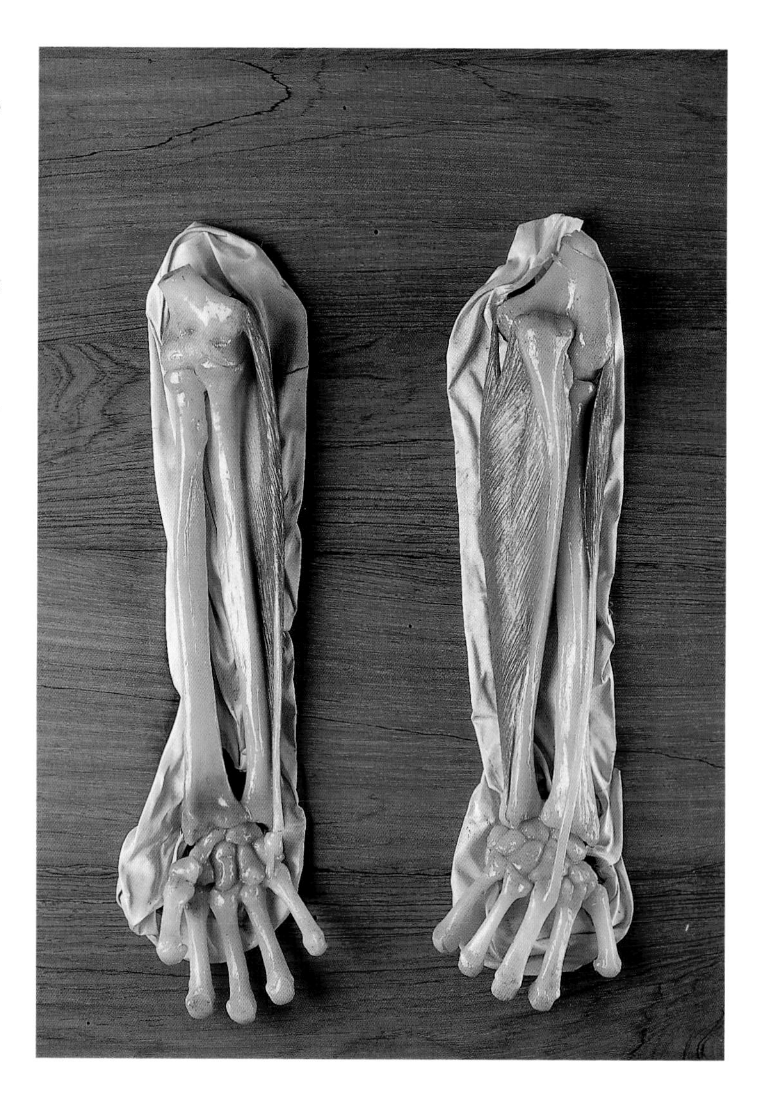

A muscle which contributes to extension at the wrist joint

Darstellung eines Muskels, der an der Streckung des Handgelenkes beteiligt ist

Muscle de la loge latérale de l'avant-bras (long extenseur radial du carpe)

[XXVI, 550]

Musculus extensor digitorum, Musculus extensor carpi ulnaris

Display showing muscles bringing about extension of the fingers and sideways movement of the hand

Darstellung von Muskeln zur Streckung der Finger und Seitwärtsbewegung der Hand

Muscles de la loge postérieure de l'avant-bras (extenseur des doigts et extenseur ulnaire du carpe)

[XXVI, 555]

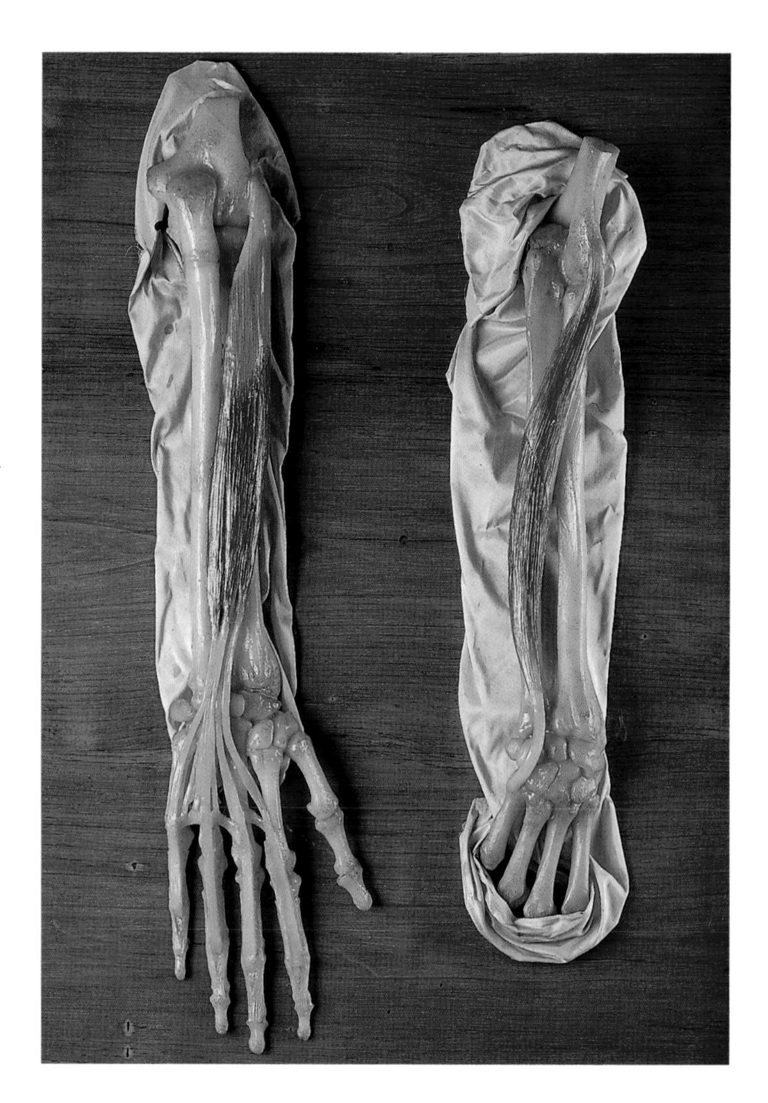

The extensor muscle of the index finger

Darstellung des Streckmuskels des Zeigefingers

Muscle extenseur propre de l'index

[XXVI, 547]

Musculi interossei manus, Musculus abductor pollicis brevis, Musculus flexor pollicis brevis, Musculus adductor pollicis, Musculus abductor digiti minimi

Display showing the deep muscles in the hollow of the hand which bring about fine movements of the fingers

Darstellungen von tiefen Muskeln der Hohlhand für feinmotorische Fingerbewegungen

Muscles courts de la main (intrinsèques) pour les mouvements fins des doigts

[XXVII, 545]

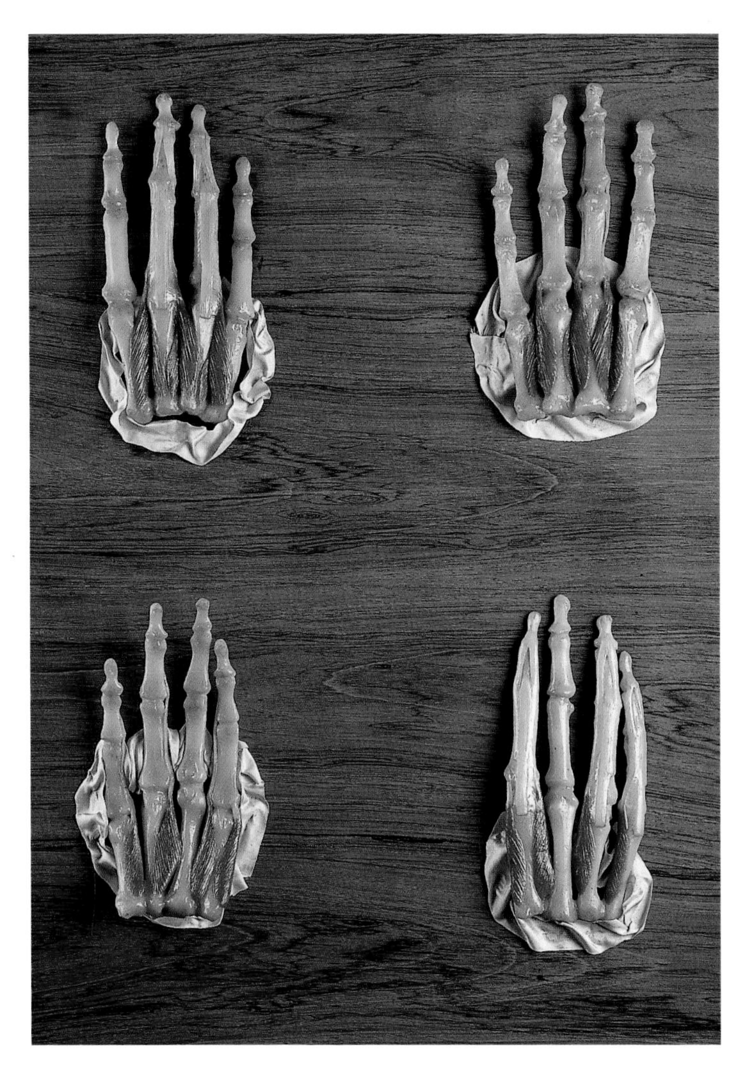

Display showing muscles which contribute to fine movements of the fingers

Darstellung von Muskeln, die an den Fingerbewegungen beteiligt sind

Muscles courts de la main (muscles interosseux)

[XXVII, 544]

Musculi digitorum manus I et V, Musculus flexor digiti minimi brevis, Musculus opponens pollicis

Display showing the muscles of the ball of the thumb and little finger

Darstellungen von Muskeln des Daumen- und Kleinfingerballens

Muscles courts pour le pouce (éminence thénar), et pour le petit doigt (éminence hypothénar)

[XXVI, 543]

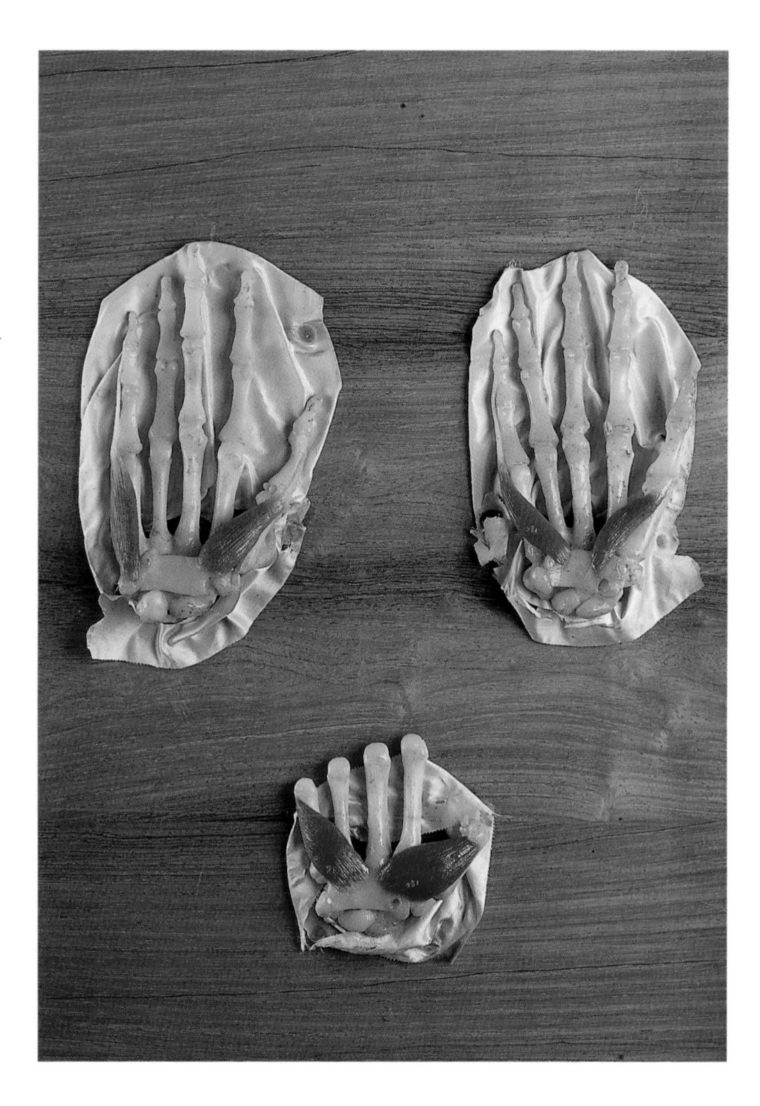

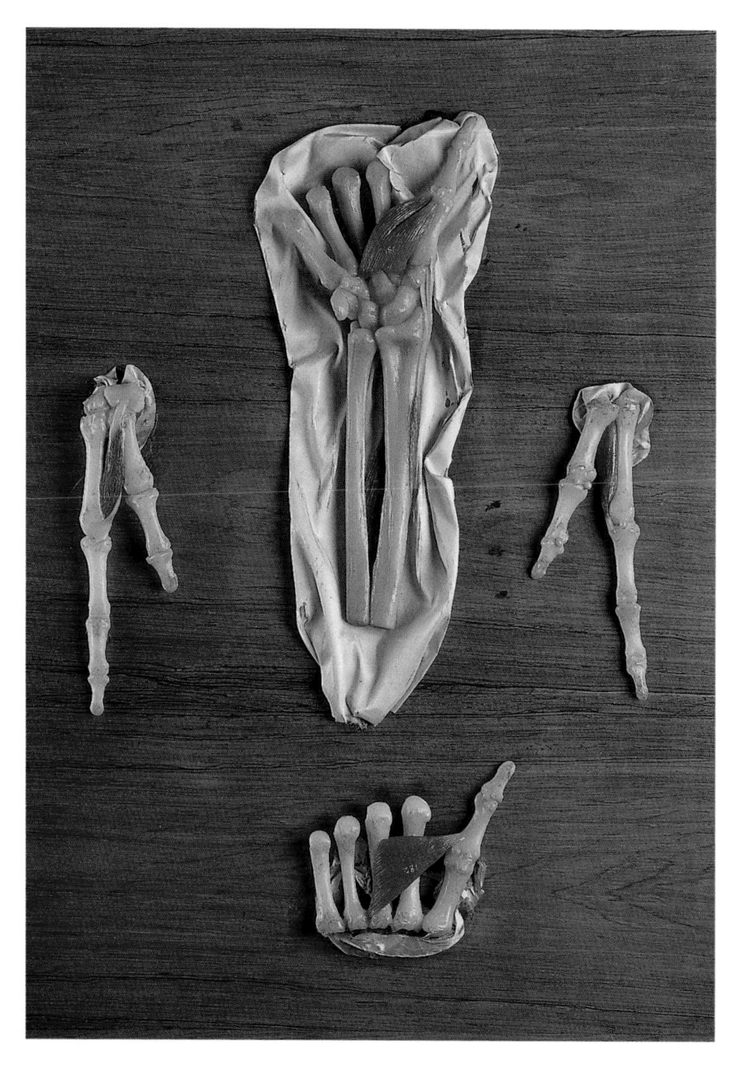

Display showing the muscles responsible for fine movements of the index finger and thumb

Darstellungen von Muskeln für die feinmotorischen Bewegungen von Zeigefinger und Daumen

Muscles courts de la main pour le pouce et l'index

[XXVII, 546]

Display showing the joints of the shoulder, elbow and wrist with muscle insertions and tendons

Darstellung von Muskeln, Muskelansätzen und Sehnen von Schultergelenk, Ellenbeuge und Handbeugeseite

Bourses et gaines synoviales annexées aux muscles et tendons des articulations de l'épaule, du coude et de la main

[XXVII, 407]

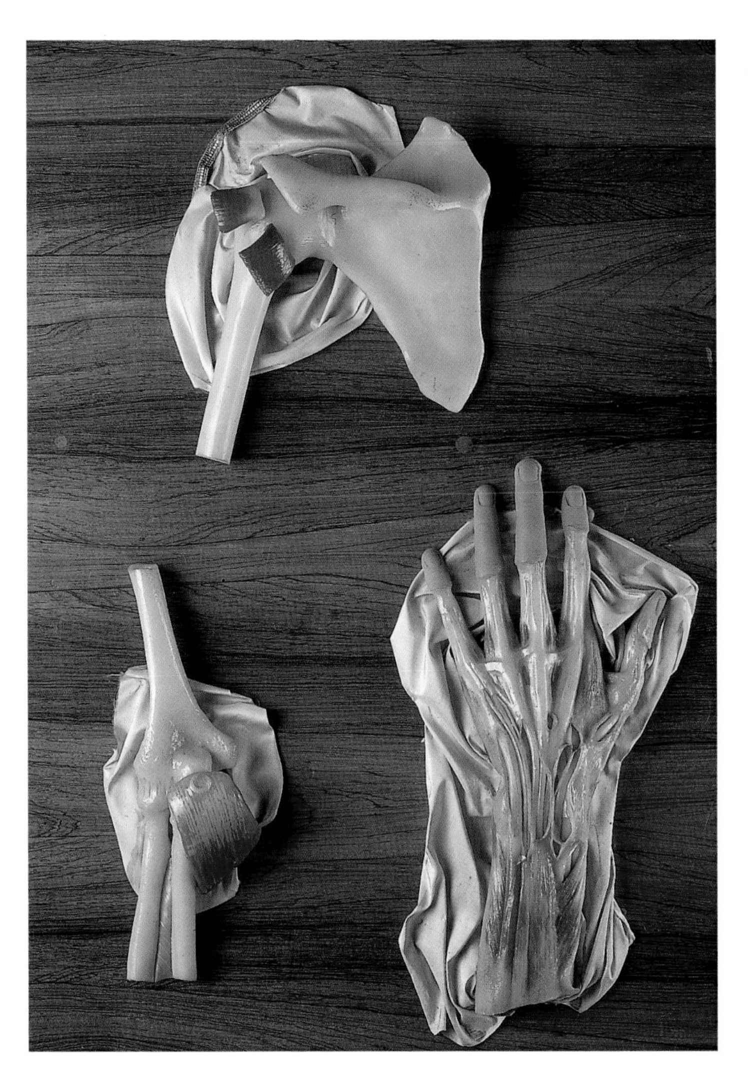

Display showing the joints of the shoulder, elbow and wrist with muscle insertions and tendons

Darstellung von Muskeln, Muskelansätzen und Sehnen von Schultergelenk, Ellenbeuge und Handstreckseite

Bourses et gaines synoviales annexées aux muscles et tendons des articulations de l'épaule, du coude, et de la main

[XXVII, 406]

A large muscle running downwards from the lumbar vertebral column viewed from in front

Ansicht des großen Lendenmuskels von vorn

Muscle psoas vu de l'avant

[XXIX, 535]

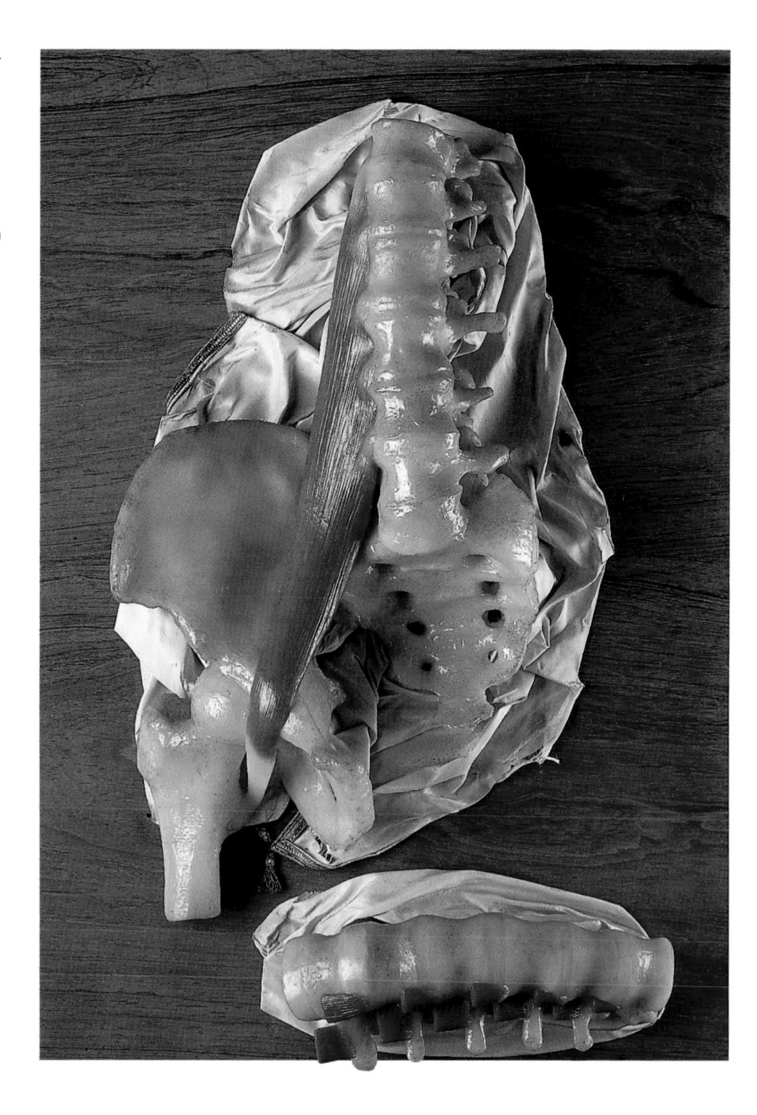

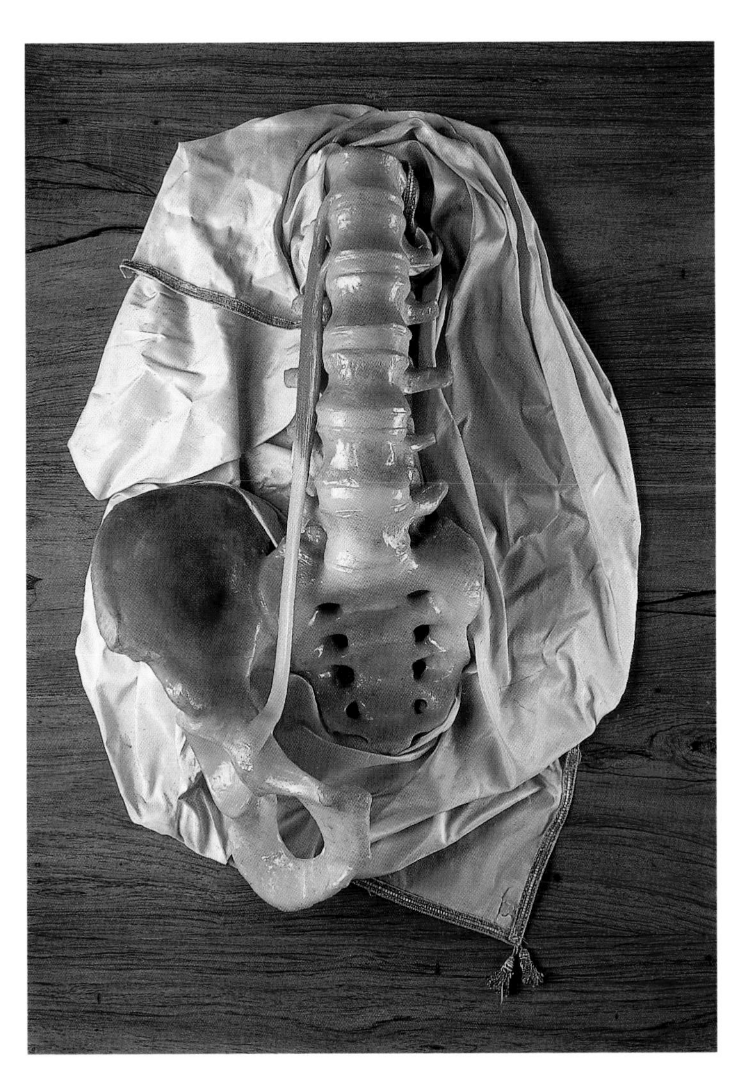

A small muscle running downwards from the lumbar vertebral column seen from in front

Ansicht des kleinen Lendenmuskels von vorn

Muscle petit psoas (inconstant) vu de l'avant

[XXV, 452]

A muscle of the posterior abdominal wall seen from behind

Blick von hinten auf einen Muskel der hinteren Bauchwand

Muscle carré des lombes vu de l'arrière

[XXV, 459]

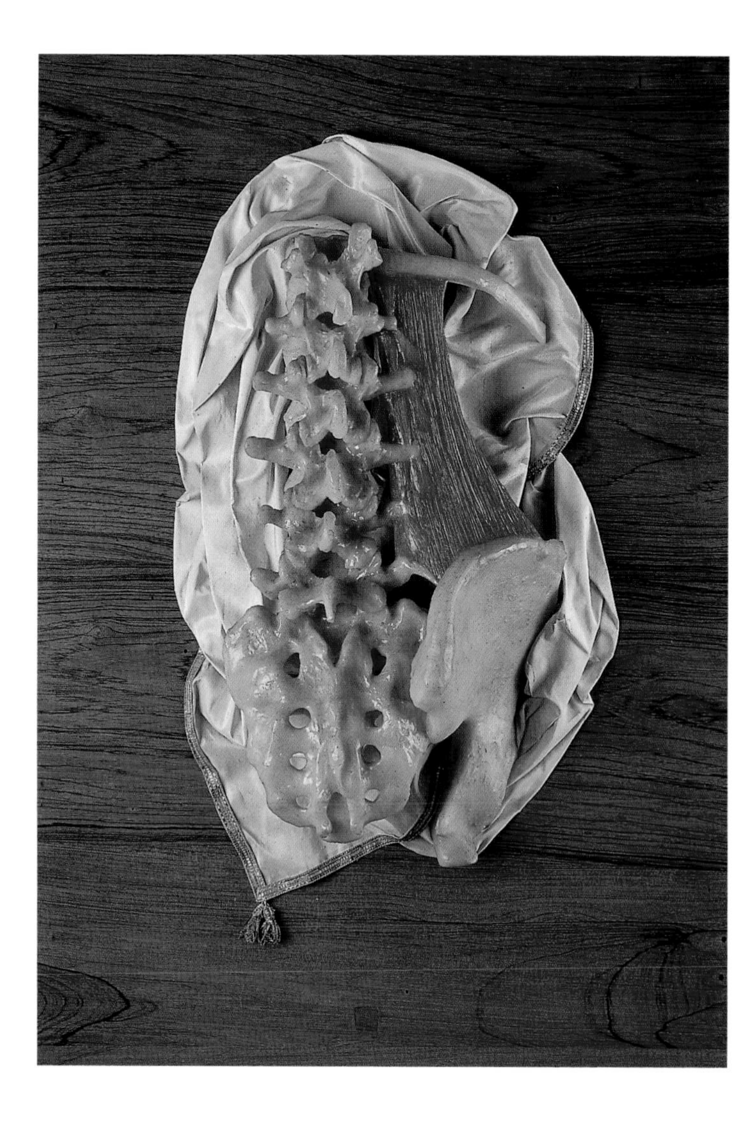

View of the muscles of the floor of the pelvis from below (upper specimen) and from above (lower specimen), also showing the circular course of the muscles of the external sphincter

Ansicht des muskulären Beckenbodens von unten (oberes Präparat) und von oben (unteres Präparat) mit Darstellung der zirkulär verlaufenden Fasern des äußeren Schließmuskels

Muscles du plancher du bassin (périnée) : vue du dessous (en haut) avec muscle sphincter de l'anus, vue du dessus (en bas)

[XXV, 448]

Musculi faciales, Musculi colli, Musculi thoracis, Musculi abdominis, Musculi membri superioris, Musculi membri inferioris

Whole body specimen displaying both the superficial and deep layers of muscles

Ganzkörperpräparat zur Darstellung oberflächlicher und tiefer Schichten der Muskulatur

Préparation du corps entier représentant des muscles superficiels et profonds

[XXX, 958]

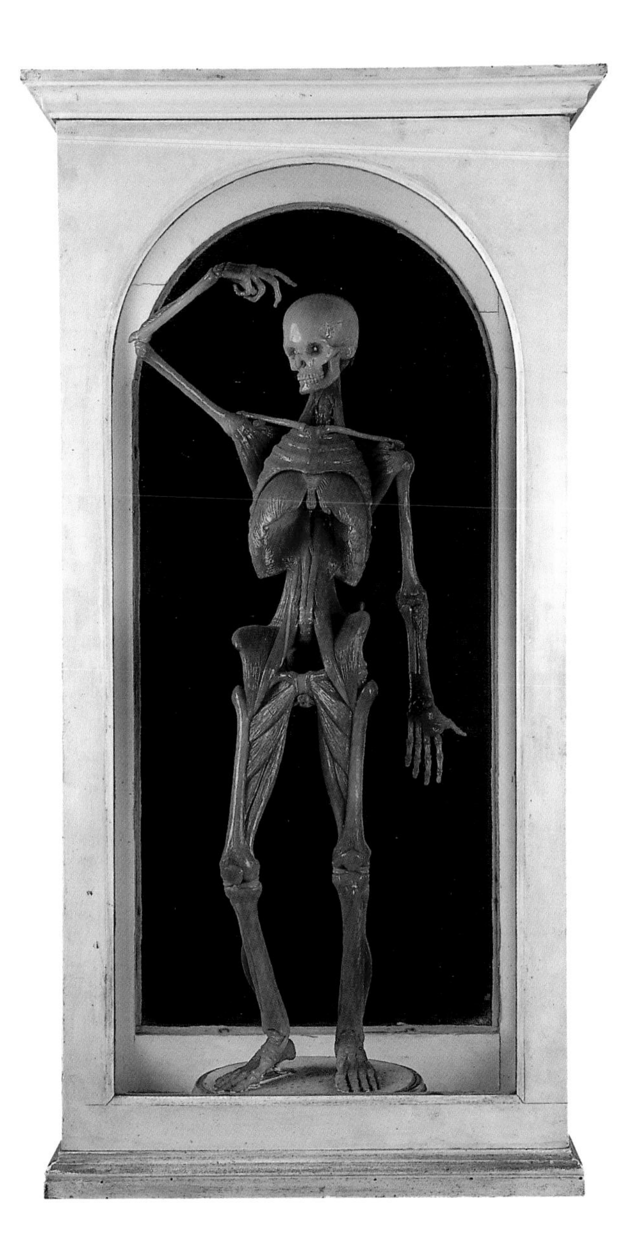

Whole body specimen displaying the deep layer of muscles

Ganzkörperpräparat zur Darstellung tiefer Schichten der Muskulatur

Préparation du corps entier représentant des muscles profonds

[XXIX, 959]

Several views of the bony pelvis and pelvic floor

Verschiedene Darstellungen des knöchernen Beckens mit Beckenboden

Muscles du plancher du bassin (périnée)

[DEP. 12]

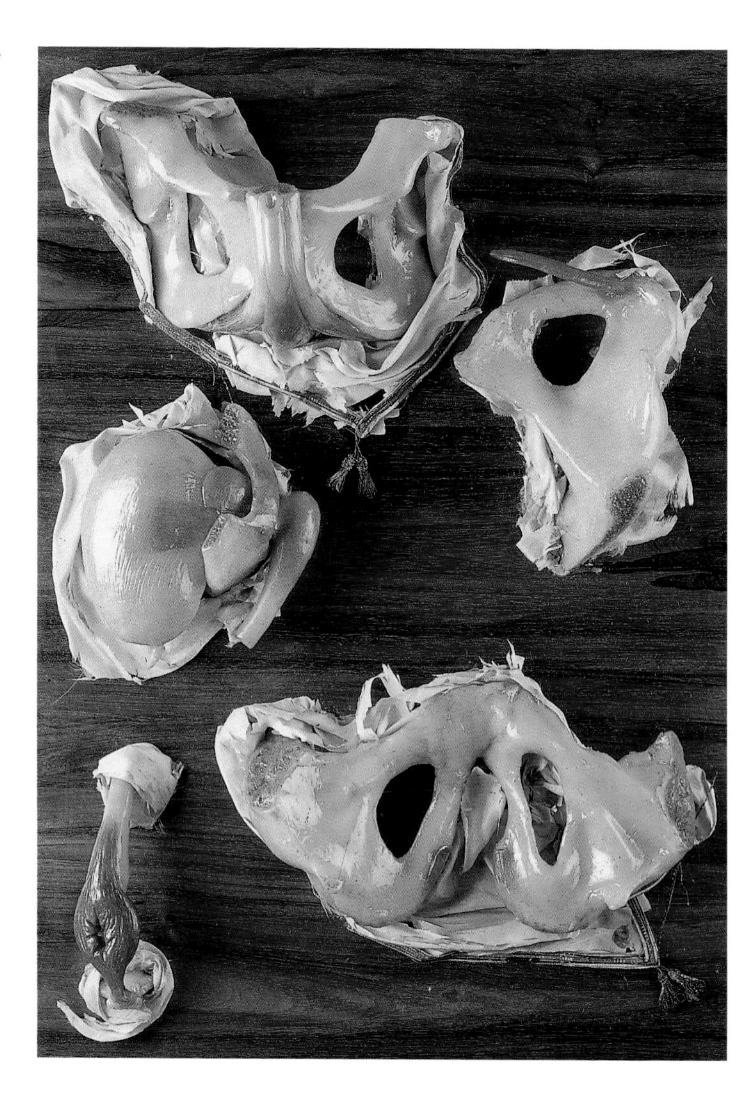

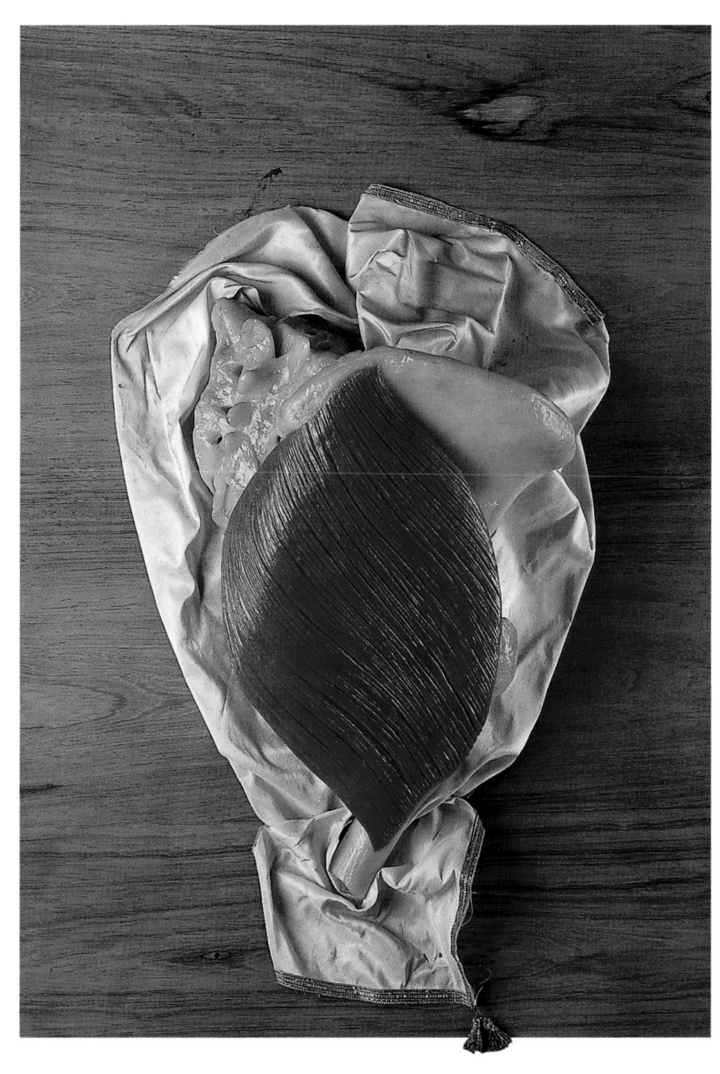

The large outermost muscle of the buttock seen from the side

Ansicht des großen äußeren Gesäßmuskels von der Seite

Muscle superficiel de la région fessière (grand fessier) vu du côté droit

[XXVI, 542]

Musculus glutaeus medius

Display showing part of the deep muscles of the buttock

Darstellung eines Teiles der tiefen Gesäßmuskulatur

Muscle le plus profond de la région fessière (petit fessier)

[XXVI, 566]

◄
Display showing part of the muscles of the middle layer of the buttock

Darstellung eines Teiles der mittleren Schicht der Gesäßmuskulatur

Muscle moyen fessier vu de l'avant et de l'arrière

[XXVI, 567]

Musculus tensor fasciae latae

A muscle attached to the inside of the wing of the iliac bone

Darstellung eines Muskels an der Innenseite der Darmbeinschaufel

Muscle iliaque attaché sur la face interne du grand bassin

[XXVI, 534]

◄

Display showing a superficial muscle which has a spreading insertion into the tendinous sheet at the side of the thigh

Darstellung eines oberflächlichen Muskels, der in die Sehnenplatte am seitlichen Oberschenkel einstrahlt

Muscle superficiel de la région fessière (tenseur du fascia lata)

[XXVI, 536]

Musculus obturatorius externus

A deeply placed muscle of the thigh

Darstellung eines tiefliegenden Oberschenkelmuskels

Muscle adducteur de la cuisse (pectiné)

[XXVI, 537]

◀
A muscle which contributes to rotation at the hip joint

Darstellungen eines Muskels, der an Drehbewegungen im Hüftgelenk beteiligt ist

Muscle obturateur externe rotateur de la cuisse (pelvitrochantérien)

[XXV, 843]

Musculus adductor brevis

A muscle on the inside of the thigh

Darstellung eines Muskels der Oberschenkelinnenseite

Muscle adducteur situé dans la région interne de la cuisse (long adducteur)

[XXVI, 533]

◄
A muscle which connects the pubic bone with the femur

Darstellung eines Muskels, der das Schambein mit dem Oberschenkelknochen verbindet

Muscle adducteur tendu entre le pubis et le fémur (court adducteur)

[XXVI, 532]

Musculus adductor magnus

A muscle on the inside of the thigh

Darstellung eines Muskels der Oberschenkelinnenseite

Muscle adducteur situé dans la région interne de la cuisse (grand adducteur)

[XXVI, 541]

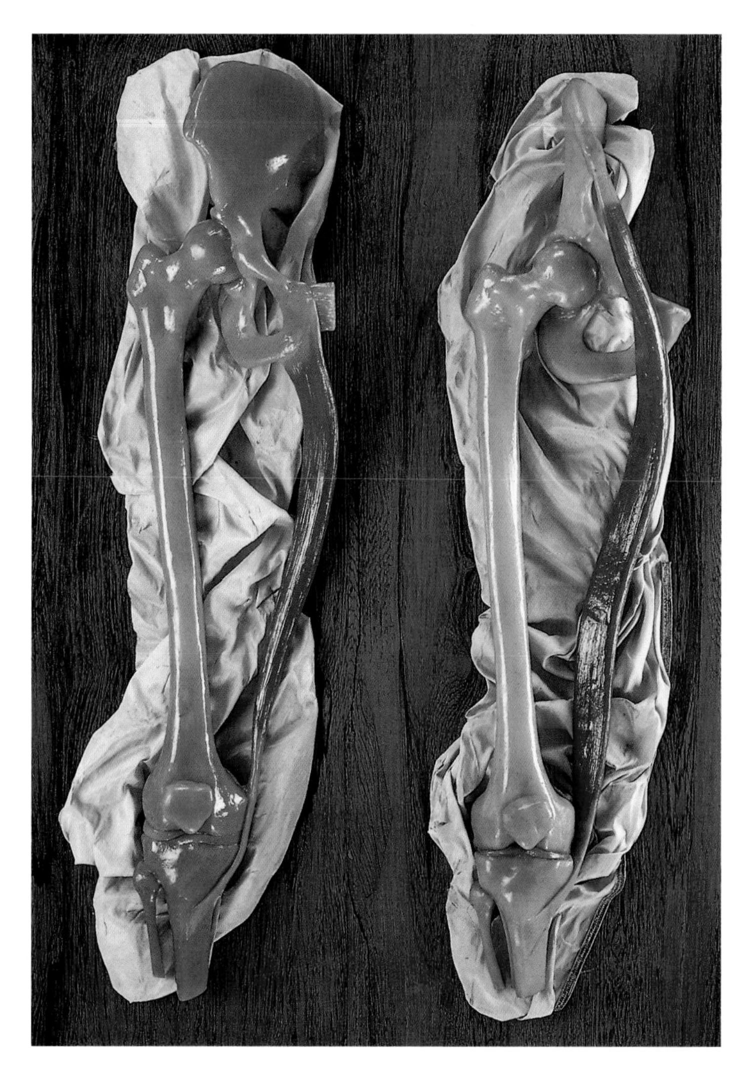

Display showing muscles of the thigh

Darstellung von Oberschenkel-muskeln

Muscles de la cuisse tendus entre le bassin et le tibia (gracile à gauche, et couturier à droite)

[XXVI, 580]

Musculus rectus femoris

Two views of a muscle at the front of the thigh

Muskel an der Oberschenkelvorderseite in zwei Ansichten

Portions du muscle quadriceps de la région antérieure de la cuisse (droit fémoral)

[XXVI, 582]

Muscles at the inner side and back of the thigh

Muskeln der Oberschenkelinnen- und -rückseite

Muscles profonds de la région postérieure de la cuisse (ischio-jambiers)

[XXVI, 583]

Muscles at the back of the thigh

Muskeln der Oberschenkelrückseite

Muscles superficiels de la région postérieure de la cuisse (ischio-jambiers)

[XXVI, 581]

Display showing the deeper parts of the four-headed muscle of the thigh

Darstellung der tieferen Anteile des vierköpfigen Oberschenkelmuskels

Chefs profonds du muscle quadriceps de la région antérieure de la cuisse (vastes)

[XXVI, 540]

Musculus peronaeus longus, Musculus peronaeus brevis, Musculus extensor digitorum longus, Musculus triceps surae

Superficial muscles of the lower leg and foot seen from the outer side

Oberflächliche Muskeln des Unterschenkels und Fußes in der Ansicht von der Außenseite

Muscles et tendons de la jambe et du pied

[XXVII, 614]

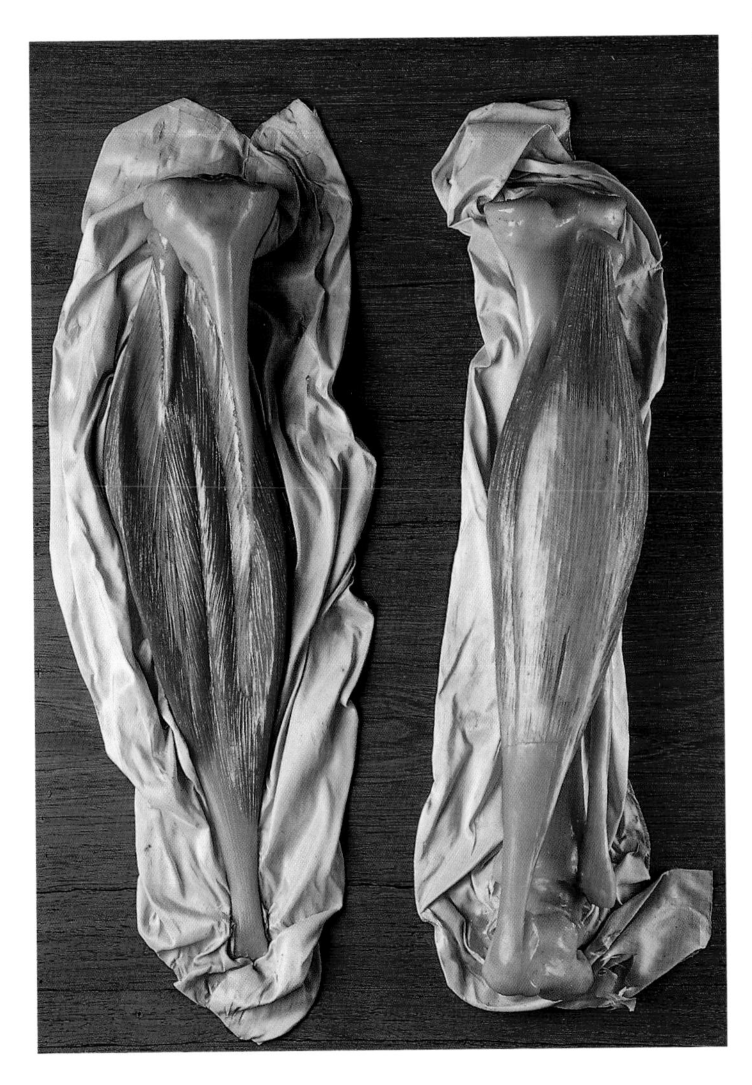

The large muscles of the calf with the tendon of Achilles

Darstellung des großen Wadenmuskels mit Achillessehne

Chef profond du muscle triceps du mollet (soléaire), et tendon d'Achille

[XXVI, 563]

The long extensor muscle of the toes

Ansicht des Zehenstreckers

Muscles antérieurs de la jambe (tibial antérieur et extenseur des orteils)

[XXVI, 527]

A long muscle on the outer side of the lower leg which helps to maintain the tension of the arch of the foot

Darstellung eines langen Muskels an der Außenseite des Unterschenkels, der an der Verspannung des Fußgewölbes beteiligt ist

Muscle latéral de la jambe (long fibulaire) élévateur du bord externe du pied

[XXVI, 529]

Musculi abductores et flexores hallucis sive digiti minimi

Display showing muscles which contribute to movement of the toes

Darstellung von Muskeln, die an Zehenbewegungen beteiligt sind

Muscles courts du pied moteurs des orteils

[XXVI, 561]

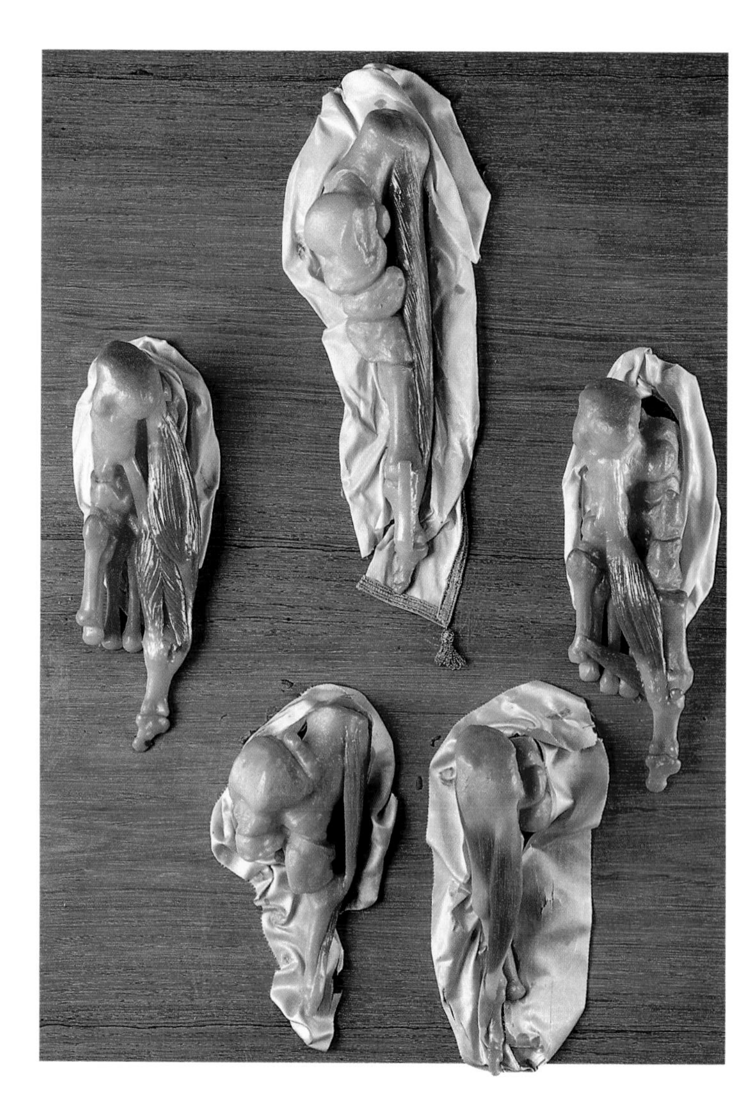

Display showing muscles and tendons in the different layers of the sole of the foot

Darstellung von Muskeln und Sehnen verschiedener Schichten der Fußsohle

Plans musculaires et fibreux de la plante du pied

[XXVI, 558]

Display showing muscles and tendons in the different layers of the sole of the foot

Darstellung von Muskeln und Sehnen verschiedener Schichten der Fußsohle

Muscles et tendons de la plante du pied

[XXVI, 562]

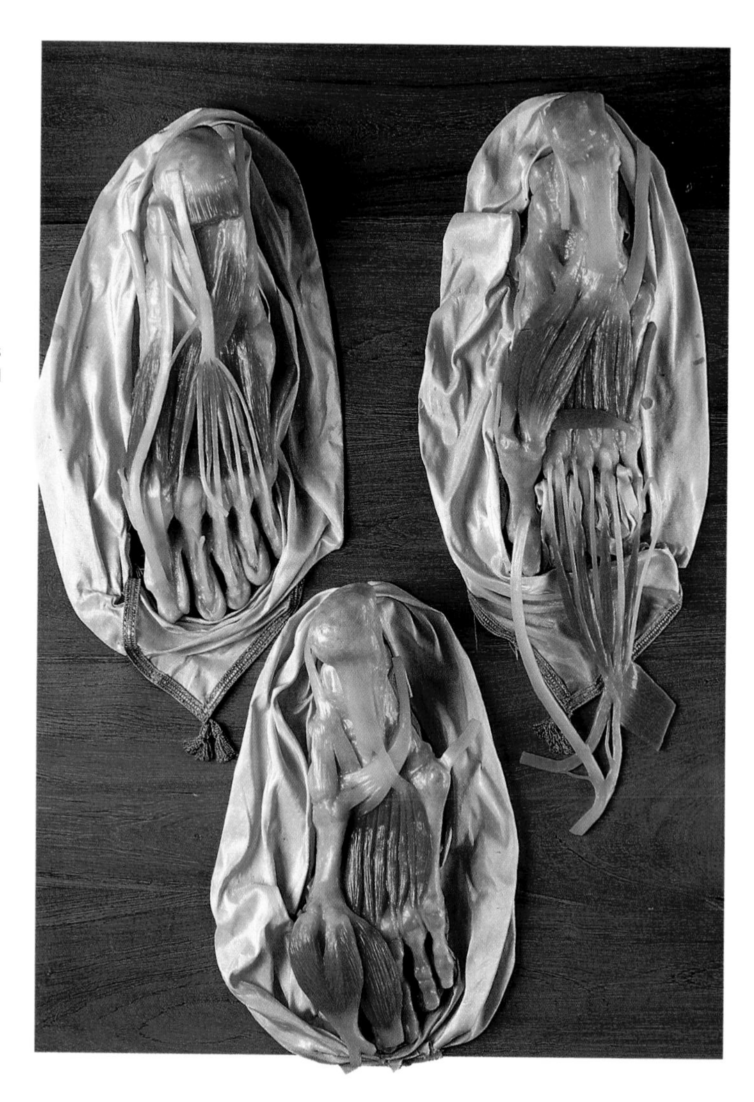

Display showing muscles and tendons in the different layers of the sole of the foot

Darstellung von Muskeln und Sehnen verschiedener Schichten der Fußsohle

Muscles et tendons de la plante du pied

[XXVI, 560]

The superficial tendons and ligaments of the foot

Darstellung der oberflächlichen Sehnen und Bandverbindungen des Fußes

Plans fibreux et gaines tendineuses de la cheville et du pied

[XXVI, 423]

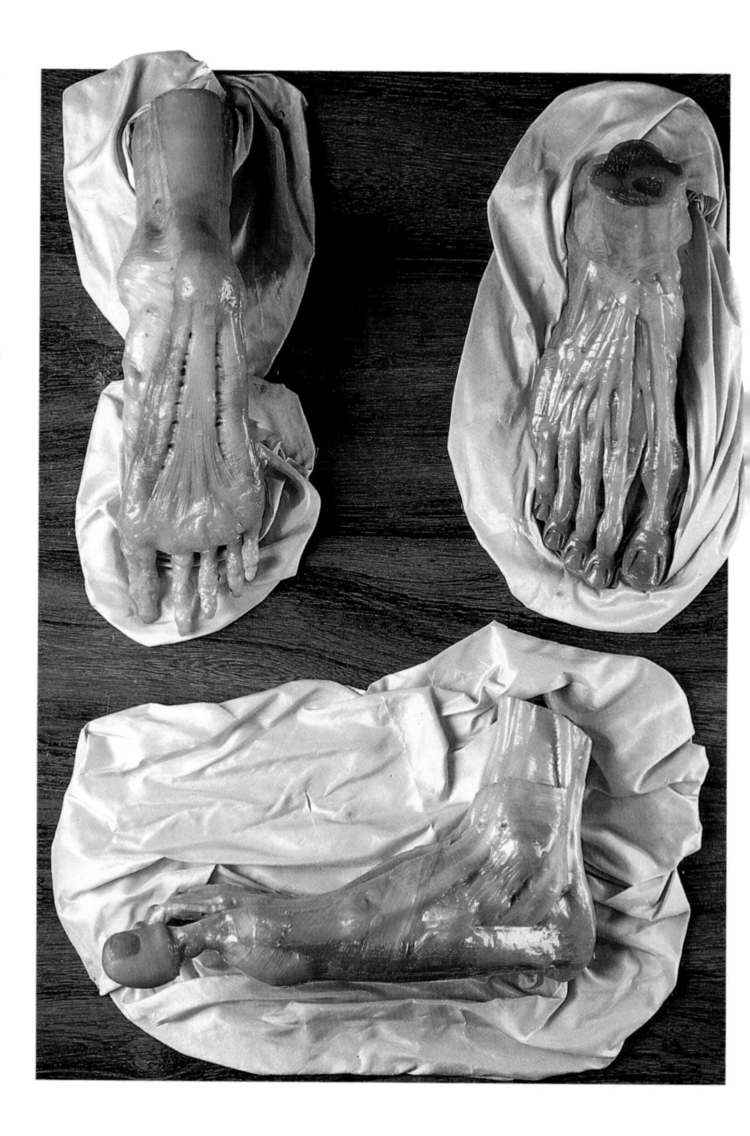

Display showing the tendons in the back of the foot (upper specimen) and sole of the foot (lower specimen)

Darstellung der Sehnen des Fußrückens (oberes Präparat) und der Fußsohle (unteres Präparat)

Tendons du dos du pied (en haut), et de la plante du pied (en bas)

[XXVI, 425]

Musculi faciales, Musculi colli, Musculi thoracis, Musculi abdominis, Musculi membri superioris, Musculi membri inferioris

Whole body specimen displaying the superficial muscles

Ganzkörperpräparat zur Darstellung der oberflächlichen Muskulatur

Préparation du corps entier représentant des muscles superficiels

[XXX, 955]

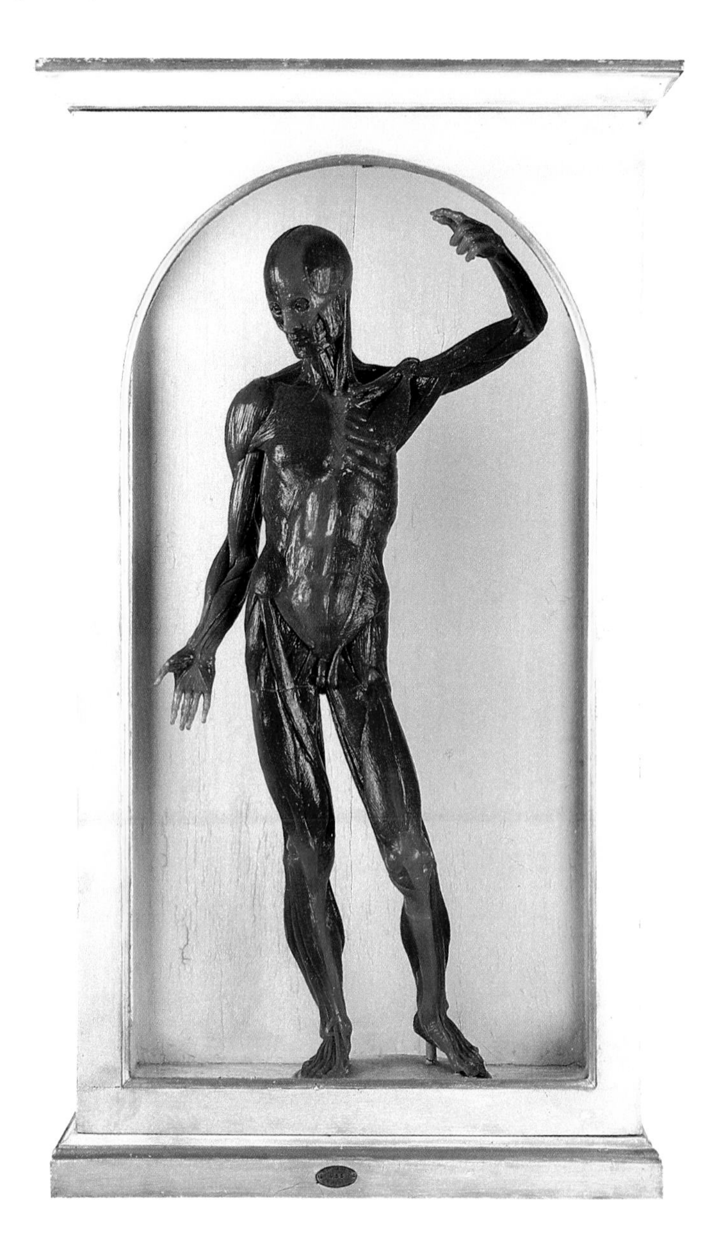

Whole body specimen with the deep layers of muscles

Ganzkörperpräparat zur Darstellung tiefer Schichten der Muskulatur

Préparation du corps entier représentant des muscles profonds

[XXV, 441]

Musculi faciales, Musculi colli, Musculi thoracis, Musculi abdominis, Musculi membri superioris, Musculi membri inferioris

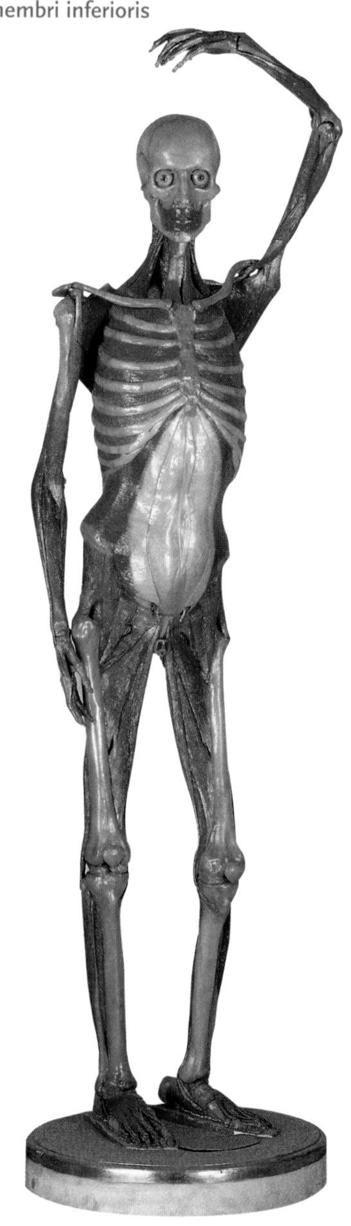

Whole body specimen with the deep layers of muscles

Ganzkörperpräparat zur Darstellung tiefer Schichten der Muskulatur

Préparation du corps entier représentant des muscles profonds

[XXV, 443]

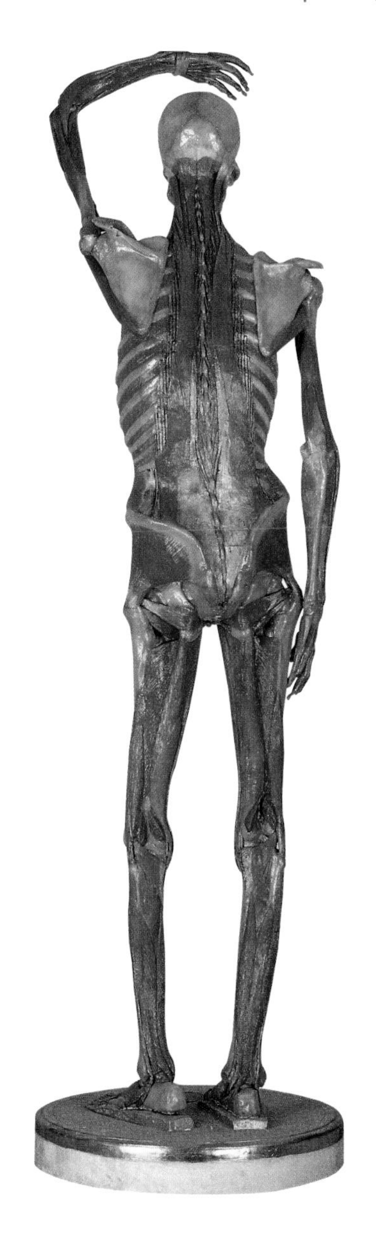

Whole body specimen with the deep layers of muscles

Ganzkörperpräparat zur Darstellung tiefer Schichten der Muskulatur

Préparation du corps entier représentant des muscles profonds

[XXV, 443]

Systema cardiovasculare

Heart and Blood Circulation

Herz und Kreislauf

Cœur, artères et veines

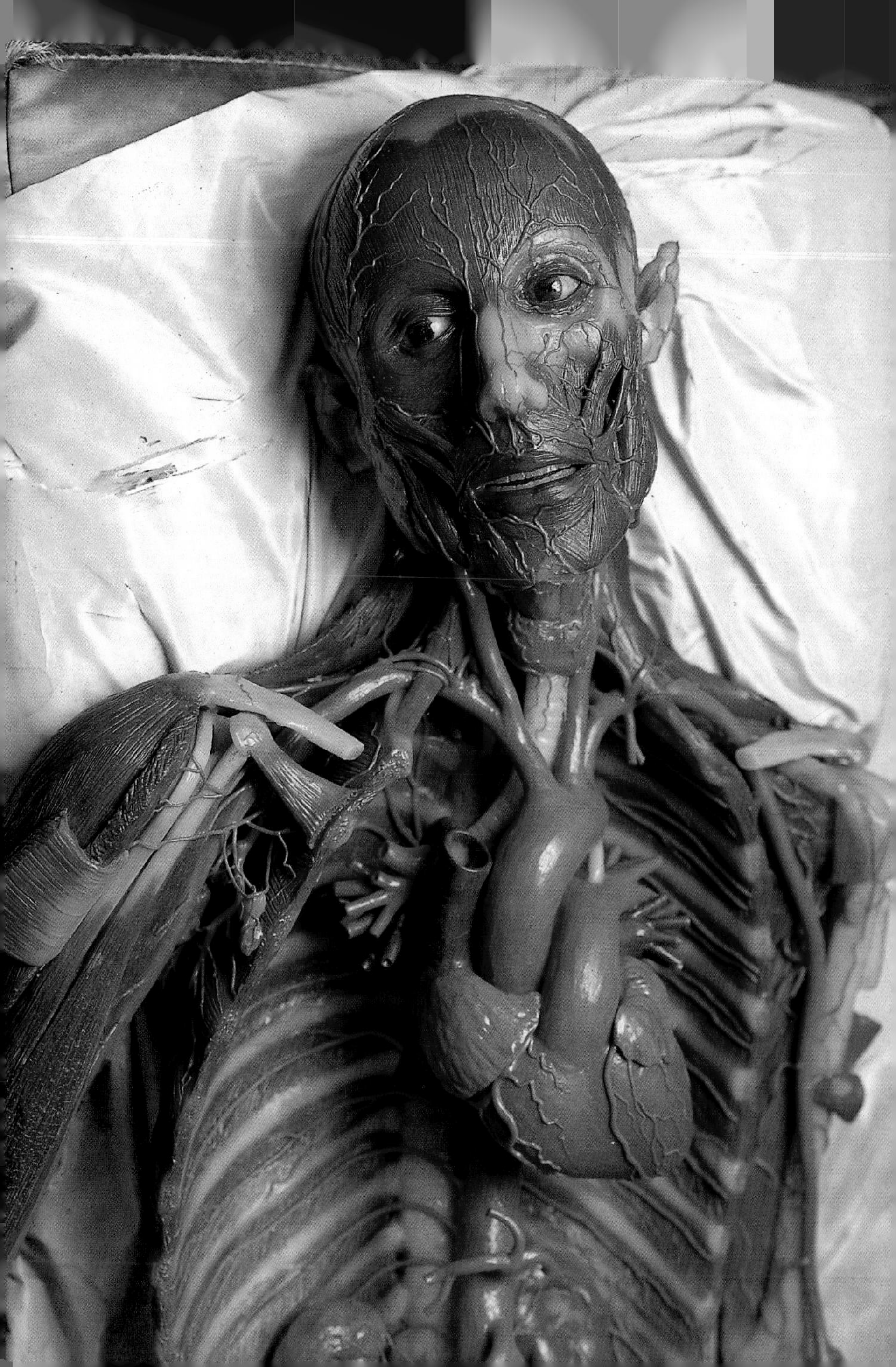

Heart and Blood Circulation

Of all those organs which are essential to life, the heart has long occupied a unique position in our cultural history, having been accepted as the seat of the soul, and of love, feeling and the spirit. In radical contrast to this view, these wax models bear impressive witness to the state of knowledge in topographical anatomy during the 17th century. As early as 1628, the English physician William Harvey (1578–1657) discovered the relationship between the heart and circulation. Harvey's revolutionary theory of the blood circulation upset medical ideas which had been taken for granted for centuries, but it was generally accepted by the end of the 18th century.

The selection of specimens was related to the corresponding woodcuts and copperplate illustrations of the 16th and 17th centuries. These models represent many aspects of the cardiovascular system, such as the position of the heart between the lungs within the thorax, its relationship to the diaphragm, the pericardium with its nerves and vessels and the internal structure of both atria and ventricles. In some of the models one can look through openings in the anterior wall of the heart and see the position of the atrioventricular and semilunar valves on both sides of the organ.

The great vessels entering and leaving the heart and the vessels supplying the head and neck are shown in their relationship to the muscles, and the arteries and veins are clearly distinguished by their coloring. The whole distribution of the arterial system, including the large branches on the anterior abdominal wall, in the limbs and between the dissected muscles of the pelvis, can be easily followed.

Page · Seite · Page
255:

Detail from pages 260–261

Detail der Seiten 260–261

Détail des pages 260–261

Herz- und Blutkreislauf

Das Herz nimmt kulturgeschichtlich unter allen lebenswichtigen Organen eine Sonderstellung ein, da es lange Zeit als Sitz der Seele, der Liebe, der Gefühle und des Geistes angesehen wurde. Die Wachspräparate zeugen im Gegensatz zu dieser Betrachtungsweise eindrucksvoll vom Kenntnisstand der makroskopischen Anatomie des 17. Jahrhunderts. Bereits 1628 hatte der englische Anatom William Harvey (1578–1657) das Zusammenspiel zwischen Herz und Kreislauf entdeckt. Harveys revolutionäre Kreislauftheorie, die ein über Jahrhunderte gültiges Medizinkonzept ins Wanken gebracht hatte, war im ausgehenden 18. Jahrhundert allgemein anerkannt.

Die Auswahl der Präparate des Herzens orientiert sich an entsprechenden Holzschnitten und Kupferstichen des 16. und 17. Jahrhunderts. An diesen Modellen werden die vielfältige Aspekte des Herzens und des Kreislaufsystems herausgearbeitet, wie etwa die topographische Lage des Herzens im Brustkorb zwischen den Lungenflügeln, die Lagebeziehung zum Zwerchfell, der Herzbeutel mit seinen Gefäßen und Nerven, das Innenrelief der beiden Herzvorhöfe und -kammern. Die Gefäßversorgung des Herzens mit Arterien, Venen und Lymphgefäßen sowie die Nervenversorgung werden in unterschiedlichen Perspektiven dargestellt. In einzelnen Modellen kann der Betrachter durch die aufgeschnittene Herzvorderwand in das Innere des rechten und linken Herzens blicken und die Position der Segel- und Taschenklappen einsehen.

Die großen herznahen Blutgefäße sowie die versorgenden Gefäße der Hals- und Kopfregion sind in ihrer natürlichen Lage in den Muskellogen wiedergegeben, wobei die arteriellen von den venösen Gefäßen durch unterschiedliche Farbgebung klar unterschieden sind. Der gesamte Verlauf der arteriellen Gefäße, einschließlich größerer Verzweigungen an der vorderen Bauchwand, den Extremitäten sowie dem Becken zwischen den aufpräparierten Muskelgruppen, läßt sich klar verfolgen.

Le cœur et l'appareil circulatoire

Le cœur occupe une place toute particulière, car il a longtemps été considéré dans l'histoire des civilisations comme le siège de l'âme et de l'esprit, de l'amour et des émotions. En opposition à ces conceptions, les modèles en cire témoignent de façon remarquable de l'état des connaissances exactes de l'anatomie macroscopique du XVIIe siècle. En 1628, l'anatomiste anglais William Harvey (1578–1657) découvre l'interdépendance du cœur et de l'appareil circulatoire ; sa théorie révolutionnaire de la circulation du sang, ébranlant un concept médical multiséculaire, est globalement admise à la fin du XVIIIe siècle.

Le choix des préparations du cœur se réfère à des gravures correspondantes, sur bois ou sur cuivre, des XVIe et XVIIe siècles. Sur les modèles en cire ressortent de nombreux aspects de la morphologie du cœur et des vaisseaux : topographie du cœur à l'intérieur de la cavité thoracique entre les deux poumons, position par rapport au diaphragme, disposition du péricarde accompagné de vaisseaux et nerfs, morphologie des cavités des oreillettes et des ventricules. La vascularisation artérielle, veineuse et lymphatique du cœur, ainsi que l'innervation, sont représentées sous différents angles. Sur des modèles du cœur isolé, une ouverture de la paroi cardiaque antérieure permet de découvrir les reliefs des cavités droites et gauches et la position des valvules cardiaques.

Les gros vaisseaux de la base du cœur, ainsi que les troncs vasculaires de la tête et du cou, sont représentés dans leur position anatomique en relation avec les loges musculaires. La distinction entre artères et veines est clairement donnée par une coloration spécifique, rouge ou bleue, de chaque type de vaisseau. Le trajet des artères entre les groupes musculaires disséqués, ainsi que le trajet des grosses branches collatérales, peut être suivi sur toute leur longueur au niveau de la paroi abdominale, du bassin, et des membres.

Page · Seite · Page 259:

Detail from pages 474–475

Detail der Seiten 474–475

Détail des pages 474–475

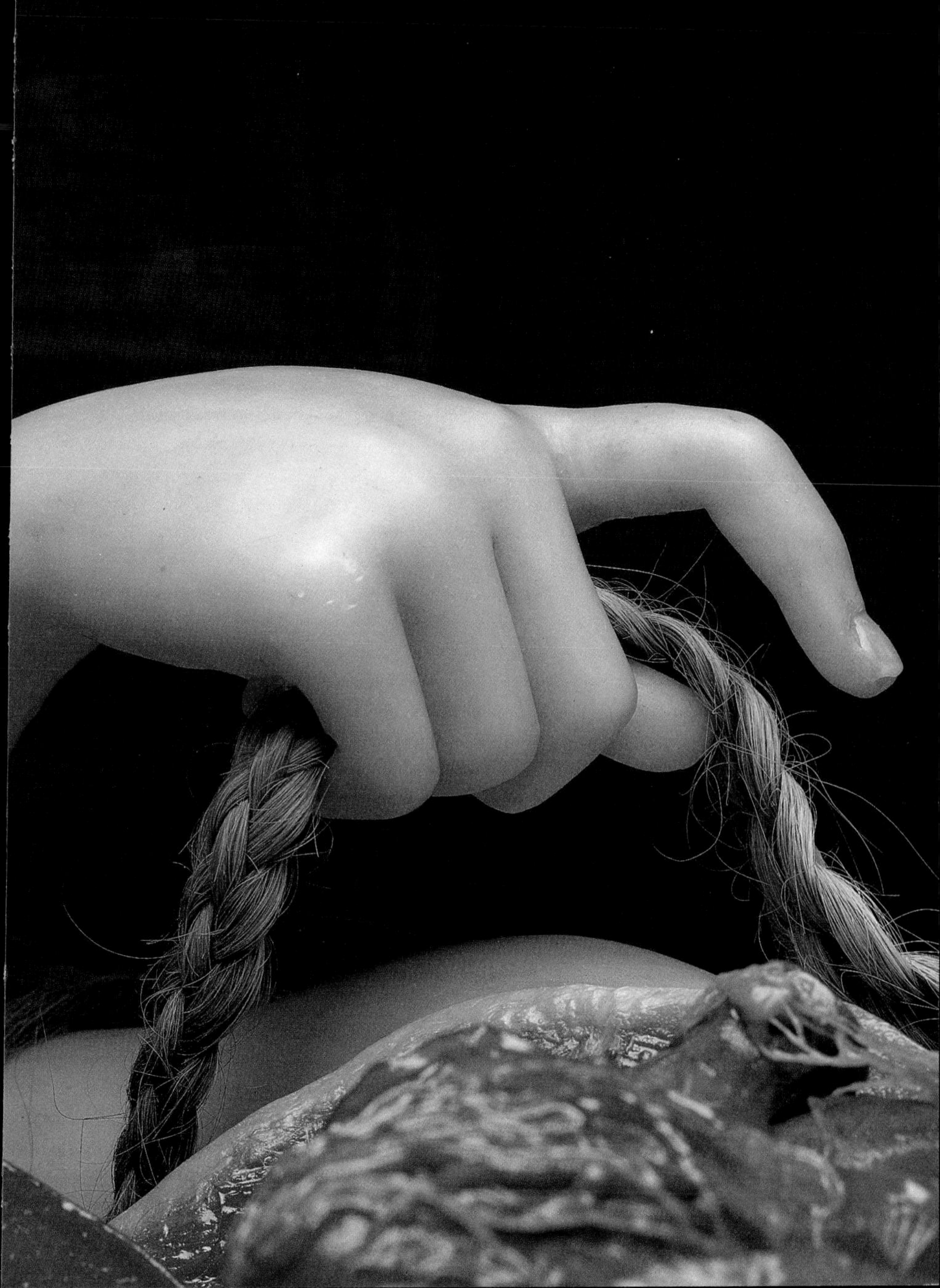

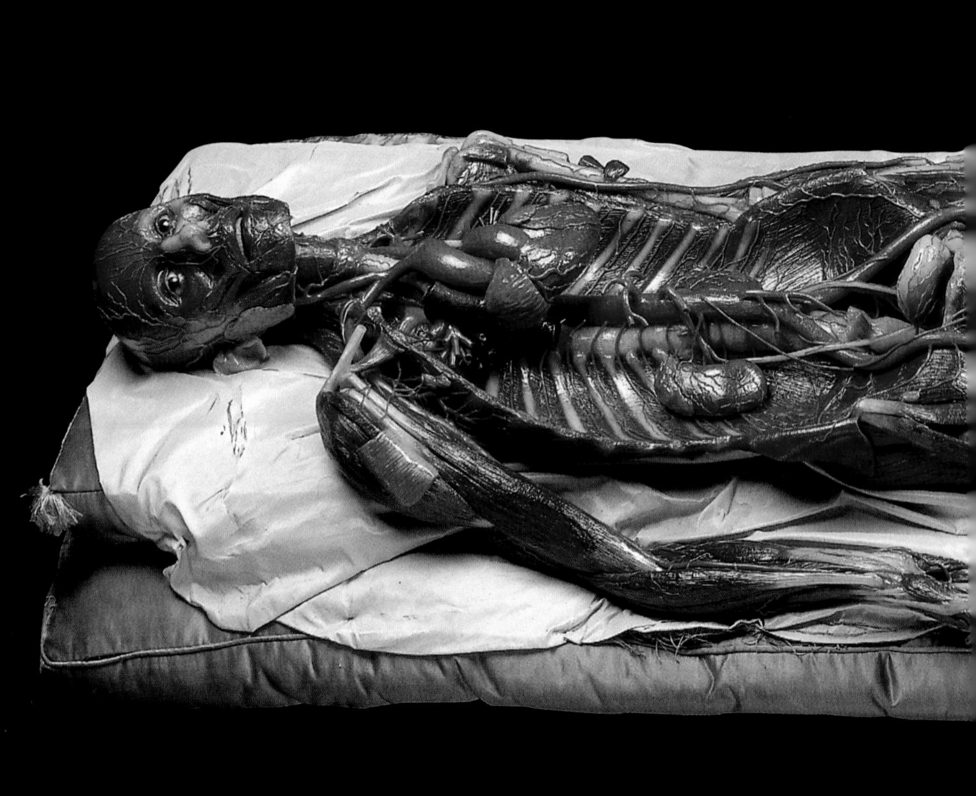

Arteriae

Whole body specimen with the arteries displayed

Ganzkörperpräparat mit Darstellung der Arterien

Préparation du corps entier représentant les artères

[XXV, 446]

Thoracic cavity laid open to show the heart in its pericarxdium, the thymus gland and the great vessels. The lungs have been displaced sideways.

Eröffneter Brustraum mit Blick auf das Herz im Herzbeutel, die Thymusdrüse sowie die großen Gefäße und die zur Seite gedrängten Lungenflügel

Cavité thoracique ouverte : péricarde, gros vaisseaux de la base du cœur destinés au cou et aux membres supérieurs, poumons réclinés

[XXIX, 754]

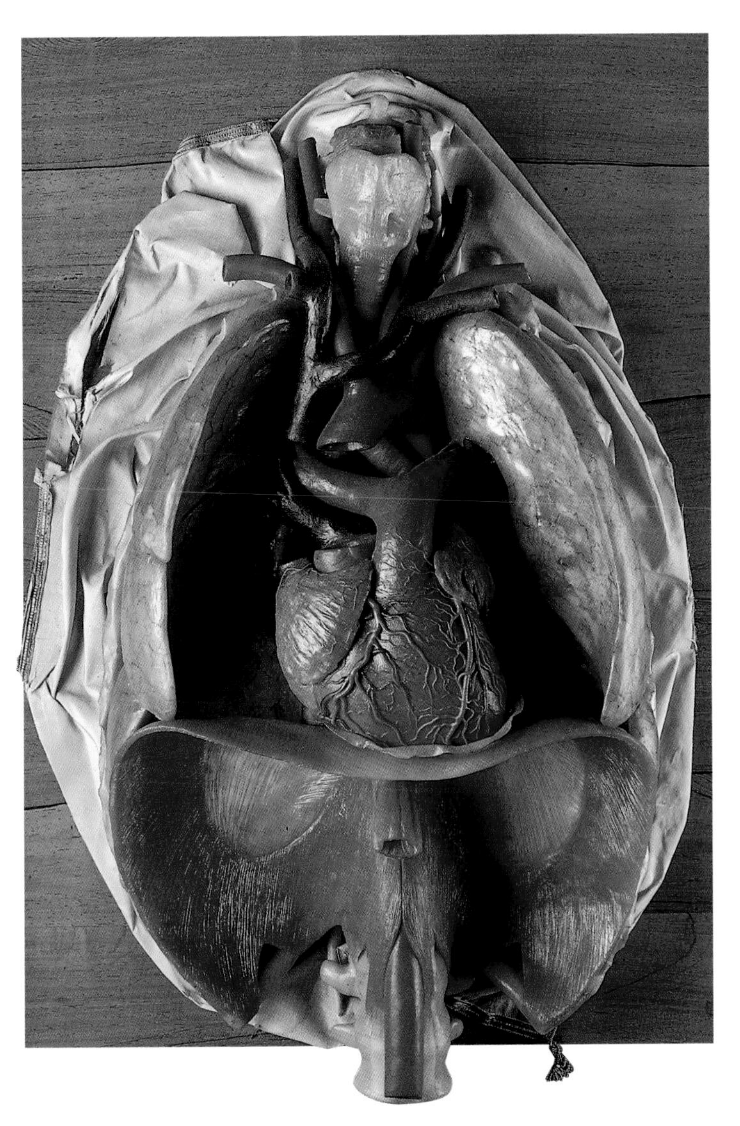

The heart displayed in its natural position in the thoracic cavity. The pericardium has been removed and the lungs displaced sideways.

Darstellung des Herzens in seiner natürlichen Lage im Brustraum. Der Herzbeutel ist enfernt, die Lungenflügel sind zur Seite gedrängt.

Cœur en position naturelle : péricarde enlevé, poumons réclinés

[XXIX, 770]

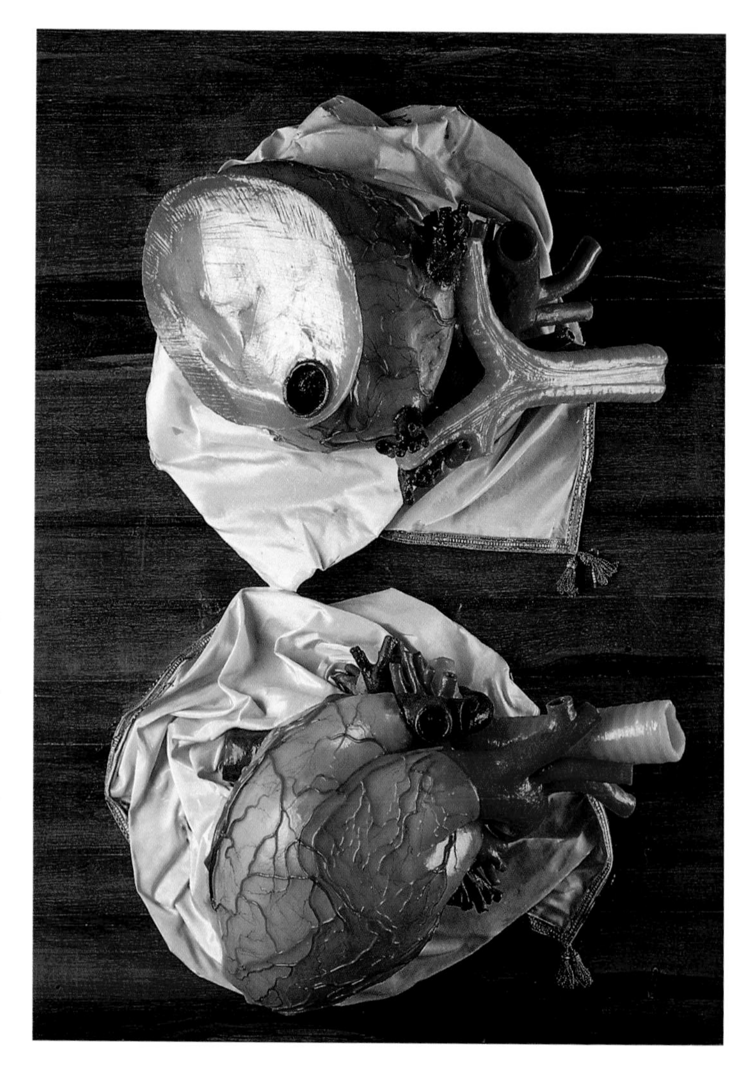

Specimen showing the heart with the aorta, the trachea and the main bronchi. In the upper specimen a part of the diaphragm has been left behind. The specimen is displayed as if viewed from the inside of the abdominal cavity. The lower specimen shows the anterior surface of the heart.

Präparate des Herzens mit Hauptschlagader, Luftröhre und Hauptbronchien. Am oberen Präparat ist ein Teil des Zwerchfells belassen. Die Ansicht entspricht einem Blick aus der Bauchhöhle. Das untere Präparat zeigt die Vorderfläche des Herzens.

Cœur et gros vaisseaux de la base, trachée et bronches principales ; rapports du cœur avec une partie du diaphragme, vue du bas et de l'arrière (en haut) ; cœur et artères coronaires vus de l'avant (en bas)

[XXVII, 625]

Specimens representing various aspects of the heart. The upper specimen shows the anterior surface of the heart with the coronary arteries.

Präparate des Herzens in verschiedenen Ansichten. Das obere Präparat zeigt die Vorderfläche des Herzens mit Herzkranzgefäßen.

Cœur et gros vaisseaux de la base, artères coronaires vus de l'avant (en haut) et de la droite (en bas)

[XXVII, 624]

Upper specimen: anterior aspect of the heart seen from the left. A branch of the left coronary artery. Lower specimen: base of the heart lying on the diaphragm

Oberes Präparat: Ansicht des Herzens von links vorne. Blick auf einen Ast der linken Herzkranzarterie. Unteres Präparat: Ansicht auf die dem Zwerchfell aufliegende Herzbasis

Artères et veines du cœur (coronaires) vus de l'avant (en haut) et de l'arrière (en bas)

[XXVII, 623]

▶
Detail from pages | Detail der Seiten | détail des pages 332–333

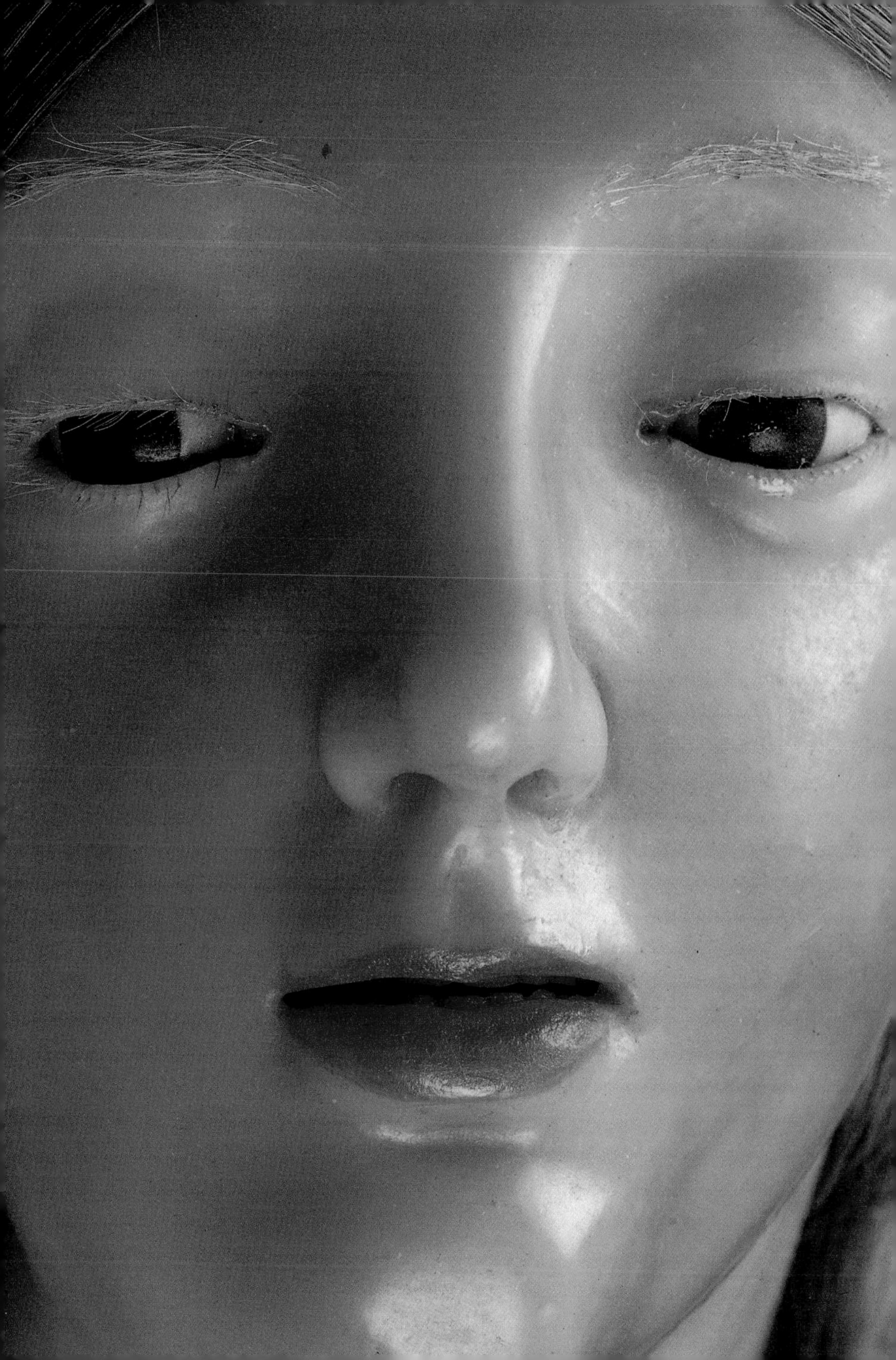

Cor, Arcus aortae, Arteria pulmonalis, Valvulae semilunares

The heart seen from above and from the left. Above, the trachea can be seen, in front of which lies the aortic arch, and in front of that the pulmonary artery. Lower specimens: a view into the pulmonary artery

Blick von oben links auf das Herz. Oben im Bild die Luftröhre, davor der Bogen der Hauptschlagader, vor diesem die Lungenarterie. Die unteren Präparate bieten Einblick in die Lungenarterie.

Cœur et gros vaisseaux (en haut) ; morphologie interne de l'artère pulmonaire et de ses valvules (en bas)

[XXVII, 620]

View from in front into the opened right and left ventricles, below the aortic arch

Blick von vorn in die eröffnete rechte und linke Herzkammer, unten der Aortenbogen

Ventricules droit et gauche ouverts (en haut) ; crosse de l'aorte isolée (en bas)

[XXVII, 619]

Various specimens of the heart with the ventricles laid open (above) and the atria (below). The ligamentous structure (depicted in yellow) connecting the pulmonary artery with the arch of the aorta is a remnant of the fetal circulation.

Verschiedene Präparate des Herzens mit Eröffnung der Herzkammern (oberes Präparat) und der Herzvorhöfe (unteres Präparat). Gelb dargestellt ist die bandartige Verbindung zwischen Lungenarterie und Aortenbogen, ein Relikt des fötalen Kreislaufes.

Cœur ouvert : ventricules (en haut) et oreillettes (en bas) ; en jaune (en haut) : ligament artériel, reliquat de la circulation fœtale entre l'artère pulmonaire et la crosse de l'aorte

[XXVII, 631]

The upper and lower specimens depict the heart valves and the internal form of the ventricular muscle. The two specimens in the middle show the atrial musculature.

Oberes und unteres Präparat: Darstellung der Herzklappen und des Innenreliefs der Kammermuskulatur. Die beiden Präparate in der Mitte zeigen die Muskulatur der Vorhöfe.

Valves et valvules du cœur, reliefs des ventricules (en haut et en bas), reliefs musculaires des oreillettes (au milieu)

[XXVII, 617]

Various views of the musculature of the atria

Verschiedene Darstellungen der Muskulatur der Vorhöfe

Architecture des fibres musculaires des oreillettes (myocarde)

[XXVII, 634]

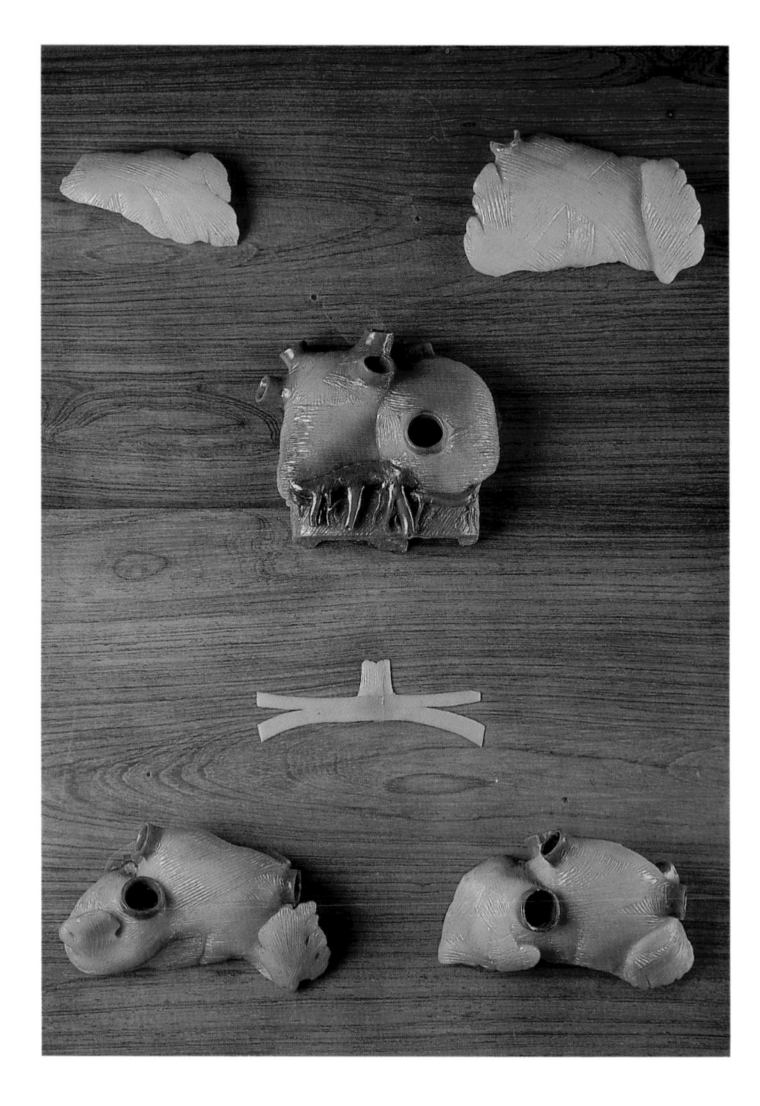

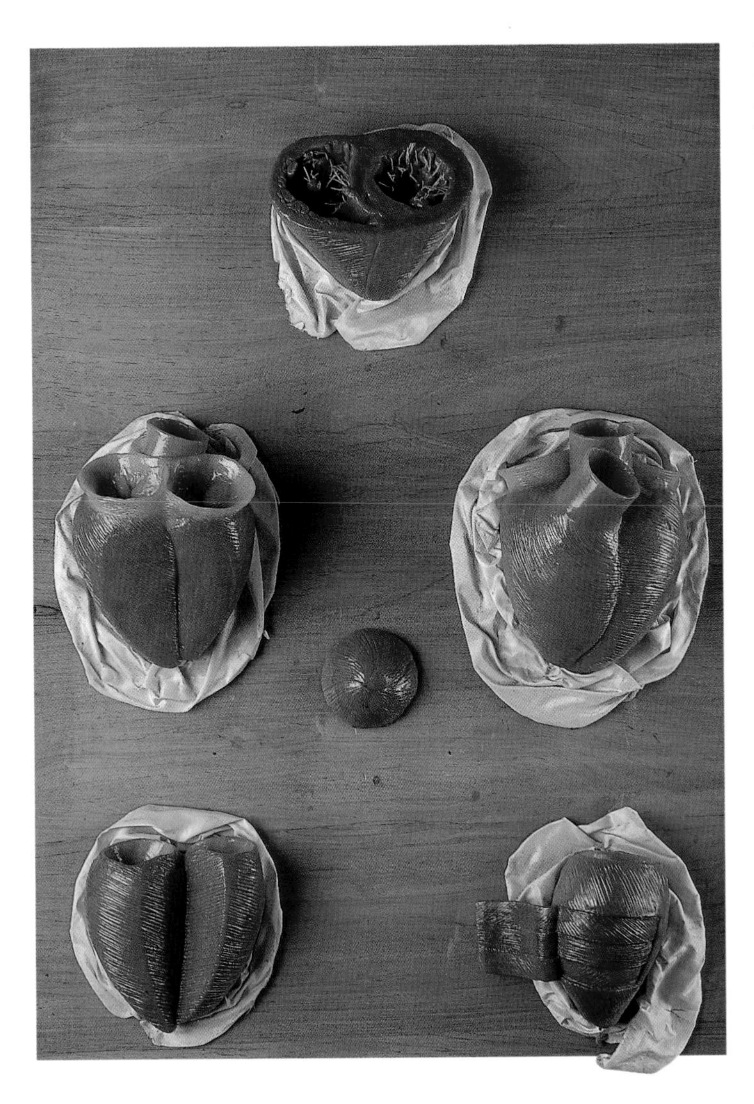

Various views showing the arrangement of the fibers of the ventricular muscle

Verschiedene Darstellungen der Anordnung der Fasern der Kammermuskulatur

Architecture des fibres musculaires des ventricules (myocarde)

[XXVII, 621]

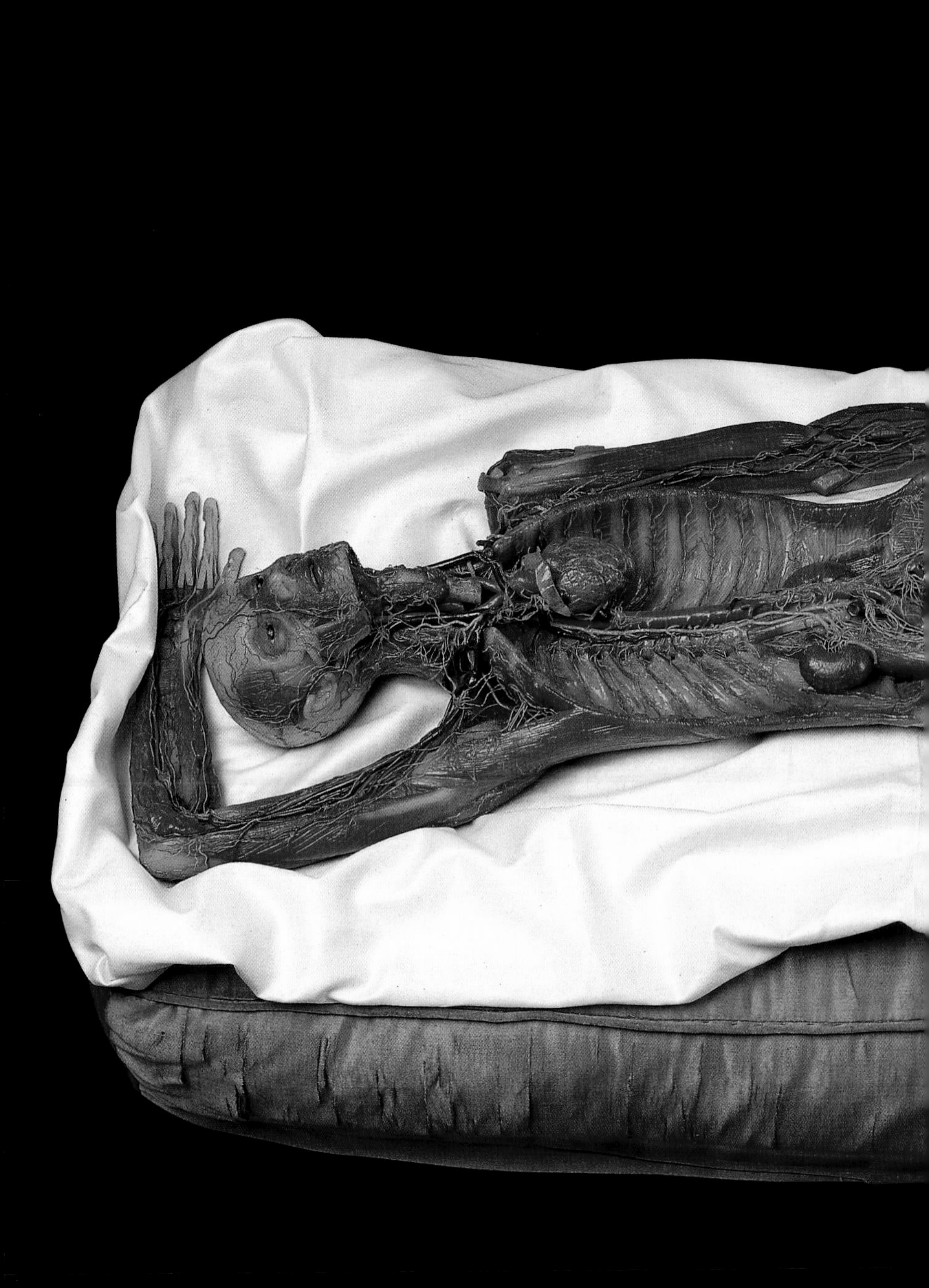

Vasa sanguinea, Vasa lymphatica

Whole body specimen showing sections of the vascular system

Ganzkörperpräparat zur Darstellung von Abschnitten des Gefäßsystems

Préparation du corps entier représentant différents éléments de l'appareil circulatoire

[DEP. 15, 11]

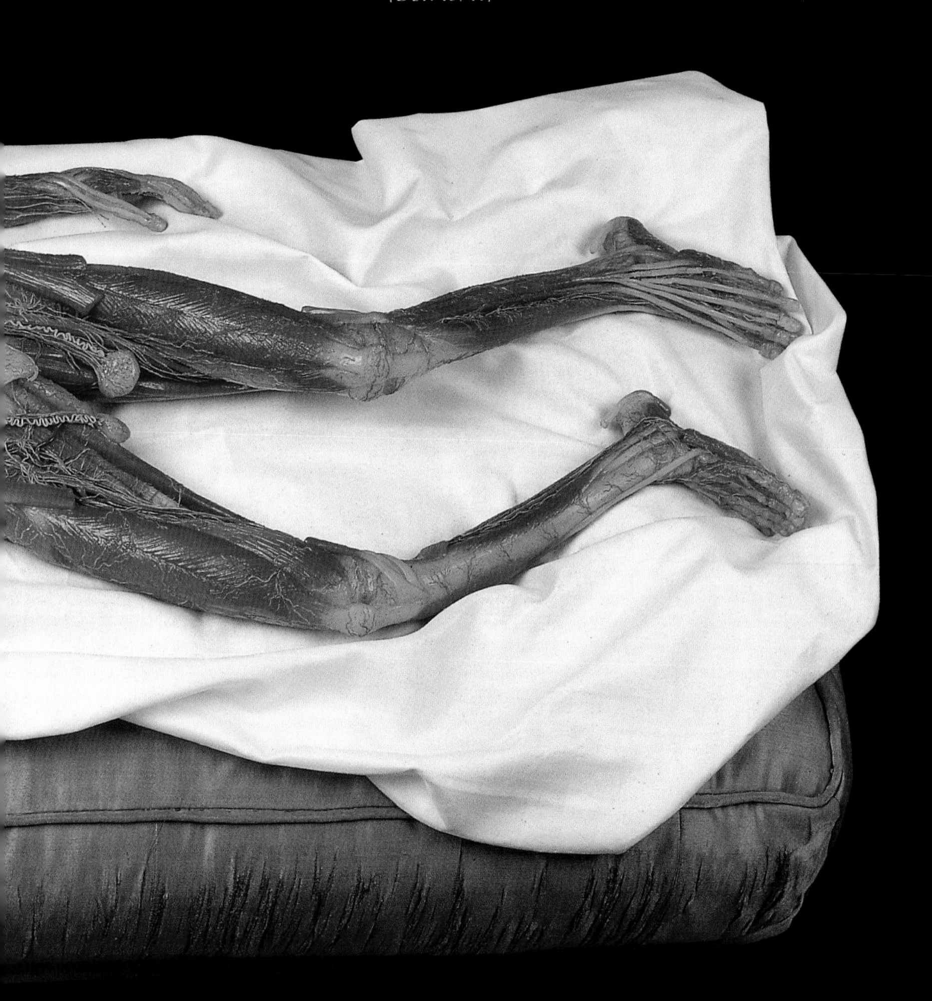

Various views depicting the ventricular muscle. The two smaller specimens represent individual ventricles.

Verschiedene Darstellungen der Herzkammermuskulatur. Bei den kleineren Präparaten handelt es sich um Darstellungen einzelner Kammern

Architecture des fibres musculaires des ventricules

[XXVII, 645]

Demonstration of the fetal heart showing the bypass between the pulmonary artery and the aortic arch, and also between the two atria (lower specimen)

Darstellungen des fötalen Herzens mit Kurzschlußverbindungen von Lungenarterie und Aortenbogen sowie den beiden Vorhöfen (untere Präparate)

Cœurs de fœtus : communication entre l'artère pulmonaire et la crosse de l'aorte, communication entre les deux oreillettes (en bas)

[XXVII, 632]

Thoracic cavity laid open and seen from in front. On the left the heart can be seen in the pericardium. Behind are lying the branches of the pulmonary arteries and veins and the bronchi.

Eröffneter Brustkorb in der Ansicht von vorne. Links im Bild ist das Herz im Herzbeutel erkennbar; dahinter liegend die Äste der Lungenarterien und -venen sowie der Bronchien.

Cage thoracique ouverte ; en arrière du péricarde : artère et veines pulmonaires droites, bronches droites, le poumon droit étant enlevé

[XXIX, 878]

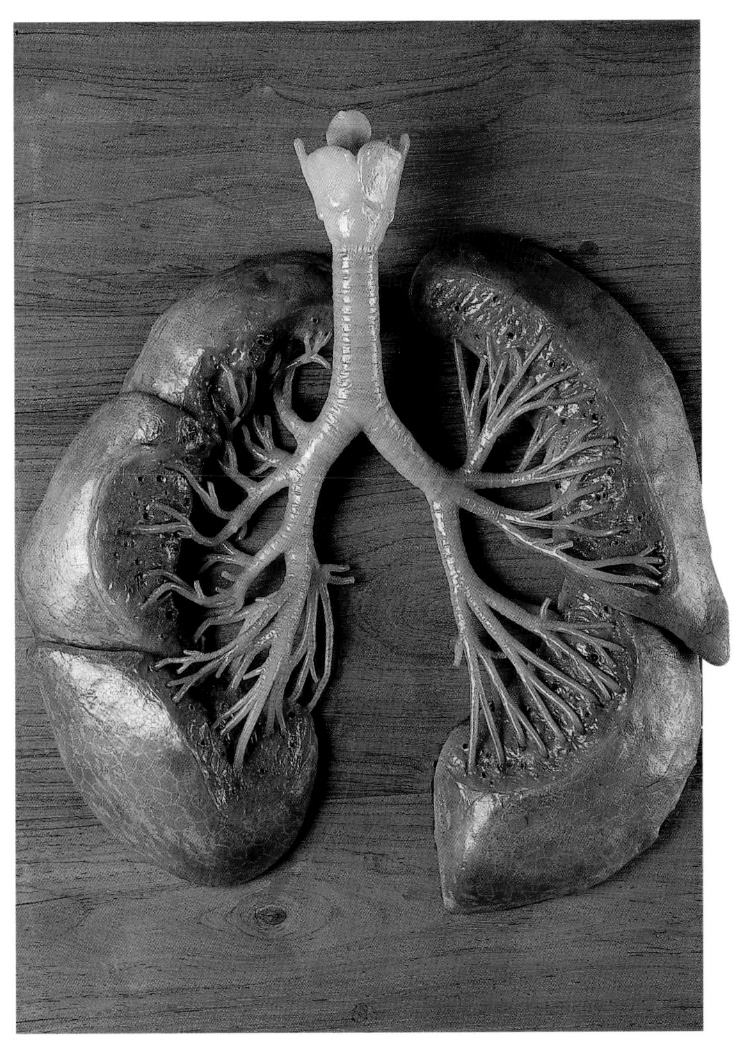

Demonstration of the respiratory tract, showing the larynx, trachea, bronchi and their subdivisions

Darstellung der Atemwege mit Kehlkopf, Luftröhre, Bronchien und ihren Verzweigungen

Appareil respiratoire : larynx, trachée, arbre bronchique disséqué dans les poumons

[XXIX, 769]

The left and right lungs with their respective two (left) or three lobes (right) seen from behind. The aorta and its intercostal branches can also be seen.

Blick von hinten auf den linken und den rechten Lungenflügel mit den zwei (links) bzw. drei Lungenlappen (rechts) sowie auf die Hauptschlagader mit den abgehenden Zwischenrippenarterien

Viscères du thorax vus de l'arrière : lobes pulmonaires (deux à gauche et trois à droite), aorte et artères intercostales

[XXIX, 772]

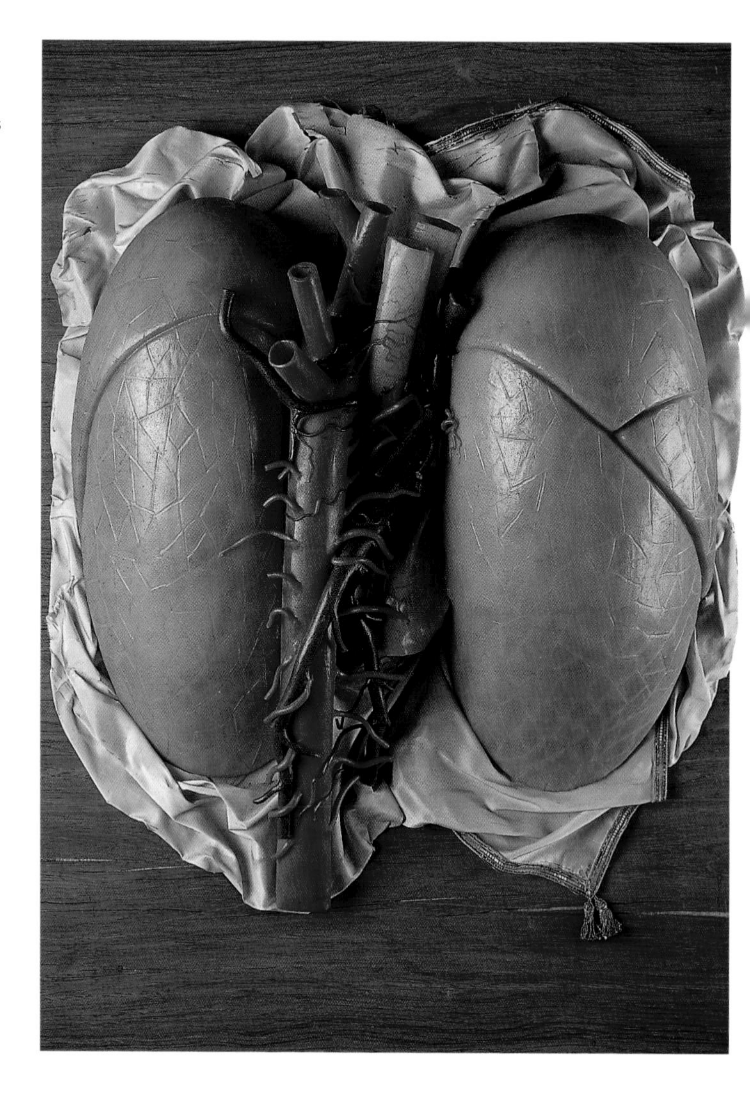

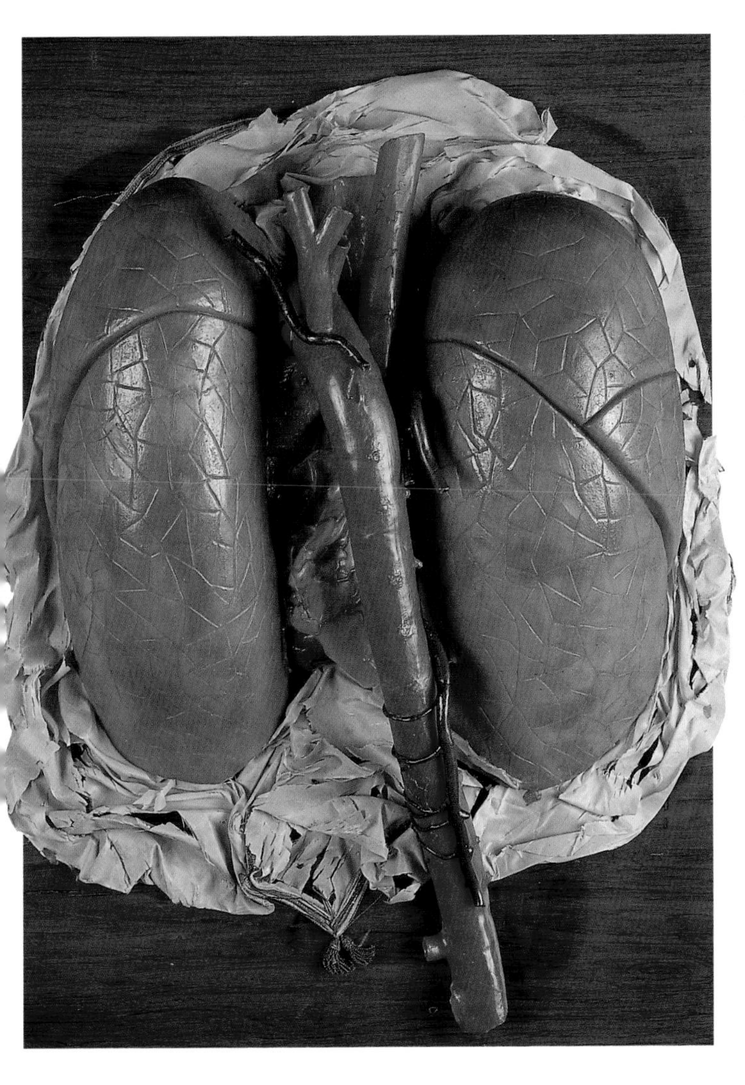

Specimen with both lungs and the thoracic aorta seen from behind

Präparat der beiden Lungenflügel und der Brustschlagader in der Ansicht von hinten

Viscères du thorax vus de l'arrière : poumons et aorte

[DEP. 9]

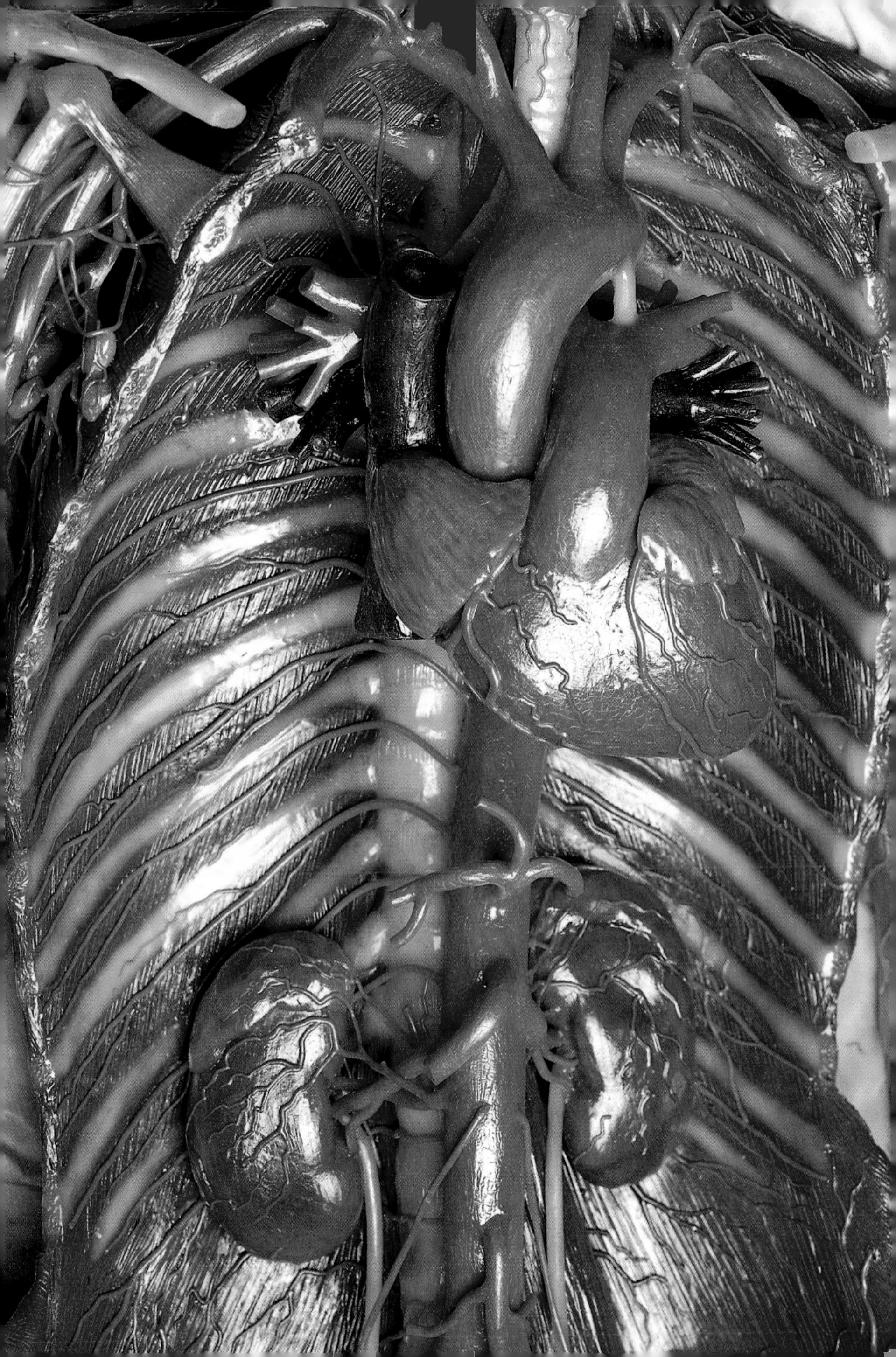

Various views of the aortic arch. The coronary arteries arise from the aortic bulb

Verschiedene Darstellungen des Aortenbogens. Die Herzkranzgefäße haben ihren Ursprung im Bereich der Auftreibung

Crosse de l'aorte et ses branches collatérales ; les artères coronaires naissent du bulbe aortique

[XXVII, 629]

◀
Detail from page | Detail von Seite | détail de la page 260–261

The aorta and the large arteries in the neck region

Darstellung der Hauptschlagader und der großen Arterien im Bereich des Halses

Crosse de l'aorte et principales artères du cou

[XXVII, 627]

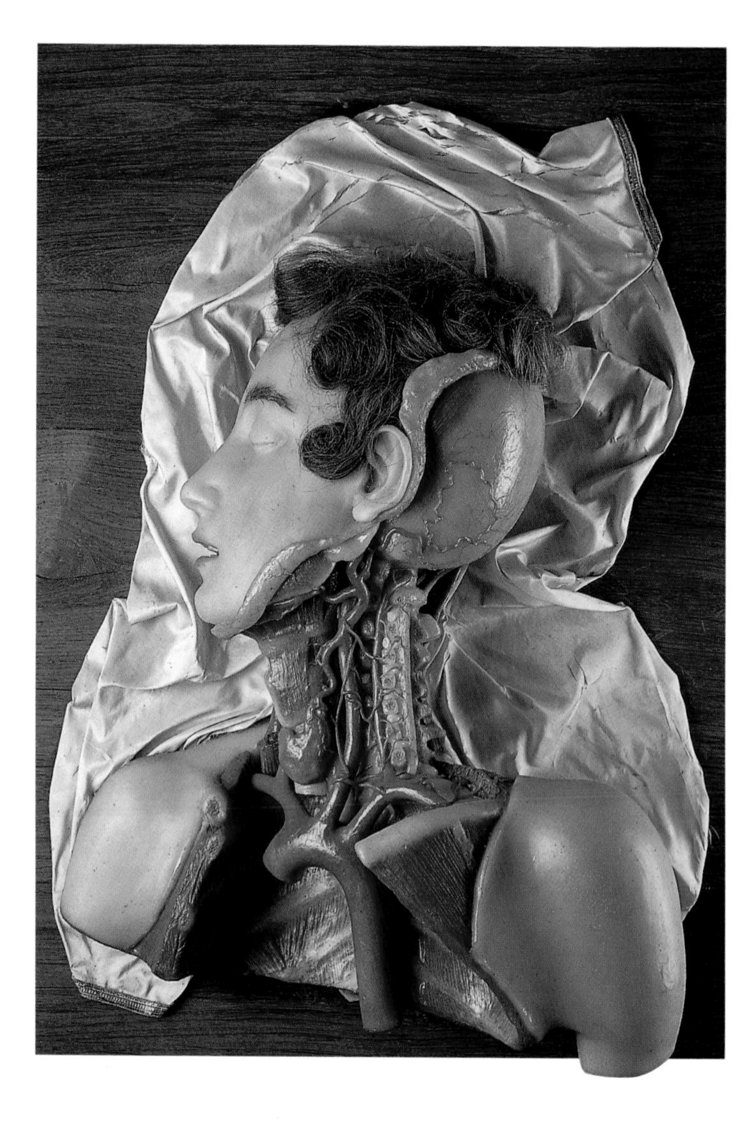

Arcus aortae, Arteria brachiocephalica, Arteria carotis communis, Arteria subclavia, Arteria carotis externa, Arteria vertebralis

Specimen showing the arteries of the head and neck. On the left side of the model the carotid artery with its branches can be seen taking origin from the aortic arch.

Präparat der Arterien im Bereich des Kopfes und Halses. Links im Bild ist die vom Aortenbogen abzweigende Halsschlagader mit ihren Aufzweigungen dargestellt.

Arbre artériel de la tête et du cou : ramifications de l'artère carotide (à gauche)

[XXVII, 594]

Vena cava superior, Vena brachiocephalica, Vena jugularis interna et externa, Sinus sagittalis superior, Sinus transversus, Sinus sigmoideus, Plexus venosus vertebralis internus

The veins of the head and neck

Darstellung der Venen im Bereich des Kopfes und Halses

Arbre veineux de la tête et du cou

[XXVII, 593]

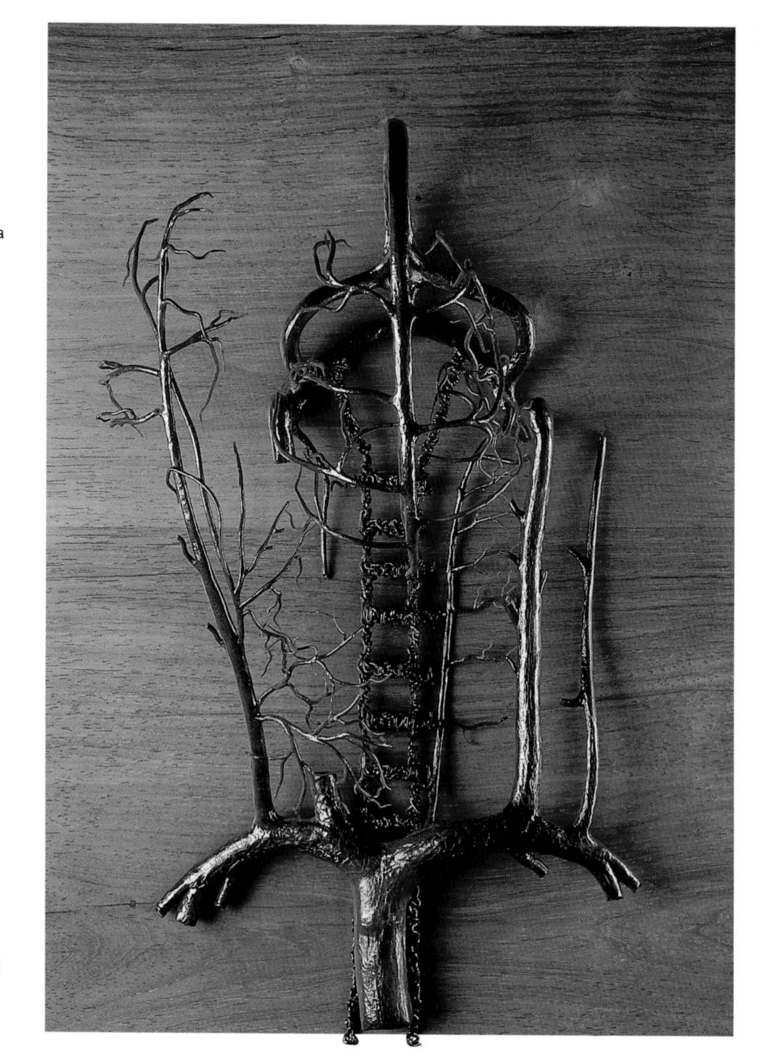

►
The veins in the region of the neck and face

Darstellung von Venen im Bereich von Gesicht und Hals

Veines de la tête et du cou

[XXVII, 611]

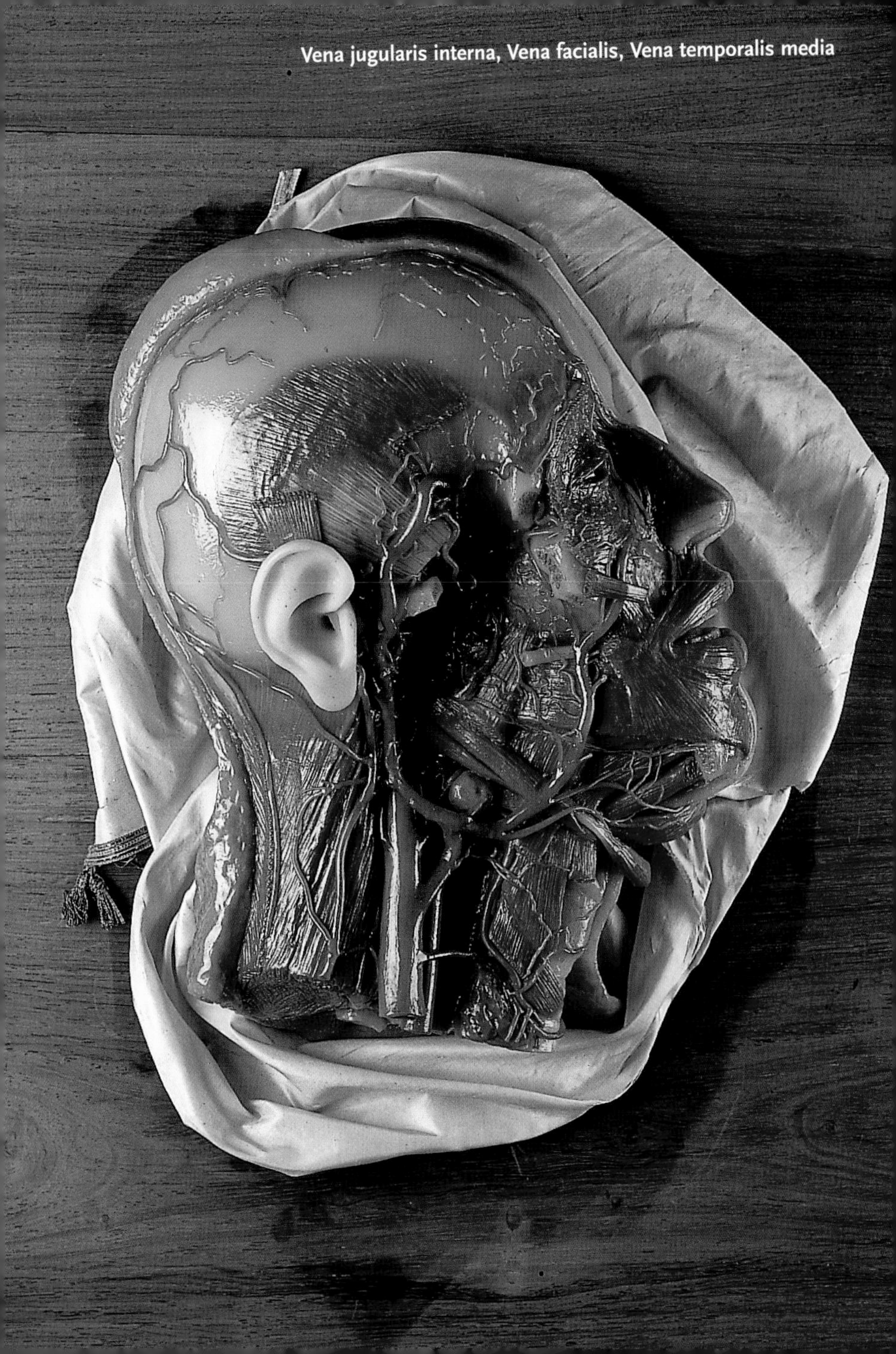

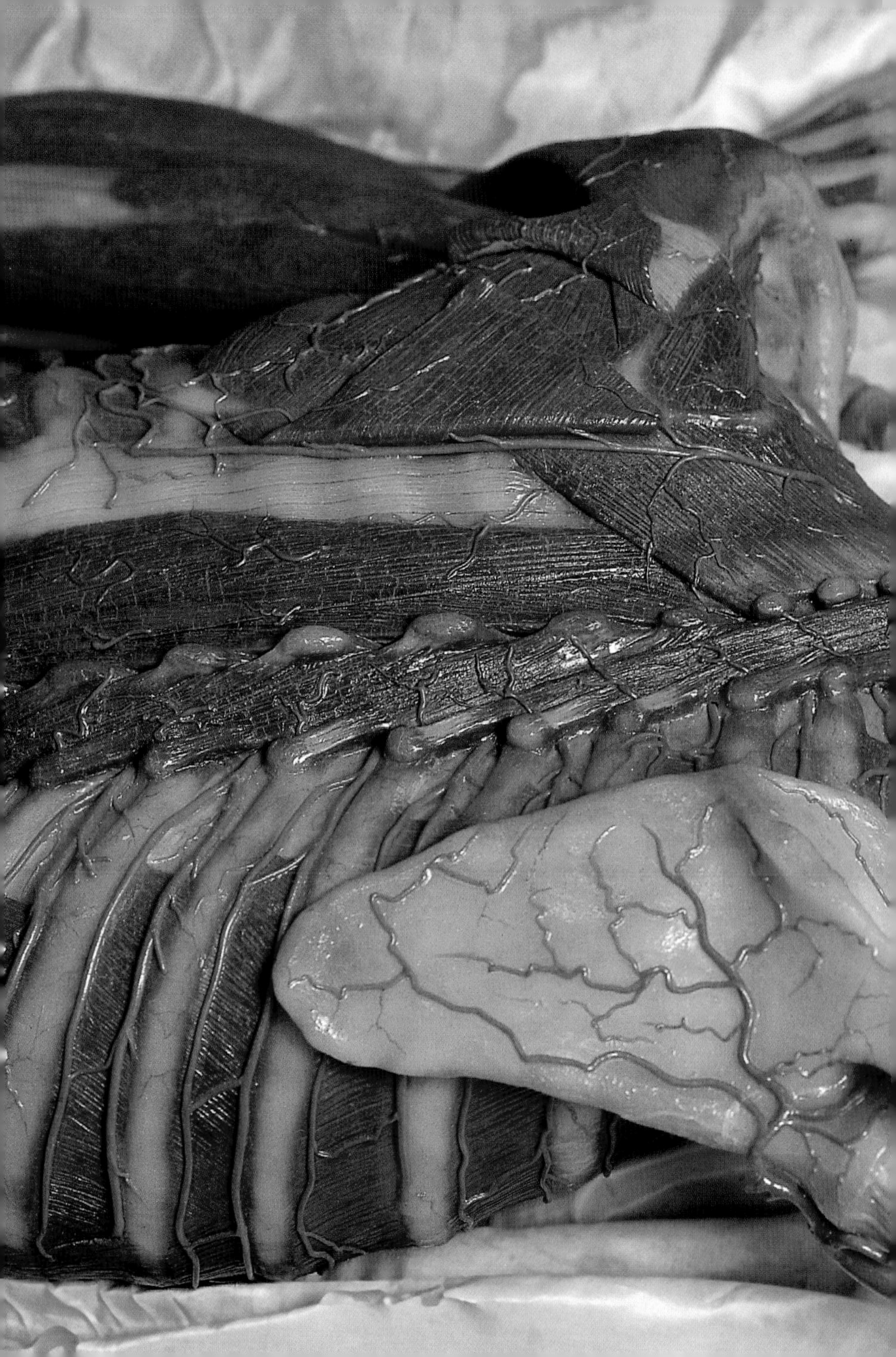

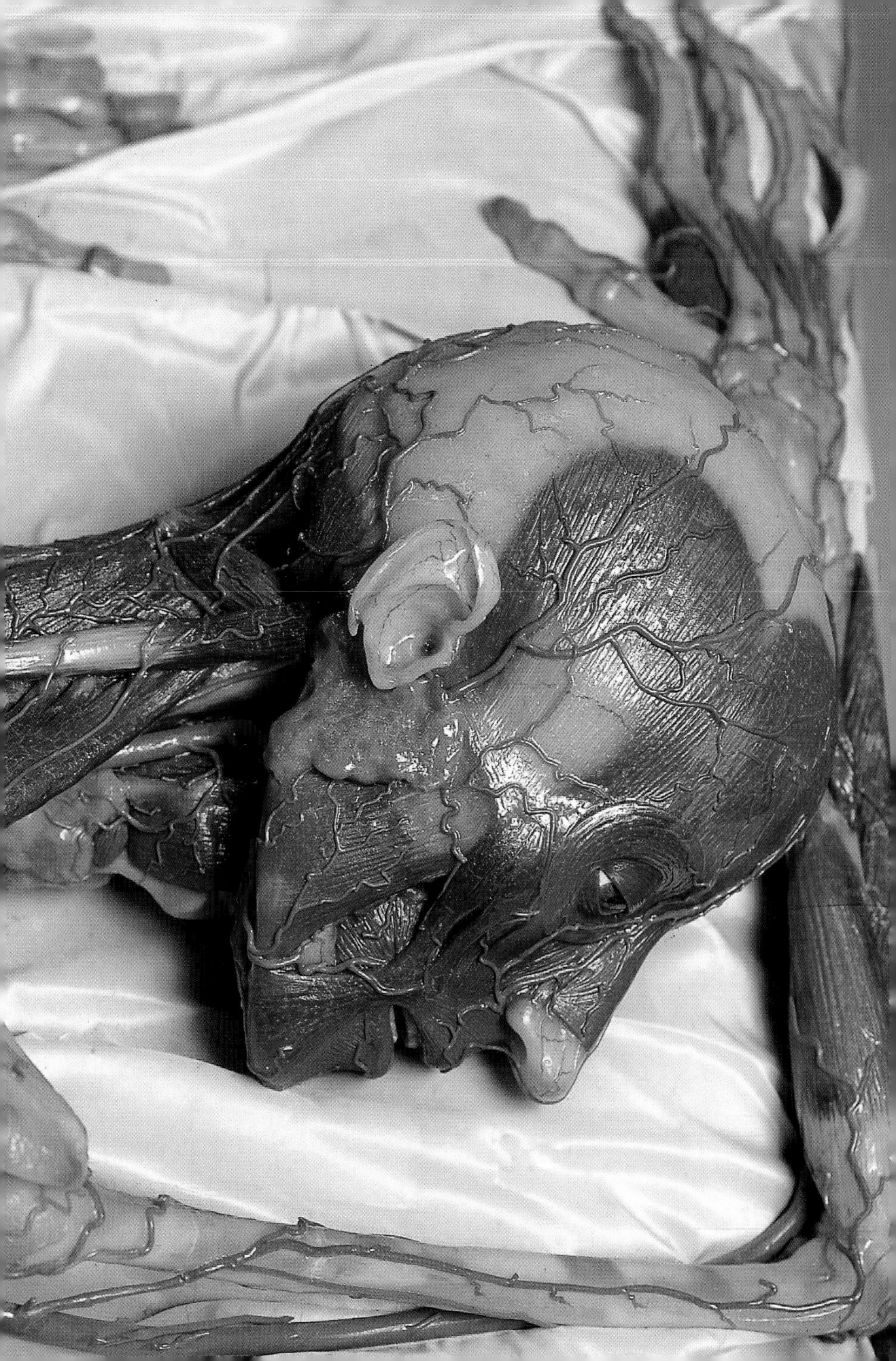

Arteria facialis, Arteria temporalis superficialis, Arteria transversa facei, Arteria occipitalis, Vena retromandibularis, Vena facialis, Vena jugularis externa

The superficial arteries and veins of the head and neck

Darstellung von oberflächlich verlaufenden Arterien und Venen im Bereich von Kopf und Hals

Artères et veines superficielles de la tête et du cou

[XXVII, 626]

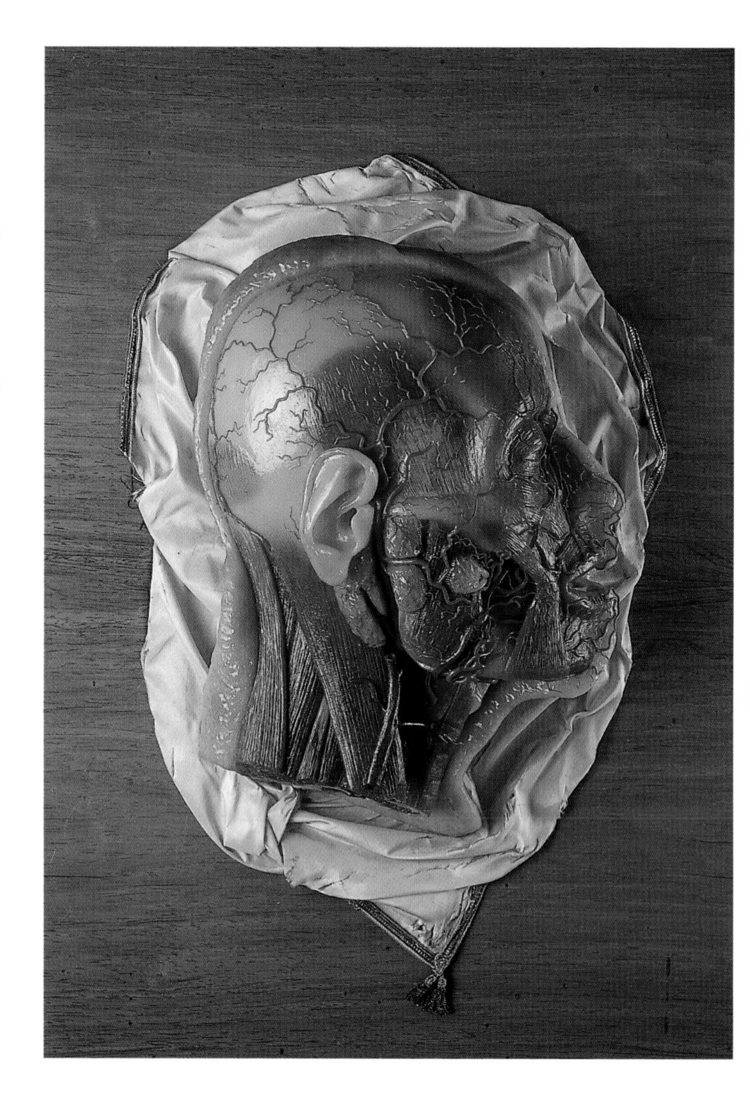

Pages · Seiten · Pages 288–289:

Detail from pages 466–467

Detail der Seiten 466–467

Détail des pages 466–467

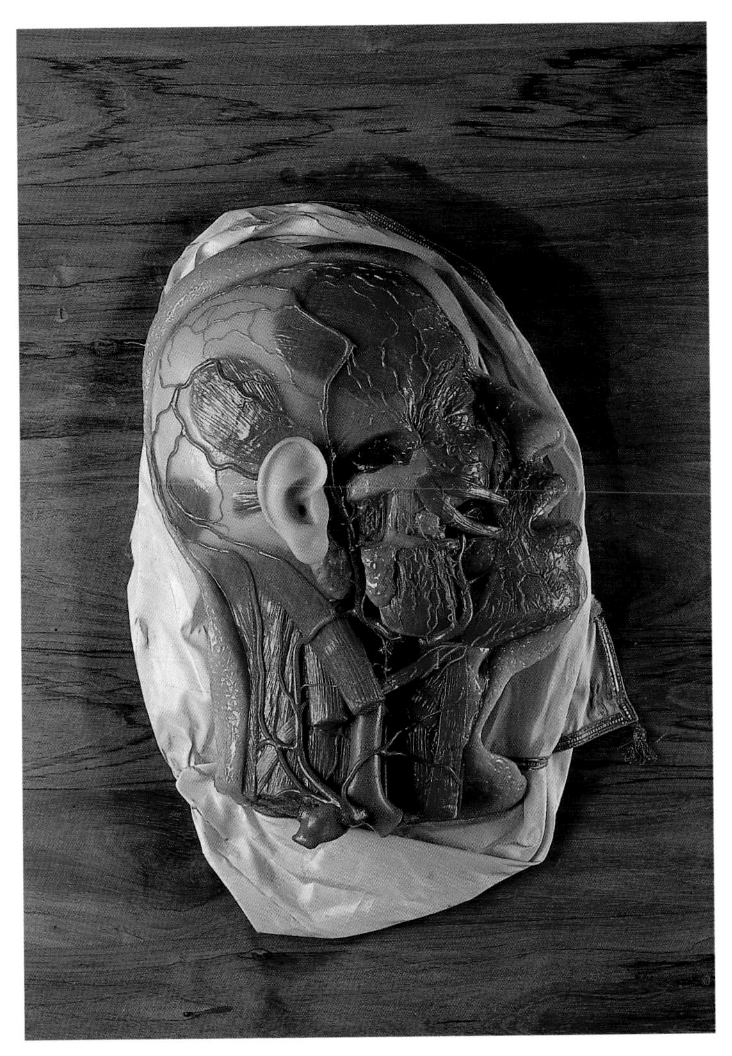

The veins in the region of head and neck

Darstellung von Venen im Bereich von Kopf und Hals

Veines de la tête et du cou

[XXVII, 612]

The superficial veins of head and neck. The parotid gland and its duct can be seen lying in front of the ear.

Oberflächlich verlaufende Venen und Arterien im Bereich von Kopf und Hals. Vor dem Ohr liegt die Ohrspeicheldrüse mit ihrem Ausführungsgang.

Artères et veines de la tête et du cou ; glande parotide et conduit excréteur (de Stenon) en avant du pavillon de l'oreille

[XXVII, 787]

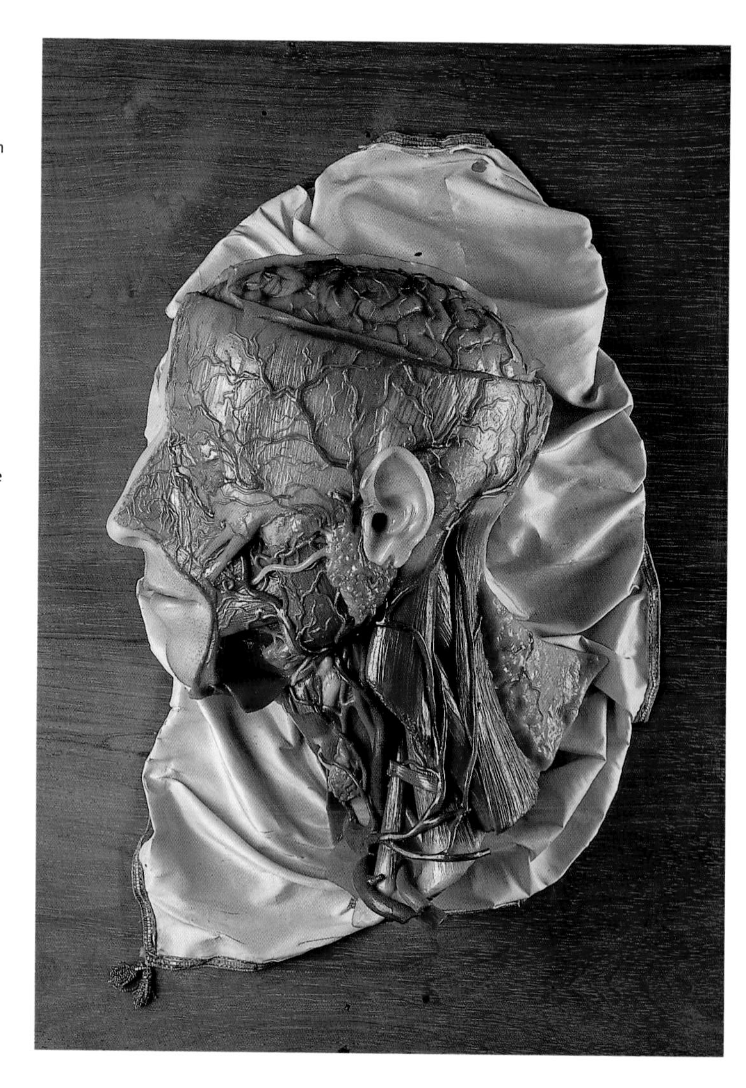

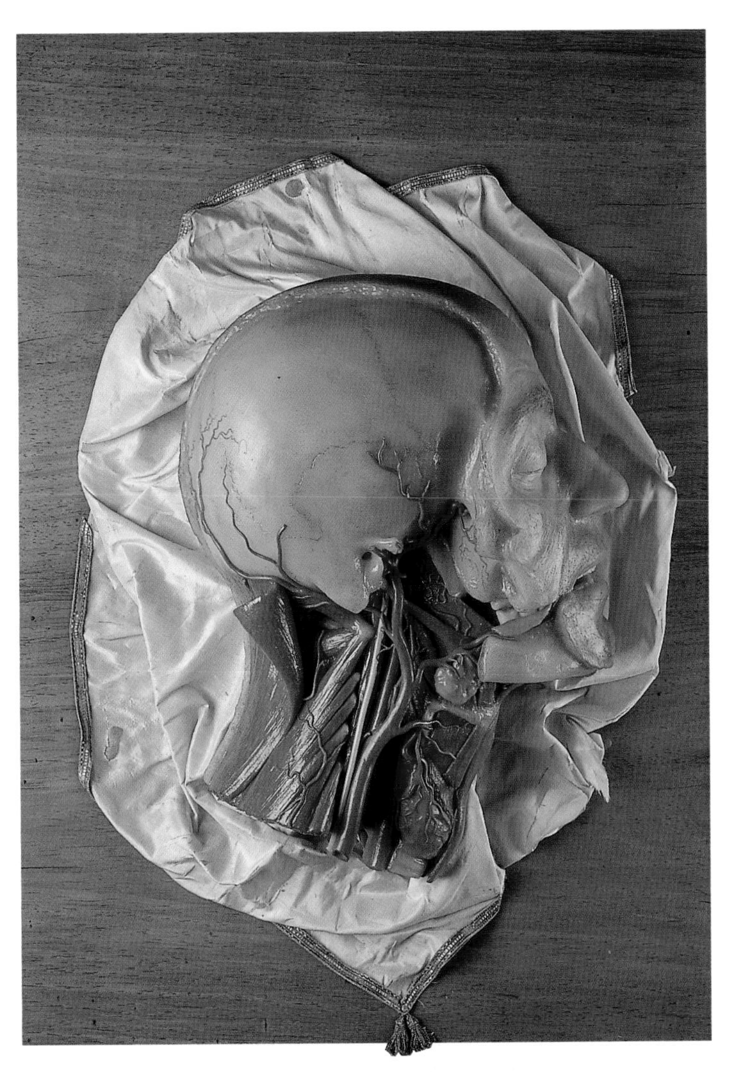

The carotid artery and the vagus nerve

Darstellung der Halsschlagader und des Vagus-nerven

Artères carotides, nerf vague

[XXVII, 628]

The arteries of the anterior thoracic and abdominal walls

Darstellung der Arterien der vorderen Brust- und Bauchwand

Artères des parois antérieures du thorax et de l'abdomen

[XXV, 616]

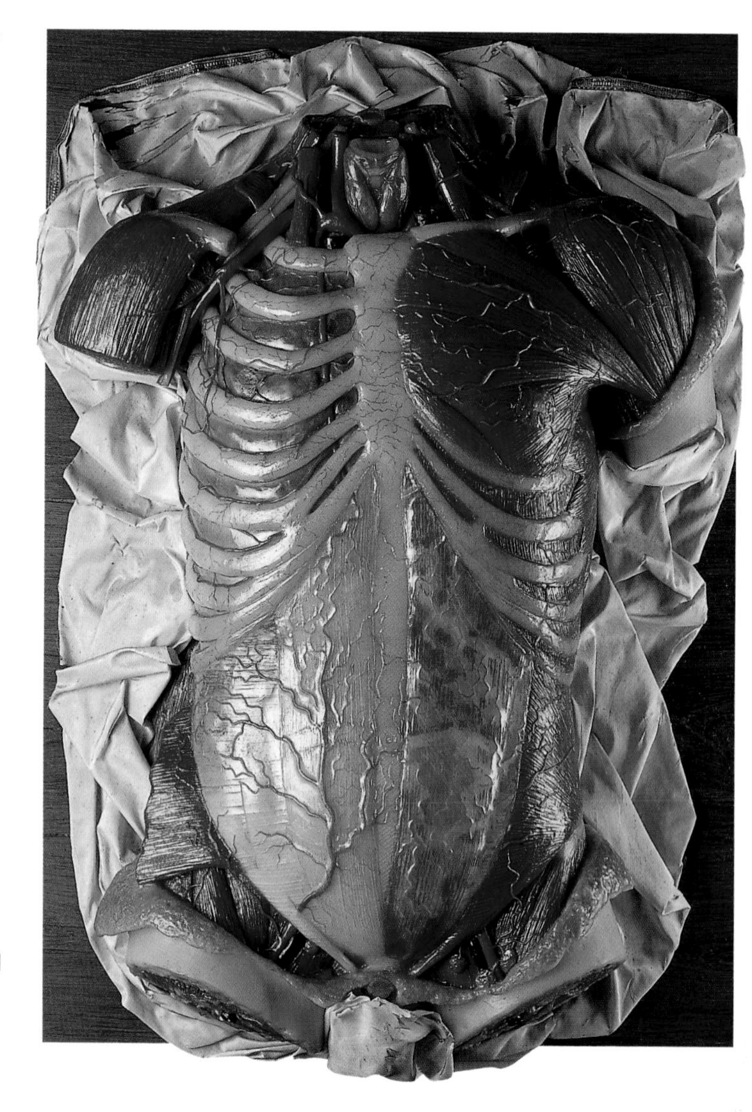

▶
Detail from pages | Detail der Seiten | détail des pages 320–321

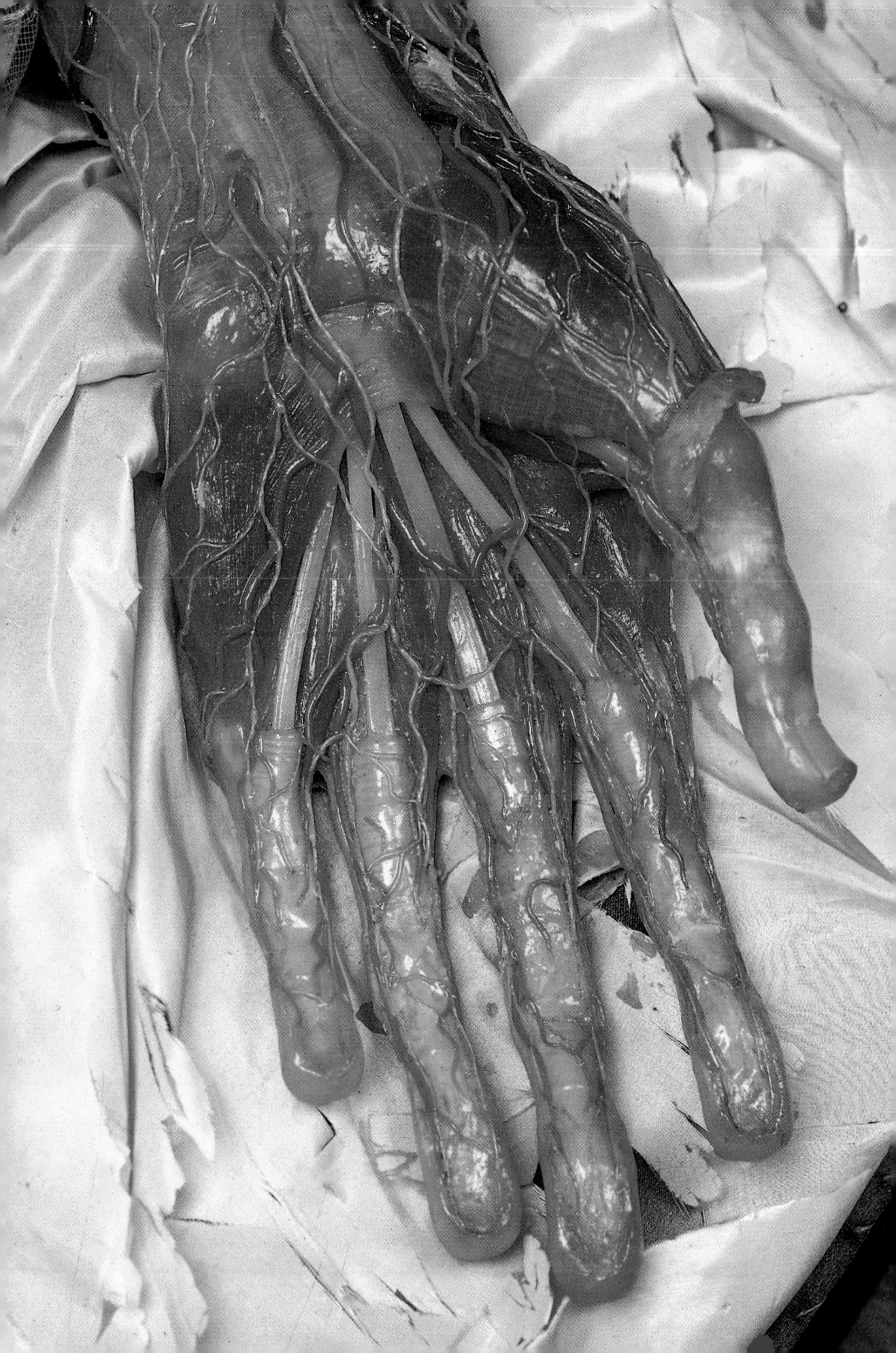

The superficial veins and cutaneous nerves of the arm

Darstellung des oberflächlichen Venen und der Hautnerven des Armes

Veines et nerfs superficiels du membre supérieur vus de l'avant

[XXIX, 641]

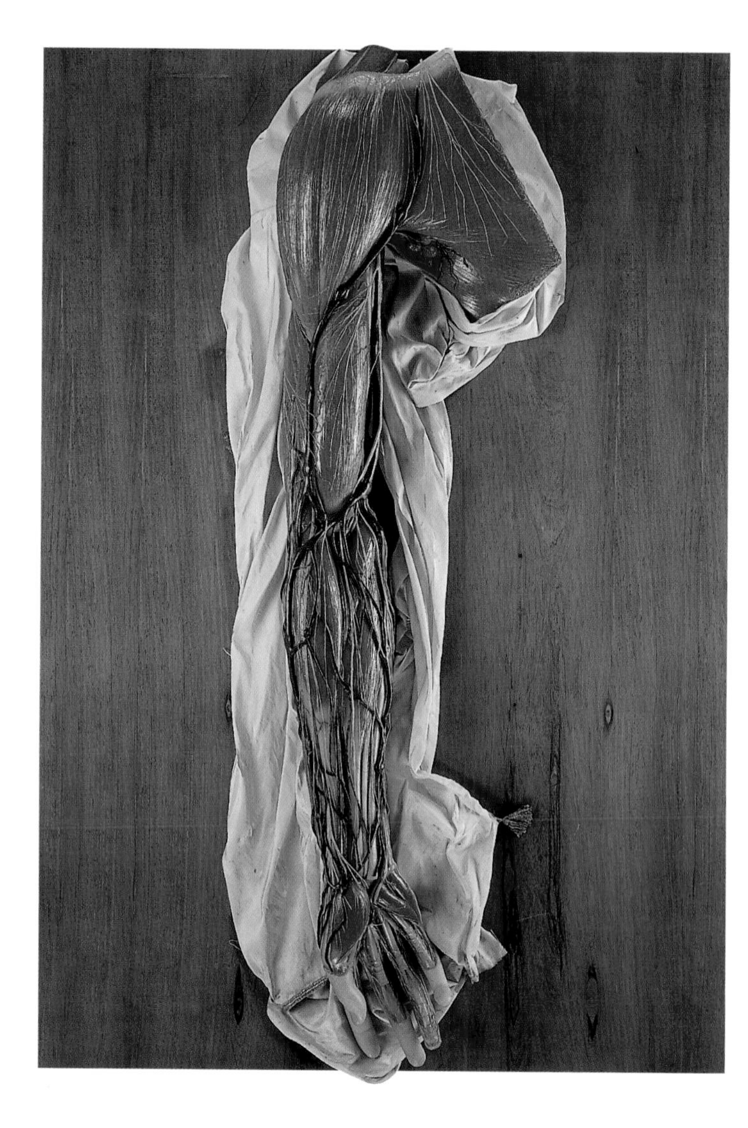

The muscles, vessels and nerves of the arm

Darstellung von Muskeln, Gefäßen und Nerven des Armes

Artères, veines, et nerfs du membre supérieur

[XXVII, 639]

The deep arteries of the upper arm and forearm

Darstellung von tiefen Arterien des Ober- und Unterarmes

Artères du membre supérieur vues de l'arrière

[XXVII, 644]

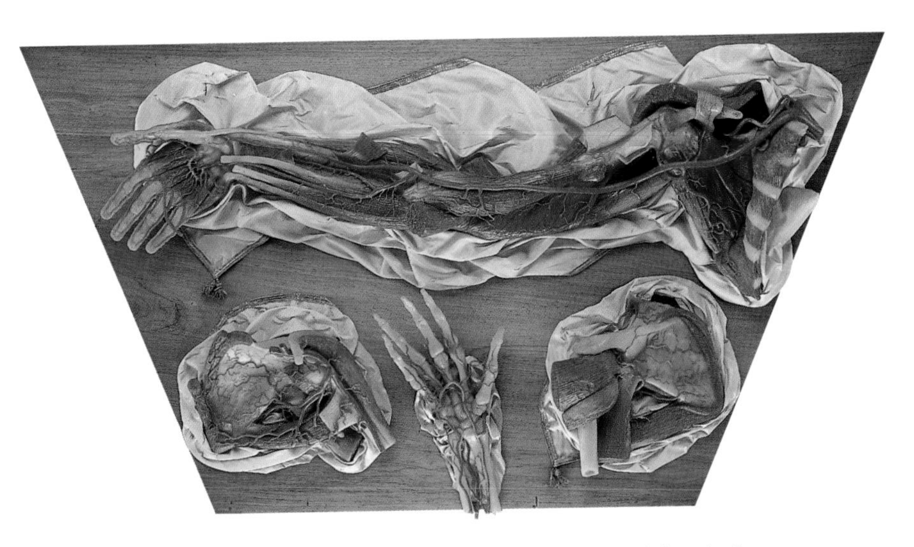

The arteries supplying the shoulder girdle, arm and hand

Darstellung der arteriellen Versorgung des Schultergürtels, des Armes und der Hand

Artères de l'épaule, du bras, de l'avant-bras, et de la main

[XXVII, 637]

The arteries, veins and nerves of the arm

Darstellung der Arterien, Venen und Nerven des Armes

Artères, veines, et nerfs du membre supérieur projetés sur le squelette

[XXVII, 640]

The arteries associated with the pelvic floor

Darstellung der Arterien im Bereich des Beckenbodens

Artères de la région fessière profonde et du plancher du bassin

[XXVII, 608]

The arteries in the region of the buttock after removal of the large gluteal muscle

Darstellung der Arterien im Gesäßbereich nach Entfernung des großen Gesäßmuskels

Artères de la région fessière profonde, le muscle grand fessier étant enlevé

[XXVII, 605]

View from in front of the organs and arteries of a male pelvis. The rectum has been ligated above, and the urinary bladder displaced to one side to show the entry of the ureter.

Blick von vorn auf Organe und Arterien eines männlichen Beckens. Der Enddarm ist oben abgebunden, die Harnblase zur Seite gedrängt, so daß die Mündung des Harnleiters sichtbar wird.

Artères et viscères du bassin masculin ; ligature du côlon sigmoïde, abouchement de l'uretère dans la vessie réclinée

[XXVII, 609]

The deep course of the femoral artery

Darstellung der Schenkelarterie in einer tieferen Schicht

Artère fémorale et ses branches à la partie supérieure de la cuisse vues de l'avant

[XXVII, 607]

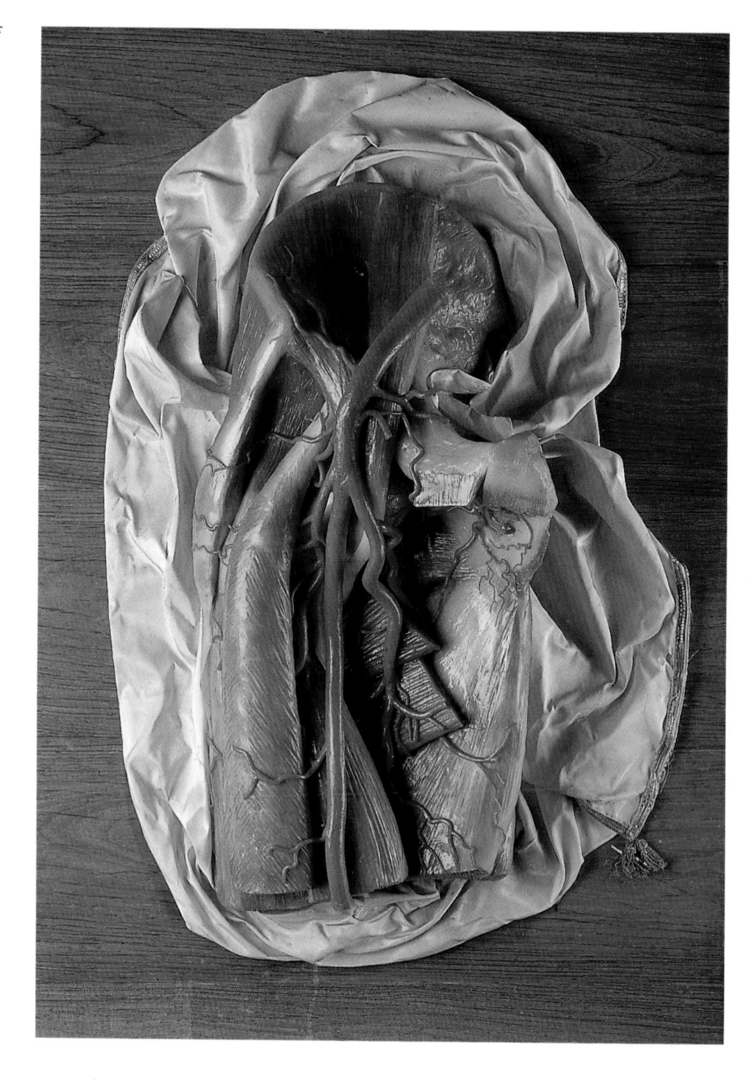

►
Detail from page | Detail von Seite | détail de la page 260–261

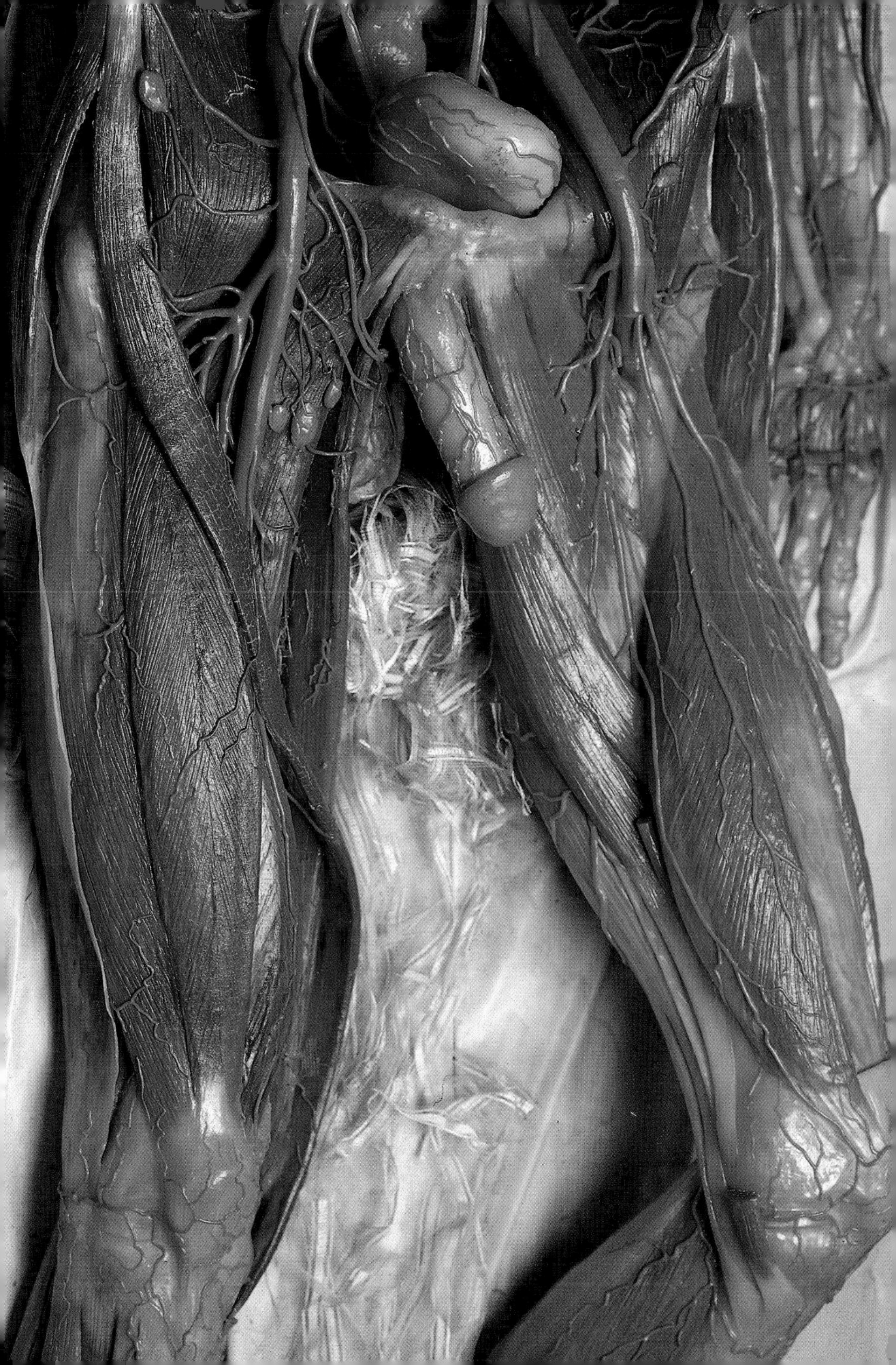

Arteria femoralis

The femoral artery

Darstellung der Schenkelarterie

Artère fémorale et ses rapports avec les muscles de la cuisse vue de l'avant

[XXVII, 603]

The thigh seen from behind, showing the sciatic nerve and popliteal artery

Ansicht des Oberschenkels von hinten mit Darstellung des Ischiasnerven und der Kniekehlenarterie

Artères fessières et poplitée, nerf sciatique dans les régions profondes de la fesse, postérieure de la cuisse, et du genou

[XXVII, 604]

The thigh seen from behind. The superficial muscles have been removed to show the deep course of the arteries and nerves.

Unterschenkel in der Ansicht von hinten. Zur Darstellung der in der Tiefe verlaufenden Arterien und Nerven sind oberflächliche Muskeln entfernt.

Artères et nerfs profonds de la jambe visibles après ablation du muscle triceps du mollet

[XXVII, 615]

The arteries of the sole of the foot

Darstellung der Arterien der Fußsohle

Artères de la plante du pied

[XXVII, 613]

View of the superficial veins of the lower limb discharging into the femoral vein near the inguinal ligament

Blick auf die Einmündung der oberflächlich liegenden Beinvene in die Schenkelvene im Bereich des Leistenbandes

Abouchement d'une veine superficielle de la cuisse (grande saphène) dans la veine fémorale en dessous du pli de l'aine

[XXI, 402]

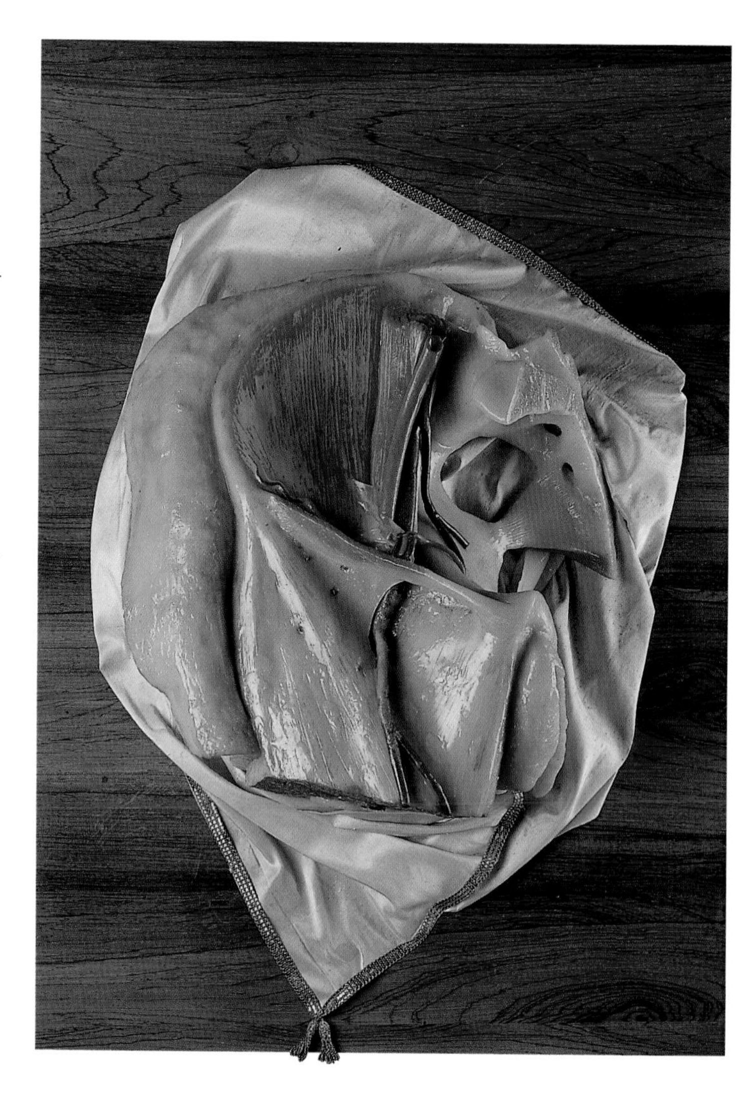

Specimen showing the arterial supply of the lower limb

Darstellung der arteriellen Versorgung des Beines

Arbre artériel du membre inférieur projeté sur le squelette

[XXVII, 642]

The blood vessels of the lower limb

Darstellung der Blutgefäße des Beines

Artères et veines du membre inférieur projetées sur le squelette

[XXVII, 635]

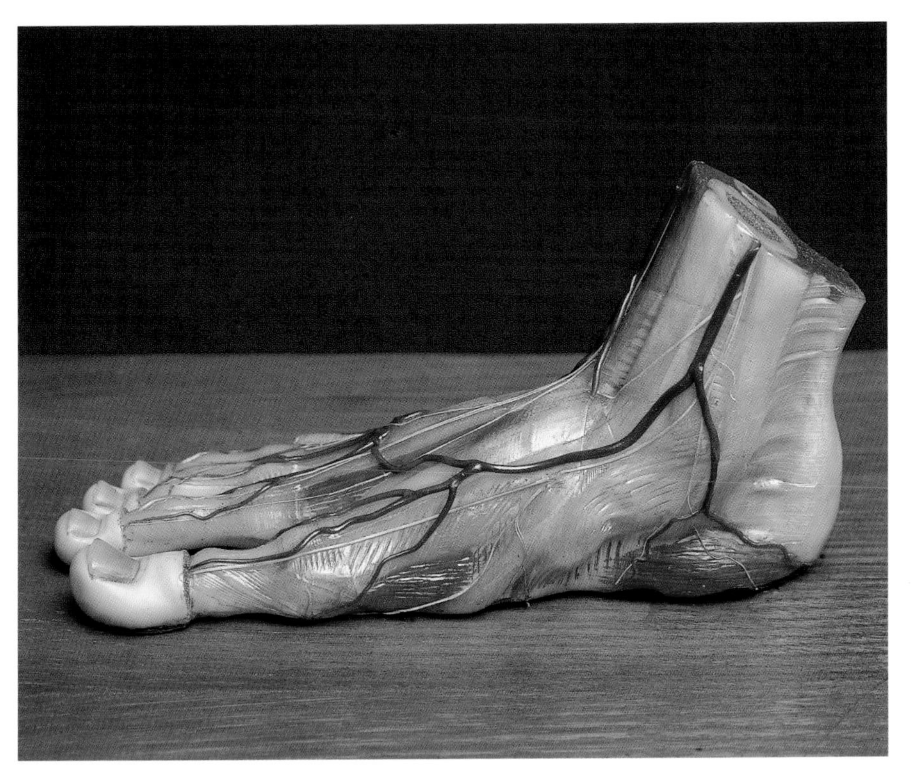

The superficial veins and cutaneous nerves of the foot

Darstellung von oberflächlichen Venen und Hautnerven des Fußes

Veines et nerfs superficiels du dos du pied et de la cheville

[DEP. 13]

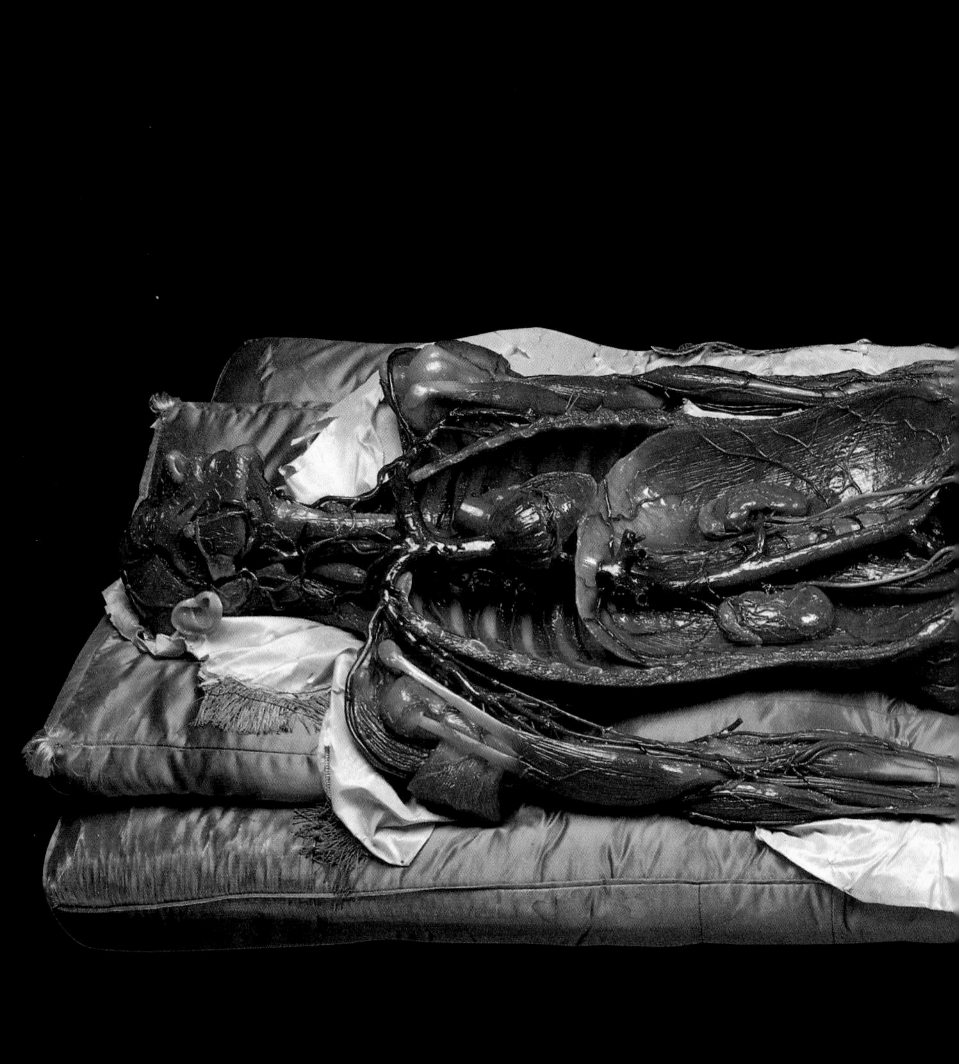

Venae

Whole body specimen with the veins displayed

Ganzkörperpräparat mit Darstellung der Venen

Préparation du corps entier représentant les grandes voies veineuses

[XXV, 526]

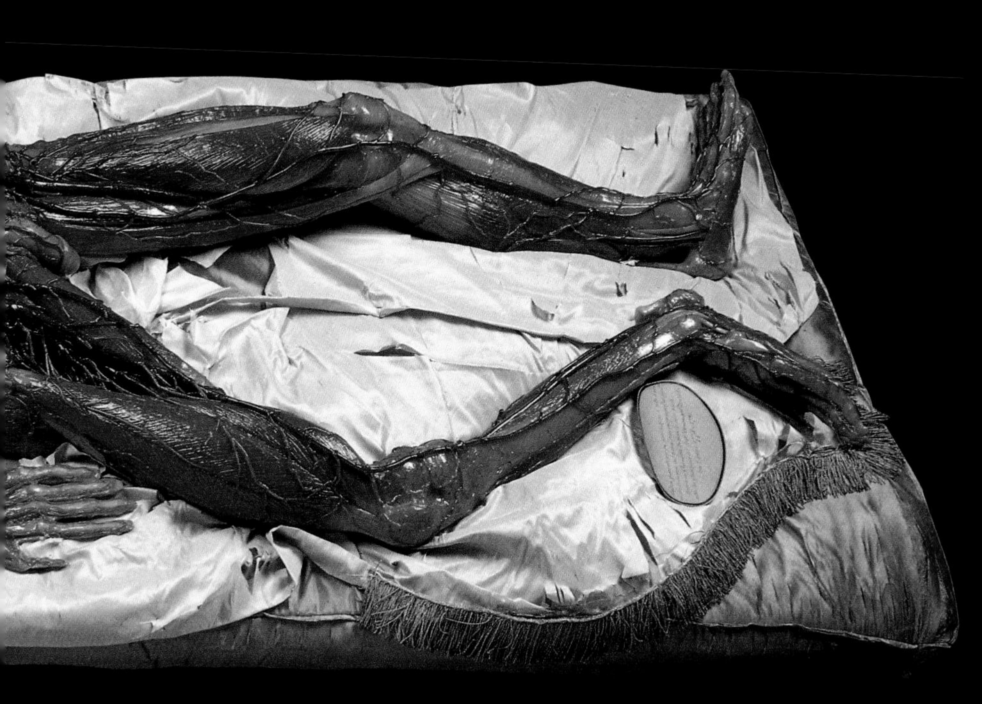

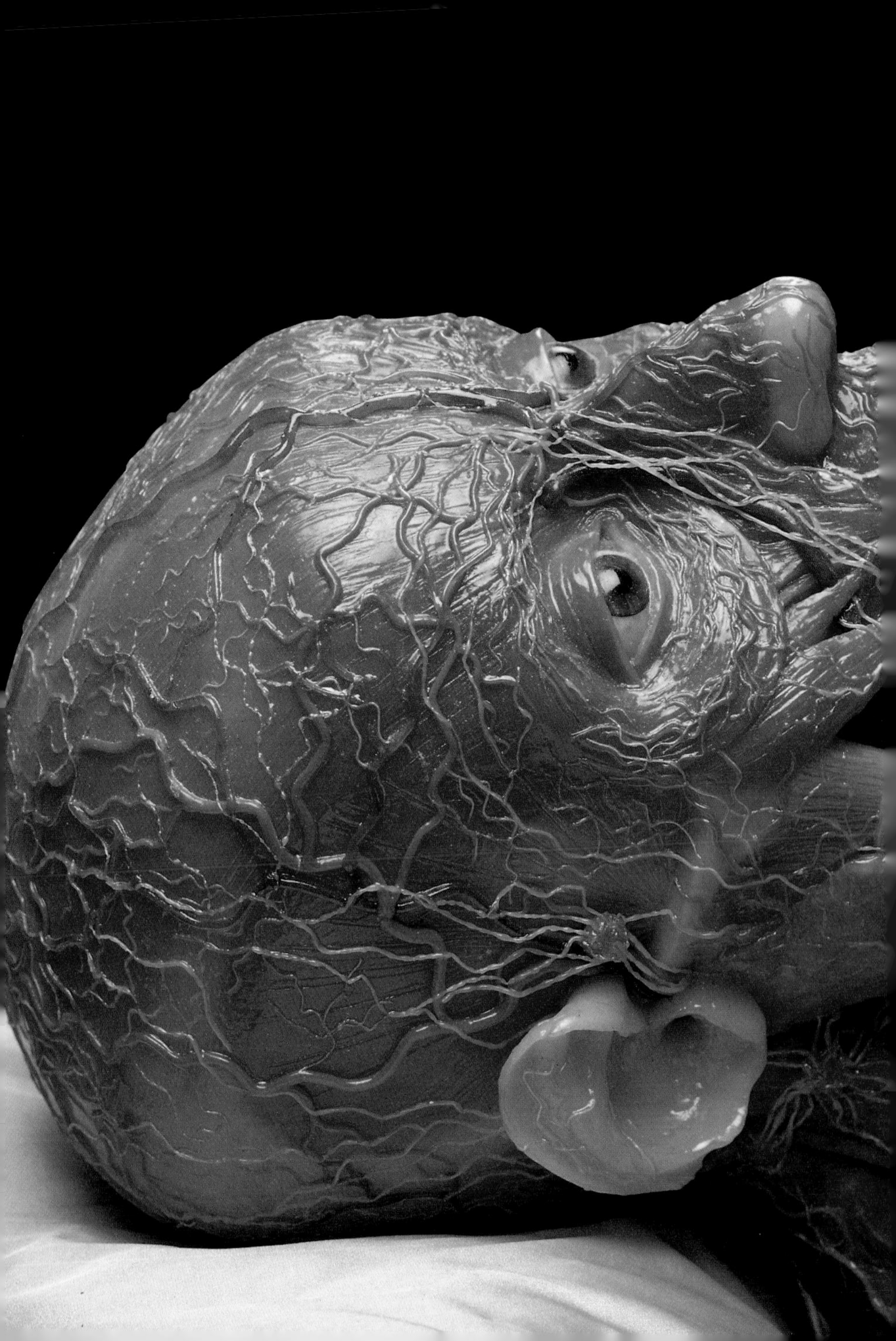

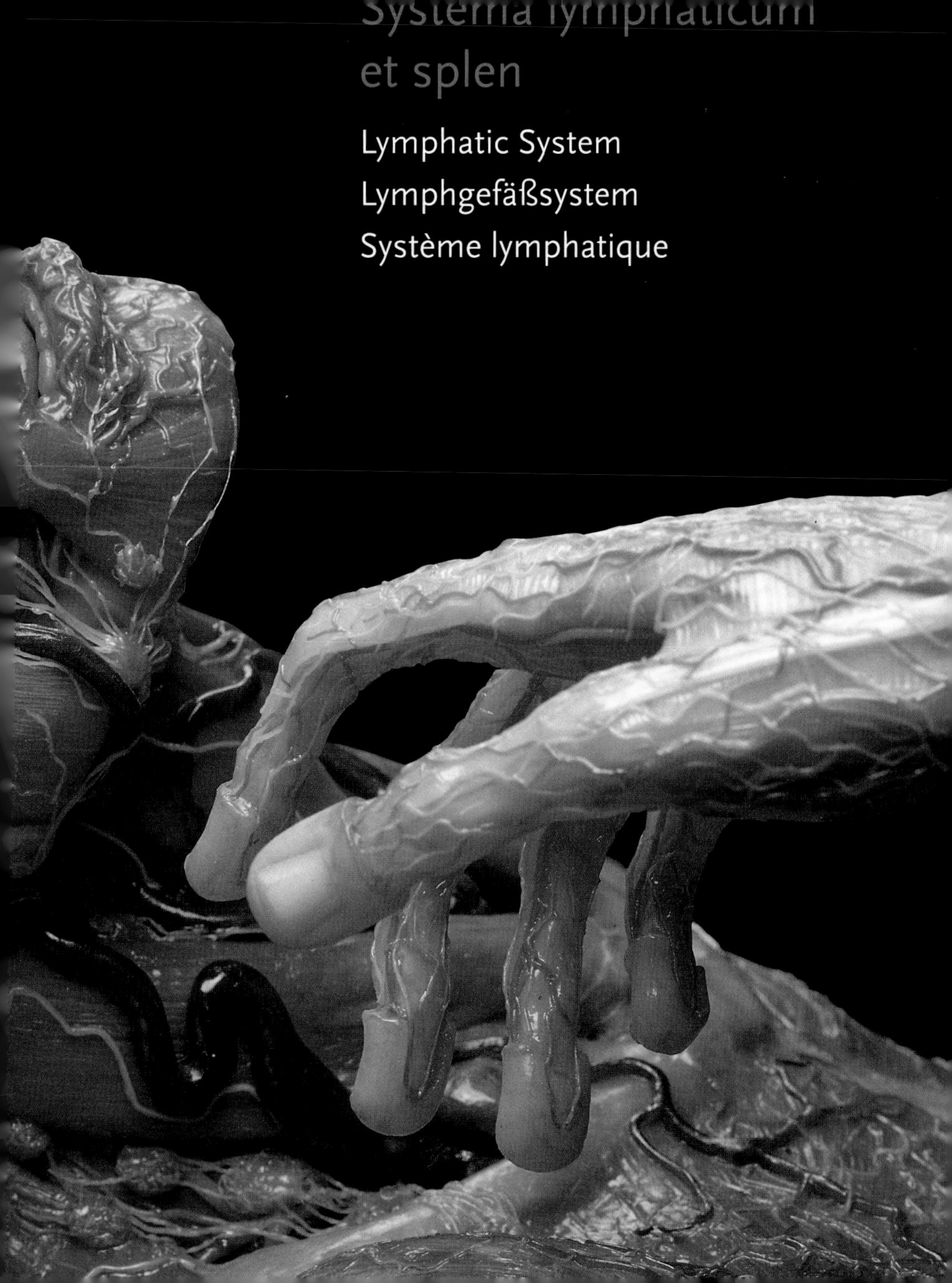

Systema lymphaticum et splen

Lymphatic System
Lymphgefäßsystem
Système lymphatique

Lymphatic System

The lymphatic vessels, and especially the lymphatic organs such as the thymus, spleen and lymph nodes, play a central role in the immune system of our entire organism. In the 17th century Gasparo Aselli (1582–1626) had already observed the lymphatic vessels in the mesentery of a dog dissected immediately after feeding. Following the ingestion of food fats in the form of very fine droplets in the delicate lymph vessels of the intestine, the latter turned white and thus became visible. At the end of the 18th century the Italian Anatomist, Paolo Mascagni (1752–1815), used a sophisticated mercury injection technique to demonstrate that all organs possess their own widely branching lymph vessels with their associated regional lymph nodes. He described and named most of the vessels and nodes which are familiar to us today.

The models shown here summarize most impressively what was then the most up-to-date anatomical knowledge of that time. Without any doubt, these specimens must count as outstandingly graphic achievements.

Lymphatisches System

Das Lymphgefäßsystem und insbesondere die lymphatischen Organe, wie beispielsweise Thymus, Milz und Lymphknoten, nehmen eine zentrale Stellung in dem komplexen Abwehrsystem unseres Gesamtorganismus ein. Bereits Gasparo Aselli (1582–1626) entdeckte im 17. Jahrhundert die Lymphgefäße im Darmgekröse eines unmittelbar nach der Mahlzeit sezierten Hundes. Durch die Aufnahme der Nahrungsfette in Form feinster Tröpfchen in die sehr feinen Darmlymphgefäße waren diese weißlich gefärbt und dadurch sichtbar. Ende des 18. Jahrhunderts gelang es dem italienischen Anatom Paolo Mascagni (1752–1815) mittels einer raffiniert eingesetzten Quecksilber-Injektionstechnik zu zeigen, daß alle Organe über ein eigenständiges und weit verzweigtes Lymphgefäßsystem mit regionalen Lymphknoten verfügen. Er beschrieb und benannte die meisten der uns heute bekannten Lymphgefäße und Lymphknoten.

Die hier gezeigten Modelle fassen diese damals hochaktuellen anatomischen Erkenntnisse zusammen. Zweifelsohne gehören diese Präparate zu den herausragenden darstellerischen Leistungen.

Pages · Seiten · Pages 316–317:

Detail from pages 320–321

Detail der Seiten 320–321

Détail des pages 320–321

Système lymphatique

Le système lymphatique, et plus particulièrement les organes lymphatiques tels que la rate, les nœuds ou ganglions lymphatiques, et chez l'enfant le thymus, occupent une place primordiale dans la défense de l'organisme. Au XVIIe siècle, Gasparo Aselli (1582–1626) avait découvert les vaisseaux lymphatiques dans le mésentère d'un chien disséqué pendant la digestion, le transport de fines gouttelettes de graisse alimentaire par les vaisseaux lymphatiques de l'intestin les rendant lactescents et donc visibles. A la fin du XVIIIe siècle, l'anatomiste italien Paoli Mascagni (1782–1815) réussit à démontrer, grâce à une technique d'injection élaborée de mercure, que tous les organes sont tributaires d'une même circulation lymphatique, largement ramifiée, disposant de nœuds ou ganglions lymphatiques dans chaque région ; il a décrit et dénommé la plupart des vaisseaux et nœuds lymphatiques connus de nos jours.

Les connaissances anatomiques de cette époque, toujours valables actuellement, sont résumées de manière impressionnante par cette série de modèles en cire ; ces préparations font sans aucun doute partie des œuvres les plus représentatives de la collection.

Systema lymphaticum, Venae epifasciales

Whole body specimen with the superficial veins and lymphatic vessels displayed

Ganzkörperpräparat mit Darstellung oberflächlicher Venen und Lymphgefäße

Préparation du corps entier représentant les veines et les vaisseaux lymphatiques superficiels

[XXVIII, 740]

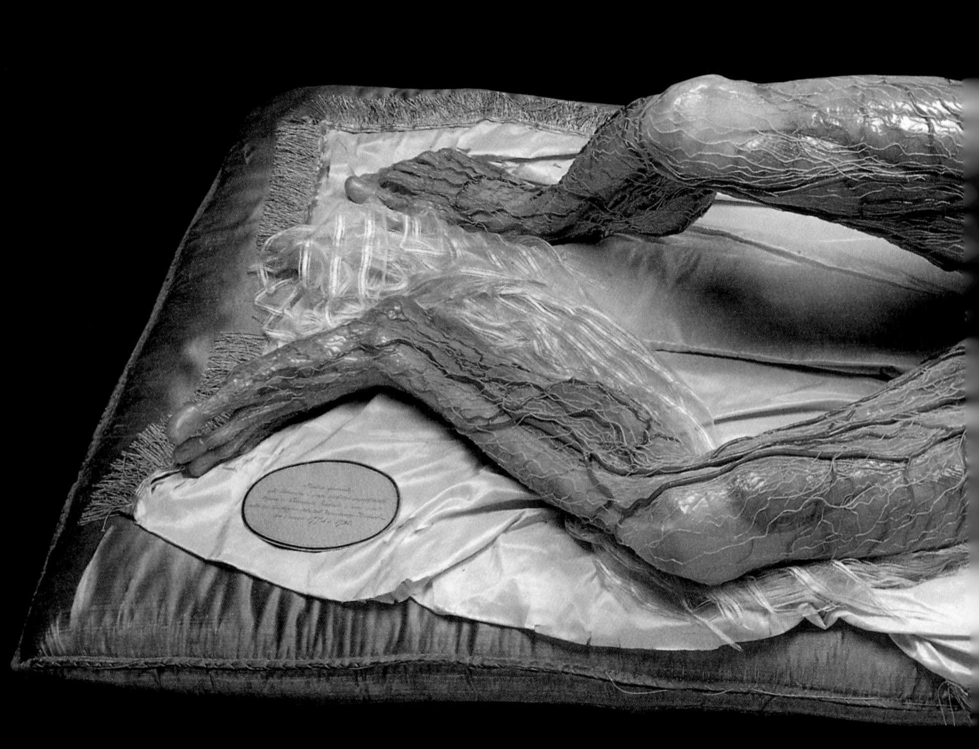

Systema lymphaticum

Whole body specimen showing the lymphatic vessels

Ganzkörperpräparat mit Darstellung des Lymphgefäßsystems

Préparation du corps entier représentant les grandes voies lymphatiques

[XXVII, 646]

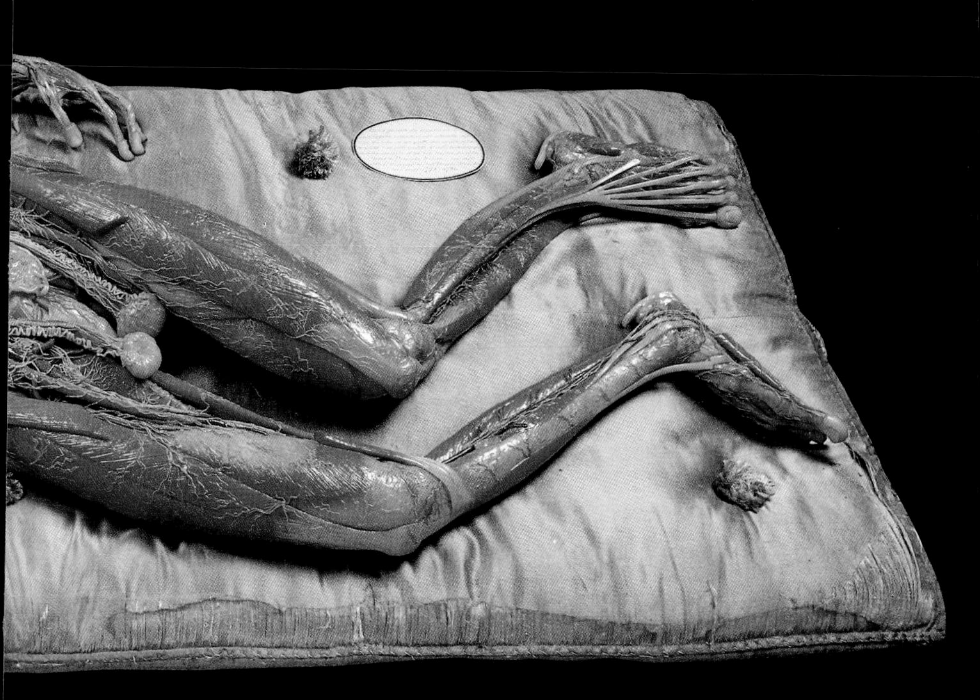

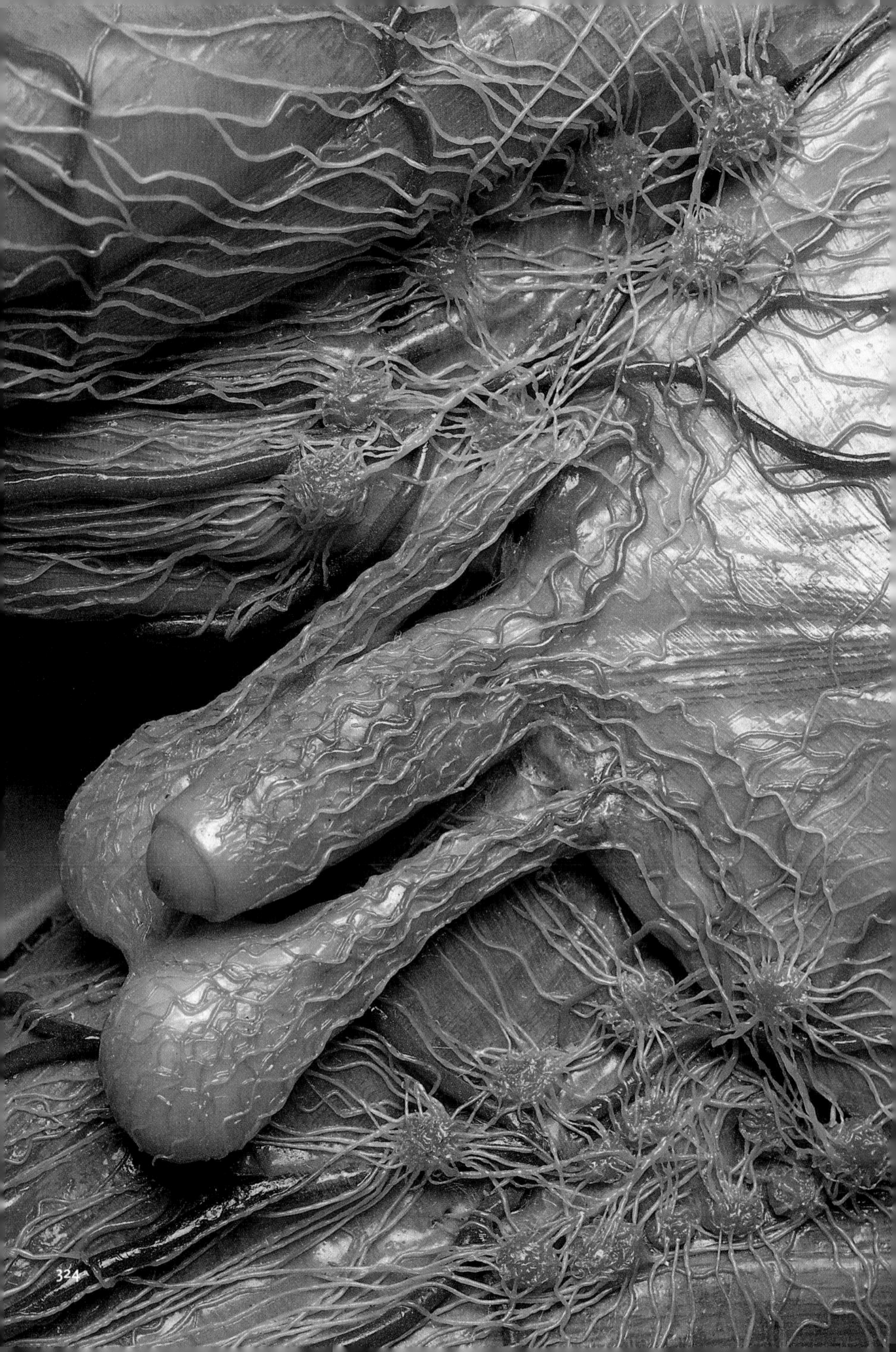

Different views of lymph nodes with efferent and afferent lymphatic vessels

Lymphknoten in verschiedenen Darstellungen mit zu- und abführenden Lymphgefäßen

Nœuds lymphatiques (ganglions) avec vaisseaux lymphatiques afférents et efférents

[XXVII, 416]

◂
Detail from pages | Detail der Seiten | détail des pages 320–321

Vasa lymphatica, Nodi lymphatici submandibulares, Nodi lymphatici submentales, Nodi lymphatici retroauriculares, Nodi lymphatici cervicales superficiales

Lymphatic vessels and nodes of the head and neck. The skull has been laid open to reveal the pia-arachnoid and the brain. The lymphatic vessels of the pia-arachnoid as displayed in the specimen do not exist.

Lymphgefäße und Lymphknoten im Bereich von Gesicht, Hinterhaupt und Hals. Lymphgefäße in der weichen Hirnhaut, wie sie hier im Präparat am eröffneten Schädel dargestellt sind, existieren nicht.

Vaisseaux et nœuds lymphatiques de la face, de la nuque, et du cou. Les lymphatiques représentés au contact du cerveau n'existent pas.

[XXVII, 414]

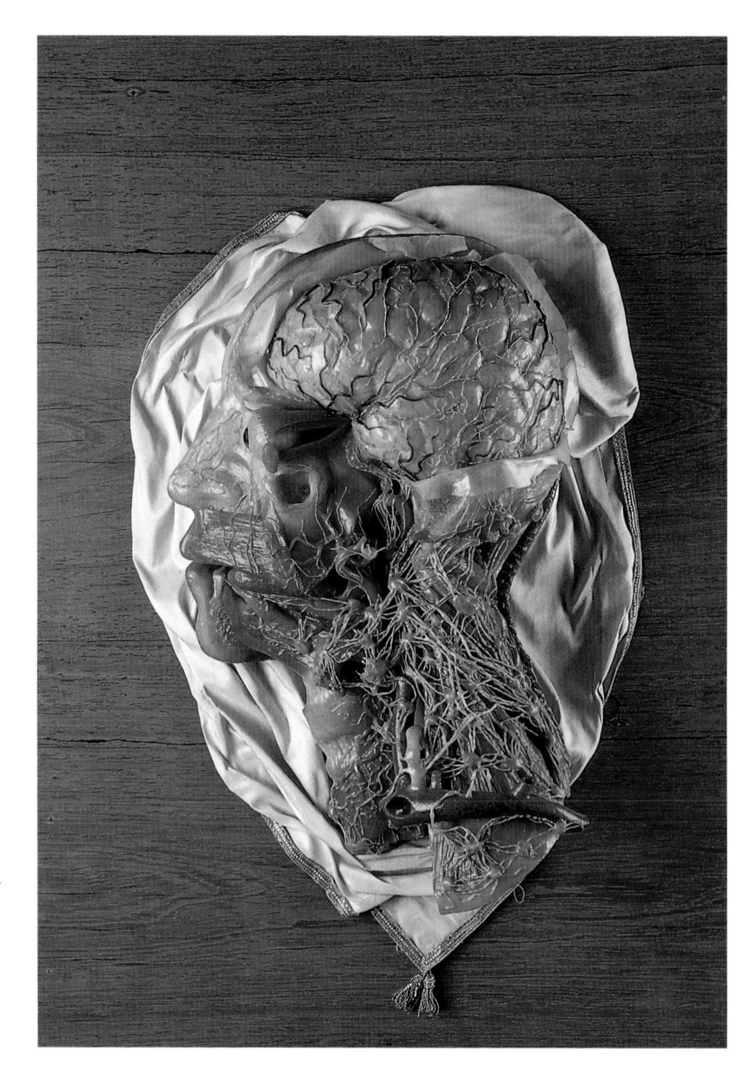

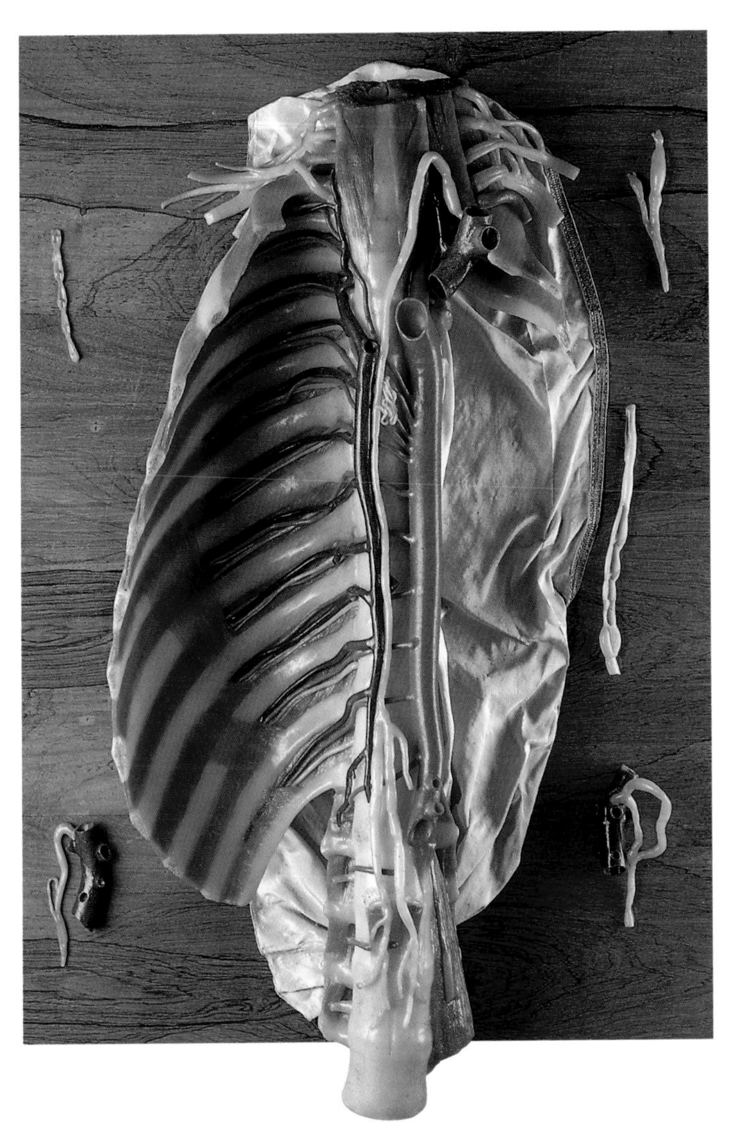

View into the thoracic cage to show the thoracic duct (the main collecting lymphatic vessel), its tributaries and its opening into the great vein on the left side of the neck

Einblick in den Brustkorb mit Darstellung des Milchbrustganges (großes Lymphsammelgefäß), seinen Zuflüssen und seiner Einmündung in die linke große Halsvene

Conduit thoracique (canal collecteur lymphatique principal) : origine et trajet intra-thoracique, abouchement dans la veine jugulaire gauche

[XXVII, 852]

Ductus thoracicus, Vasa lymphatica pulmones superficiales

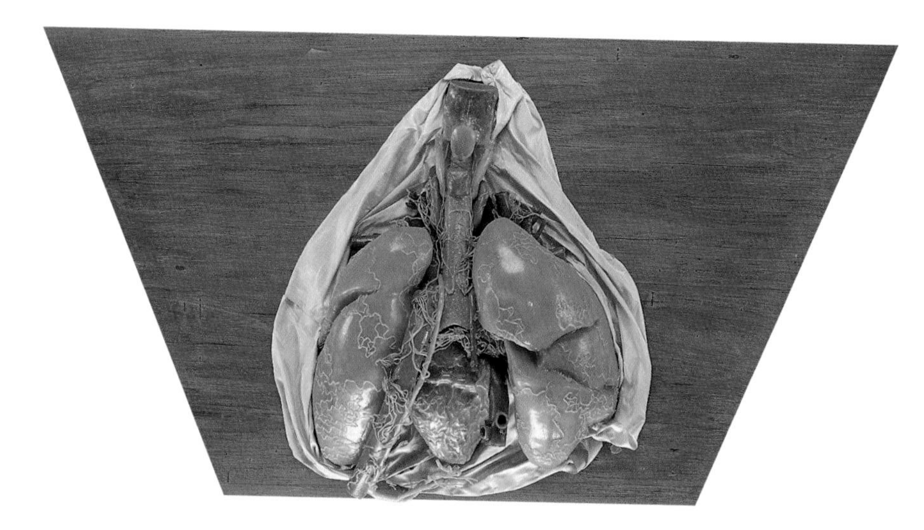

View from behind into the thoracic cavity to show the course of the lymphatic vessels and the thoracic duct

Ansicht von hinten auf die im Brustkorb verlaufenden Lymphgefäße sowie den Milchbrustgang

Conduit thoracique et vaisseaux lymphatiques des viscères du thorax vus de l'arrière

[XXIX, 862]

►
Detail from pages |
Detail der Seiten |
détail des pages
322–323

Lymphatic vessels of the liver and on the internal surface of the breastbone. The thoracic duct can be discerned as a pale cord.

Darstellung der Lymphgefäße der Leber und der Innenfläche des Brustbeins. Als heller Strang ist der Milchbrustgang erkennbar.

Vaisseaux lymphatiques du foie et de la face profonde de la paroi antérieure du thorax

[XXX, 894]

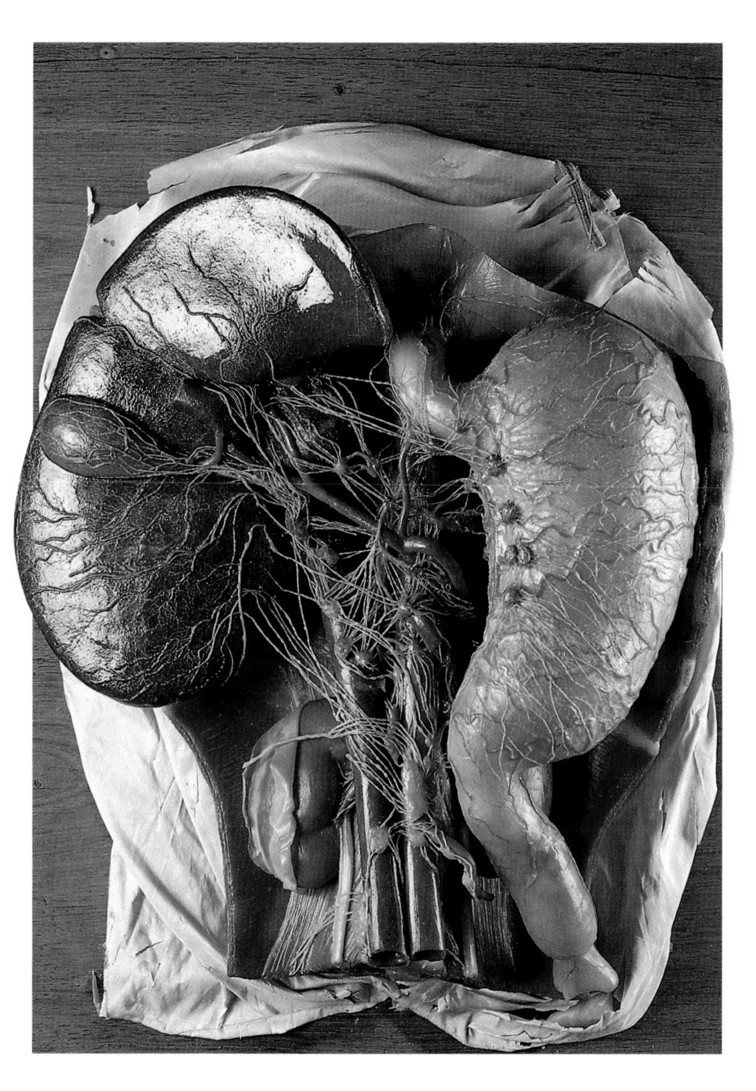

Lymphatic vessels and nodes around the organs in the upper part of the abdominal cavity

Lymphgefäße und Lymphknoten im Bereich der Oberbauchorgane

Vaisseaux et nœuds lymphatiques des viscères de la partie haute de l'abdomen (foie, estomac)

[XXX, 851]

Vasa lymphatica thoracalis et abdominalis

Whole body specimen showing the lymphatic vessels of the thoracic and abdominal cavities

Ganzkörperpräparat mit Darstellung des Lymphgefäßsystems in Brust- und Bauchraum

Préparation du corps entier représentant les veines et les vaisseaux lymphatiques des viscères du thorax et de l'abdomen

[XXIX, 745]

Lymphatic vessels on the anterior surface of the stomach

Darstellung der Lymphgefäße auf der Magenvorderfläche

Vaisseaux lymphatiques de la face antérieure de l'estomac ; pancréas et rate

[XXX, 854]

Lymph nodes and lymphatic vessels on the posterior surface of the stomach, and the thoracic duct receiving the main lymphatic tributaries

Darstellung der Lymphgefäße und Lymphknoten auf der Magenrückfläche mit Milchbrustgang und einmündenden Lymphgefäßstämmen

Vaisseaux lymphatiques de la face postérieure de l'estomac ; origine du conduit thoracique, rate

[XXX, 850]

Lymph nodes and lymphatic vessels of the small intestine

Darstellung der Lymphknoten und Lymphgefäße des Dünndarmes

Vaisseaux et nœuds lymphatiques de l'intestin grêle et du mésentère

[XXIX, 833]

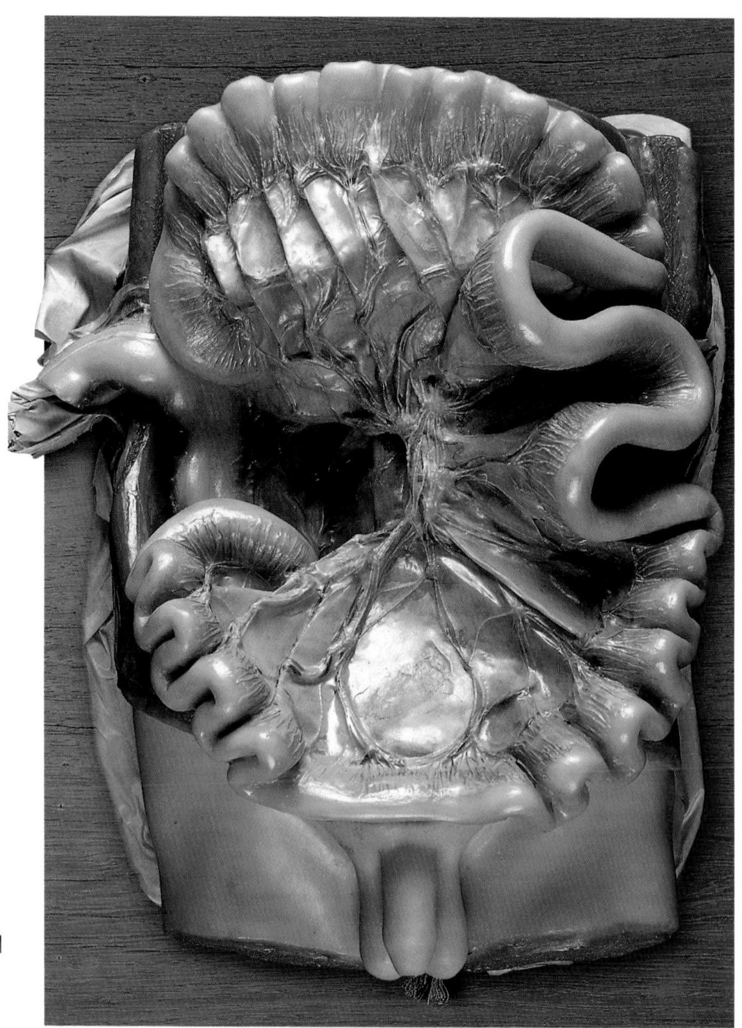

▶
Detail from pages | Detail der Seiten | détail des pages 338–339

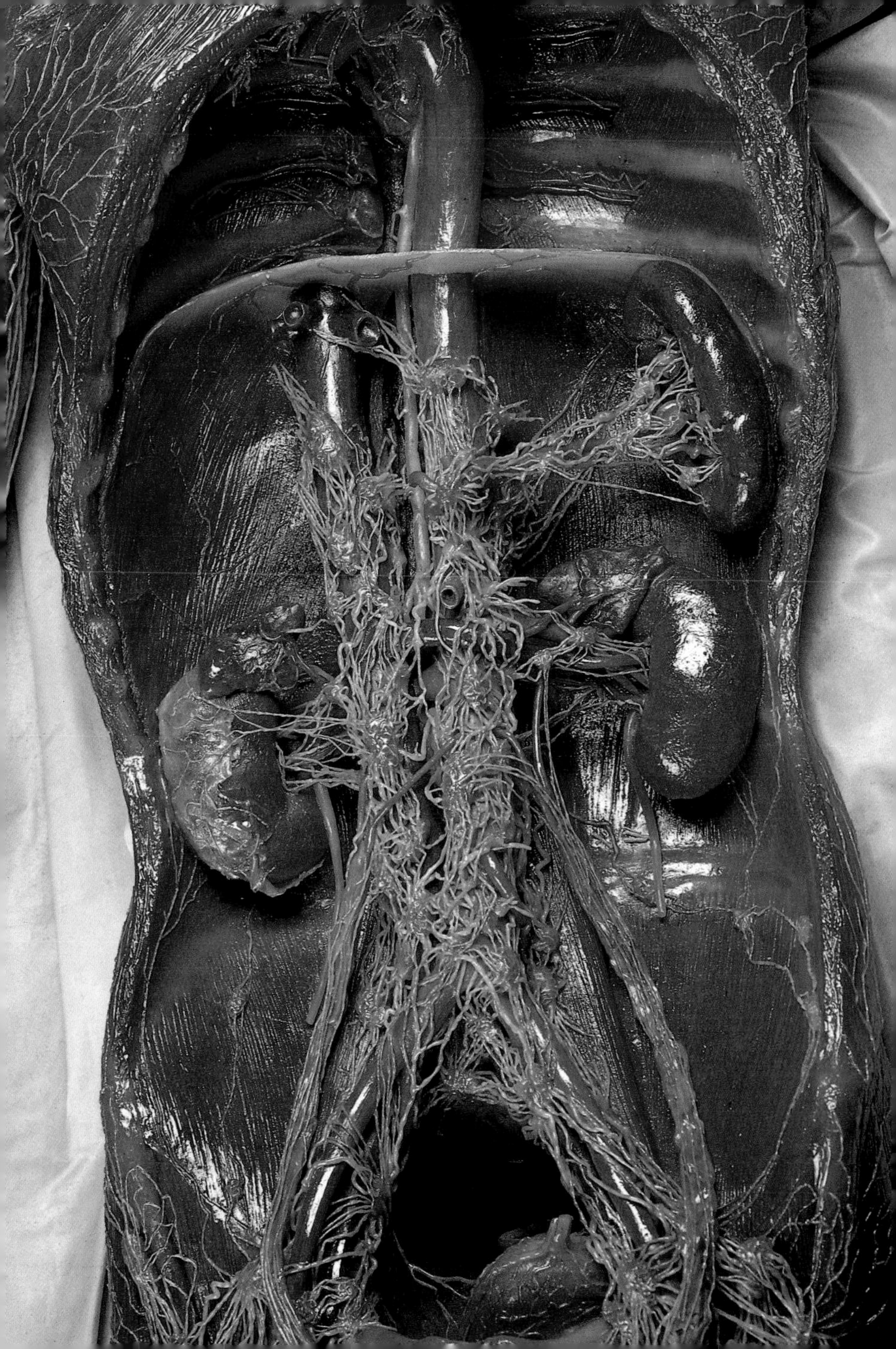

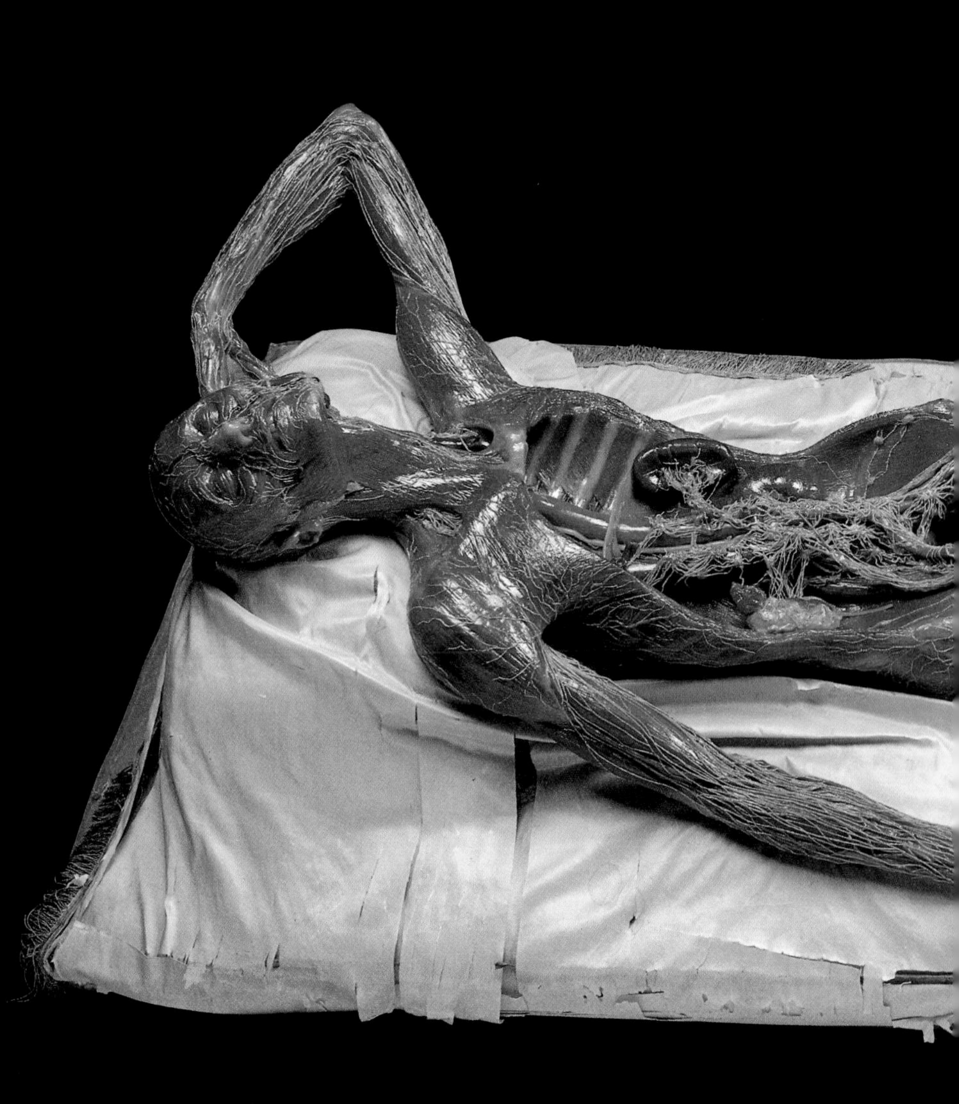

Systema lymphaticum

Whole body specimen showing the lymphatic vessels

Ganzkörperpräparat mit Darstellung des Lymphgefäßsystems

Préparation du corps entier représentant les veines et les vaisseaux lymphatiques

[XXVIII, 739]

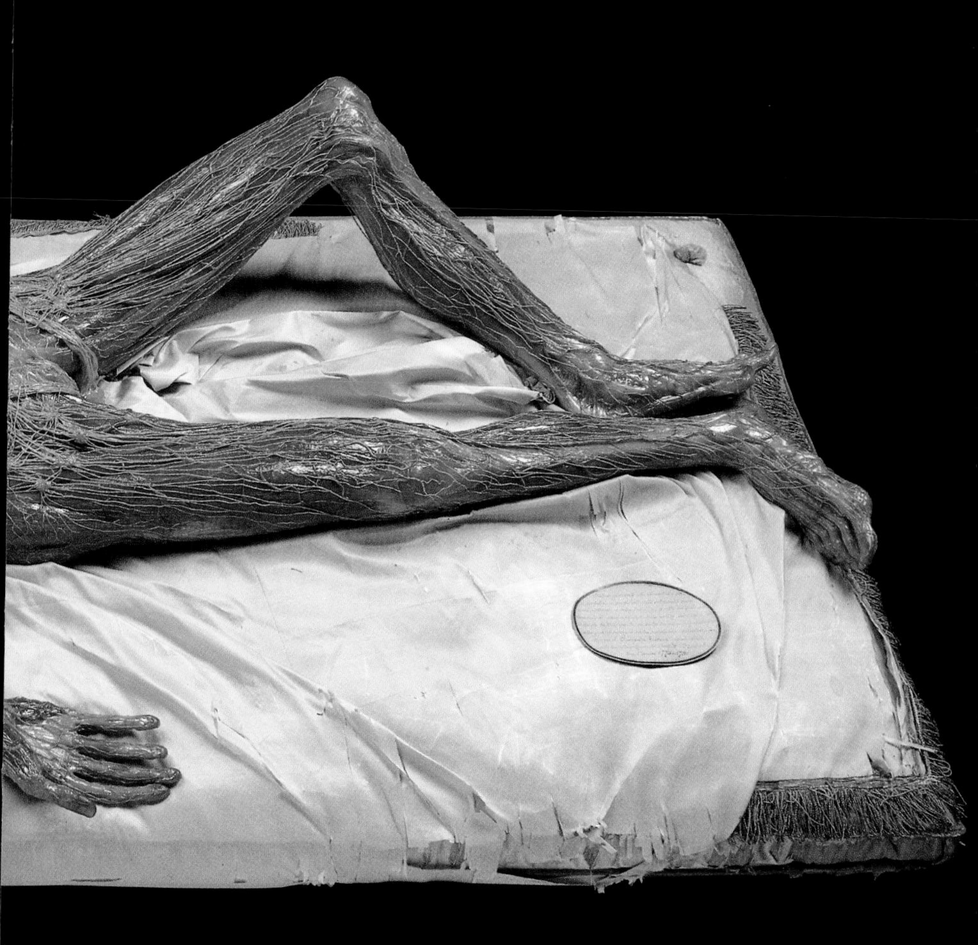

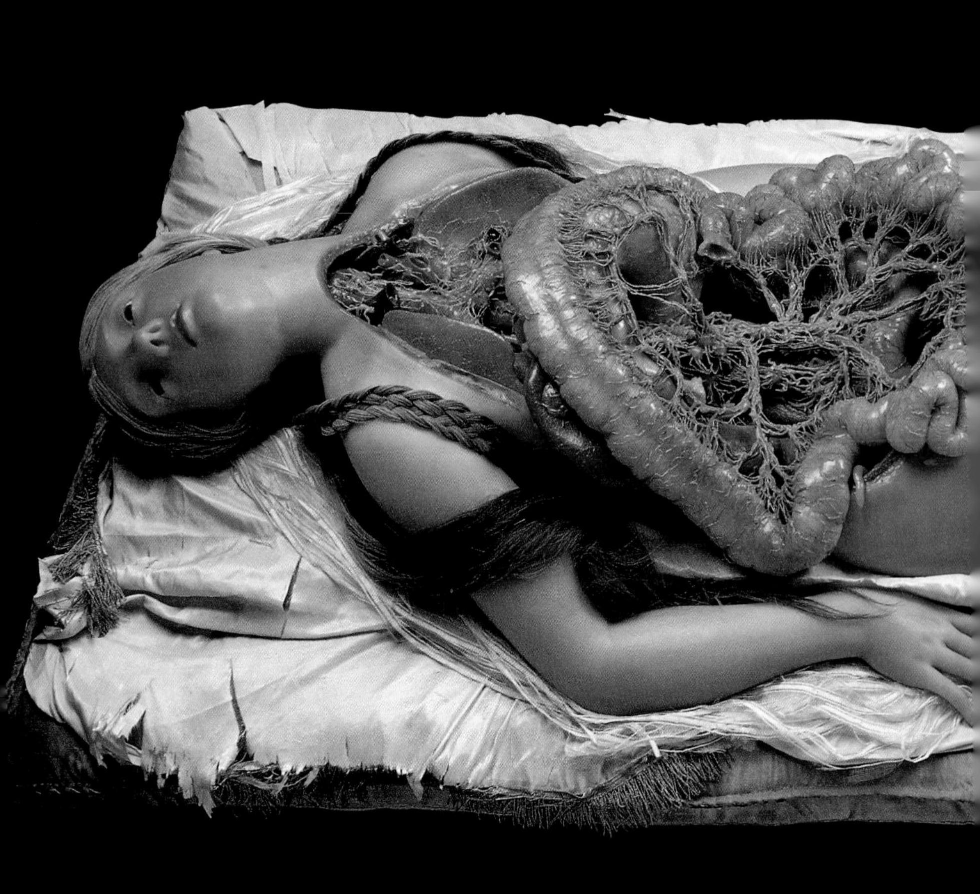

Vasa mesenterica

Whole body specimen showing the vessels and lymph nodes within the abdominal cavity

Ganzkörperpräparat mit Darstellung der Gefäße und Lymphknoten der Bauchhöhle

Préparation du corps entier représentant les veines et les vaisseaux et nœuds lymphatiques des viscères de l'abdomen et du thorax

[XXIX, 747]

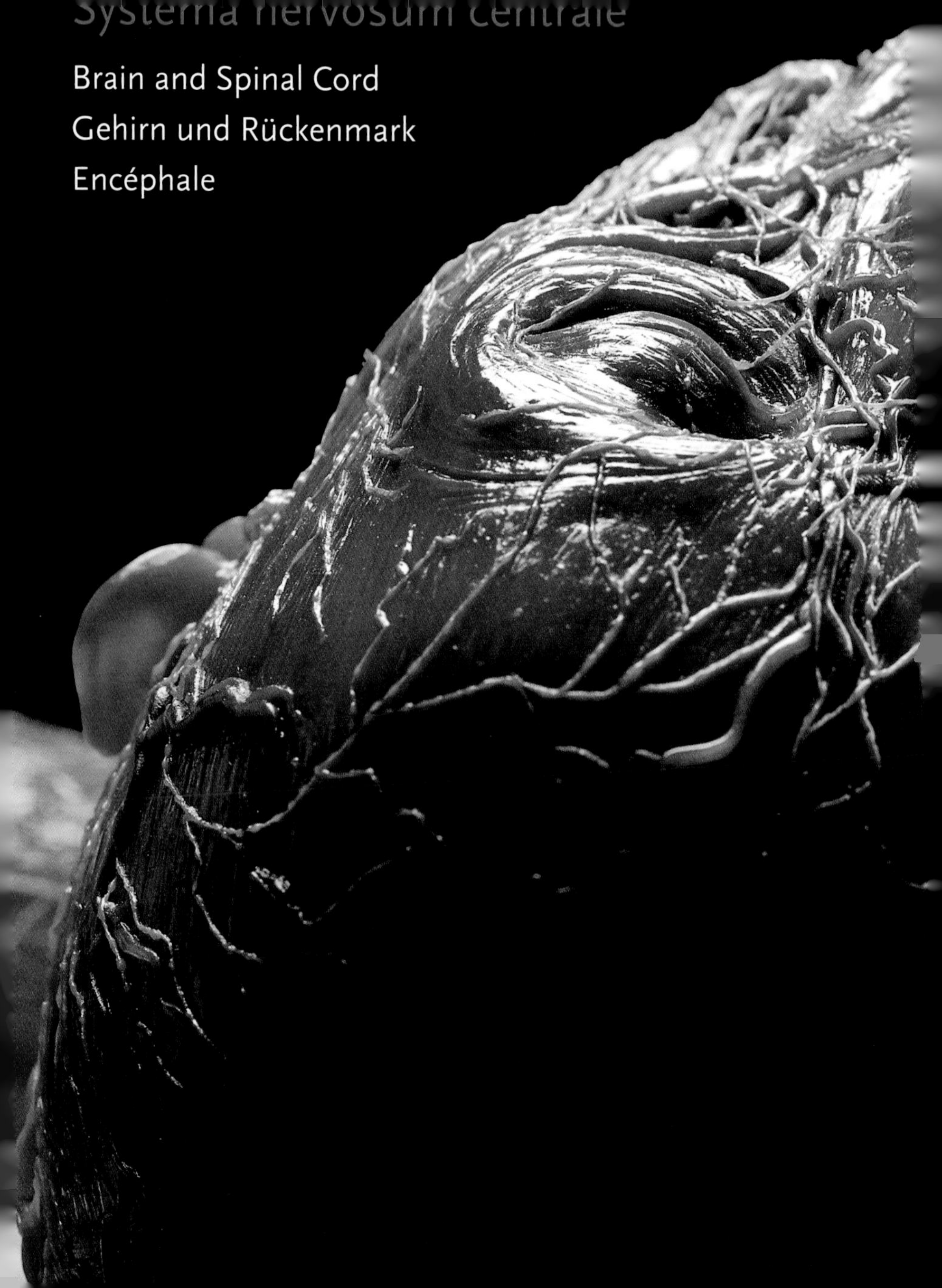

Systema nervosum centrale

Brain and Spinal Cord

Gehirn und Rückenmark

Encéphale

The Brain

The uniqueness of every human being is founded in his soul, i. e. his mental, religious, cultural, social and motor capabilities, which are carried out and controlled by the neuronal networks in our brains. The nerve cells are the morphological basis of the complex activities of the brain, their processes (dendrites, axons) and specialist sites of contact (synapses) serve to conduct and transmit information. Whereas two centuries ago only the vaguest concepts of their significance existed, we, by applying our modern imaging techniques, are able today to obtain detailed morphological and functional information about the individual brain without opening the skull. According to contemporary medical knowledge, the complete and irreversible cessation of the function of the cerebrum and brainstem as the site of complex and vital cerebral processes is the criterion of the individual's death. The following illustrations display the various aspects of the anatomy of the brain: an intact infant's skull after removal of the skin and muscles of the head showing the still patent anterior fontanelle, a skull laid open to reveal the connective tissue coverings of the brain (the hard and soft meninges) with their arteries and veins, the vessels on the upper and lower surfaces of the brain, including the emerging nerves and their course through the openings in the base of the skull. In contrast to these still relevant insights, the cerebral gyri bear little relationship to the original specimen, since in most models they are merely stylized. In most of the sections the visible border between the dark outer cortex and the internal white matter has been artificially emphasized.

Of great importance is the display of the points of attachment of the cranial nerves to the brain stem, as well as the various fluid-filled internal chambers (ventricles) of the brain, the position and extent of which can be very well studied in the specimens. These cerebral ventricles used to be of great interest, because they were thought during the 17th century to be the site of the spirit and the soul.

Many of the models are anatomically correct and impress one also by their great artistic merit. However, there are also certain exhibits seen from unusual points of view, the significance and meaning of which may not always be easy for the non-professional to grasp.

Pages · Seiten · Pages 342–343:

Detail from pages 338–339

Detail der Seiten 338–339

Détail des pages 338–339

►

G. G. Zumbo: Anatomist's model of a head (cf. pp. 13, 18, 348)

Präparat eines Kopfes (vgl. S. 13, 18, 348)

Préparation anatomique d'une tête (cf. p. 13, 18, 348)

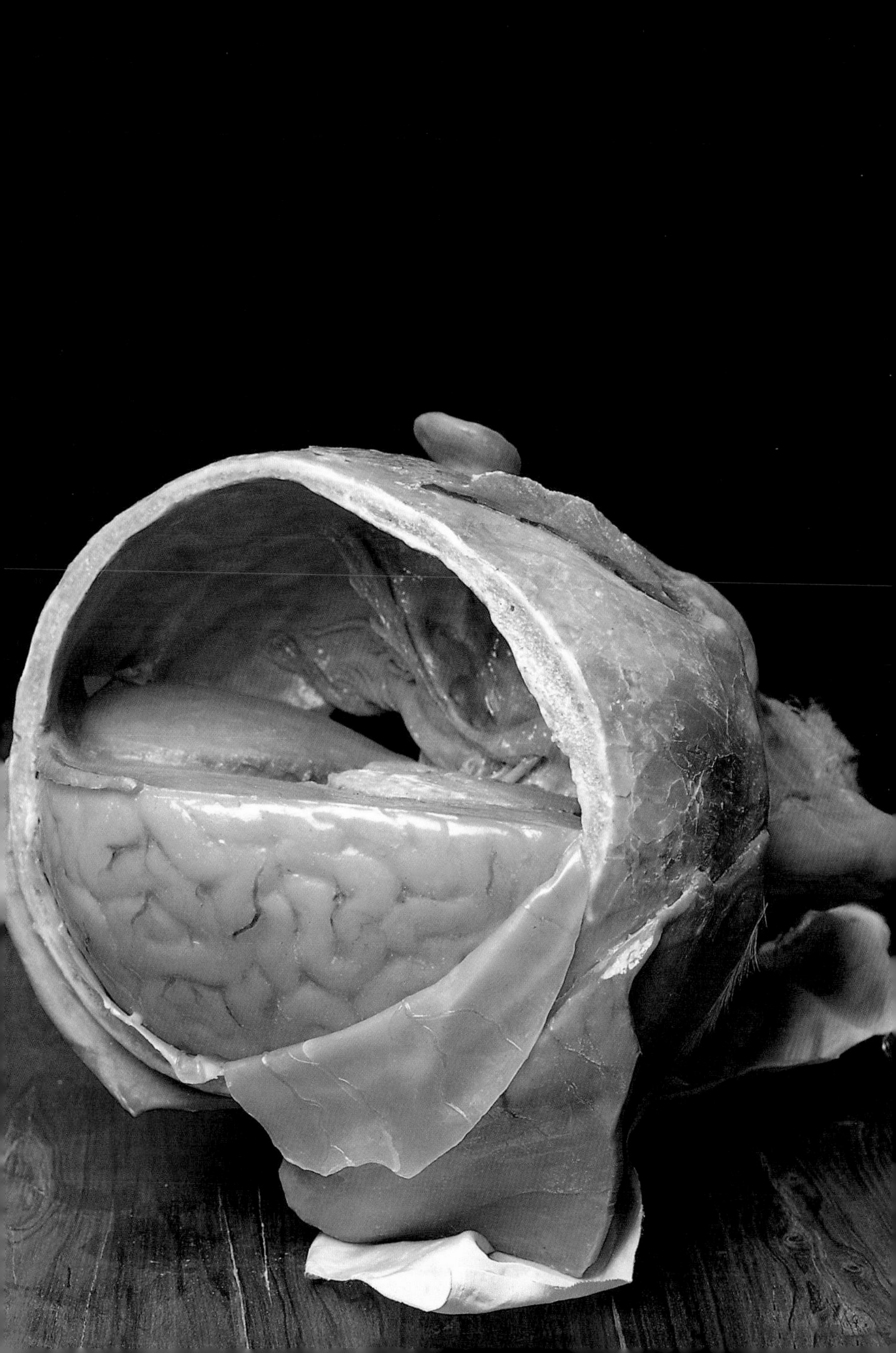

Das Gehirn

Die Einzigartigkeit eines jeden Menschen ist begründet in seiner Seele, d. h. in seinen geistigen, religiösen, kulturellen, sozialen und motorischen Fähigkeiten, die von den neuronalen Netzwerken in unserem Gehirn getragen und gesteuert werden. Die morphologische Grundlage dieser komplexen Hirnleistungen bilden die Nervenzellen. Mit ihren Ausläufern und über spezielle Kontake, den Synapsen, dienen sie der Informationsleitung und -übertragung. Während vor 200 Jahren nur vage Vorstellungen über ihre Funktionsweise bestanden, können wir heute mit modernen bildgebenden Verfahren am ungeöffneten Gehirn detaillierte Aussagen über die komplexe Gestalt des individuellen Gehirns machen, und funktionelle Teilleistungen erfassen. Nach den heutigen medizinischen Erkenntnissen gilt der endgültige Ausfall der Funktion von Großhirn und Stammhirn als den Regionen komplexer und lebensnotwendiger Hirnleistungen als Todeszeichen des Menschen.

Auf den nachfolgenden Tafeln sind die verschiedensten anatomischen Ansichten zur Hirnanatomie zu sehen: ein intakter Säuglingsschädel nach Wegnahme der Kopfhaut und Kopfmuskeln mit der großen vorderen, noch offenen Fontanelle, der eröffnete Schädel mit der Aufsicht auf die Bindegewebshüllen des Gehirns (harte und weiche Hirnhaut) mit versorgenden arteriellen und venösen Blutgefäßen, die Gefäße auf der Hirnoberfläche und Hirnunterfläche, Hirnhälften mit austretenden Hirnnerven einschließlich ihrer Verläufe durch die Knochenkanäle des Schädels. Im Gegensatz zu diesen bis heute gültigen Erkenntnissen haben die in den Modellen gezeigten Hirnwindungen nur wenig oder keinen Bezug zum Originalpräparat. In den meisten Schnittpräparaten ist die sichtbare Grenze zwischen der dunkleren Hirnrinde und dem weißlichen Hirnmark herausgearbeitet.

Thematische Schwerpunkte bilden die Darstellung der Austrittsstellen der Hirnnerven am Hirnstamm sowie die verschiedenen flüssigkeitsgefüllten Innenräume (Ventrikel) des Gehirns, deren Lage und Ausdehnung an den Präparaten sehr gut studiert werden können. Diese Hirnventrikel waren von großem allgemeinen Interesse, weil man im 17. Jahrhundert in ihnen den Sitz des Geistes und der Seele vermutete.

Zahlreiche Modelle sind anatomisch korrekt und beeindrucken gleichzeitig durch ihre hohe künstlerische Qualität. Daneben finden sich aber auch Exponate in ungewöhnlicher perspektivischer Ansicht, deren Bedeutung und Aussage dem Laien nur schwer zugänglich ist.

C'est son esprit qui donne à chaque être humain son caractère unique; il capacités spirituelles, religieuses, culturelles, soziales ou motrices. Les cellules nerveuses sont la base morphologique de ces fonctions cérébrales complexes ; par leurs prolongements et à travers des zones de contact spécifiques, les synapses, elles véhiculent et transmettent les informations. Il est possible aujourd'hui, par des techniques d'imagerie, de mettre en évidence non seulement des détails de l'architecture complexe du cerveau d'un individu mais encore des localisations fonctionnelles particulières. Dans l'état actuel des connaissances médicales, l'arrêt total et définitif des fonctions du cerveau et du tronc cérébral signe la mort de l'individu.

Les vues les plus variées de l'anatomie du cerveau sont représentées dans les planches suivantes. Sur un crâne de nouveau-né, après ablation de la peau et des muscles de la tête, la grande fontanelle apparait encore ouverte, et les méninges cérébrales (dure-mère et pie-mère) sont pourvues de leur vascularisation artérielle et veineuse. Plus loin, les vaisseaux de la convexité et de la base du cerveau sont au contact des structures nerveuses. D'un hémi-cerveau partent les nerfs crâniens, dont le trajet peut être suivi jusque dans les canaux osseux de la base du crâne.

La représentation des circonvolutions cérébrales n'a que peu ou pas de relation avec la réalité anatomique, contrairement à de nombreux autres détails encore valables aujourd'hui ; les circonvolutions ont souvent été stylisées. Dans la plupart des coupes, la limite entre la substance blanche et la substance grise du cerveau a été mise en évidence.

Une attention particulière a été portée à la représentation des points d'émergence des nerfs crâniens au niveau du tronc cérébral, et à la représentation des cavités cérébrales, ou ventricules, contenant le liquide cérébro-spinal, dont la disposition et le développement peuvent être bien étudiés sur ces modèles. Les ventricules cérébraux ont suscité un grand intérêt car ils étaient considérés, au XVIIe siècle, comme le siège de l'esprit et de l'âme.

De nombreux modèles sont les témoins fidèles de la réalité anatomique et sont remarquables par leur grande qualité artistique. A côté de ces pièces de qualité, il en existe quelquesunes dont les vues perspectives inhabituelles en rendent l'interprétation difficilement accessible au non initié.

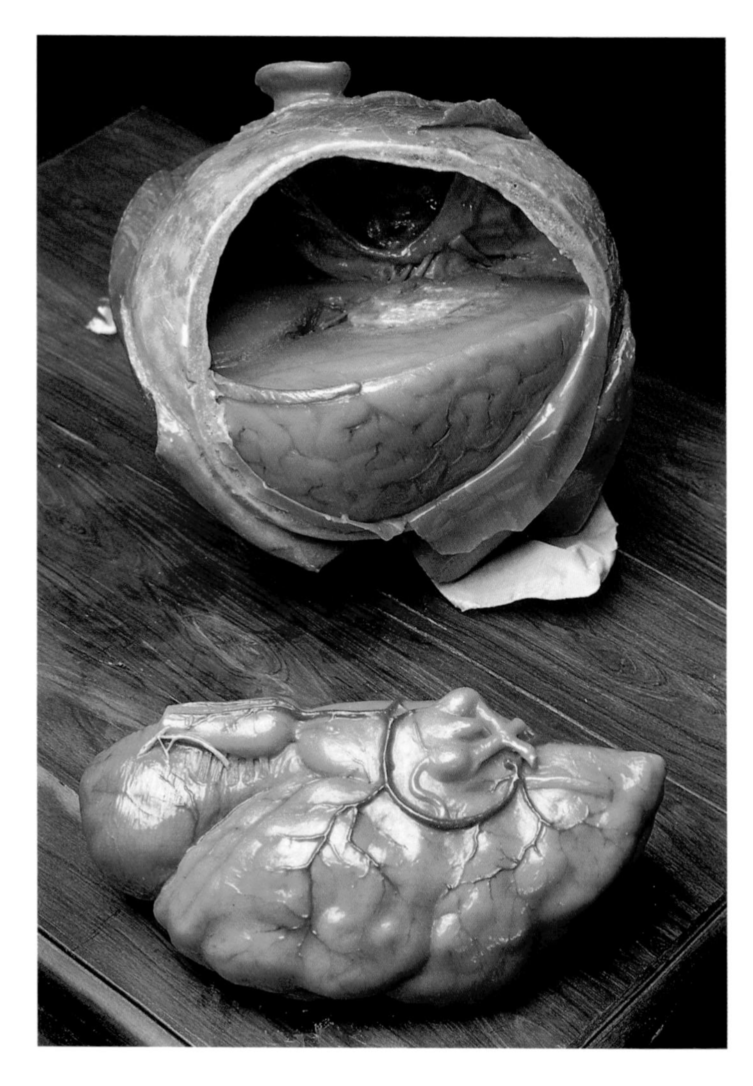

Specimen of a head. The calvaria has been removed and half of the brain can be taken out.

Präparat eines Kopfes. Das Schädeldach ist entfernt und eine Hirnhälfte herausnehmbar.

Tête ; ouverture de la voûte du crâne permettant d'enlever un hémisphère cérébral

[SALA ZUMBO, D1]

Specimen showing the dura mater with its arteries. The left half of the brain has been removed.

Präparat mit Blick auf die harte Hirnhaut mit ihren Arterien, die linke Hirnhälfte ist entfernt

Méninges (duremère) et artères méningées (hémisphère gauche du cerveau enlevé)

[XXVIII, 679]

Internal aspects of the calvaria (upper specimen) and base of skull after removal of the brain. View of the dura mater and its blood vessels. In the midline of the upper specimen one can see the sickle-shaped falx of the cerebrum.

Innenansicht von Schädeldach (oberes Präparat) und Schädelbasis nach Entnahme des Gehirns. Blick auf die harte Hirnhaut und ihre Blutgefäße. In der Mittellinie des oberen Präparates ist die Großhirnsichel erkennbar.

Voûte (en haut) et base du crâne (en bas) en vue endocrânienne recouvertes par la dure-mère avec ses vaisseaux ; la faux du cerveau est visible sur la ligne médiane de la voûte.

[XXVIII, 591]

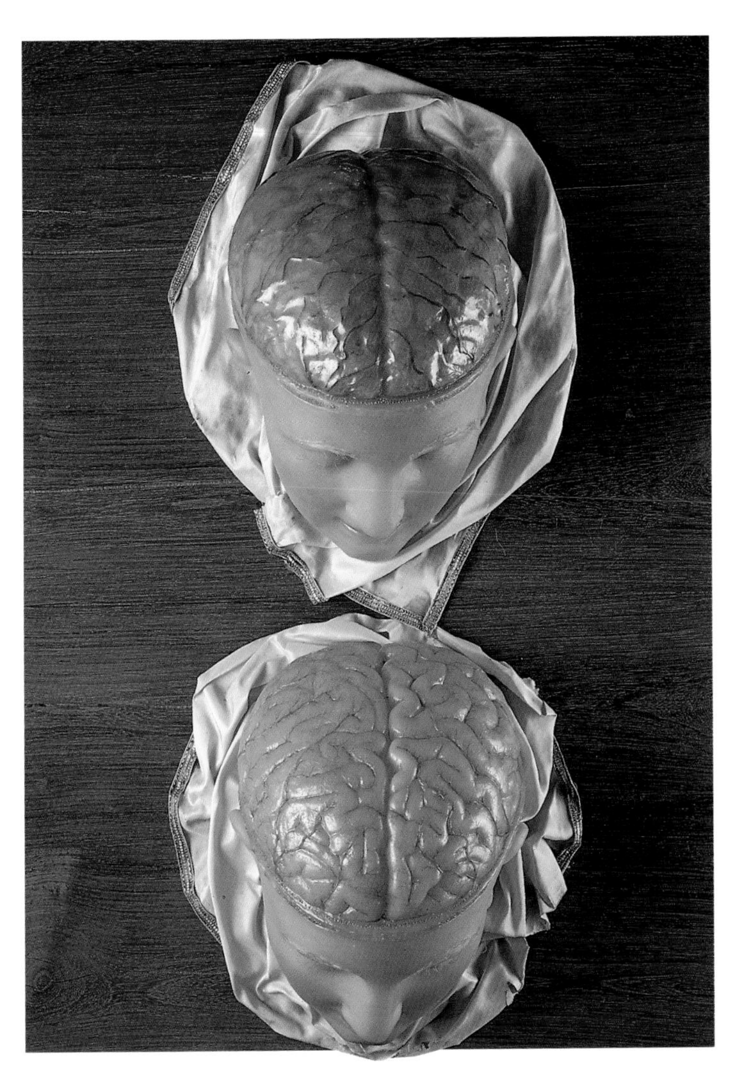

View of the brain after removal of the calvaria and dura mater. In the upper specimen the soft meninges with their vessels can be seen. The convolutions of the brain surface can be clearly seen in the lower specimen after removal of the meninges.

Blick auf das Gehirn nach Entfernung des Schädeldaches und der harten Hirnhaut. Im oberen Präparat ist die weiche Hirnhaut mit Gefäßen dargestellt. Das Windungsrelief des Großhirns wird nach Entfernung der weichen Hirnhaut deutlich (unteres Präparat).

Cerveau après ouverture de la voûte du crâne et de la dure-mère ; les circonvolutions cérébrales apparaissent après résection de la pie-mère et de ses vaisseaux (en bas)

[XXVIII, 592]

The calvaria of a fetus seen from above, showing the anterior fontanelle still open (left specimen) and the dura mater after removal of the calvaria (right specimen). Below: the calvaria seen from the inside

Blick von oben auf das Schädeldach eines Föten mit offener vorderer Fontanelle (linkes Präparat) und auf die harte Hirnhaut nach Enfernung des Schädeldaches (rechtes Präparat). Unteres Präparat: Innenansicht des Schädeldaches

Voûte du crâne du fœtus avec la fontanelle antérieure ouverte (en haut à gauche) ; dure-mère après ablation de la voûte (en haut à droite). En bas : vue endocrânienne de la voûte

[XXVIII, 590]

The venous sinuses of the dura mater

Darstellungen der venösen Blutleiter in der harten Hirnhaut

Conduits veineux limités par la duremère (sinus veineux)

[XXVII, 595]

Sinus durae matris

The venous sinuses of the dura mater

Darstellungen der in der harten Hirnhaut gelegenen venösen Blutleiter

Conduits veineux limités par la duremère (sinus veineux)

[XXVII, 596]

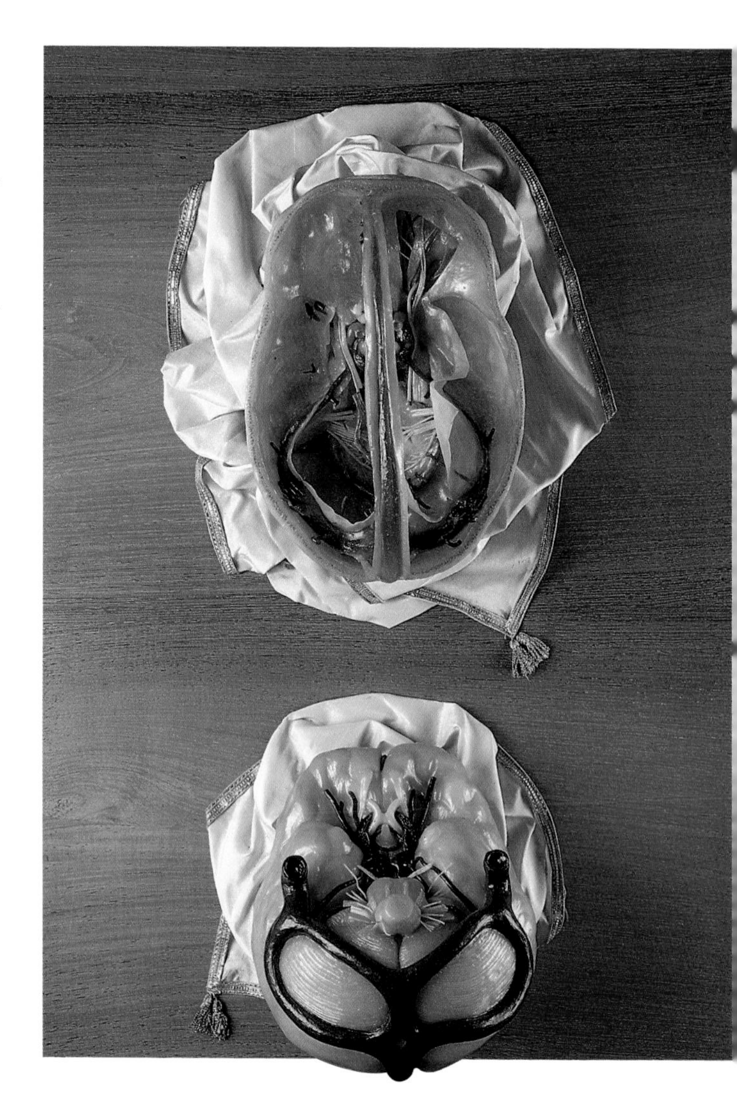

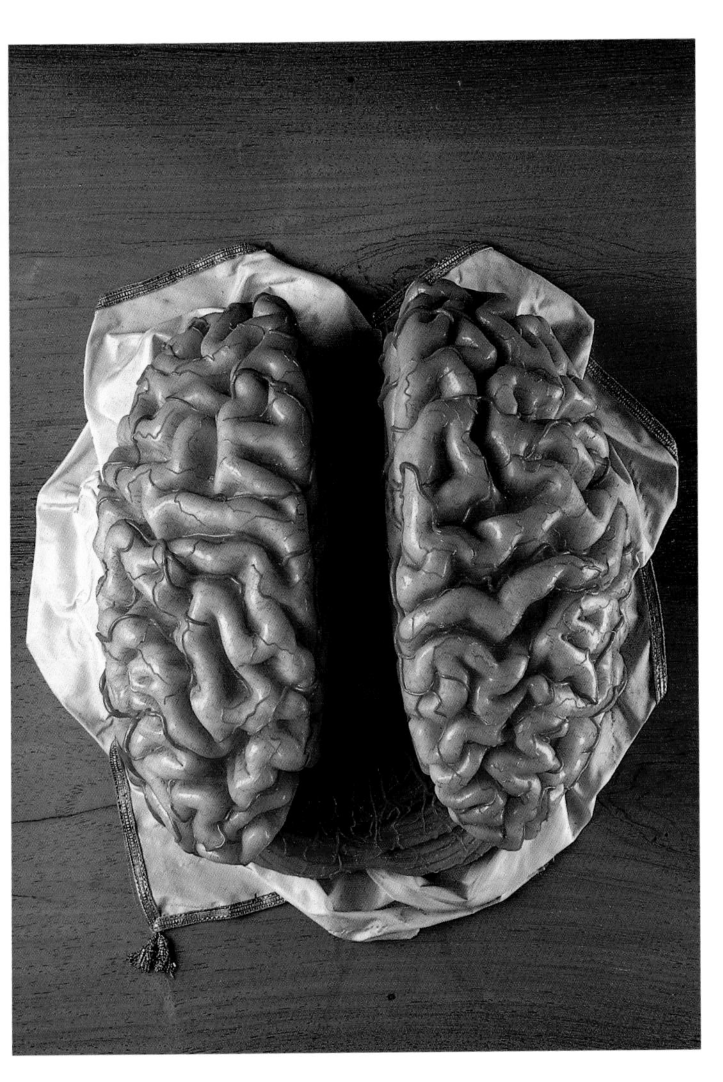

Specimen of the brain with its superficial arteries. The cerebrum has been divided in the midline and the two halves pushed aside to reveal the cerebellum (below).

Präparat des Gehirns mit oberflächlichen Arterien. Das Großhirn ist in der Mittellinie durchtrennt, die beiden Hirnhälften sind zu den Seiten verschoben, so daß der Blick auf das Kleinhirn (unten) freigegeben wird.

Artères de la convexité du cerveau ; après section médiane du cerveau, le déplacement latéral des hémisphères dégage la face supérieure du cervelet (en bas)

[XXVIII, 602]

Arteria carotis interna, Arteria cerebri media et anterior, Arteria cerebelli posterior inferior

Venae cerebri superficiales, Sinus sagittalis superior, Confluens sinuum, Sinus transversus

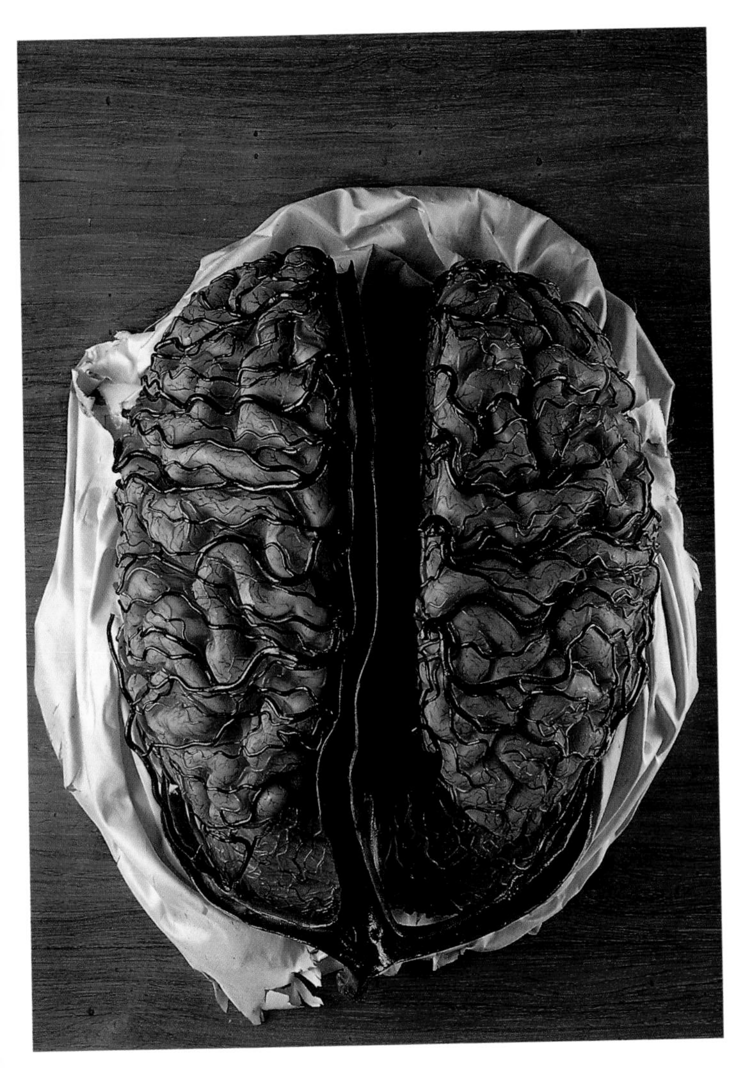

The superficial veins of the brain and their openings into the superior sagittal sinus within the falx of the cerebrum

Darstellung der oberflächlichen Hirnvenen und ihrer Mündungen in den venösen Blutleiter der Großhirnsichel

Veines superficielles de la convexité du cerveau s'abouchant dans les sinus veineux de la dure-mère

[XXVIII, 598]

◄

Specimen showing the various arteries at the base of the brain

Darstellung verschiedener Arterien an der Hirnbasis

Artères de la base de l'encéphale

[XXVIII, 601]

View of the arteries and veins at the base of the brain

Blick auf Arterien und Venen an der Hirnbasis

Artères et veines de la base de l'encéphale

[XXVIII, 597]

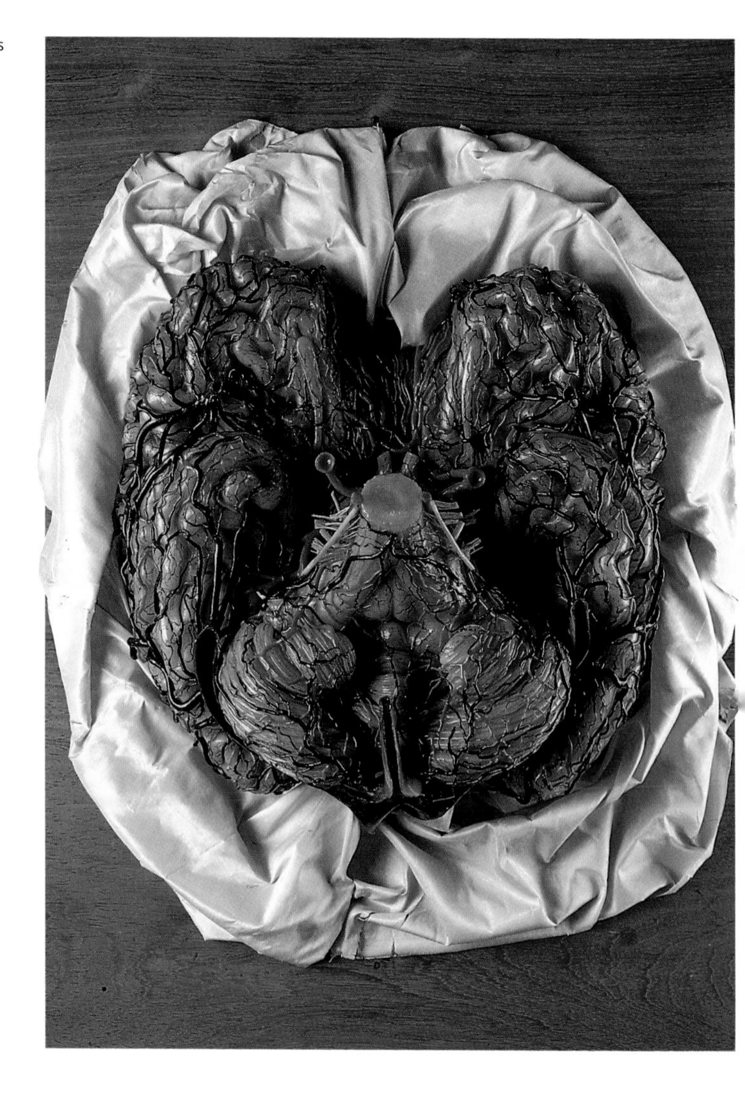

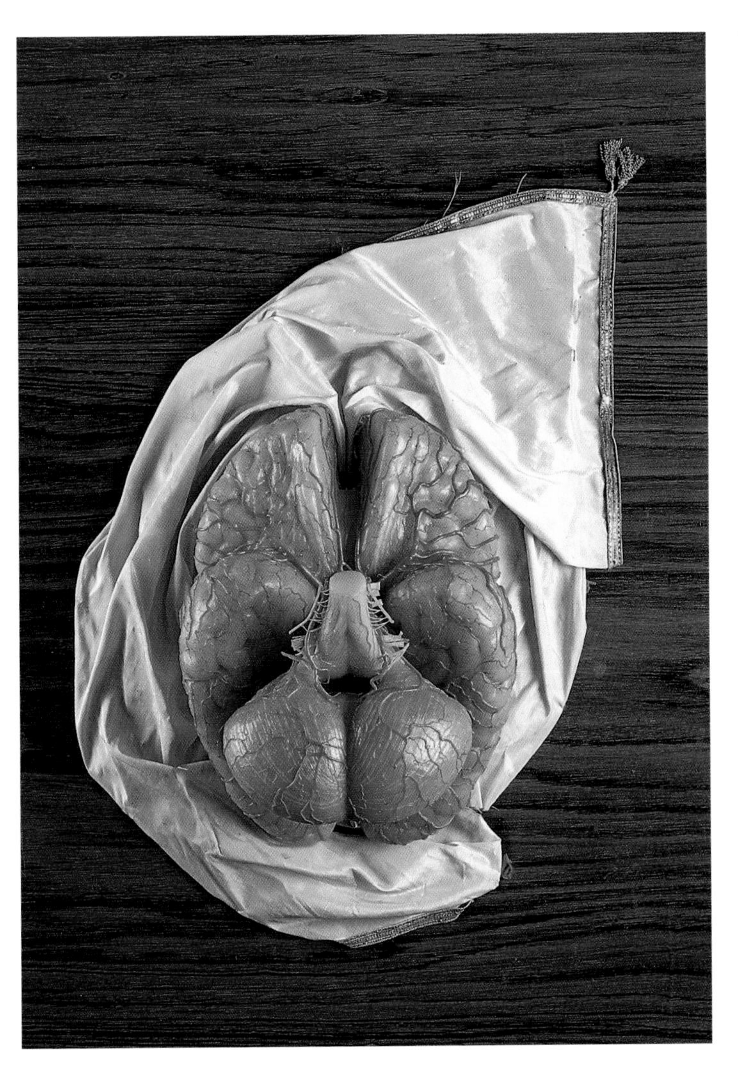

View of the arteries at the base of the brain

Darstellung von Arterien der Hirnbasis

Artères de la base de l'encéphale

[XXVIII, 675]

Plexus choroideus ventriculi lateralis, Circulus arteriosus Willisii

View of the arteries at the base of the brain. The lower specimen shows the confluence of the arteries which supply the brain.

Darstellung der Arterien an der Hirnbasis. Das untere Präparat zeigt den Zusammenfluß der hirnversorgenden Arterien.

Artères de la base du cerveau (en haut) ; arbre artériel montrant les anastomoses de la base (cercle de Willis) (en bas)

[XXVIII, 682]

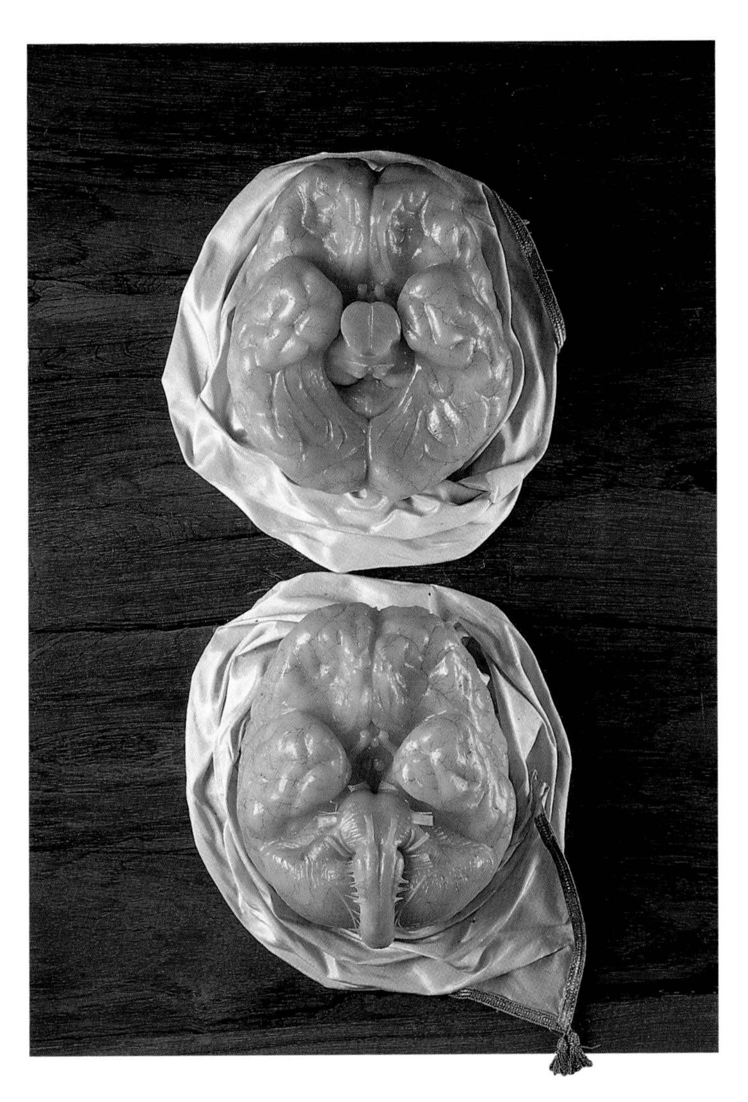

Base of the brain. In the upper specimen the main posterior lobes of the brain can be seen after the removal of the cerebellum.

Ansichten der Hirnbasis. Im oberen Präparat werden nach Entfernung des Kleinhirns die Hinterhauptslappen des Großhirns erkennbar.

Base de l'encéphale ; ablation du cervelet découvrant le lobe occipital (en haut)

[XXVIII, 719]

Thalamus

View of the base of the brain. The cerebellum and part of the brainstem have been removed. The small specimen in the middle shows the thalamus.

Ansicht der Hirnbasis. Das Kleinhirn und Teile des Hirnstammes sind entfernt. Das kleinere mittlere Präparat stellt den Thalamus dar.

Base du cerveau après ablation du cervelet (en haut) ; noyaux gris profonds du cerveau (en bas)

[XXVIII, 732]

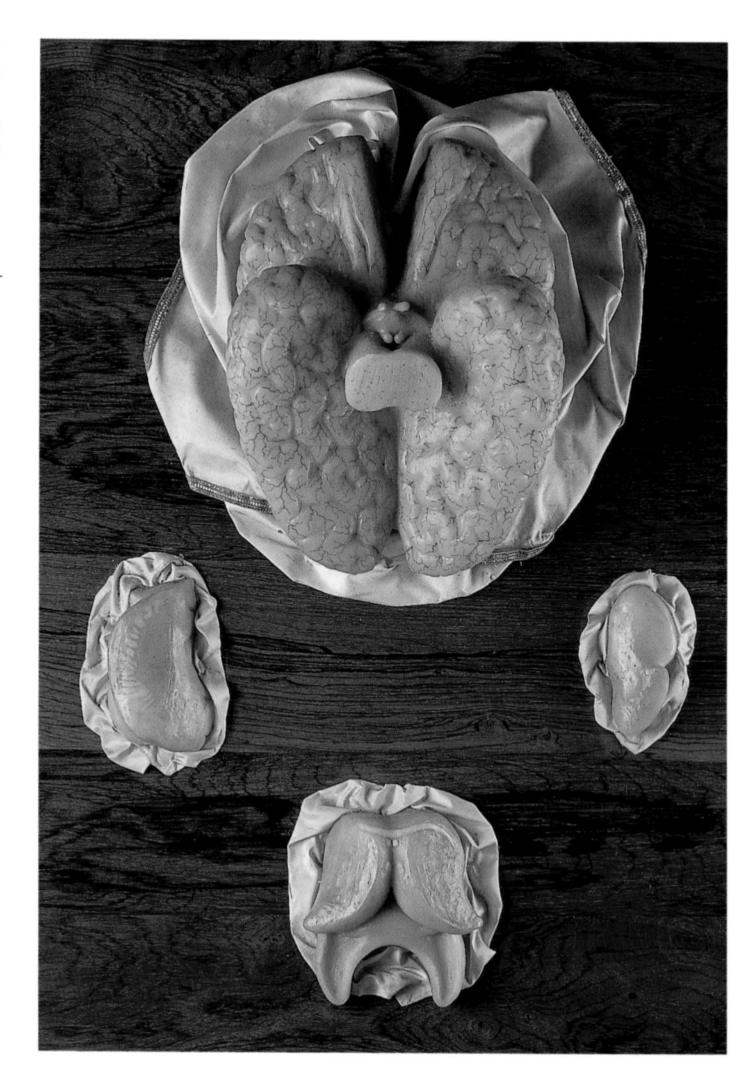

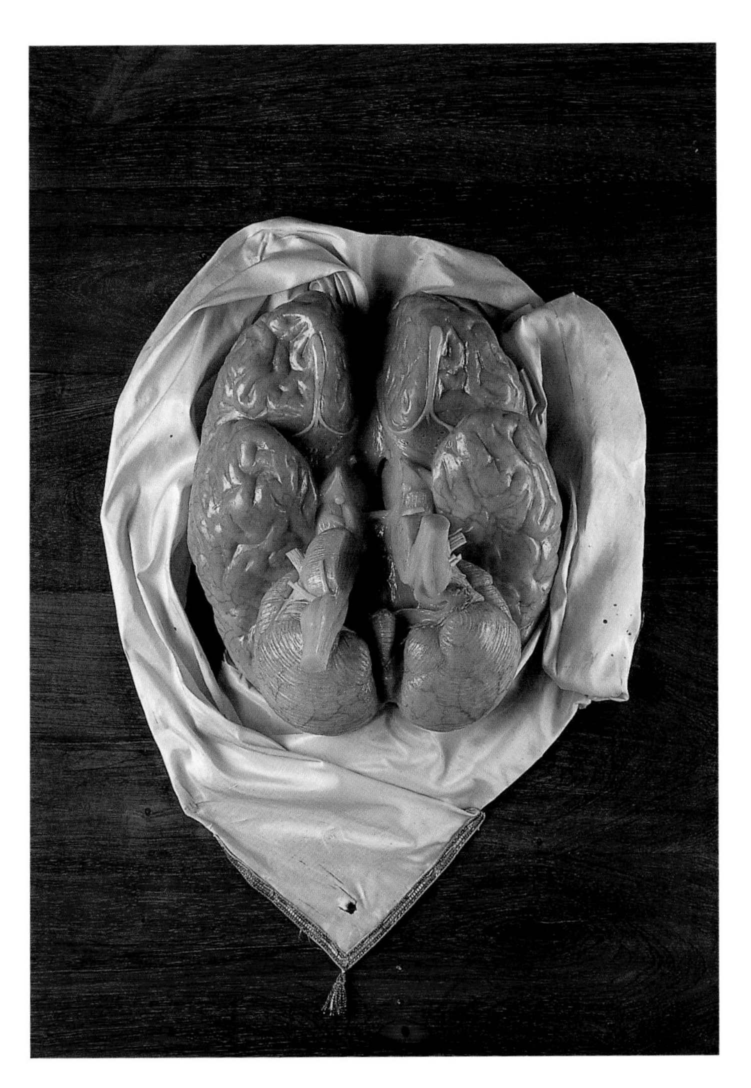

View of the base of the brain showing parts of the olfactory bulb and tract. The brainstem has been divided in the midline.

Blick auf die Hirnbasis mit Anteilen des Riechkolbens und der Riechbahn. Der Hirnstamm ist in der Mittellinie durchtrennt.

Base de l'encéphale avec tractus olfactif; section médiane du tronc cérébral

[XXVIII, 731]

View of the base of the brain. The cerebellum has been lifted up and pushed out of shape. The lower specimen shows a side view of the brainstem with the emerging cranial nerves.

Ansicht der Hirnbasis. Das Kleinhirn ist angehoben und verformt. Die unteren Präparate zeigen Seitenansichten des Hirnstammes mit austretenden Hirnnerven.

Base de l'encéphale, cervelet soulevé et déformé (en haut) ; vues latérales du tronc cérébral avec l'émergence des nerfs crâniens (en bas)

[XXVIII, 714]

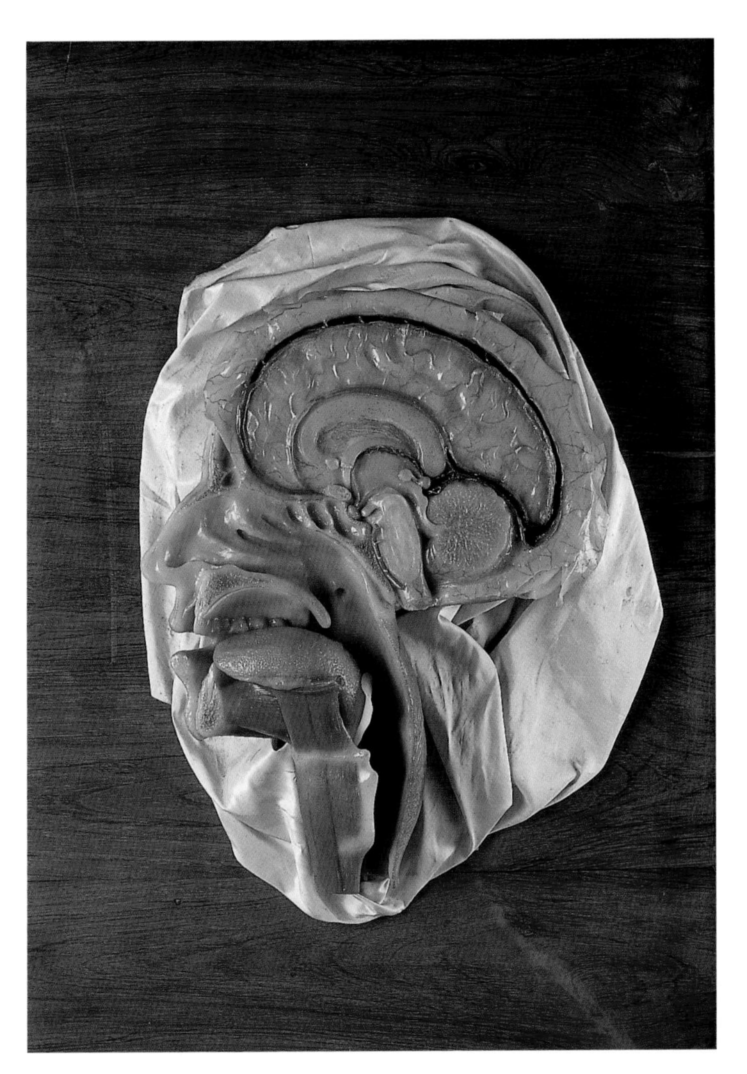

Specimen of a head which has been divided in the midline to show the relationship between cranial cavity and facial skeleton.

Präparat eines in der Mittellinie durchtrennten Kopfes, an dem die Lagebeziehungen von Hirn- und Gesichtsschädel deutlich werden.

Coupe médiane de la tête, rapports entre l'encéphale et le squelette du crâne et de la face

[XXVIII, 664]

Specimen showing the left half of the brain. The C-shaped structure below the gyri is the corpus callosum, which is the main collection of fibers connecting the two halves of the cerebrum.

Präparat einer linken Hirnhälfte. Das C-förmig gekrümmte Gebilde unterhalb der Großhirnwindungen ist der Balken, die mächtigste Nervenfaserverbindung zwischen beiden Großhirnhälften.

Hémisphère gauche du cerveau ; parmi les commissures interhémisphériques, le corps calleux en forme de C est la plus grosse

[XXVIII, 720]

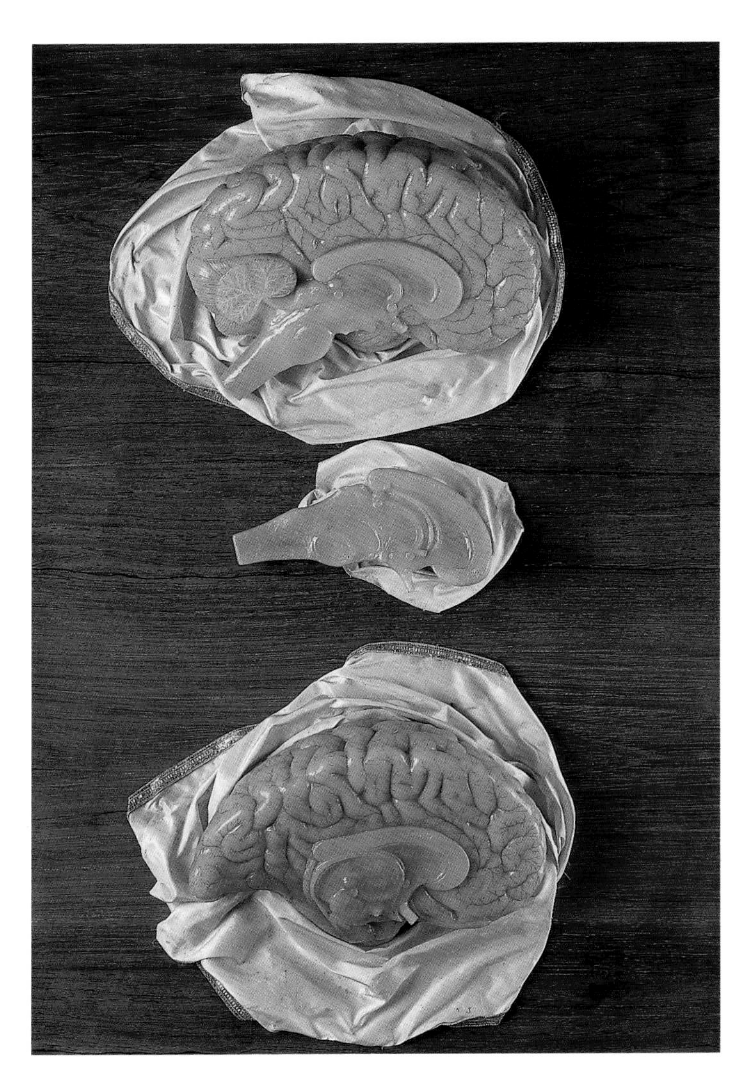

View of the left half of a brain which has been divided in the midline. In the lower specimen the brainstem and cerebellum have been removed.

Ansicht auf die linke Hirnhälfte nach Durchtrennung des Gehirns in der Mittelline. Im unteren Präparat sind Hirnstamm und Kleinhirn entfernt.

Coupes médianes de l'encéphale ; hémisphère gauche, ablation du tronc cérébral et du cervelet (en bas)

[XXVIII, 721]

Facies superior et inferior cerebri

View of the brain from above to show the right and left halves, and from below to show the brainstem and cerebellum

Ansicht des Gehirns von oben mit rechter und linker Hirnhälfte und von unten mit Hirnstamm und Kleinhirn

Cerveau vu du dessus (en haut), et du dessous avec le cervelet et le tronc cérébral (en bas)

[XXVIII, 695]

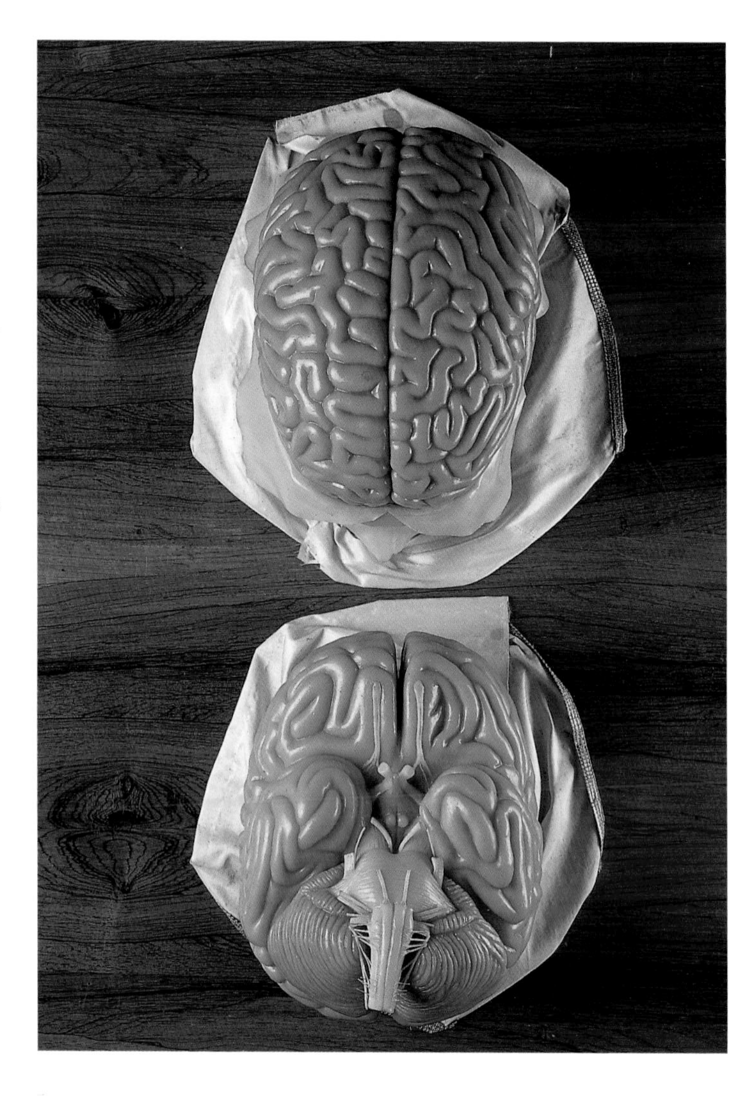

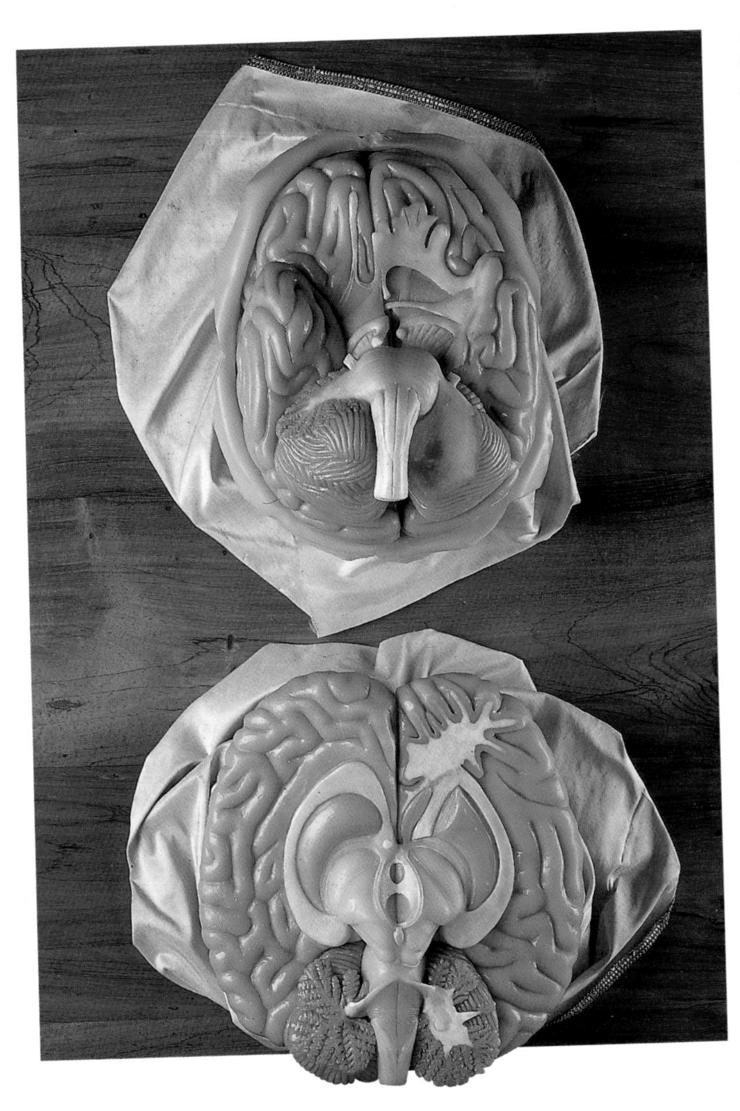

Specimens showing the white substance of the brain. In the lower specimen the halves of the brain have been separated to show the topography of the commissures in relation to the thalamus and caudate-putamen.

Präparate zur Darstellung von Nervenfaserverbindungen des Gehirns (weiße Substanz). Teile der grauen Substanz sind entfernt. Im unteren Präparat sind die Hirnhälften durchtrennt.

Préparation des faisceaux de fibres nerveuses (substance blanche) après ablation de zones de substance grise ; hémisphères séparés en bas

[XXVIII, 696]

Specimens to show the cavities of the brain (ventricles)

Präparate zur Darstellung der Hirnkammern

Ventricules cérébraux

[XXVIII, 694]

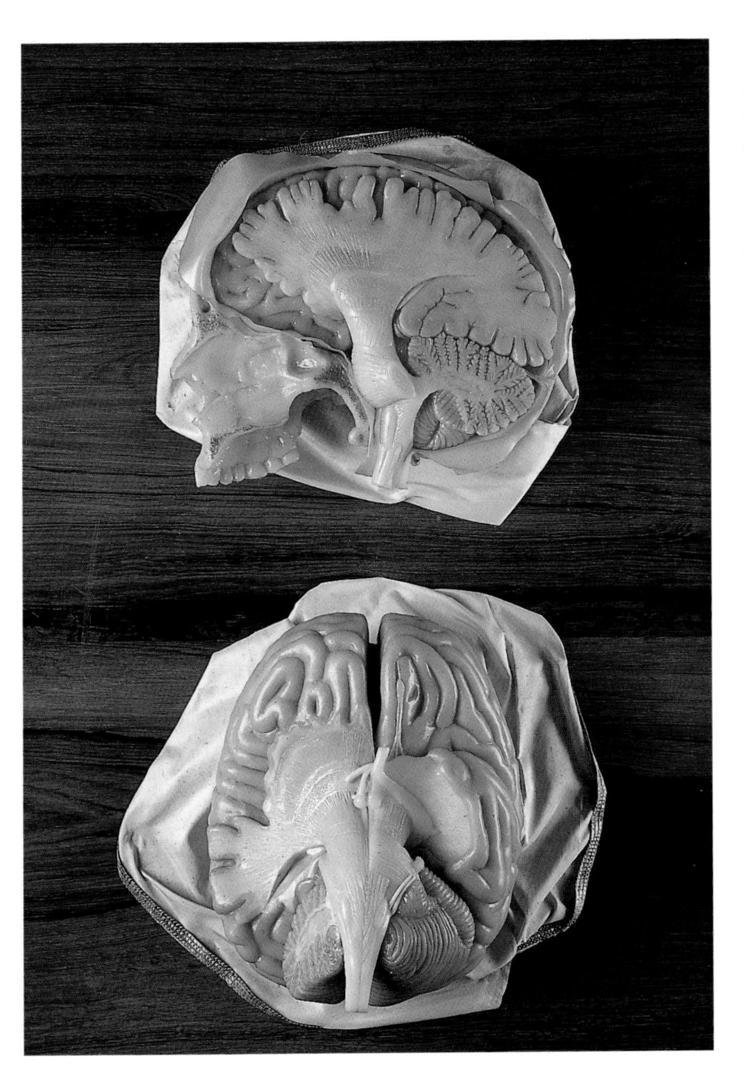

Display of the nerve fibers between the cerebrum, brainstem and spinal cord

Darstellungen von Nervenfaserverbindungen zwischen Großhirn und Hirnstamm sowie Rückenmark

Faisceaux de fibres nerveuses unissant le cerveau, le cervelet, le tronc cérébral, et la mœlle épinière

[XXVIII, 693]

Corpus callosum, Fornix, Septum pellucidum, Commissura anterior et posterior, Tentorium cerebelli

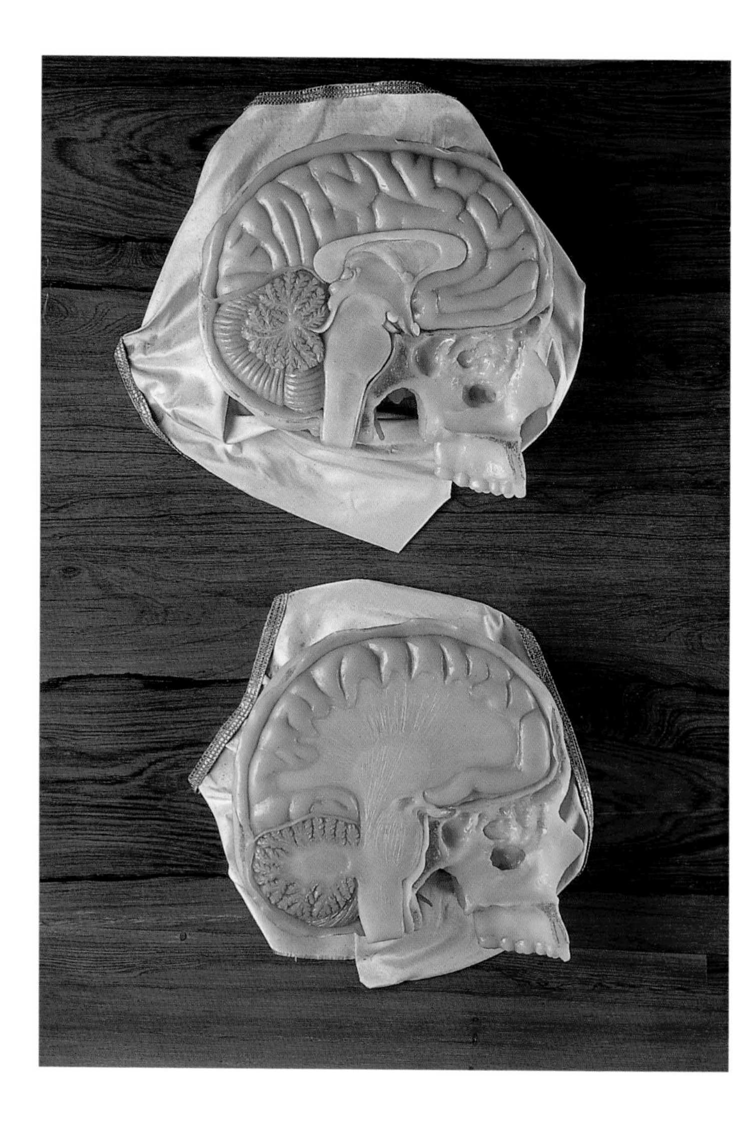

Median (above) and paramedian (below) sections through the head

Medianschnitt (oberes Präparat) und Paramedianschnitt (unteres Präparat) durch den Kopf

Encéphale dans la boîte crânienne : sections médiane (en haut) et paramédiane (en bas)

[XXVIII, 692]

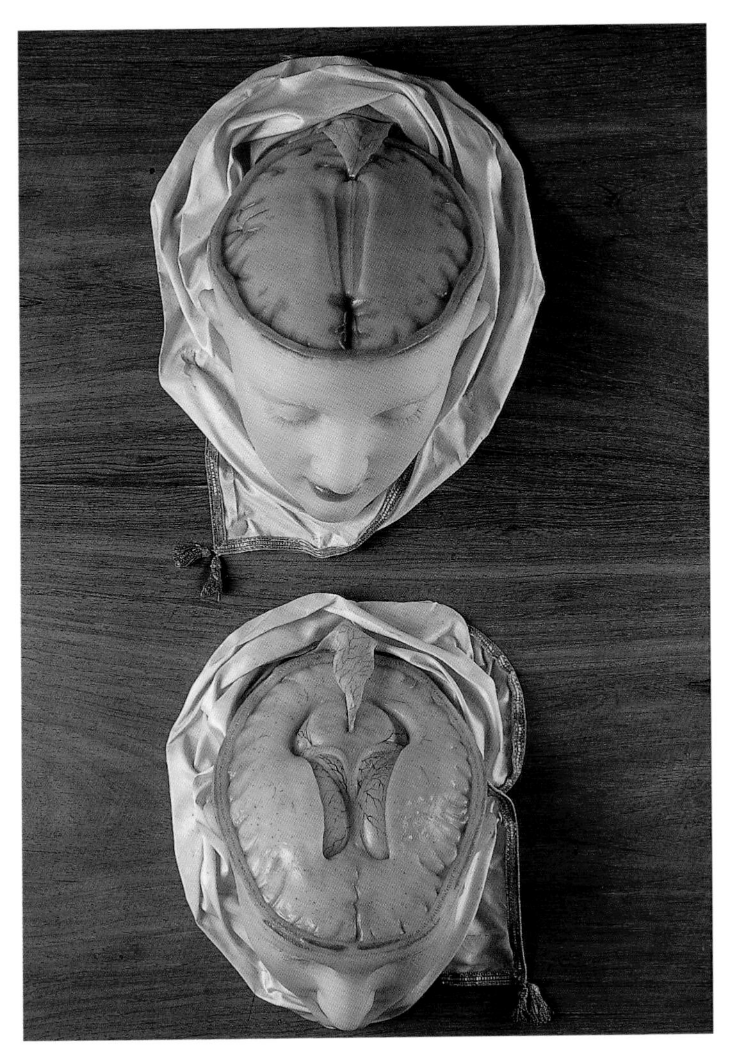

Horizontal sections through the skull a little above the corpus callosum (upper specimen) and a little below it (lower specimen). The lower section passes through the ventricles and the caudate nucleus.

Horizontalschnitte durch den Schädel etwas oberhalb des Balkens (oberes Präparat) und unterhalb des Balkens (unteres Präparat). Das untere Präparat zeigt Abschnitte der Hirnkammern und den Schweifkern.

Coupes horizontales du crâne et de l'encéphale : au dessus du corps calleux (en haut), et en dessous avec les ventricules cérébraux (en bas)

[XXVIII, 689]

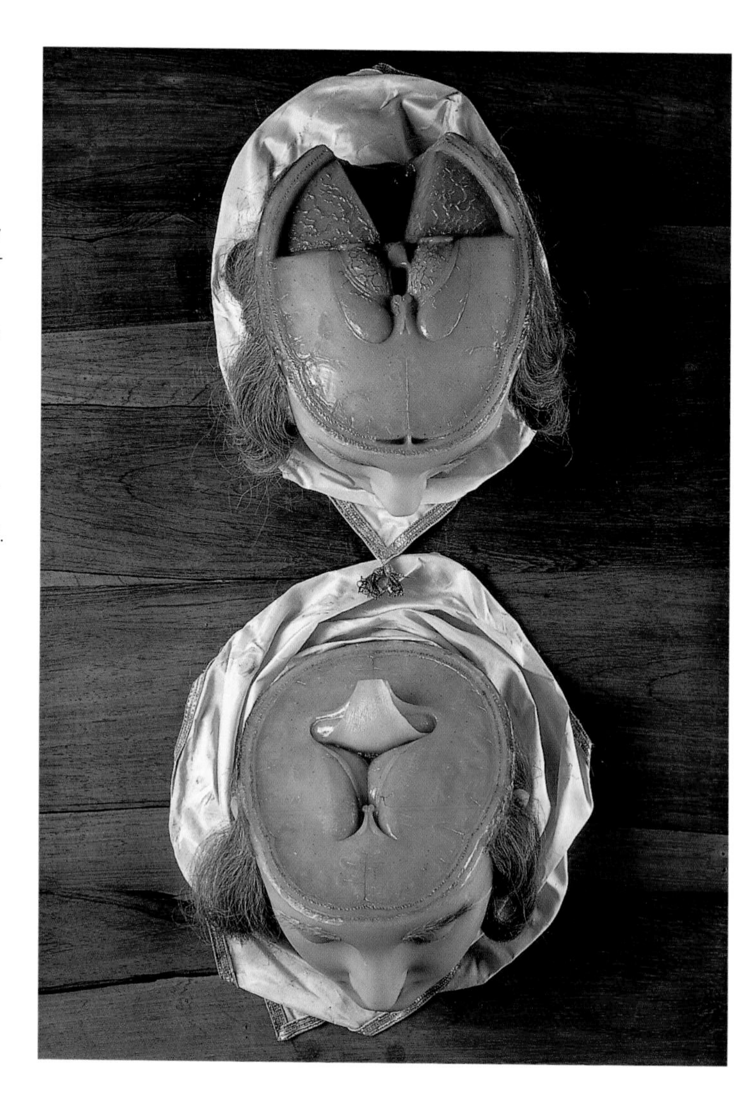

Horizontal sections at various levels through the skull. In the upper specimen one can see the caudate nucleus, and adjacent to and directly behind it, the thalamus.

Horizontalschnitte durch den Schädel auf verschiedenen Höhen, im oberen Präparat erkennt man den Schweifkern und nach hinten direkt angrenzend den Sehhügel.

Coupes horizontales du crâne et de l'encéphale : noyaux gris profonds

[XXVIII, 690]

Horizontal sections through the skull showing the caudate nucleus and thalamus. One can see sections passing through the ventricles.

Horizontalschnitte durch den Schädel mit Aufsicht auf den Schweifkern und den Thalamus. Man erkennt Abschnitte der Hirnkammern.

Coupes horizontales du crâne et de l'encéphale : noyaux gris profonds et ventricules cérébraux

[XXVIII, 688]

Horizontal section through the brain (frontal pole directed downwards). Caudate nucleus and internal cerebral vein seen from above

Horizontalschnitt durch das Gehirn (Stirnseite unten). Blick von oben auf den Schweifkern und die innere Hirnvene

Coupe horizontale du cerveau (lobe frontal vers le bas) : vue supérieure du noyau caudé et des veines profondes du cerveau

[XXVIII, 674]

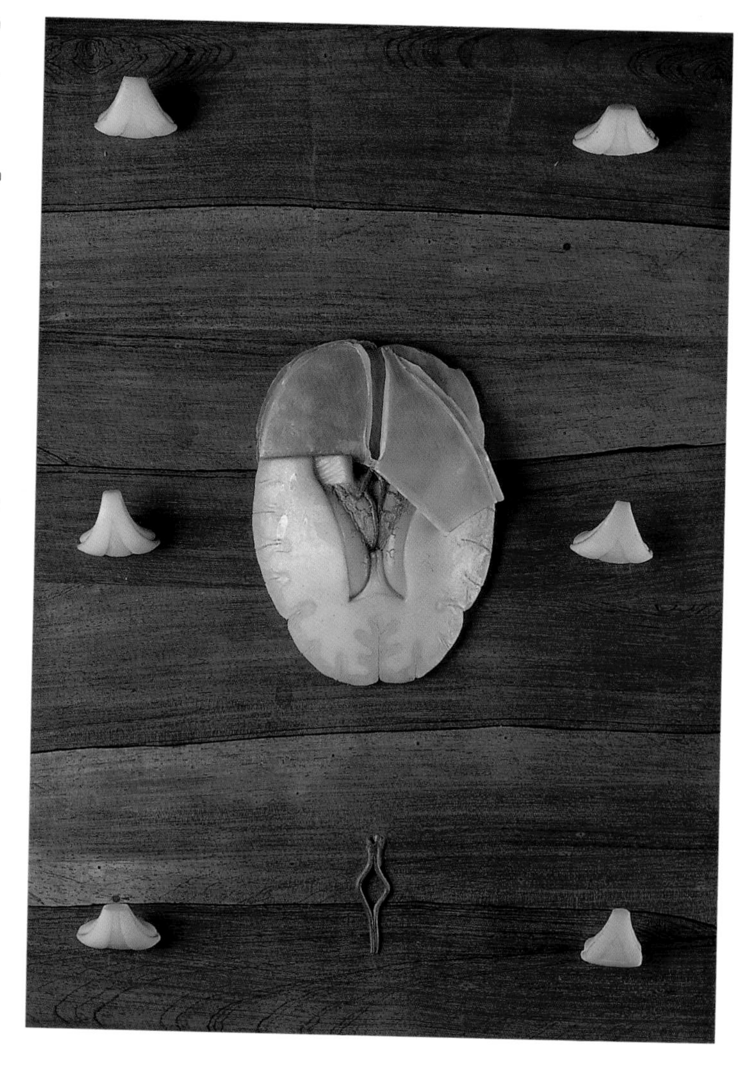

Upper specimen: hippocampus. The lower specimen shows the centrally placed cerebral nuclei such as the caudate nucleus and thalamus

Oberes Präparat: Darstellung des Hippokampus. Unteres Präparat: Darstellung zentral gelegener Kerngebiete des Gehirns wie Schweifkern und Thalamus

En haut : hippocampe disséqué. En bas : cervelet et noyaux gris profonds du cerveau : thalamus et noyau caudé

[XXVIII, 685]

Various views of the hippocampus

Verschiedene Darstellungen des Hippocampus

Ventricule latéral et hippocampe disséqués

[XXVIII, 683]

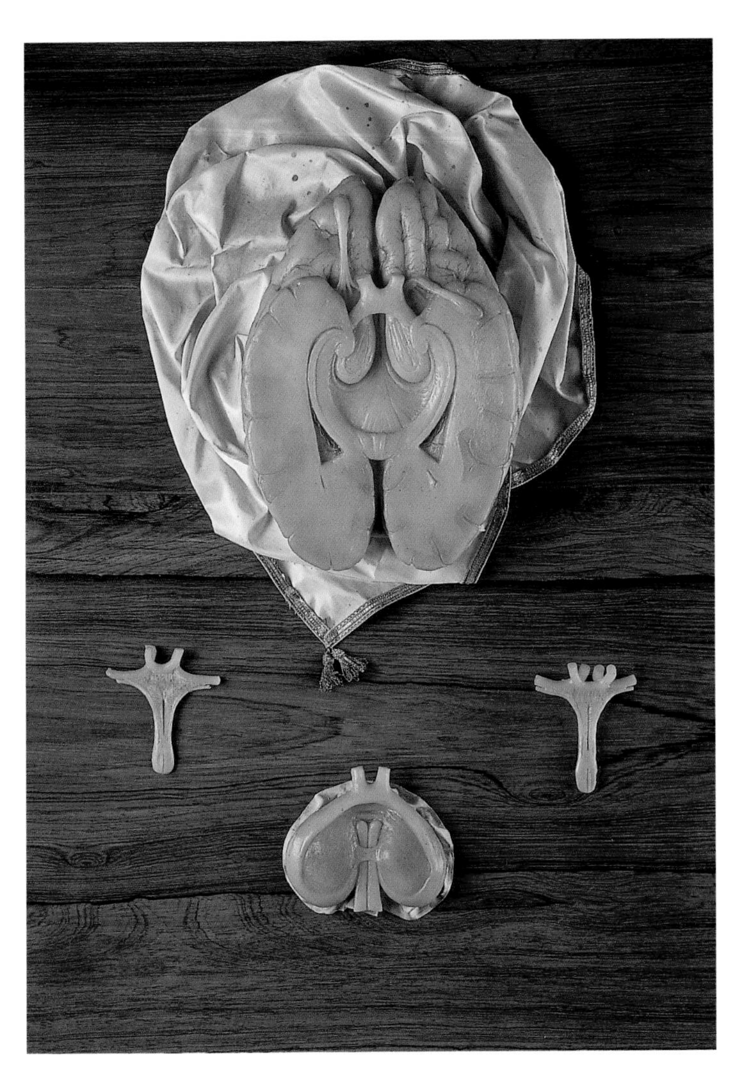

Specimens showing the connections (nerve fibers) of the hippocampus

Präparate zur Darstellung der Nervenfaserverbindungen des Hippocampus

Connexions nerveuses de l'hippocampe

[XXVIII, 729]

Nucleus caudatus, Thalamus, Glandula pinealis, Lamina quadrigemina, Cornu ammonis

Specimens from various regions of the brain. Among other things the thalamus, caudate nucleus and hippocampus can be seen.

Präparate aus verschiedenen Hirnregionen. Dargestellt sind u. a. Thalamus, Schweifkern und Hippocampus.

Diverses régions de l'encéphale : thalamus, noyau caudé, hippocampe, tronc cérébral

[XXVIII, 673]

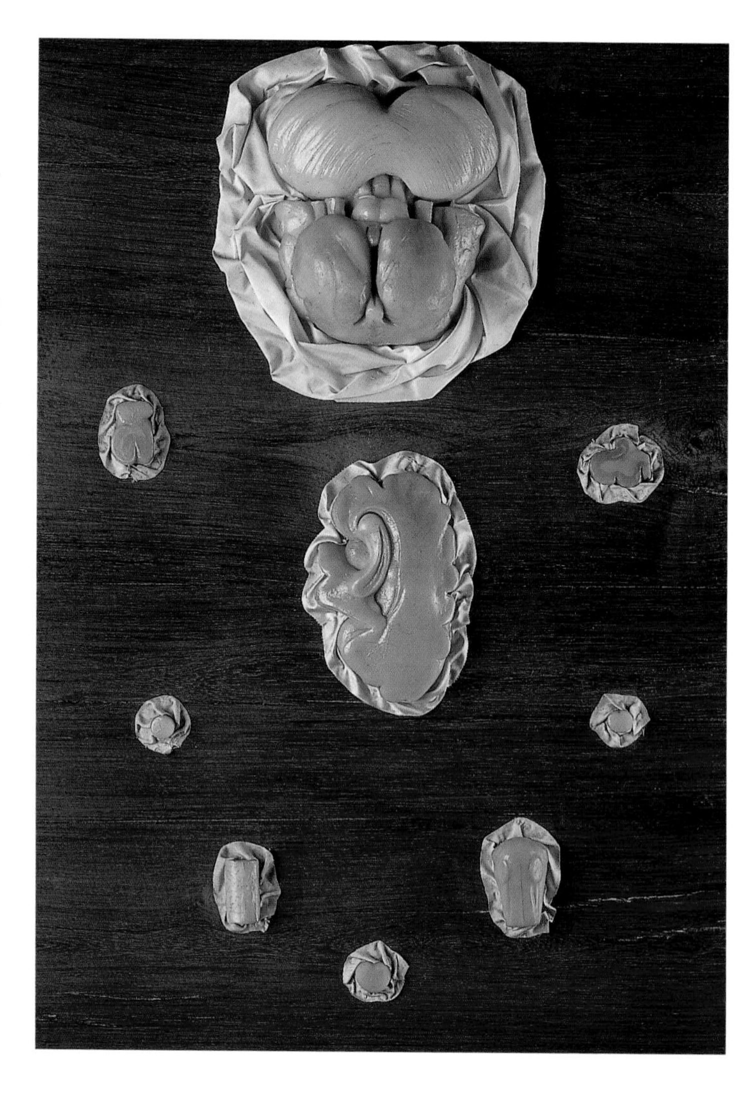

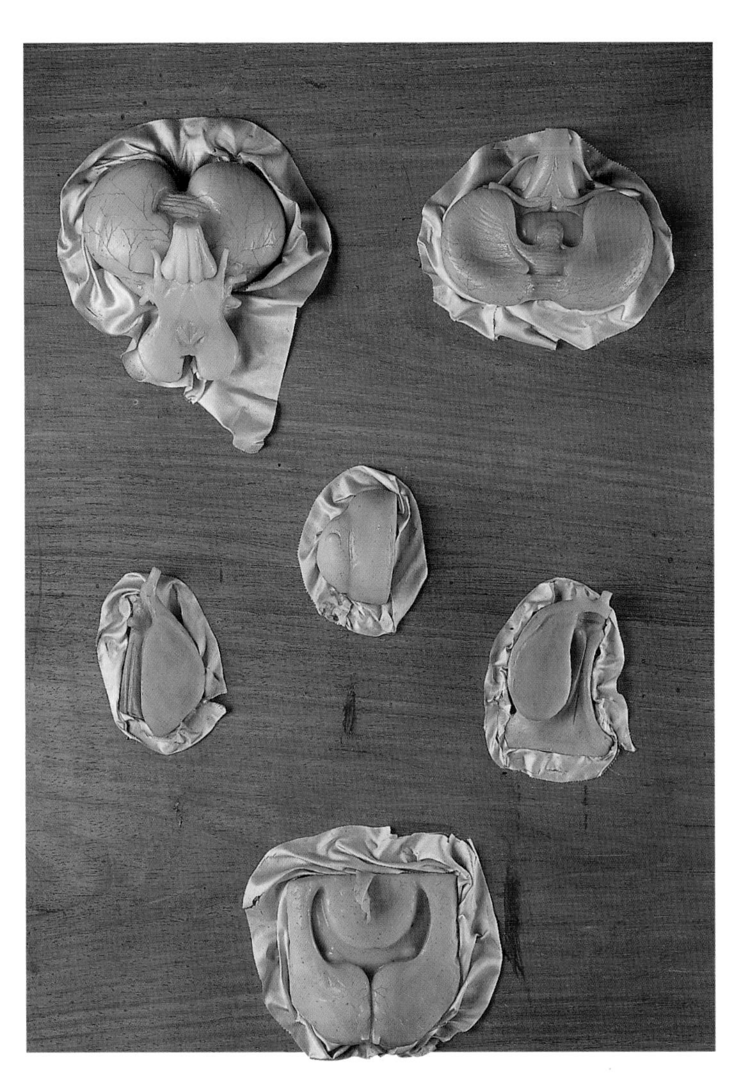

Specimens from various regions of the brain

Präparate aus verschiedenen Hirnregionen

Dissections de diverses régions de l'encéphale

[XXVIII, 681]

Cerebellum, Arbor vitae

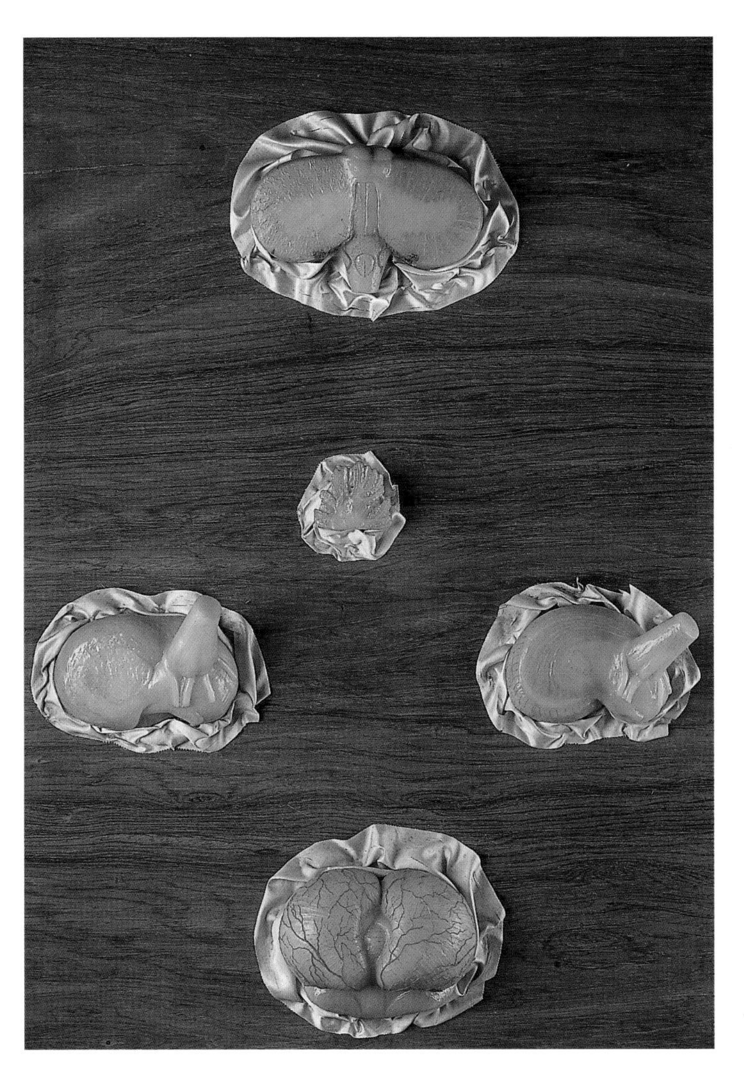

Different views of the cerebellum

Verschiedene Ansichten des Kleinhirns

Cervelet

[XXVIII, 678]

◄
Several specimens of the cerebellum

Verschiedene Präparate des Kleinhirns

Cervelet : coupes sagittales

[XXVIII, 677]

View of the inner ear showing the bony labyrinth (cochlea and three semicircular canals, above left). The other specimens represent the brainstem with cranial nerves and the cerebellum.

Darstellung des Innenohres mit der knöchernen Schnecke und den drei knöchernen Bogengängen (oben links). Die anderen Präparate zeigen den Hirnstamm mit Hirnnerven und Kleinhirn.

Oreille interne : cochlée et canaux semi-circulaires (en haut à gauche) ; sur les autres préparations : tronc cérébral, émergence des nerfs crâniens, cervelet

[XXVIII, 765]

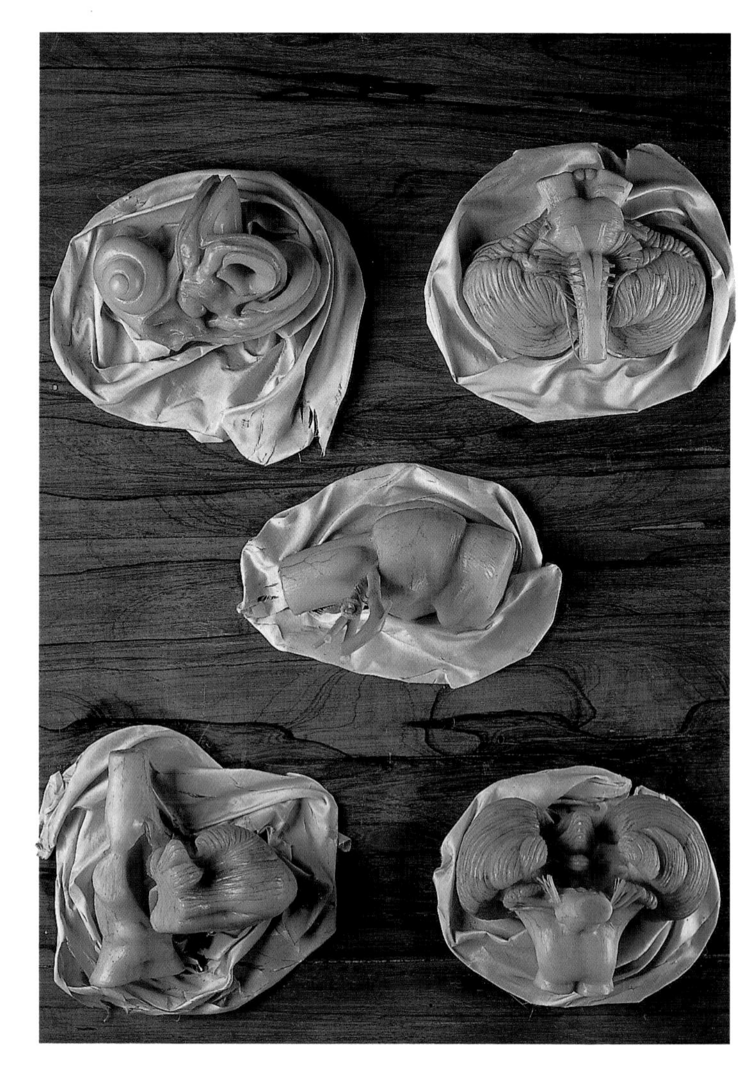

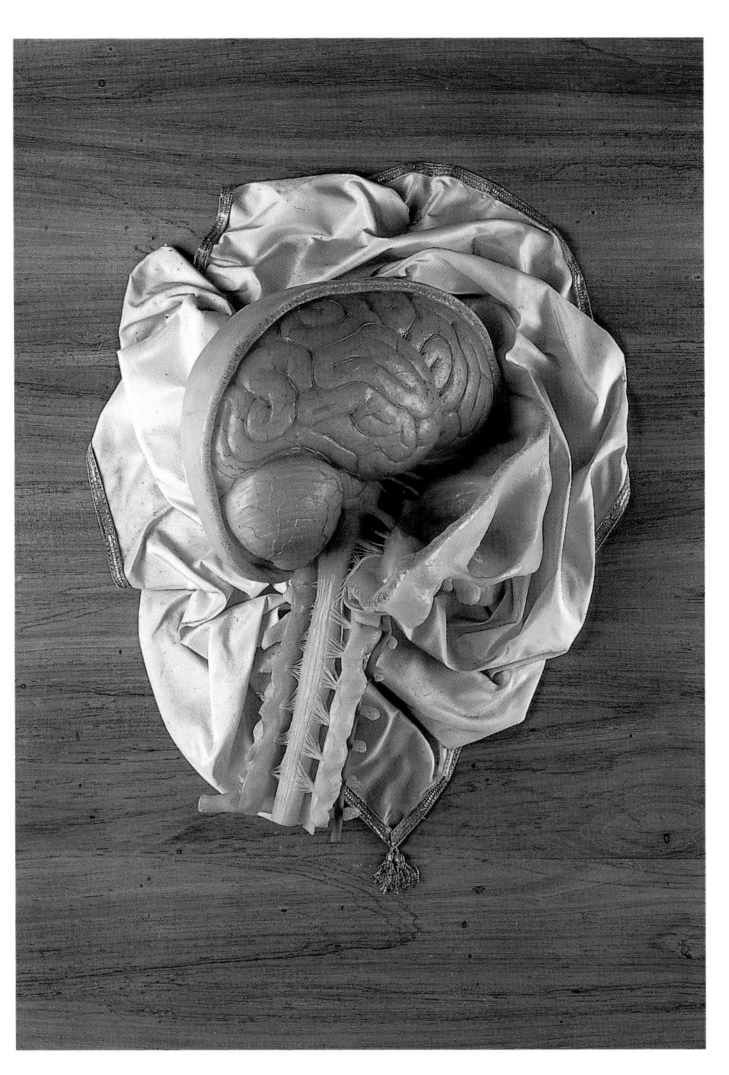

Oblique view from behind (the skull and vertebral canal have been laid open). The cerebrum and cerebellum have been lifted up to reveal the cranial nerves emerging from the brainstem.

Ansicht von schräg hinten in den eröffneten Schädel und den Wirbelkanal. Großhirn und Kleinhirn sind angehoben, um den Austritt der Hirnnerven aus dem Hirnstamm sichtbar zu machen.

Trajets intracrâniens des nerfs crâniens visibles après ouverture du crâne et du canal vertébral et après déplacement du cerveau et du cervelet (vue de l'arrière et de la droite)

[XXVIII, 728]

Various displays and views of the brainstem

Verschiedene Darstellungen und Ansichten des Hirnstammes

Tronc cérébral : coupes diverses

[XXVIII, 726]

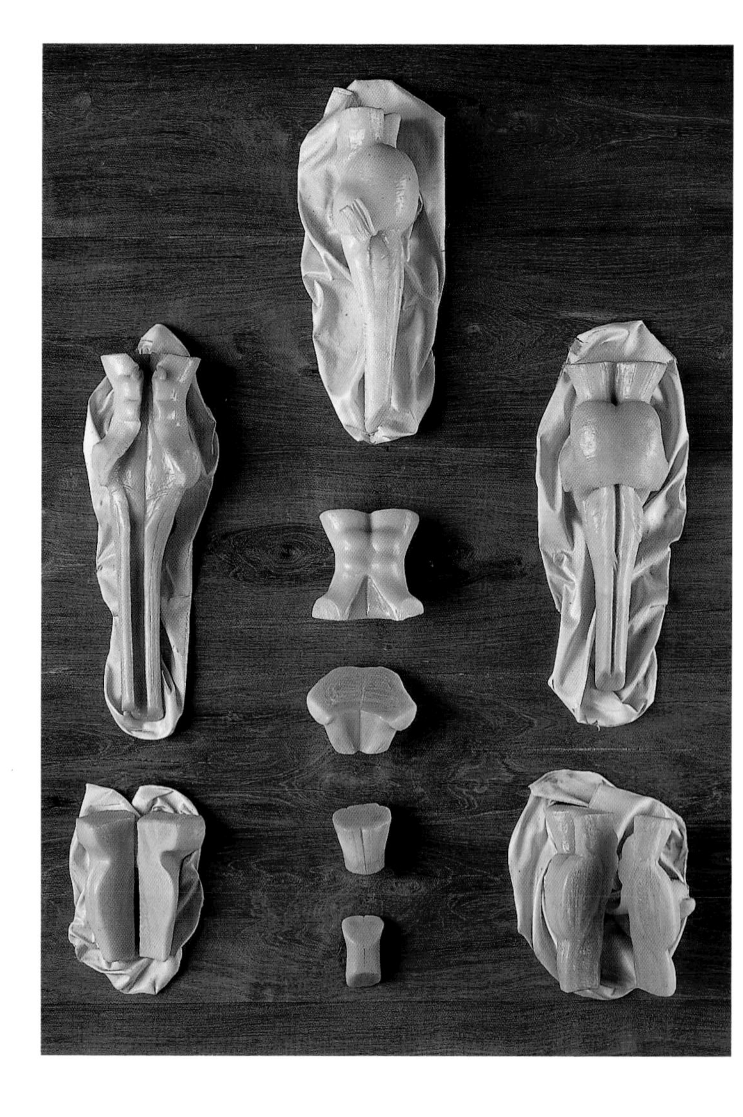

View of the spinal cord to show the spinal nerve roots emerging

Darstellung des Rückenmarks mit austretenden Nervenwurzeln

Mœlle épinière et racines des nerfs spinaux (rachidiens)

[XXIX, 736]

Medulla spinalis, Arteriae radiculares

The vertebral canal has been laid open from behind to show the spinal cord, the spinal nerve roots and the supplying blood vessels

Einblick in den eröffneten Wirbelkanal von hinten mit Rückenmark, austretenden Nervenwurzeln sowie versorgenden Blutgefäßen

Artères de la mœlle épinière et des nerfs spinaux (rachidiens) ; canal vertébral ouvert vu de l'arrière

[XXIX, 737]

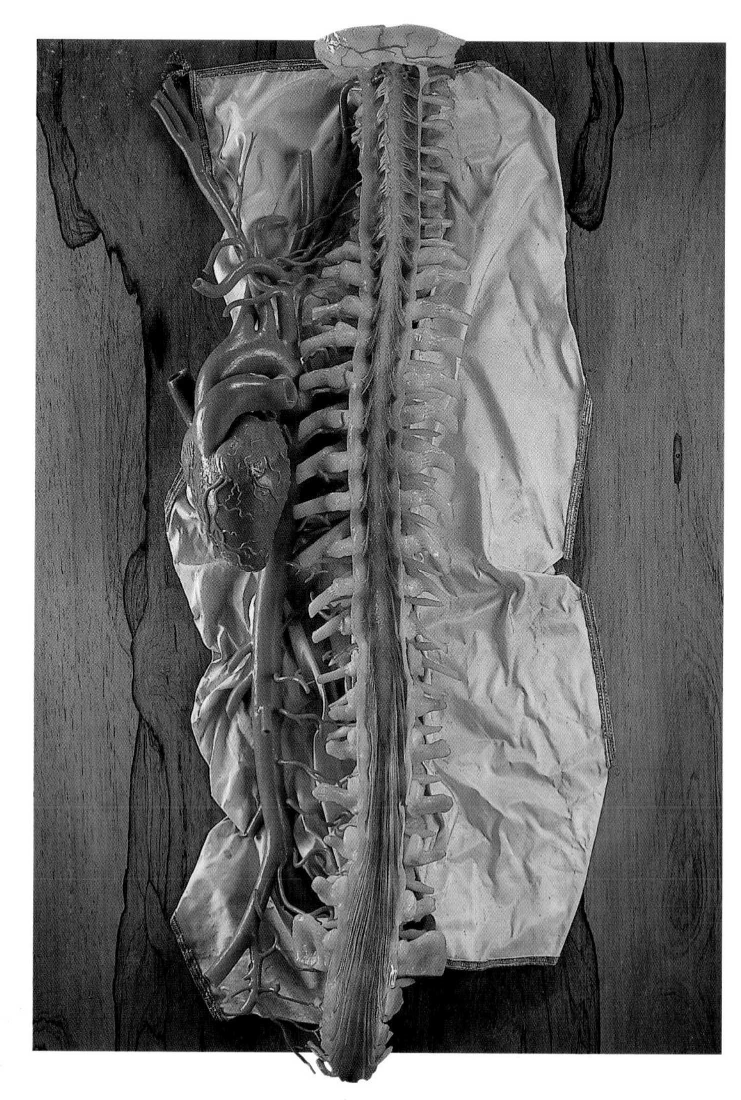

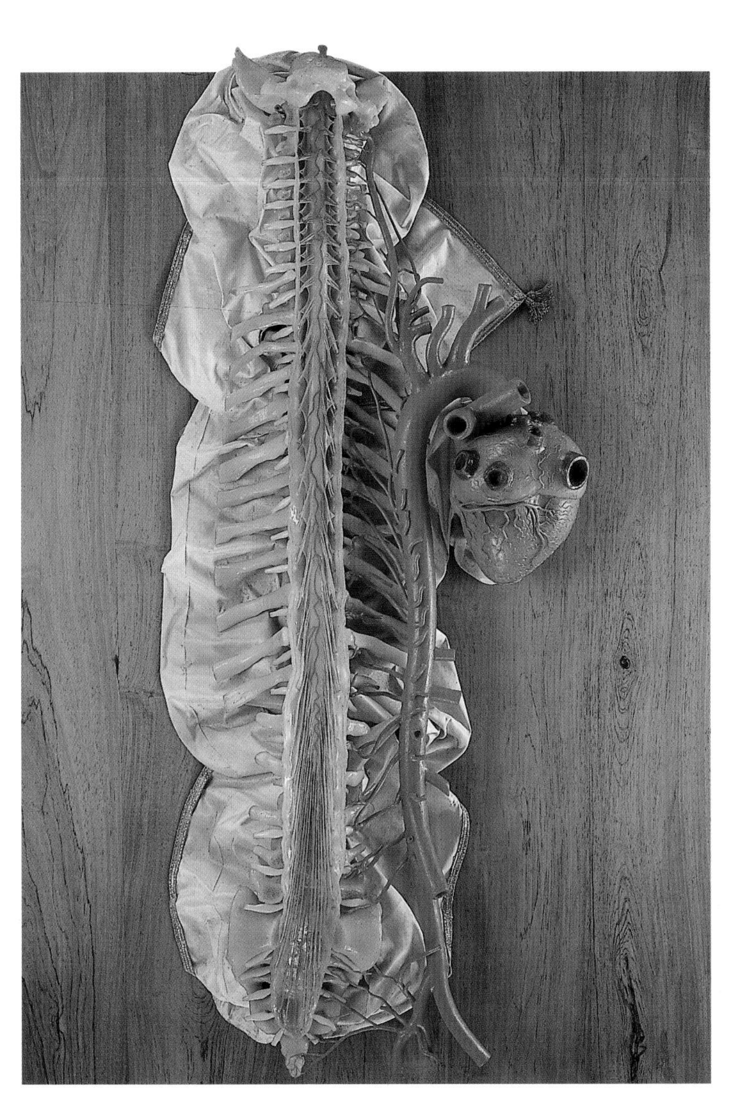

The vertebral canal has been laid open from in front to show the spinal cord, emerging spinal nerve roots and blood vessels

Einblick in den eröffneten Wirbelkanal von vorn mit Rückenmark, austretenden Nervenwurzeln sowie Blutgefäßen

Artères de la mœlle épinière et des nerfs spinaux ; canal vertébral ouvert vu de l'avant

[XXIX, 738]

Pages · Seiten · Pages 390–391:

Frontal view of the specimen from page 408

Frontalansicht des Präparates von Seite 408

Vue frontale de la préparation anatomique de la page 408

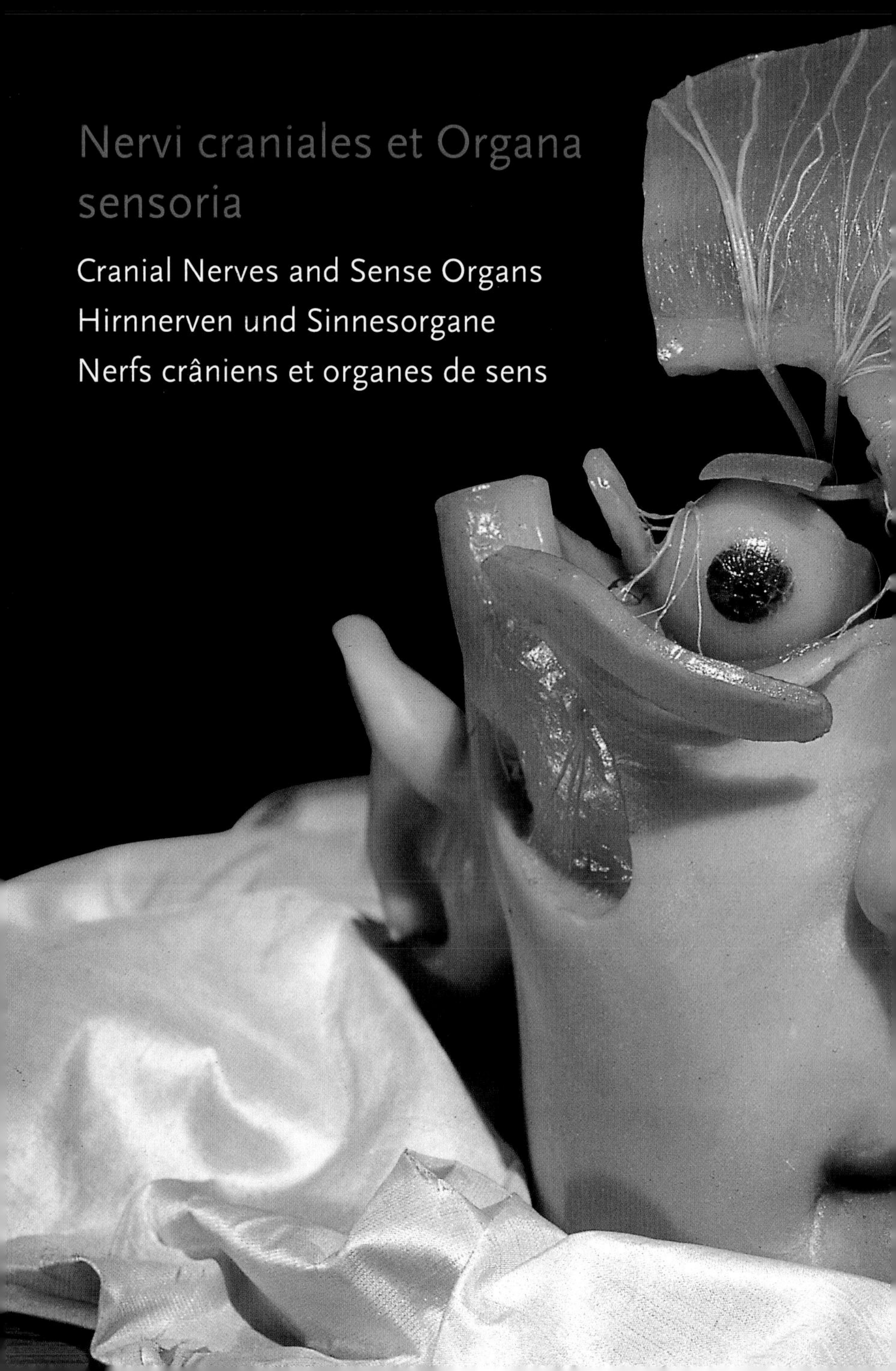

Nervi craniales et Organa sensoria

Cranial Nerves and Sense Organs

Hirnnerven und Sinnesorgane

Nerfs crâniens et organes de sens

Cranial Nerves and Sense Organs

The sense organs responsible for our perception of stimuli awakened the interest of anatomists very early on, since such special senses as hearing, sight, smell and taste have naturally always been relevant to our conscious experience.

Whereas many of the models show the detailed anatomy of the individual sense organs, other specimens display them in relationship to their central connections with the brain. This type of display is in accordance with the ideas of René Descartes (1596–1650), who interpreted the function of these organs in connection with the brain. It is indeed an essential part of cerebral function to analyze and interpret the particular stimuli which act upon our organs of special sense.

Hirnnerven und Sinnesorgane

Unsere für die Reizwahrnehmung verantwortlichen Sinnesorgane haben früh das Interesse der Anatomen geweckt, da Sinneswahrnehmungen wie Hören, Sehen, Riechen und Schmecken in unserem Bewußtsein natürlicherweise gegenwärtig sind.

Während zahlreiche Modelle im Detail die Anatomie der einzelnen Sinnesorgane dokumentieren, wird in anderen Präparaten das Sinnesorgan mit seiner nervalen Verbindung zum Gehirn gezeigt. Die Art der Darstellung lehnt sich an das Konzept von René Descartes (1596–1650) an, der die Funktionen der Sinnesorgane im Zusammenhang mit dem Gehirn interpretiert hatte. Dem Gehirn kommen dabei wesentliche Aufgaben bei der Verarbeitung der spezifischen Reize, die auf unsere Sinnesorgane einwirken, zu.

Nerfs crâniens et organes de sens

Les organes des sens, capteurs des stimulations périphériques, ont intéressé précocement les anatomistes car la perception de l'environnement par l'audition, la vue, l'odorat ou le goût est une des bases de la conscience.

L'anatomie des organes des sens isolés est démontrée en détail par de nombreux modèles en cire. Certains modèles montrent les connexions nerveuses des organes des sens avec le cerveau ; cette représentation s'appuie sur les conceptions de René Descartes (1596–1650) intégrant les fonctions des organes des sens à celles du cerveau.

▶
Detail from page | Detail von Seite | détail de la page 503

Bulbus olfactorius, Nervus opticus, Nervus oculomotorius, Nervus trochlearis, Nervus trigeminus, Nervus abducens

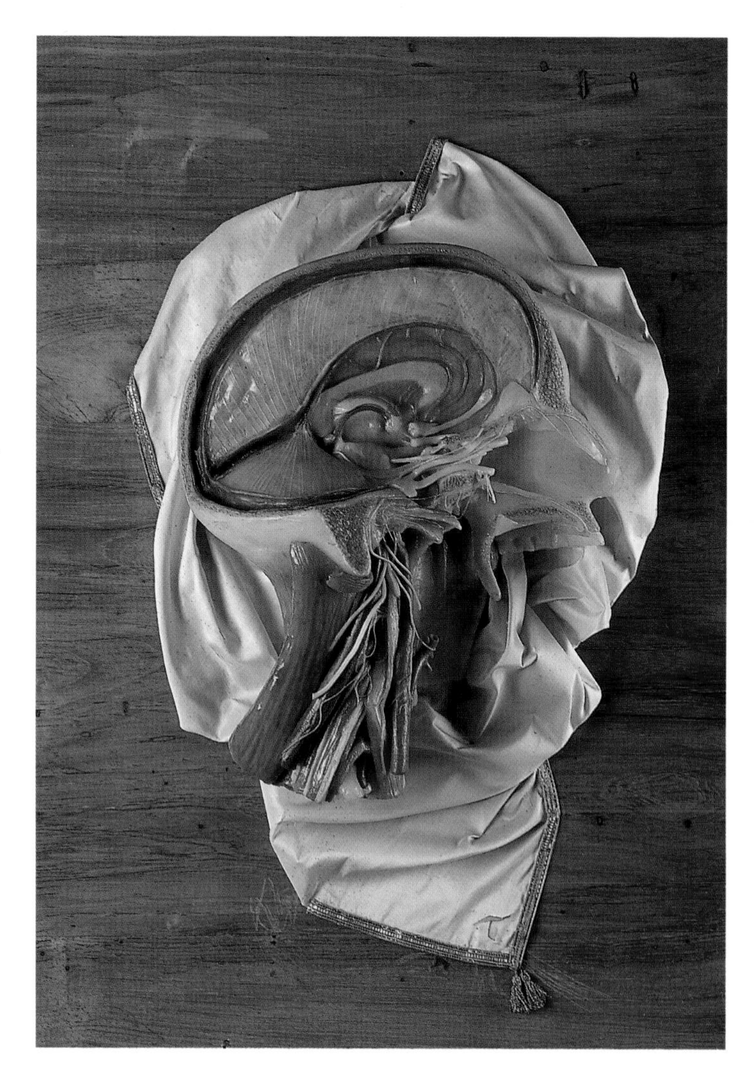

Side view of the skull laid open to display the individual cranial nerves. The bulbous thickening right at the top is part of the olfactory system. Below: the nerves supplying the eye muscles and branches of the trigeminal nerve

Ansicht von der Seite in den eröffneten Schädel. Darstellung einzelner Hirnnerven. Oben mit der kolbigen Verdickung ein Teil des Riechsystems, darunter Augenmuskelnerven und Äste des Nervus trigeminus

Nerfs crâniens vus à travers une large ouverture du crâne : système olfactif, nerfs de la motricité de l'œil, rameaux du nerf trijumeau

[XXVIII, 662]

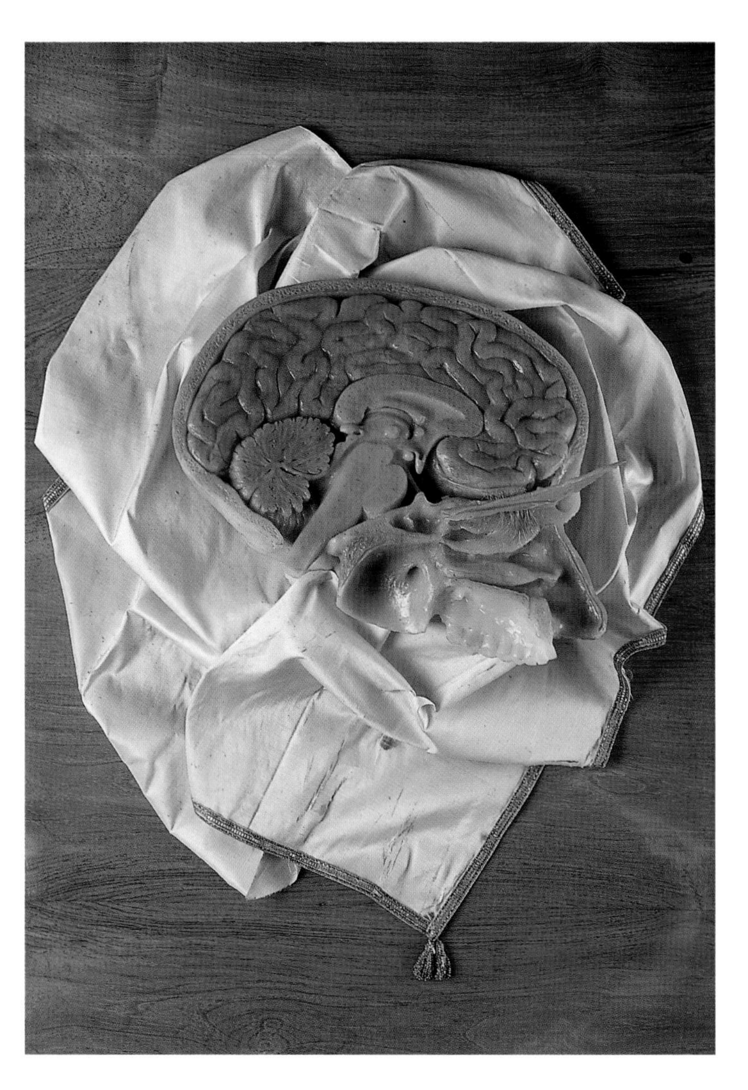

A head bisected in the midline to show the olfactory fibers

In der Mittelline durchtrennter Kopf mit Darstellung der Riechfasern

Coupe médiane du crâne : trajet des filets nerveux olfactifs

[XXVIII, 665]

The olfactory nerve fibers and the trigeminal nerve. The numerous olfactory fibers arise in the mucous membrane of the nose and enter the cranial cavity. Below: they are shown enlarged

Darstellung des Riechnerven und des Nervus trigeminus. Die zahlreichen Riechfasern, die aus der Nasenschleimhaut in den Schädel eintreten sind unten vergrößert dargestellt.

Filets nerveux olfactifs et nerf trijumeau (en haut) ; agrandissement des filets nerveux de l'olfaction issus de la muqueuse nasale et traversant la base du crâne (en bas)

[XXVIII, 663]

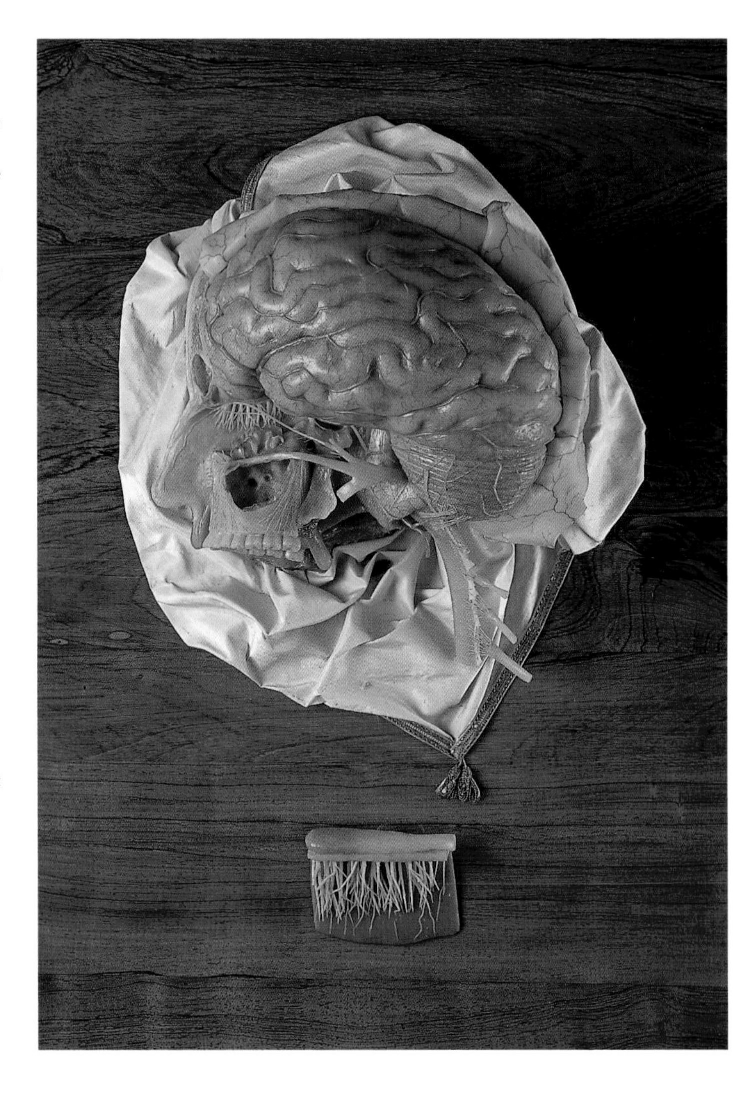

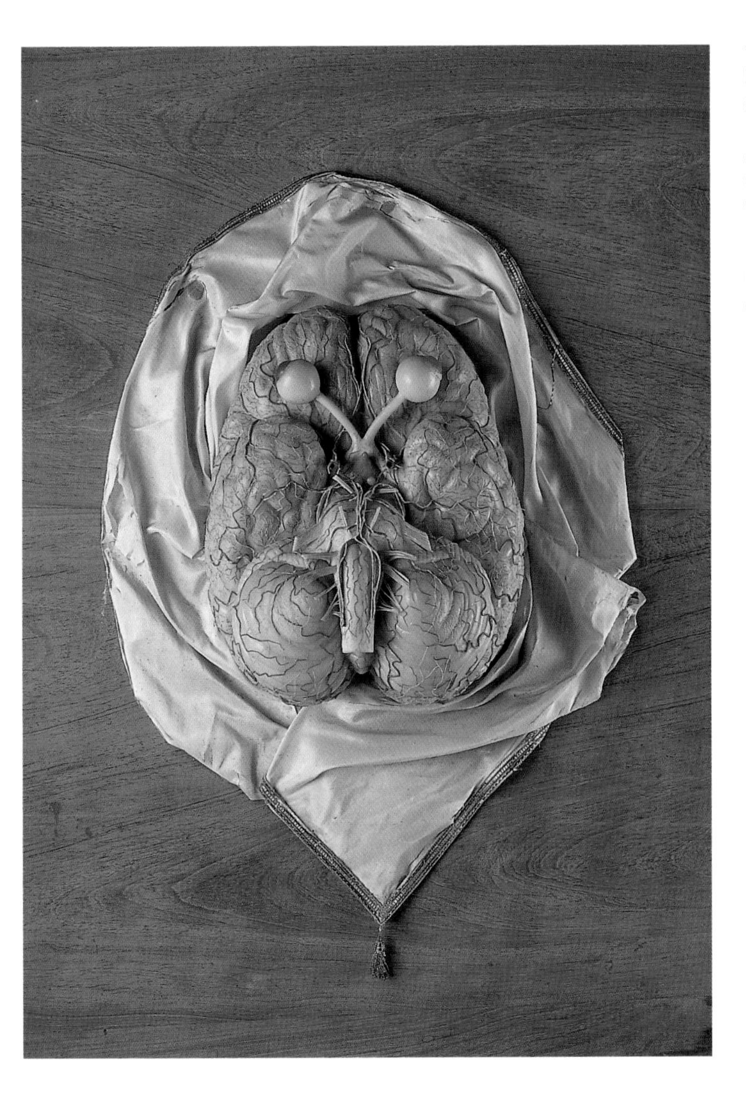

View showing the base of the brain and the optic nerve

Ansicht der Hirnbasis mit Darstellung des Sehnerven

Base de l'encéphale, nerfs optiques et globes oculaires

[XXVIII, 660]

Various dissections of the eyeball

Verschiedene Präparationen des Augapfels

Globes oculaires : rétine, iris, cristallin

[XXVIII, 708]

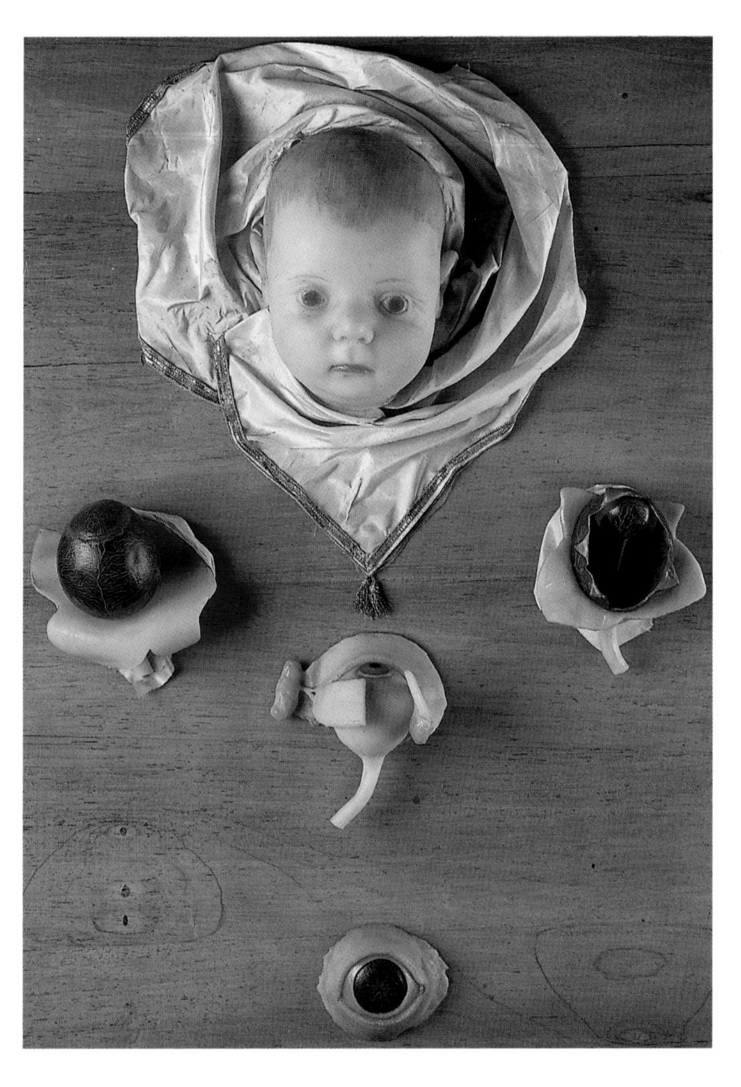

Various dissections of the eyeball of a fetus

Verschiedene Präparate von Augen eines Föten

Globe oculaire du fœtus

[XXVIII, 703]

Musculi bulbi oculi, Nervus opticus

Various presentations of the eyeball, optic nerve and external ocular muscles

Verschiedene Darstellungen von Augapfel, Sehnerv und äußeren Augenmuskeln

Globes oculaires avec nerf optique et muscles de la motricité de l'œil

[XXV, 449]

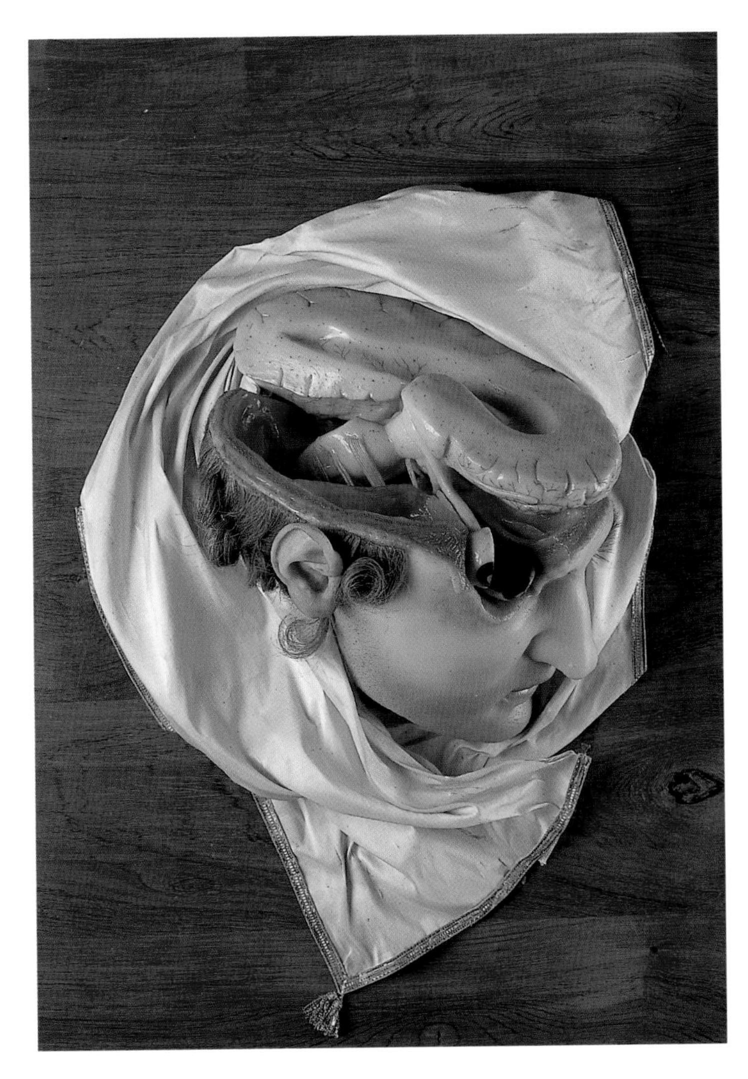

The course of the optic nerve (nerve of sight) from the eyeball to the brain

Darstellung des Sehnerven in seinem Verlauf vom Augapfel zum Gehirn

Nerf optique disséqué du globe oculaire à l'encéphale

[XXVIII, 707]

Specimens of the orbital cavity, some with the eye muscles, nerves and vessels. The upper specimen on the left shows the eyelids with the Meibomian glands (sebaceous glands).

Darstellungen der Augenhöhle teilweise mit Augenmuskeln, Nerven und Gefäßen. Im Präparat oben links sind die Augenlider mit den Meibom'schen Drüsen (Talgdrüsen) zu erkennen.

Cavités orbitaires : globe oculaire, muscles moteurs, nerfs, vaisseaux ; paupières et glandes tarsiennes (en haut à gauche)

[XXVIII, 706]

Various presentations of the eyeball and orbit. The lacrimal gland is lying against the eyeball (middle and upper specimens at the side).

Verschiedene Darstellungen von Augapfel und Augenhöhle; am Augapfel liegend die Tränendrüse (mittlere und oben seitliche Präparate)

Cavités orbitaires, globes oculaires, glandes lacrymales

[XXVIII, 705]

Nervus abducens, Arteria carotis interna

View of the brainstem and orbit. The brain has been lifted up and displaced to one side to show the course of one of the eye muscle nerves, the abducent nerve.

Blick auf Hirnstamm und Augenhöhle. Das Gehirn ist angehoben und zur Seite gedrängt, um den Verlauf eines Augenmuskelnerven, des Nervus abducens, zu demonstrieren.

Nerf abducens, un des nerfs moteur de l'œil : trajet de l'orbite au tronc cérébral (cerveau déplacé vers le haut et la gauche)

[XXVIII, 658]

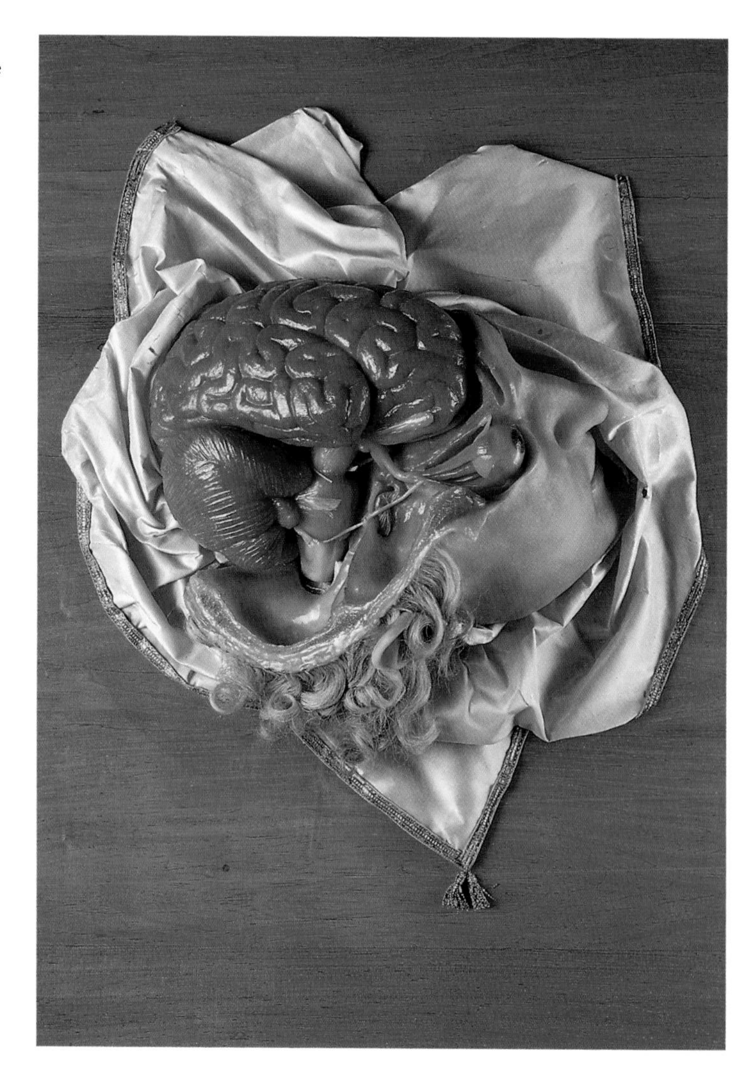

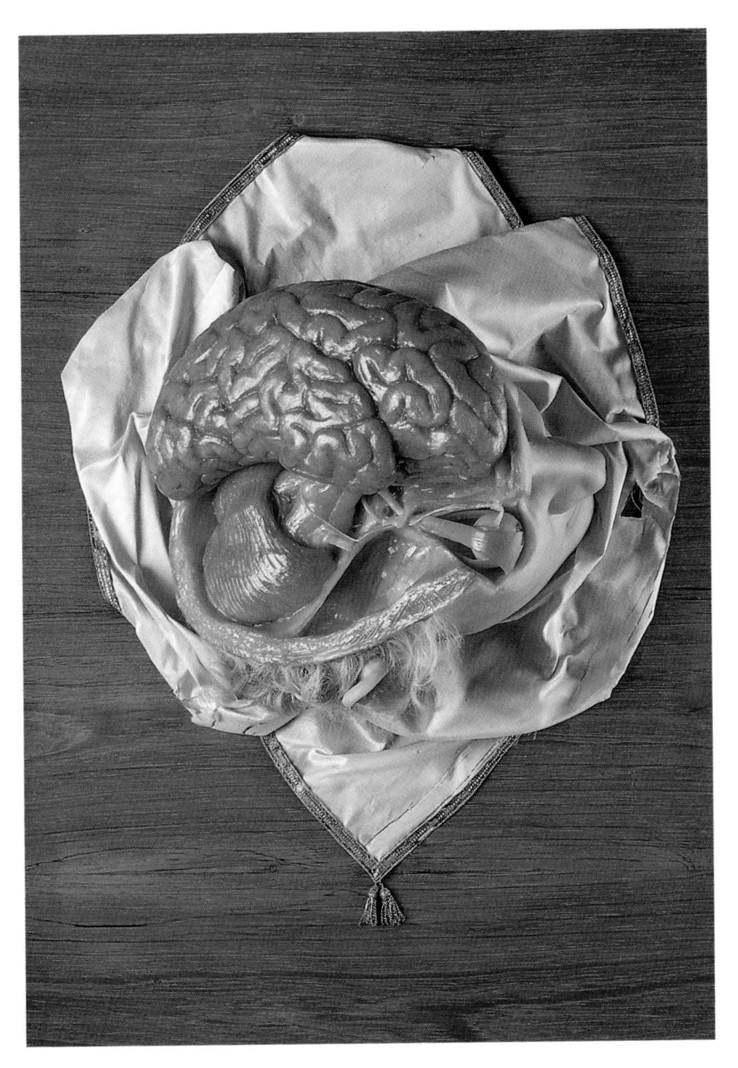

View of the brainstem and orbit. The brain has been lifted up and displaced to one side to show the course of one of the eye muscle nerves, the trochlear nerve.

Blick auf Hirnstamm und Augenhöhle. Das Großhirn ist angehoben und zur Seite gedrängt, um den Verlauf eines Augenmuskelnerven, des Nervus trochlearis, zu demonstrieren.

Nerf trochléaire, un des nerfs moteurs de l'œil : trajet de l'orbite au tronc cérébral (cerveau déplacé en haut et à gauche)

[XXVIII, 657]

View of the base of the brain showing the three subdivisions of the trigeminal nerve

Ansicht der Hirnbasis mit Darstellung der drei Äste des Trigeminusnerven

Base de l'encéphale, émergence des nerfs crâniens parmi lesquels le nerf trijumeau et ses trois branches

[XXVIII, 659]

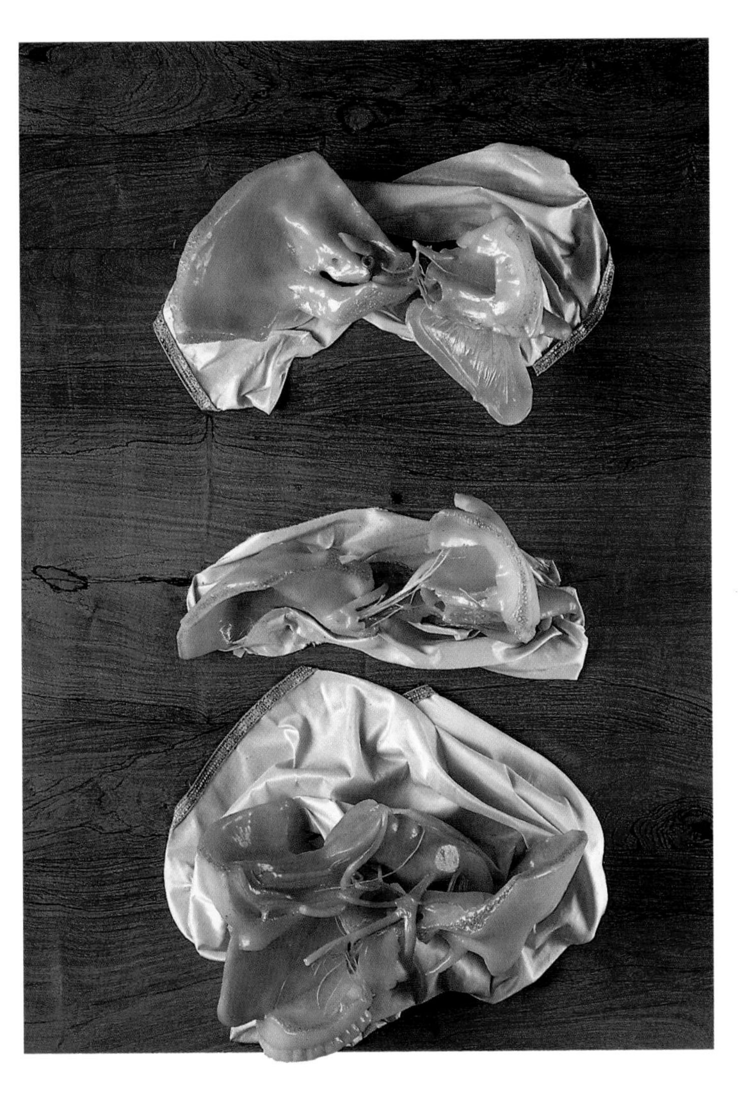

Display showing the trigeminal nerve and its branches

Darstellungen des Trigeminusnerven mit seinen Ästen

Nerf trijumeau et ses trois branches

[XXVIII, 661]

Display showing the first subdivision of the trigeminal nerve. A part of the frontalis muscle can be seen on the left, across which single branches of the trigeminal nerve are running.

Darstellung des ersten Astes des Nervus trigeminus. Links im Bild ist ein Teil des Stirnmuskels zu sehen, auf dem einzelne Trigeminusäste verlaufen.

Nerf trijumeau : branche supérieure ramifiée à la surface du muscle frontal

[XXVIII, 709]

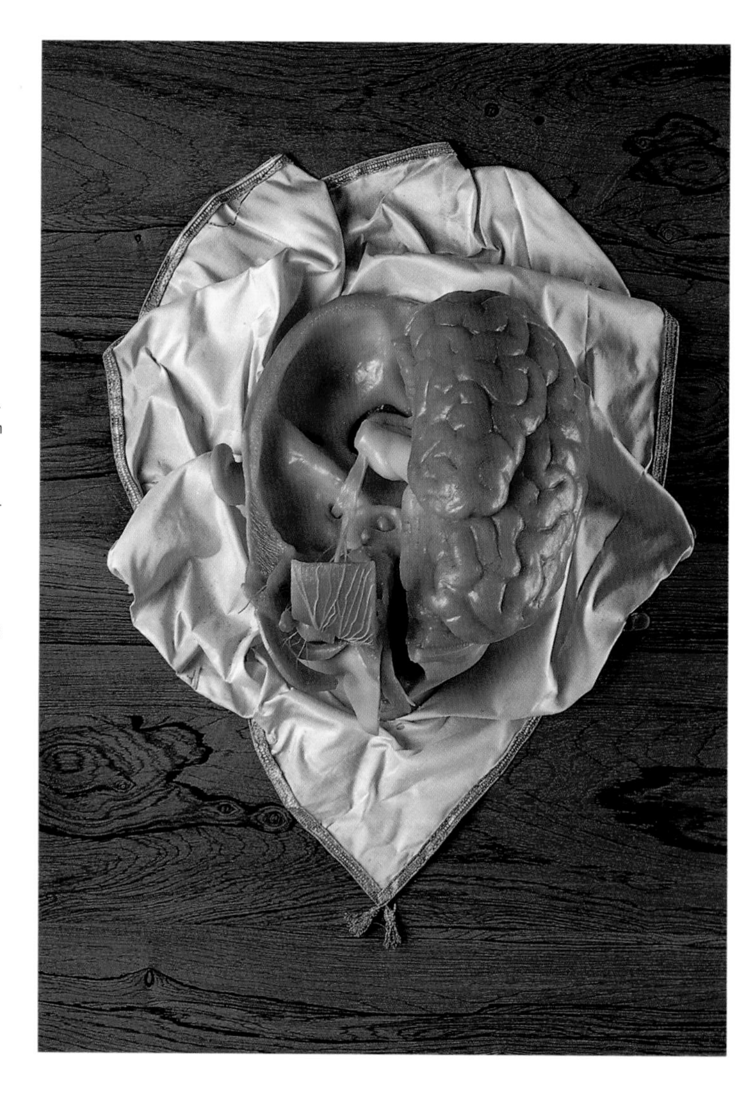

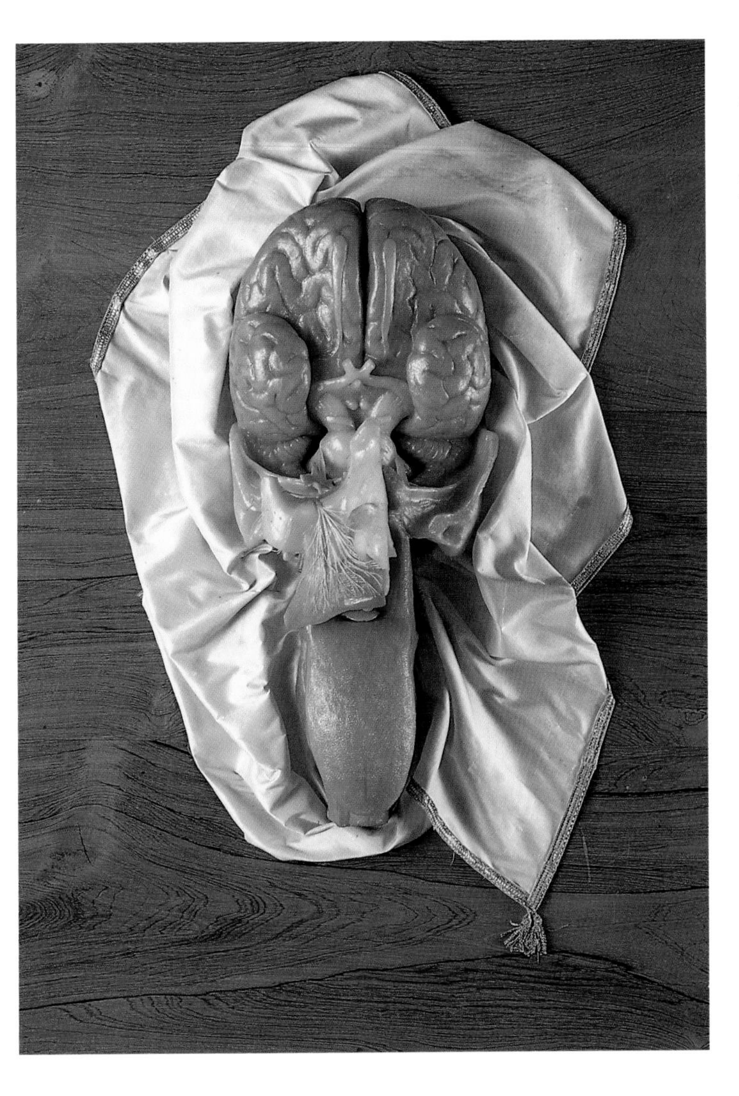

Display showing branches of the trigeminal nerve

Darstellung von Ästen des Nervus trigeminus

Nerf trijumeau : branche supérieure ramifiée dans la région du front

[XXVIII, 666]

Part of the mastica-tory muscles with their nerve supply

Teil der Kaumuskulatur mit nervöser Versorgung

Nerf trijumeau : branches pour les muscles masticateurs

[XXIX, 718]

►
Display showing the nerves and arteries of the cavity of the mouth, tongue and throat

Darstellung der Nerven und Arterien von Mundhöhle, Zunge und Rachen

Nerfs crâniens et artères pour les régions de la bouche, de la langue, et du pharynx

[XXVIII, 763]

Nervus lingualis, Nervus hypoglossus

One of the sensory nerves of the tongue

Darstellung eines sensiblen Nerven für die Zunge

Nerf trijumeau : branches pour la sensibilité de la langue

[XXVIII, 789]

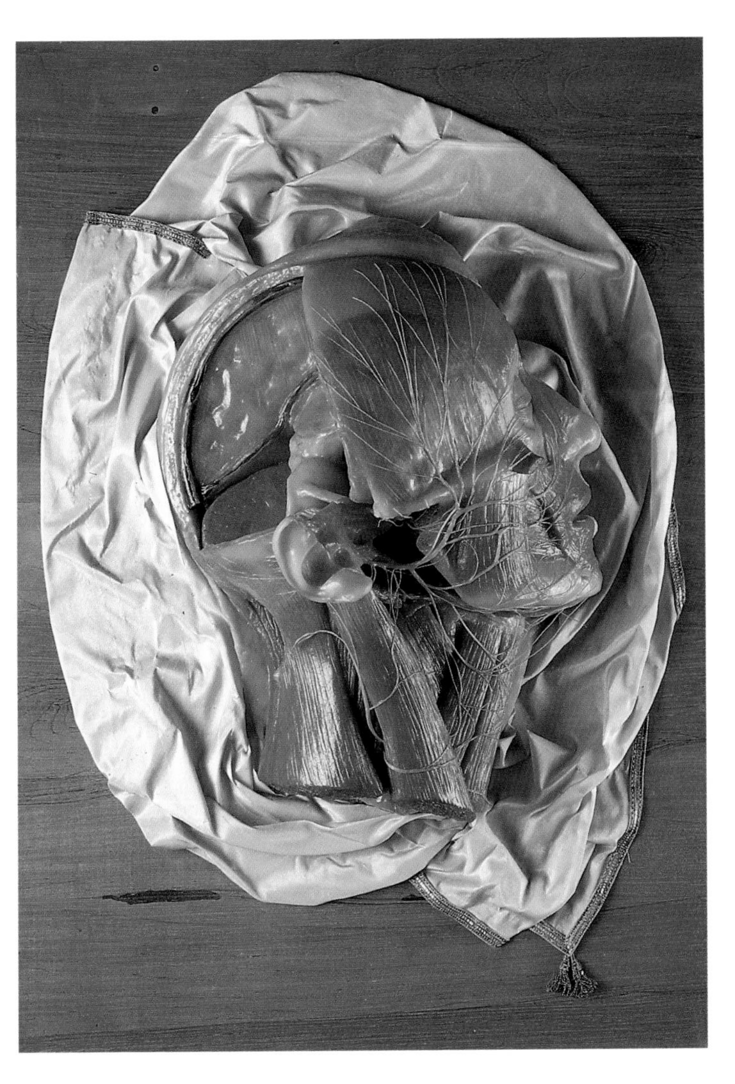

Display showing the facial nerve, which supplies the muscles of facial expression

Darstellung des Nervus facialis, der die mimische Muskulatur innerviert

Nerf facial : branches pour les muscles peauciers de la mimique

[XXVIII, 713]

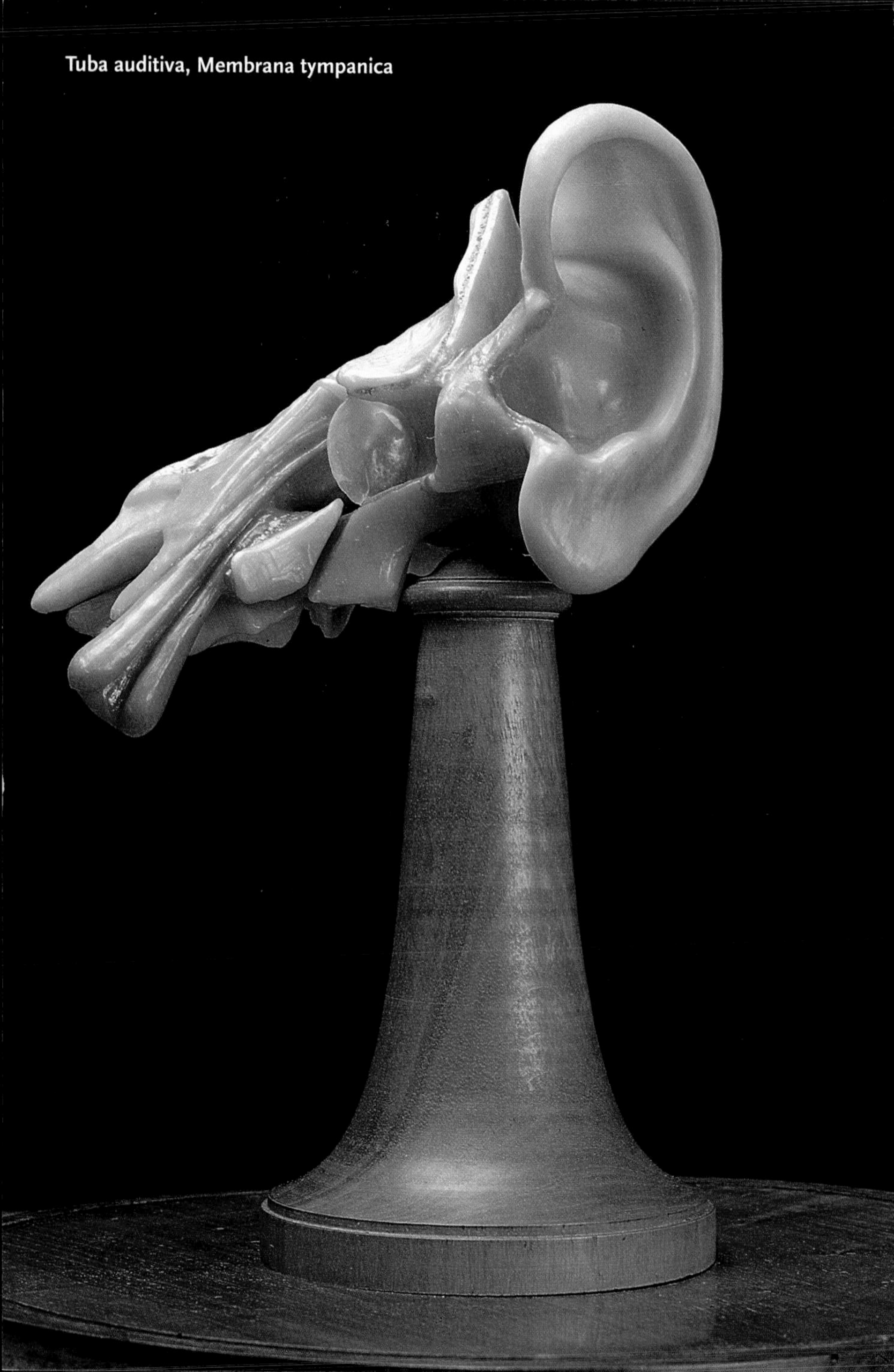
Tuba auditiva, Membrana tympanica

Various presentations of the inner ear with the cochlea and semicircular canals

Verschiedene Darstellungen des Innenohres mit knöcherner Schnecke und Bogengängen

Oreille interne : cochlée et canaux semi-circulaires

[XXVIII, 700]

◀

Specimen of the ear with the external auditory meatus, eardrum, middle ear and Eustachian tube

Präparat einer Ohrmuschel mit äußerem Gehörgang, Trommelfell, Mittelohr sowie der Ohrtrompete

Pavillon de l'oreille, conduit auditif externe, tympan, caisse du tympan, et trompe auditive (d'Eustache)

[XXVIII, 614]

Specimen of the petrous temporal bone partly chiseled away to show the organs of hearing and balance (greatly enlarged). Below on the right are the individual auditory ossicles.

Darstellung des aufgemeißelten Felsenbeines mit Gehör- und Gleichgewichtsorgan (stark vergrößert). Die Präparate unten rechts zeigen die einzelnen Gehörknöchelchen.

Organes de l'audition et de l'équilibration dans l'os temporal trépané fortement agrandis ; osselets de l'ouïe isolés (en bas à droite)

[XXVIII, 667]

The auditory ossicles and the muscles of the middle ear

Gehörknöchelchen und Muskeln des Mittelohres

Osselets de l'ouïe et muscles de l'oreille moyenne

[XXV, 499]

Display showing the organs of hearing and balance, and the motor nerve of the face passing through the petrous temporal bone

Darstellung von Gehör-, Gleichgewichts- und motorischem Gesichtsnerv im Felsenbein

Nerf de l'audition et de l'équilibration, nerf de la mimique (facial) dans l'os temporal

[XXVIII, 699]

Greatly enlarged specimen of the inner ear showing the semicircular canals and the convolutions of the cochlea

Darstellung des stark vergrößerten Innenohres mit Bogengängen und Windungen der Schnecke

Oreilles internes fortement agrandies : canaux semicirculaires, tours de spire de la cochlée

[XXVIII, 697]

Greatly enlarged specimen of the inner ear showing the semicircular canals and the convolutions of the cochlea

Darstellung des stark vergrößerten Innenohres mit Bogengängen und Windungen der Schnecke

Oreilles internes fortement agrandies : cochlée et canaux semi-circulaires ouverts (en haut)

[XXVIII, 698]

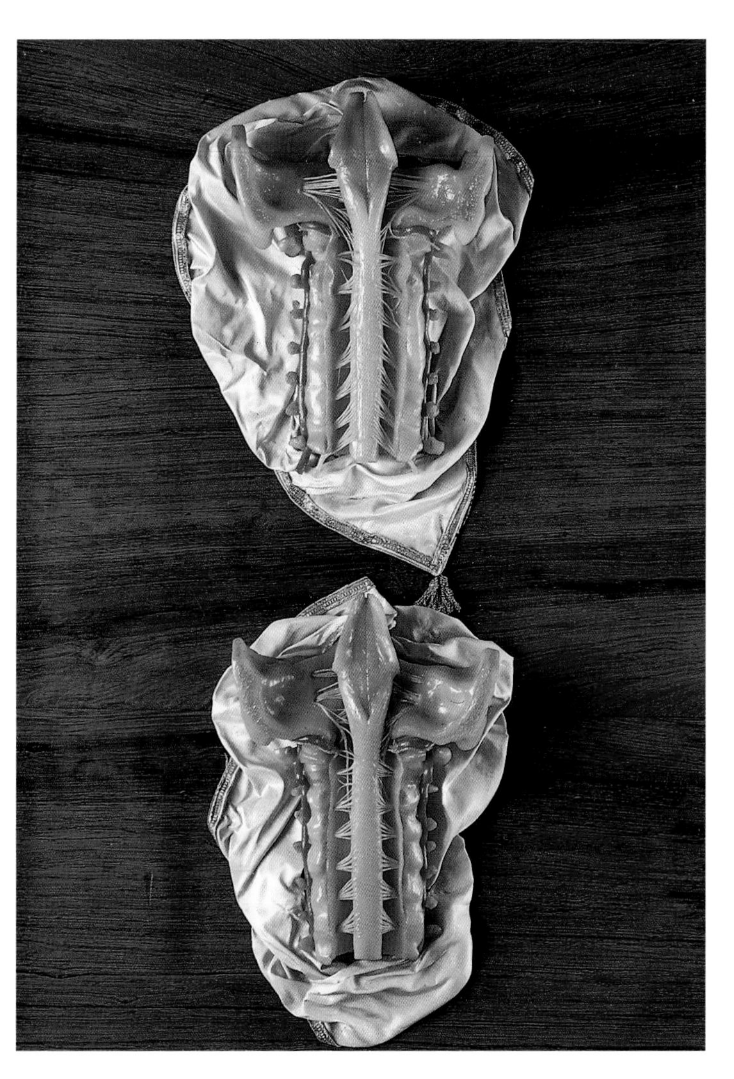

Specimen showing the brainstem and upper part of the spinal cord with the attached spinal nerves roots; seen from behind

Präparat des Hirnstammes und des oberen Rückenmarks mit austretenden Nervenwurzeln in der Ansicht von hinten

Tronc cérébral et mœlle épinière cervicale vus de l'arrière, racines de nerfs crâniens et spinaux, artère vertébrale

[XXIX, 723]

The brainstem and upper part of the spinal cord seen from behind. Variations in the course of the eleventh cranial nerve can be observed.

Hirnstamm und oberes Rückenmark in der Ansicht von hinten. Die Varianten im Verlauf des elften Hirnnerven sind sichtbar.

Tronc cérébral et mœlle épinière cervicale vus de l'arrière, variations du trajet du nerf accessoire, onzième nerf crânien

[XXIX, 724]

Display showing the tenth cranial nerve, the vagus nerve

Darstellung des zehnten Hirnnerven, der Vagusnerv

Nerf vague, dixième nerf crânien

[XXIX, 752]

Nerves in the region of the neck and thoracic cage. The vagus nerve follows an almost directly vertical course near the midline

Nerven im Bereich des Halses und des Brustkorbes. Nahezu senkrecht in der Mitte verlaufend der Vagusnerv

Nerfs du cou et des viscères du thorax, nerf vague descendant verticalement au milieu

[XXIX, 762]

The viscera of neck and thorax showing the course of the vagus nerve

Hals- und Brusteingeweide, in der Mitte der Vagusnerv

Nerf vague droit, plexus nerveux cardiaque, plexus nerveux cervical et brachial

[XXIX, 755]

Nervus vagus, Nervus laryngeus recurrens, Plexus cardiacus, Truncus sympathicus, Pars cervicalis

The neurovascular bundle of the neck

Darstellung der Gefäßnervenstraße des Halses

Nerf vague, axes nerveux et artériels du cou

[XXIX, 751]

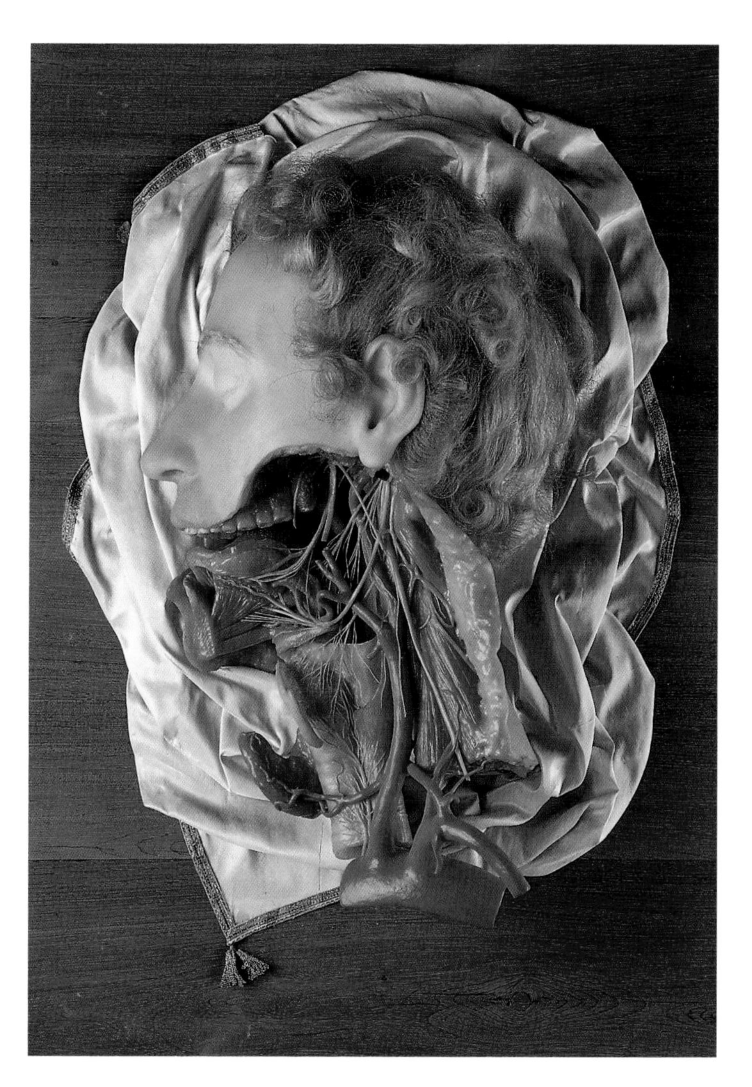

Neck dissected to show the nerves and vessels. Display showing the innervation of the tongue, floor of the mouth and larynx

Präparierter Hals mit Nerven und Gefäßen. Darstellung der Innervation von Zunge, Mundboden und Kehlkopf

Nerfs et artères du cou : innervation de la langue, du plancher de la bouche, et du larynx

[XXVIII, 761]

Nervus hypoglossus, Nervus laryngeus recurrens, Nervus vagus, Nervus laryngeus superior

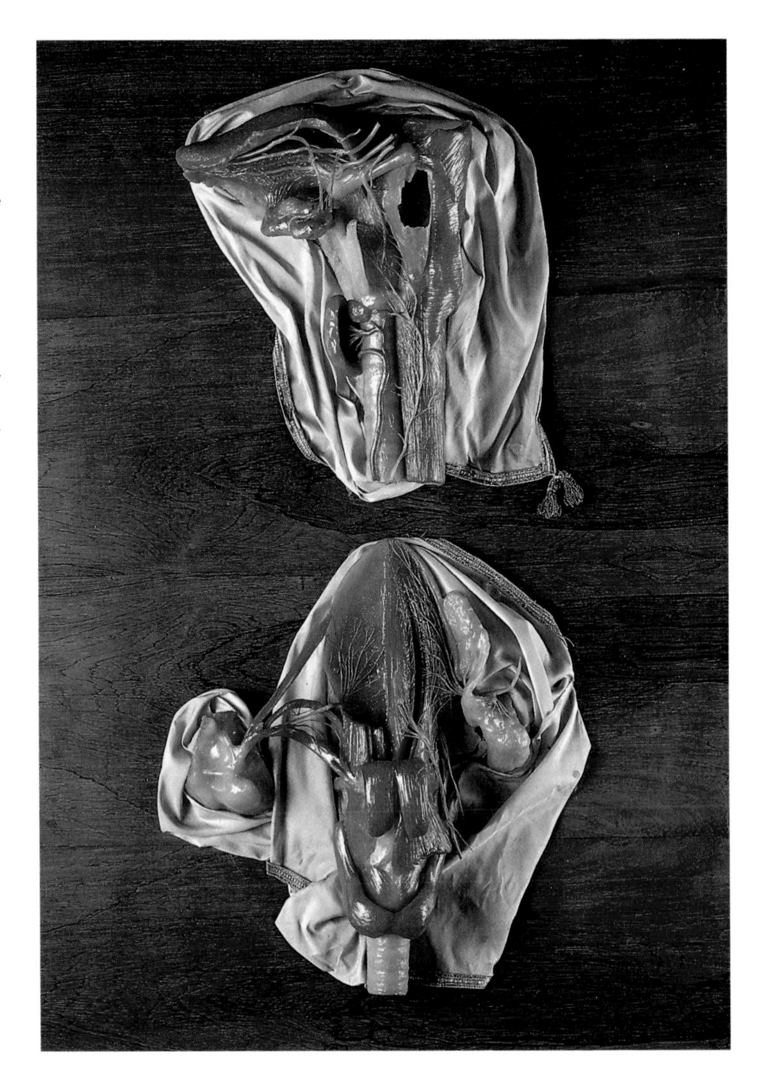

Upper specimen: display showing the innervation of the tongue. Lower specimen: the innervation of the mucous membrane and muscles of the larynx

Darstellung der Innervation der Zunge (oberes Präparat) sowie der Kehlkopfschleimhaut und -muskulatur (unteres Präparat)

Innervation de la langue (en haut), de la muqueuse et des muscles du larynx (en bas)

[XXVIII, 845]

Display showing the vagus nerve and the area innervated by it in the thoracic and abdominal cavities

Darstellung des Vagusnerven mit seinen Innervationsgebieten in Brust- und Bauchhöhle

Nerf vague : innervation des viscères du thorax et de l'abdomen

[XXIX, 797]

Display showing the vagus nerve and its area of innervated thoracic and abdominal cavities

Darstellung des Vagusnerven mit seinen Innervationsgebieten in Brust- und Bauchhöhle

Nerf vague : innervation des viscères du thorax et de l'abdomen

[XXIX, 800]

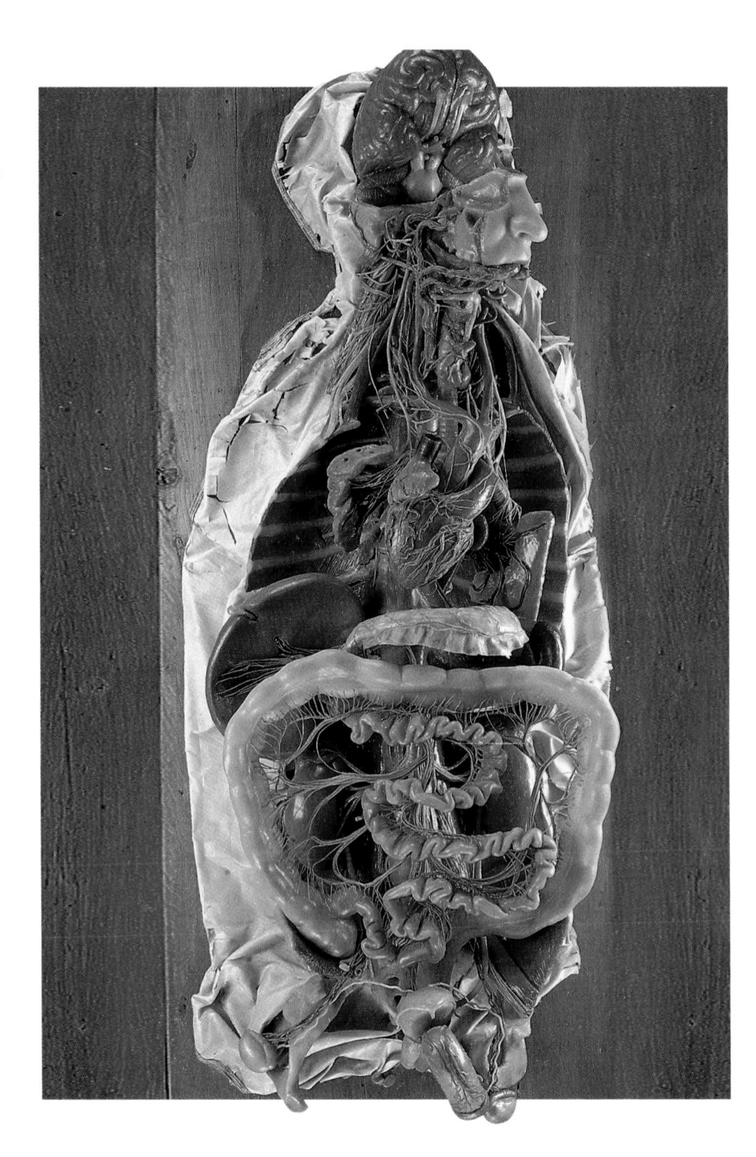

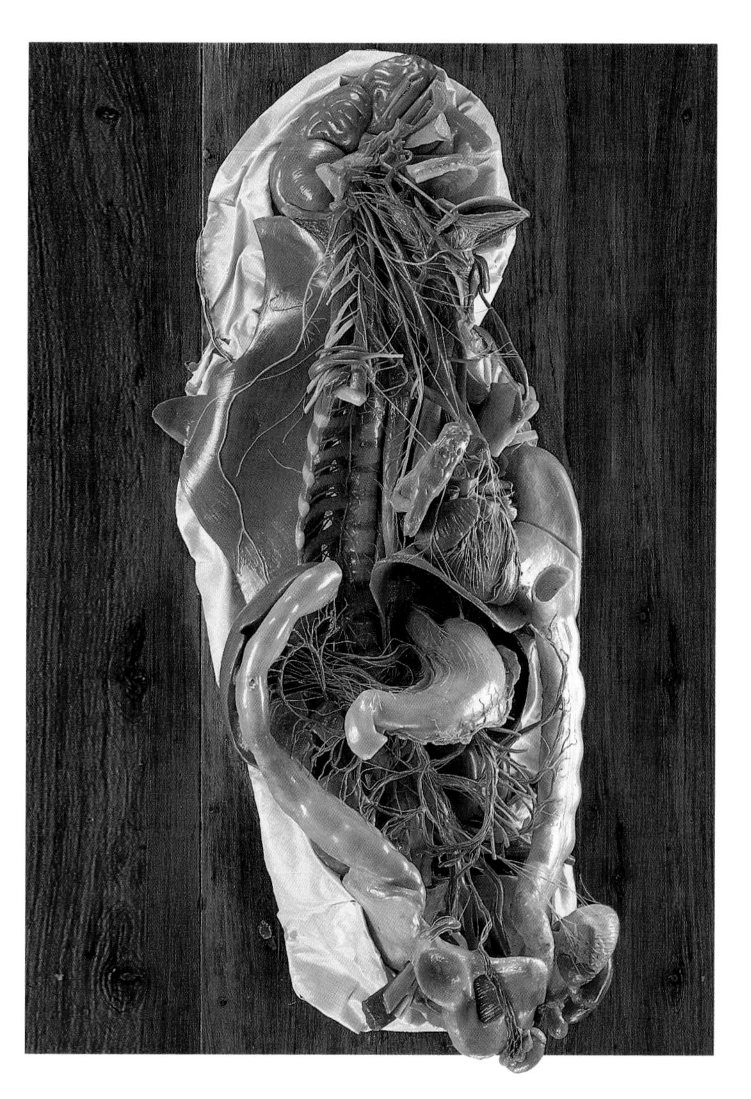

Display showing the vagus nerve and the area innervated by it in the thoracic and abdominal cavities

Darstellung des Vagusnerven mit seinen Innervationsgebieten in Brust- und Bauchhöhle

Nerf vague : innervation des viscères du thorax et de l'abdomen

[XXIX, 798]

Variations in the origin and course of the eleventh cranial nerve

Varianten im Ursprung und Verlauf des elften Hirnnerven

Nerf accessoire (spinal), onzième nerf cranien : variations de l'origine et du trajet

[XXIX, 722]

Display showing the course of the eleventh cranial nerve to the trapezius muscle

Darstellung des Verlaufes des elften Hirnnerven zum Trapezmuskel

Nerf accessoire : innervation du muscle trapèze

[XXIX, 757]

Display showing the course of the eleventh cranial nerve to the trapezius muscle

Darstellung des Verlaufes des elften Hirnnerven zum Trapezmuskel

Nerf accessoire : innervation du muscle trapèze

[XXIX, 669]

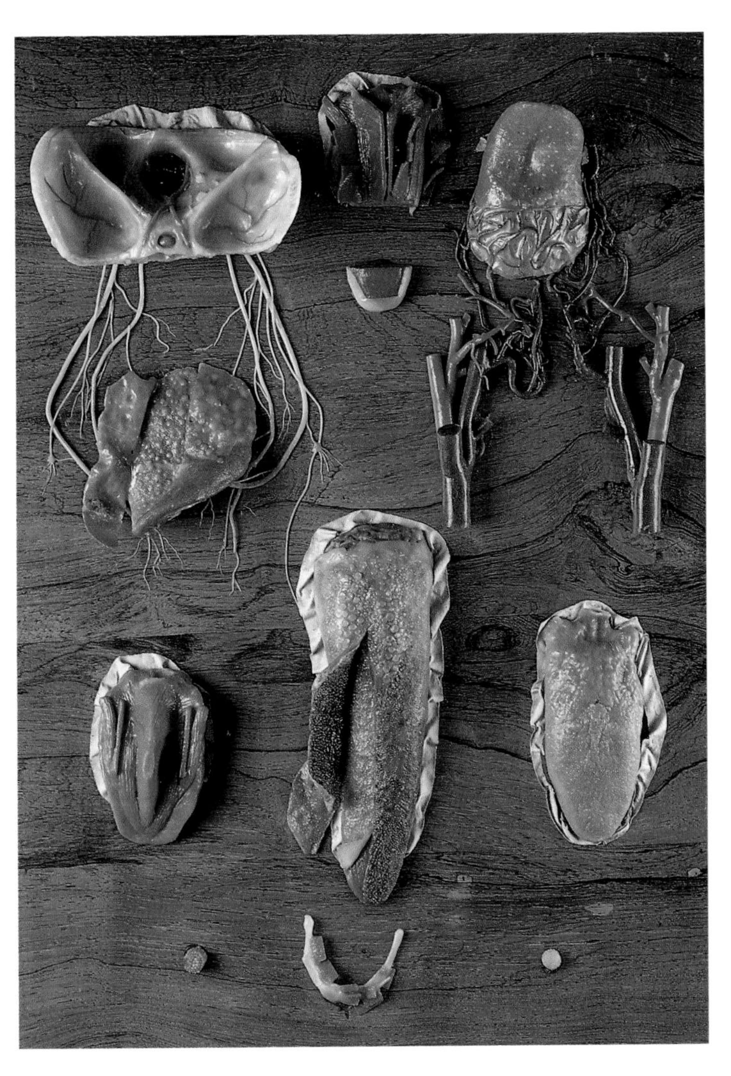

Nerves and vessels of the tongue

Nerven und Gefäße der Zunge

Nerfs et vaisseaux de la langue

[XXVIII, 704]

Nervus hypoglossus, Nervus lingualis, Nervus vagus, Nervus laryngeus superior et inferior

Display showing the innervation of the tongue; the lingual nerve running down in front of, and the hypoglossal nerve behind, the ear

Darstellung der Innervation der Zunge, des Nervus lingualis (vor dem Ohr absteigend) und Nervus hypoglossus (hinter dem Ohr absteigend)

Innervation de la langue : nerf lingual (en avant de l'oreille), nerf hypoglosse (en arrière de l'oreille)

[XXVIII, 711]

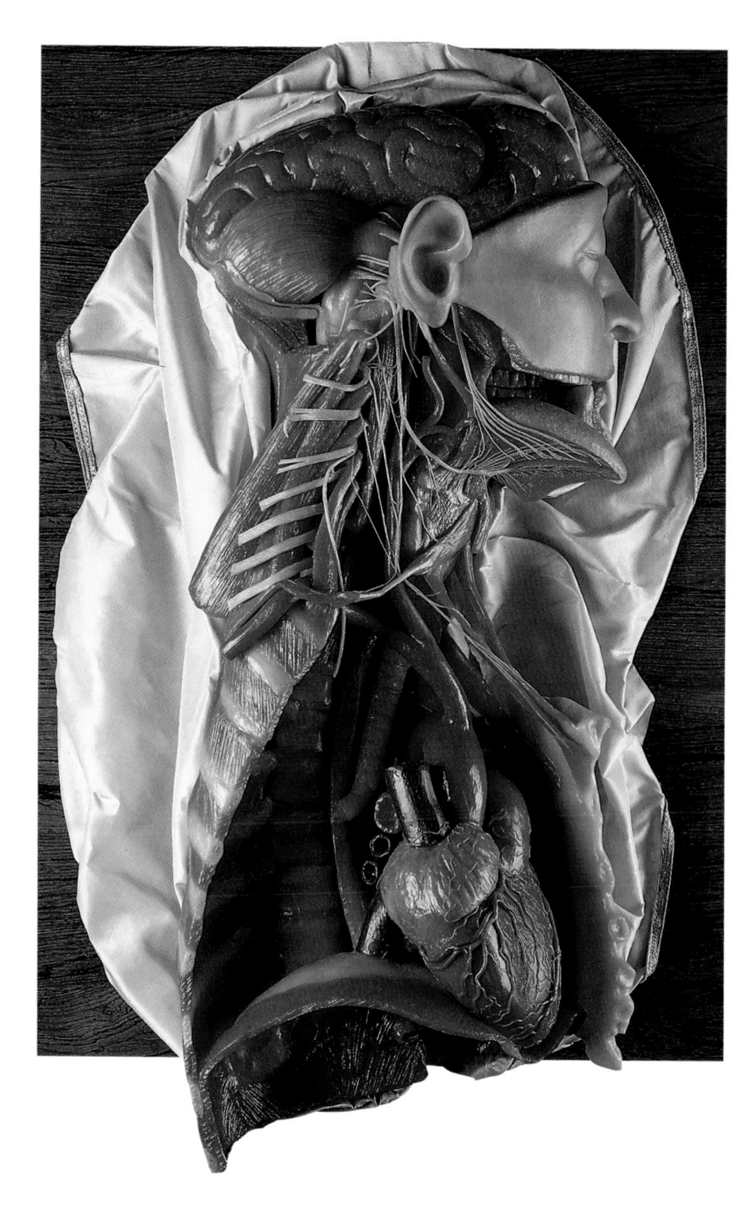

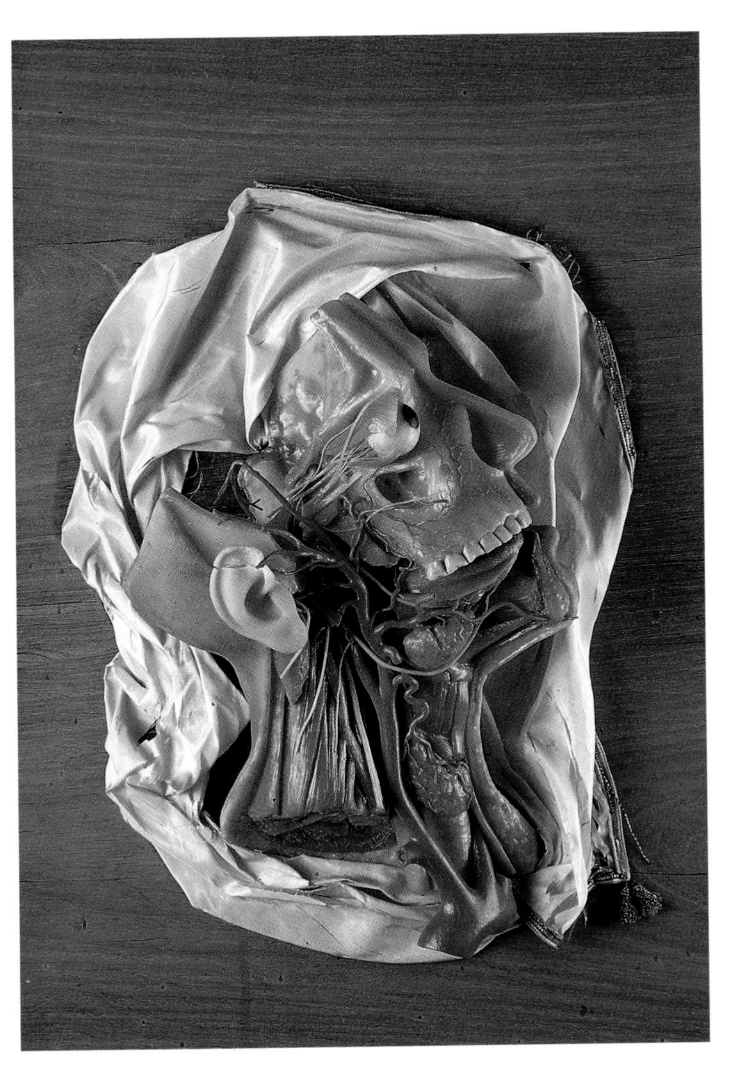

The nerves and arteries of the orbit, tongue and neck

Darstellung von Nerven und Arterien der Augenhöhle, der Zunge und des Halses

Nerfs et artères de la cavité orbitaire, de la langue, et du cou

[XXVIII, 702]

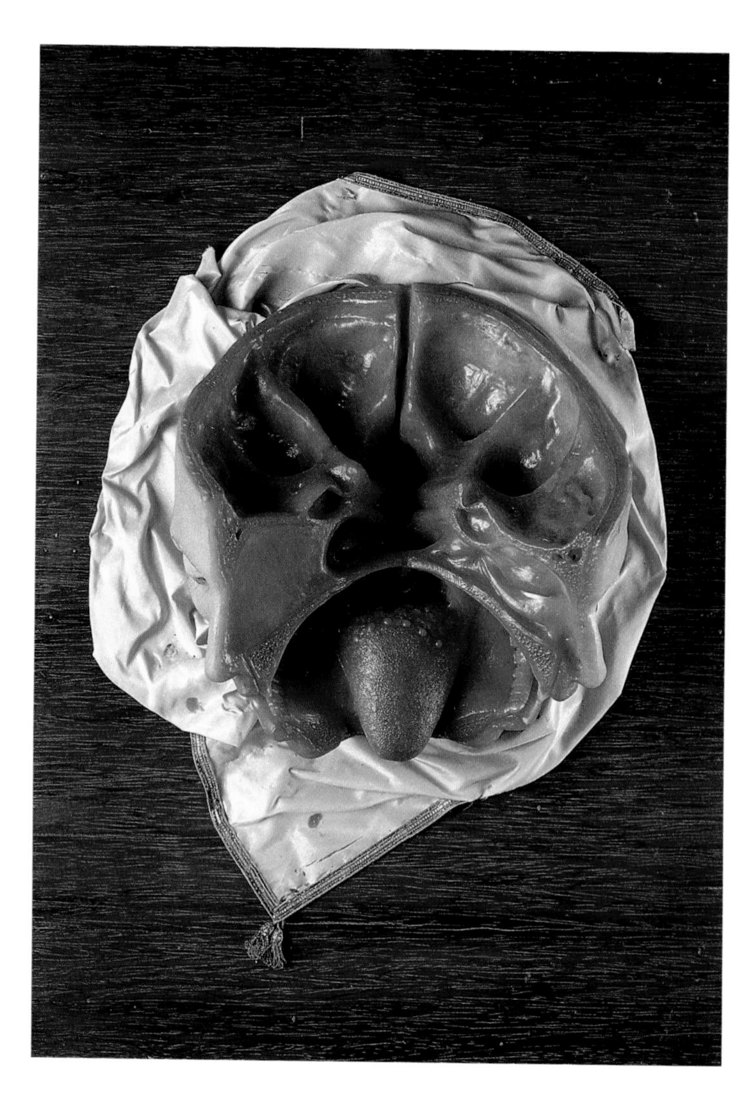

Median section of the scull. One can discern the bony roof of the oral cavity and the papillae of the tongue.

In der Mitte durchtrennter Schädel. Man erkennt das knöcherne Dach der Mundhöhle und die Papillen der Zunge.

Organes du goût : papilles de la langue (cavité buccale vue après section médiane du crâne et écartement des deux moitiés)

[XXVIII, 764]

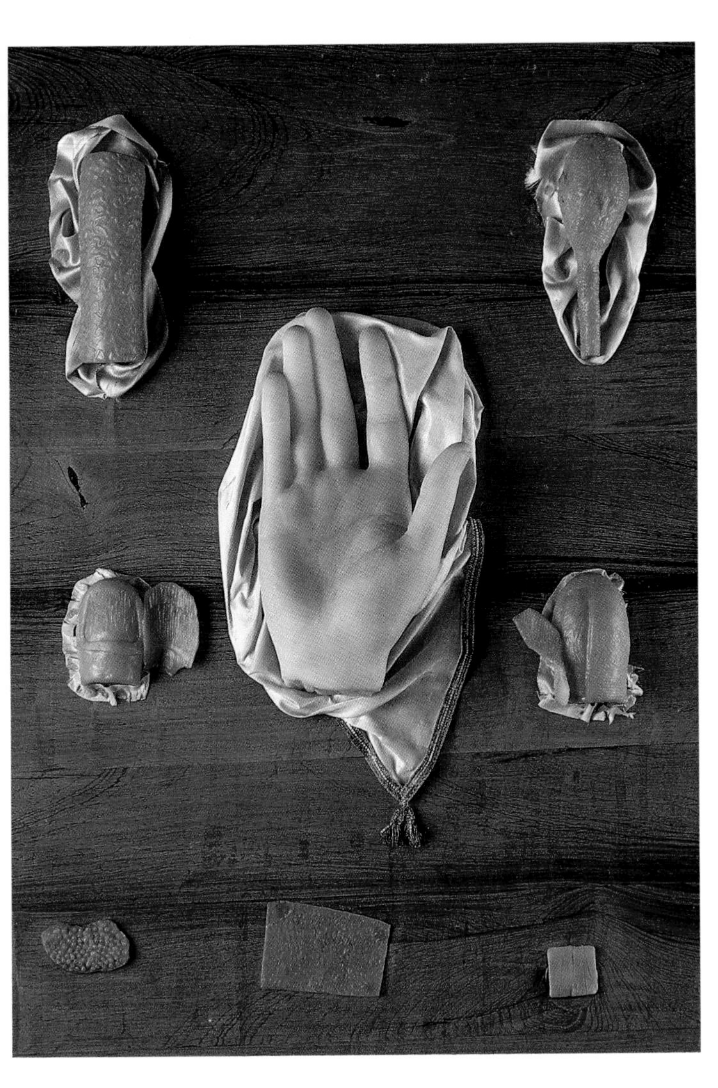

Tactile organs in the skin of the palms

Darstellung der Tastorgane der unbehaarten Haut (Leistenhaut)

Organes du tact : peau des doigts et de la paume de la main

[XXVIII, 784]

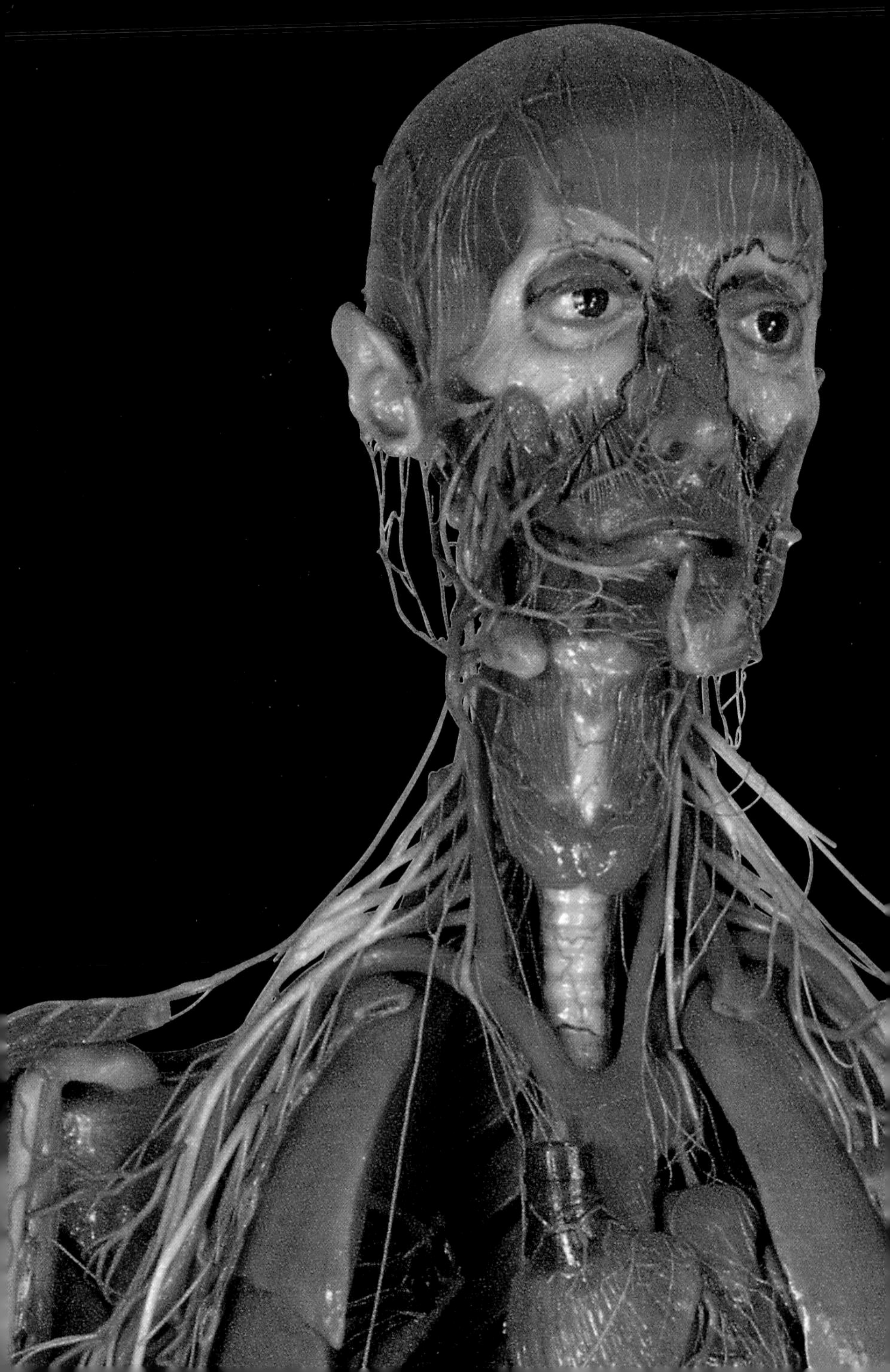

Systema nervosum peripheriale et autonomicum

Spinal Nerves and Autonomic Nervous System
Spinalnerven und autonomes Nervensystem
Nerfs spinaux et système nerveux autonome

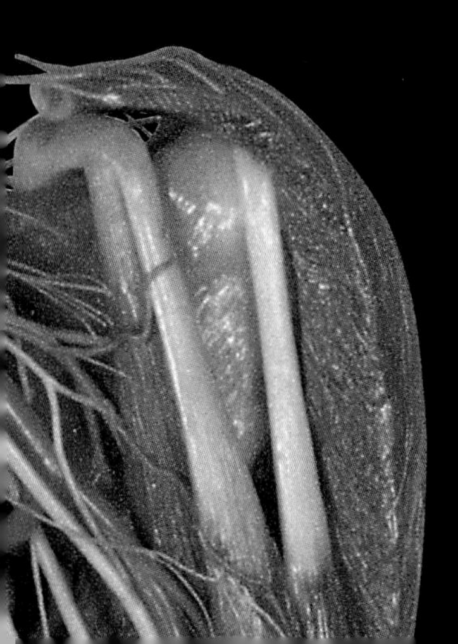

Spinal Nerves and Autonomic Nervous System

The spinal nerves – or nerves of the spinal cord – are shown in their relationship to the central nervous system, similar to the models displaying the course of the cranial nerves. The models show very clearly that the basic principle of segmentation of the spinal cord, together with the emergence of the nerve roots and ganglia and their position within the spinal canal, was already recognized by the end of the 18th century. It is striking to observe how the alternating topographical relationship between the nerves and the adjacent vertebrae is reproduced in the models. Some of the specimens display the large nerve plexuses which lie at the side of the cervical and lumbar vertebrae. It is remarkable how well they reproduce the complexity of the nerve trunks which innervate (that is to say, carry the nerve impulses for) the limbs. In the neck, thoracic and pelvic regions, elements of the sympathetic trunk as branches of the autonomic nervous system can be recognized.

In many of the models of all regions of the body one can fully understand the course and topography of the peripheral nerves. In many cases the complete course of the connective tissue septa and the muscle segmentation are displayed. Again, in other specimens, the most striking thing is the impressive way in which the fine autonomic nerve plexuses of the heart, hilum of the liver and mesentery have been reproduced.

Spinalnerven und autonomes Nervensystem

Die Spinal- oder Rückenmarksnerven werden – ähnlich den Modellen der Hirnnervenverläufe – in ihrem Bezug zum zentralen Nervensystem dargestellt. Diese Modelle zeigen deutlich, daß bereits im ausgehenden 18. Jahrhundert das Grundprinzip der Rückenmarksgliederung, die austretenden Nervenwurzeln und Ganglien sowie die Aufhängung im Wirbelkanal, erkannt war. Die topographische Beziehung der Nervenaustritte zwischen benachbarten Wirbelkörpern ist in bemerkenswerter Weise im Modell umgesetzt. Einige Präparate zeigen die großen Nervengeflechte, die im seitlichen Bereich der Hals- und Lendenwirbelsäule liegen. Sie dokumentieren eindrucksvoll, wie komplex die Nervenstämme zusammengesetzt sind und wie sich schließlich die Nerven für die Innervation (die Reizübertragung durch die Nerven) der Extremitäten formieren. Im Hals-, Brust- und Beckenbereich sind Abschnitte des Truncus sympathicus als einem der Schenkel des autonomen Nervensystems zu erkennen.

Page · Seite · Page 440–441:

Detail from page 479

Detail der Seite 479

Détail de la page 479

In einer großen Anzahl von Modellen aus allen Regionen unseres Körpers kann der Betrachter die Verläufe sowie die Topographie der peripheren Nerven nachvollziehen. Vielfach sind zur Darstellung des gesamten Verlaufes die Bindegewebssepten und Muskellogen aufgeschnitten. In anderen Präparate wiederum liegt das Hauptaugenmerk auf der Wiedergabe der sehr feinen autonomen Nervenfasergeflechte des Herzens, der Leberpforte und des Darmgekröses.

Nerfs spinaux et système nerveux végétatif

Les nerfs spinaux, ou nerfs rachidiens, sont représentés, comme les nerfs crâniens, en connexion avec le système nerveux central. Ces préparations montrent clairement qu'à la fin du XVIIIe siècle étaient connus les principes fondamentaux de la subdivision de la mœlle épinière, la disposition des racines et des ganglions des nerfs spinaux, leurs moyens de suspension dans le canal vertébral. Les rapports topographiques des nerfs spinaux entre deux vertèbres voisines sont représentés de façon remarquable. Quelques préparations montrent les grands plexus nerveux disposés de part et d'autre des segments cervical et lombaire de la colonne vertébrale ; elles montrent la complexité de la formation des troncs nerveux et la constitution des nerfs destinés aux membres. Au niveau du cou, du thorax et du bassin, peuvent être reconnus des éléments du tronc sympathique, composante du système nerveux végétatif.

Le trajet et les rapports topographiques des nerfs périphériques peuvent être suivis et analysés sur de nombreux modèles concernant toutes les régions du corps ; souvent l'ouverture des cloisons fibreuses et des loges musculaires permet de suivre un nerf sur tout son trajet. Sur certaines préparations, au niveau du cœur, du hile du foie, ou du mésentère, l'attention est attirée sur la disposition des réseaux nerveux constitués par les fines fibres nerveuses végétatives.

Nervus ischiadicus

Whole body specimen with muscles and nerves from behind

Ganzkörperpräparat zur Darstellung der Muskulatur und Nervenläufe an der Körperrückseite

Préparation du corps entier représentant les nerfs des régions postérieures

[XXIX, 748]

View of the spinal cord from behind. Left: with the dura mater. Right: showing the spinal ganglia

Blick auf das Rückenmark von hinten; links mit harter Rückenmarkshaut, rechts mit Darstellungen der Spinalganglien

Mœlle épinière et racines des nerfs spinaux vus de l'arrière ; méninges ouvertes (à gauche), méninges intactes et ganglions spinaux (à droite)

[XXIX, 717]

Below: display showing the nerve plexus of the arm. Above: the trigeminal ganglion

Darstellungen von Nervenfasergeflechten u. a. des Armnervengeflechtes (untere Präparate) und des Trigeminusganglions (obere Präparate)

Plexus nerveux : plexus brachial (en bas), ganglion du nerf trijumeau (en haut)

[XXXI, 419]

Brainstem and upper spinal cord seen from behind. The nerve roots are sometimes only shown on one side.

Hirnstamm und oberes Rückenmark in der Ansicht von hinten. Die Nervenwurzeln sind z. T. nur auf einer Seite dargestellt.

Tronc cérébral et mœlle épinière cervicale vus de l'arrière, racines des nerfs spinaux représentées d'un seul côté

[XXIX, 725]

View of the cervical nerves and of the back of the head after removal of the skin

Aufsicht auf Halsnerven und auf den Hinterkopf nach Entfernung der Kopfhaut

Nerfs de la région occipitale après section du cuir chevelu, nerfs du cou

[XXIX, 756]

Display showing the nerves at the back of the head and the neurovascular bundles in the deep region of the neck

Darstellung der Hinterhauptsnerven und der Gefäß-Nervenstraße in der tiefen Halsregion

Nerfs de la région occipitale, artères et nerfs des régions profondes du cou

[XXIX, 668]

Nervus occipitalis major

Display showing the course of a nerve at the back of the head

Darstellung des Verlaufes eines Hinterhaupts-nerven

Nerf grand occipital

[XXIX, 710]

Display showing the nerves at the back of the head

Darstellung der Hinterhaupts-nerven

Nerfs occipitaux

[XXIX, 716]

Display showing the branches of the facial nerve and the cutaneous branches of the cervical plexus

Darstellung von Ästen des Nervus facialis und von Hautästen des Halsnervengeflechtes

Rameaux du nerf facial et rameaux cutanés du plexus nerveux cervical

[XXVIII, 760]

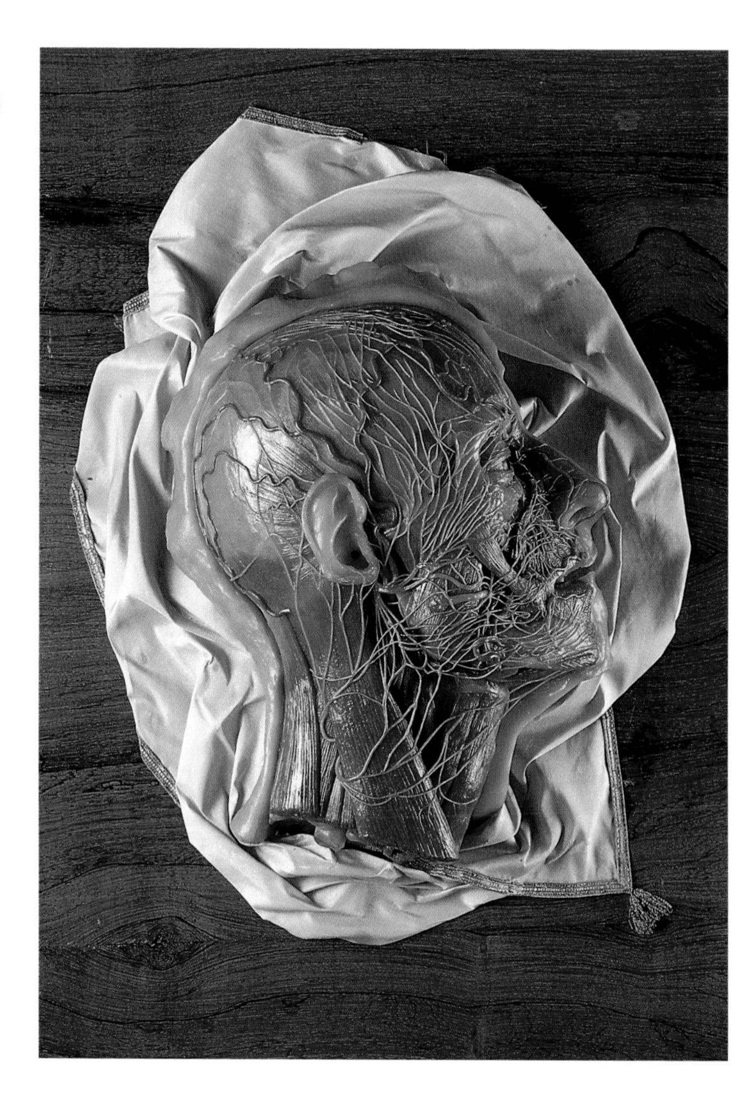

Display showing the cutaneous nerves at the side of the neck

Darstellung von Hautnerven in der seitlichen Halsregion

Plexus nerveux cervical dans la région superficielle du cou, nerfs cutanés

[XXIX, 712]

Plexus cervicalis, Nervus phrenicus, Nervus hypoglossus, Ansa cervicalis profunda, Plexus brachialis

View of the deep region of the neck and into the opened thoracic cage. The heart can be seen. Observe the nerves supplying the diaphragm running along the side of the pericardial wall.

Blick auf die tiefe Halsregion und in die eröffnete Brusthöhle mit Herz. Man beachte den Verlauf des Zwerchfellnerven entlang der seitlichen Herzbeutelwand.

Nerfs des régions profondes du cou et des viscères thoraciques ; trajet du nerf du diaphragme (phrénique) le long du péricarde

[XXIX, 753]

The nerve plexus of the upper limb

Darstellung des Armnervengeflechtes

Plexus nerveux brachial au niveau du cou

[XXIX, 813]

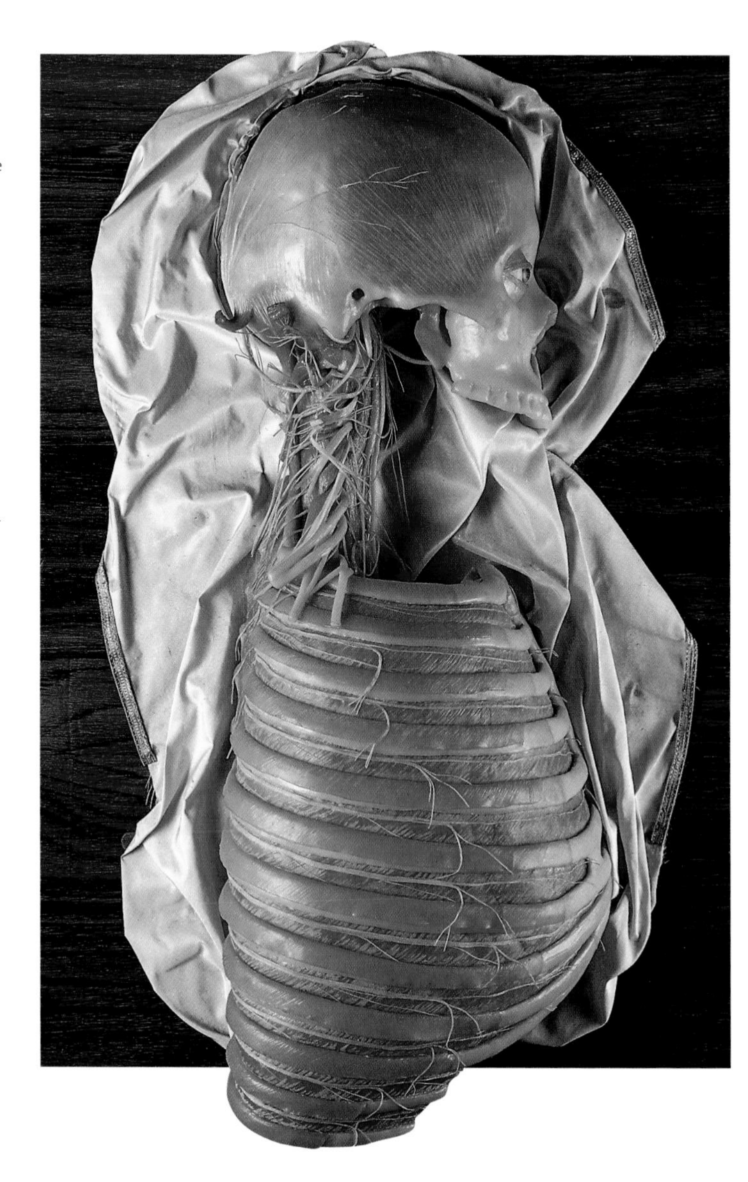

The nerves in a deeper layer of the neck. Above the thoracic cage one can see parts of the nerve plexus of the upper limb.

Nerven des Halses in einer tiefen Schicht. Oberhalb des Brustkorbs erkennt man Anteile des Armnervengeflechtes.

Nerfs des régions profondes du cou, plexus nerveux brachial au dessus de la première côte, nerfs intercostaux

[XXIX, 715]

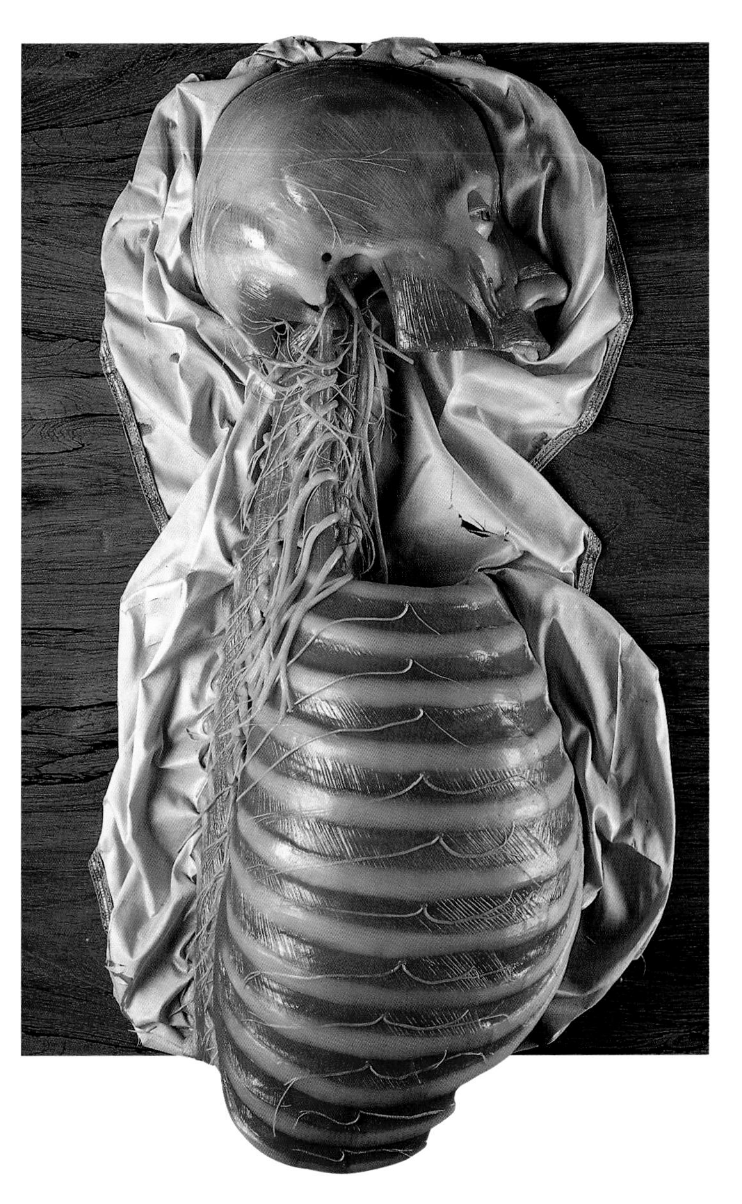

Deeper layer showing cervical plexus and plexus of the upper limb

Hals- und Armnervengeflecht in einer tiefen Schicht

Plexus nerveux cervical et brachial, rameaux des nerfs intercostaux

[XXIX, 672]

The nerves of the upper limb and side of the chest wall

Darstellung der Nerven der seitlichen Brustwand sowie des Armes

Nerfs du membre supérieur vus du côté droit

[XXIX, 807]

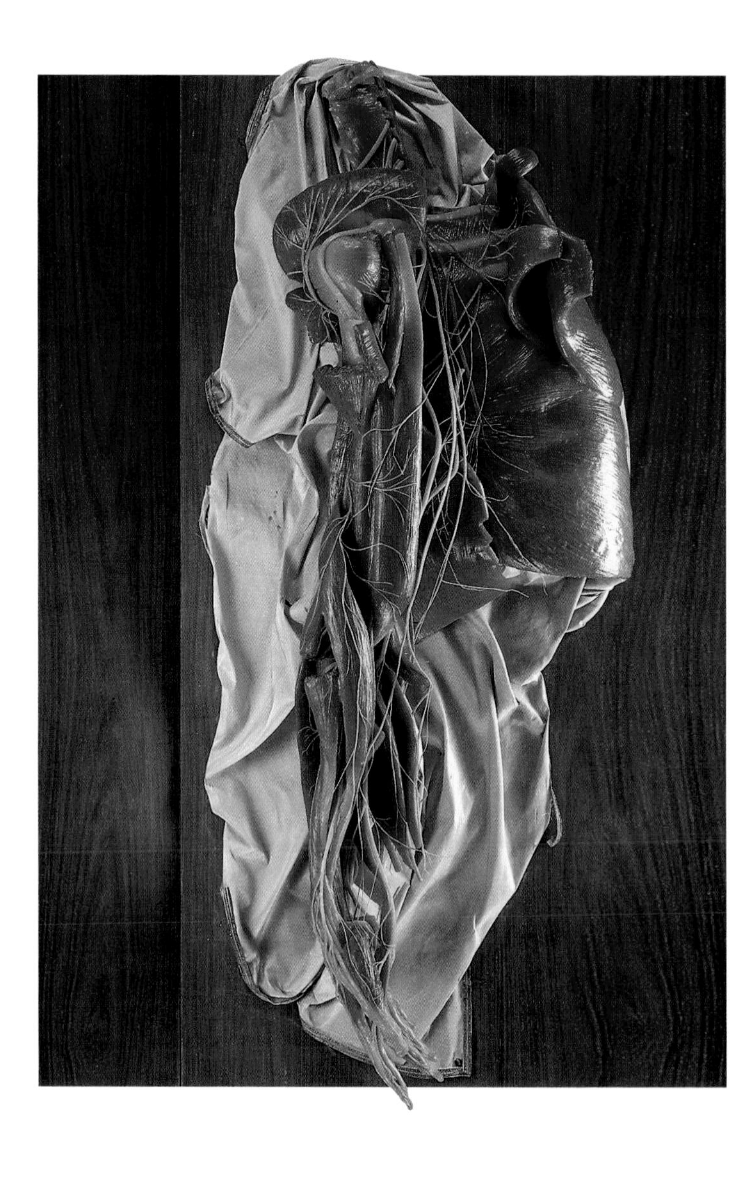

The nerves and muscles of the upper limb

Darstellung von Muskeln und Nerven des Armes

Nerfs, artères, et muscles du membre supérieur

[XXIX, 805]

Plexus brachialis

The nerves and plexus of the upper limb

Darstellung von Armnerven und Armnervengeflecht

Plexus nerveux brachial et nerfs du bras gauche

[XXIX, 810]

The skeleton of the left arm and shoulder seen from in front. The nerves and plexus of the arm, together with the associated region of the spinal cord, can also be seen.

Präparat eines linken Arm- und Schulterskelettes in der Ansicht von vorn mit Armnerven, Armnervengeflecht und zugehörigem Rückenmarksabschnitt

Plexus brachial et nerfs projetés sur le squelette du membre supérieur

[XXIX, 809]

A male pelvis divided in the midline; view of the nerve plexus of the lower limb

In der Mittellinie durchtrenntes männliches Becken mit Blick auf das Beinnervengeflecht

Plexus nerveux sacré, nerf honteux (coupe médiane du bassin masculin)

[XXIX, 781]

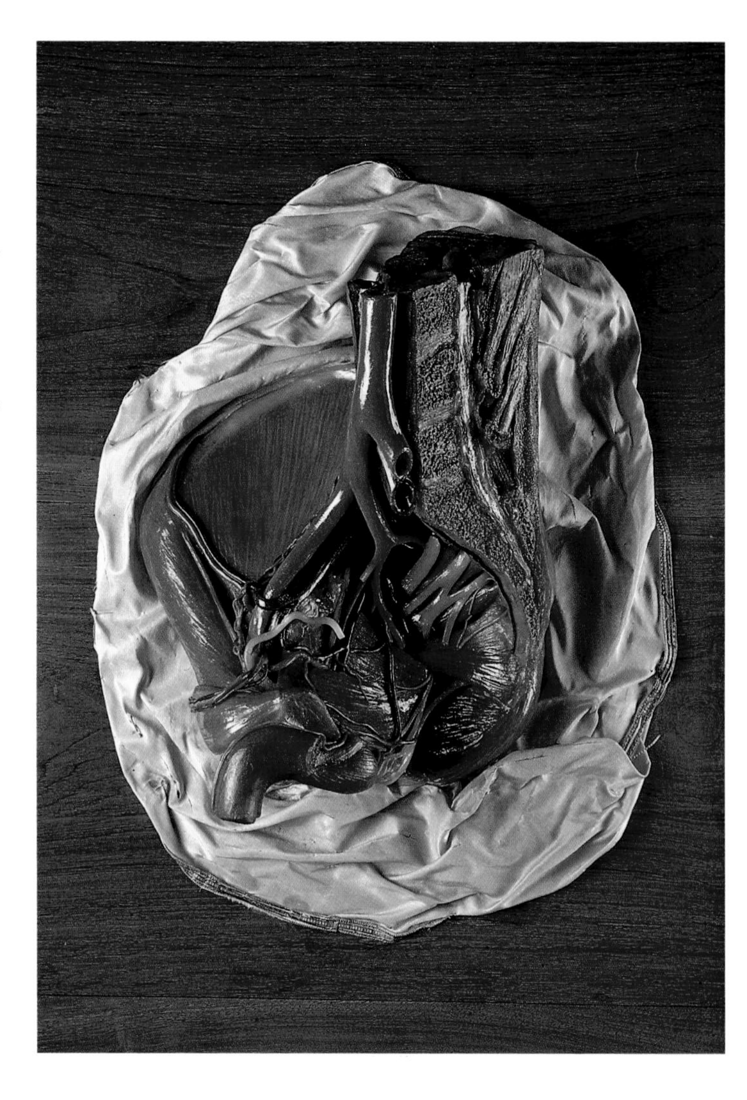

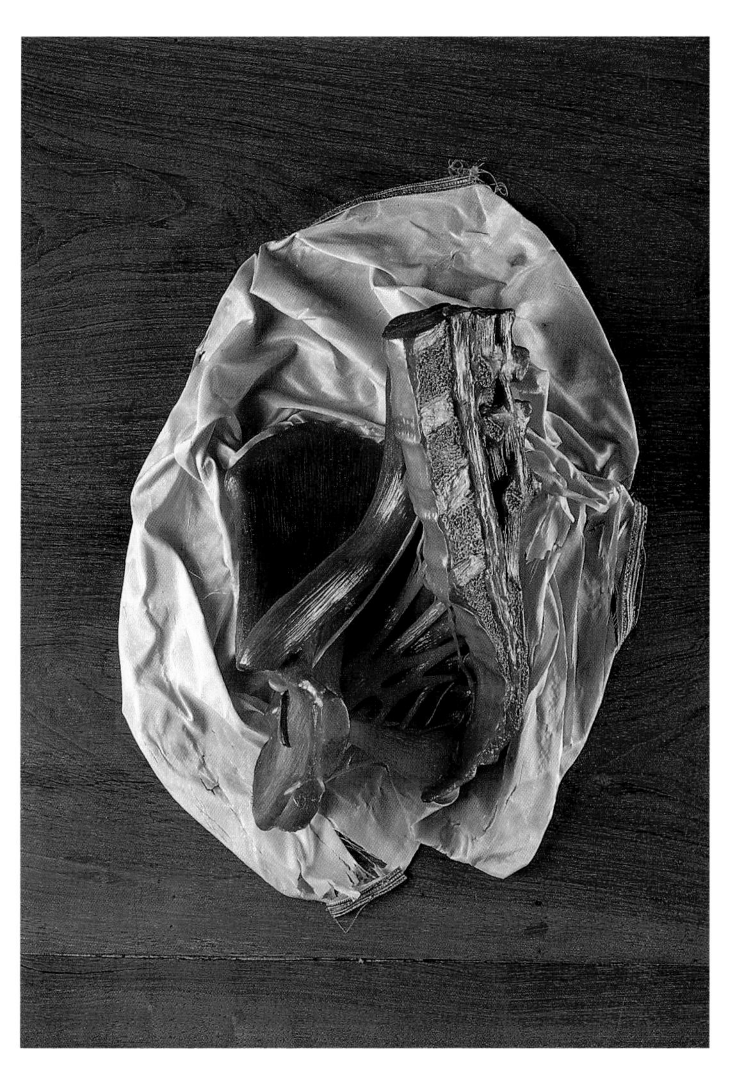

Pelvis divided in the midline; view of the nerve plexus of the lower limb

In der Mittellinie durchtrenntes Becken mit Blick auf das Beinnervengeflecht

Plexus nerveux lombo-sacré à l'origine de nerfs du membre inférieur (coupe médiane du bassin)

[XXIX, 783]

The nerves and muscles of the lower limb

Darstellung der Nerven und Muskeln des Beines

Nerfs et muscles du membre inférieur

[XXIX, 771]

Display showing the nerves supplying the muscles and skin of the sole of the foot

Darstellung der Nerven für die Fußsohlenmuskulatur und Haut

Nerfs des muscles et de la peau de la plante du pied

[XXVI, 749]

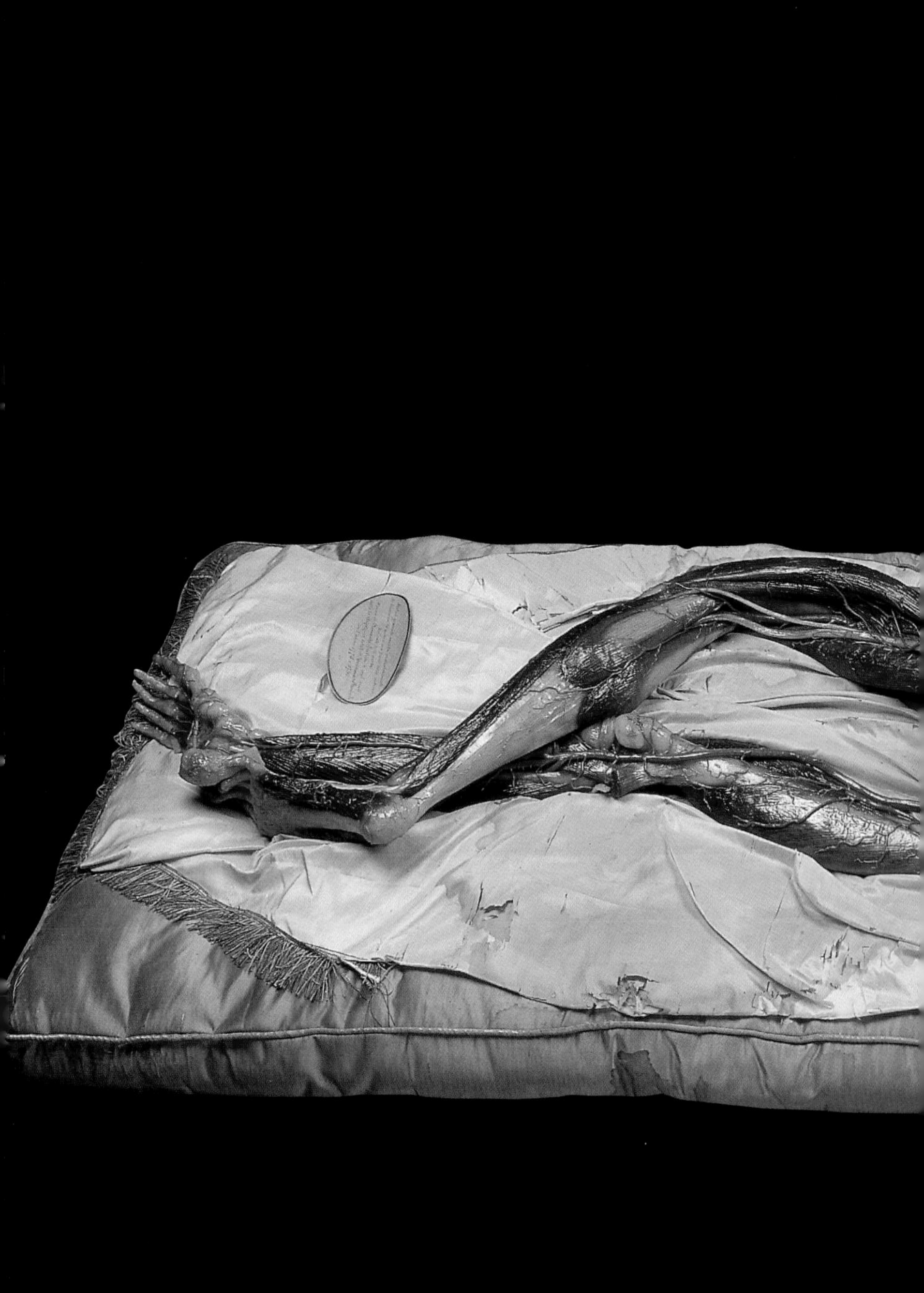

Arteriae

Whole body specimen with the arteries displayed

Ganzkörperpräparat mit Darstellung der Arterien

Préparation du corps entier représentant des nerfs et artères

[XXV, 445]

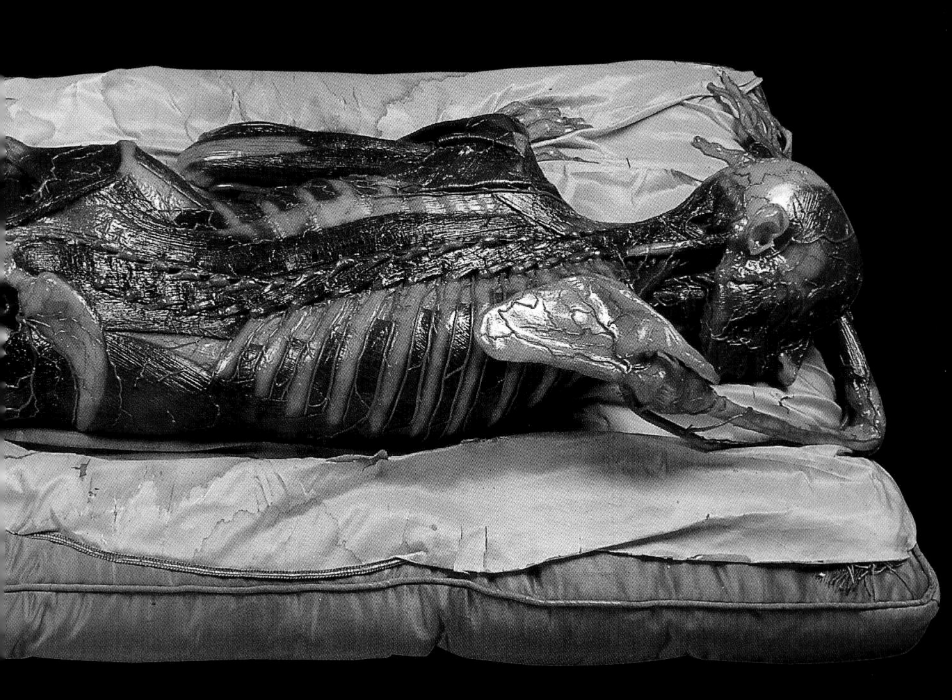

Truncus sympathicus, Plexus lumbosacralis, Plexus pelvinus

The autonomic nerve plexuses of the thoracic and abdominal cavities

Darstellung vegetativer Nervenfasergeflechte der Brust- und Bauchhöhle

Système nerveux végétatif pour les viscères du thorax, de l'abdomen, et du bassin

[XXIX, 801]

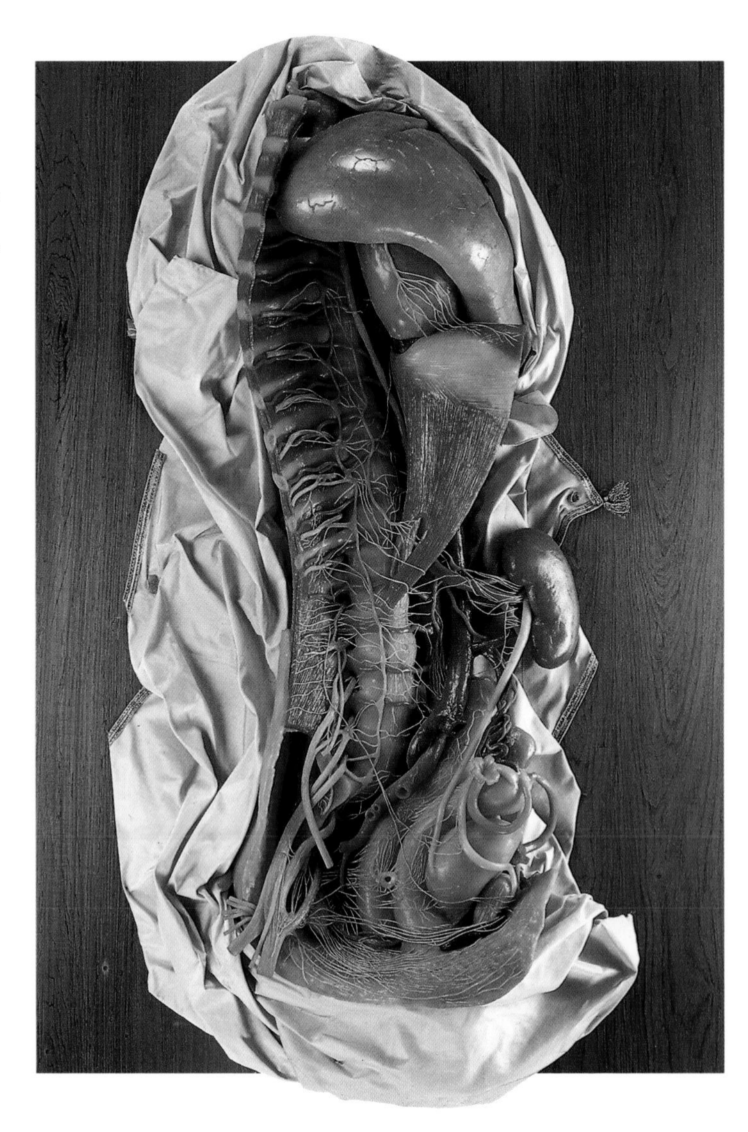

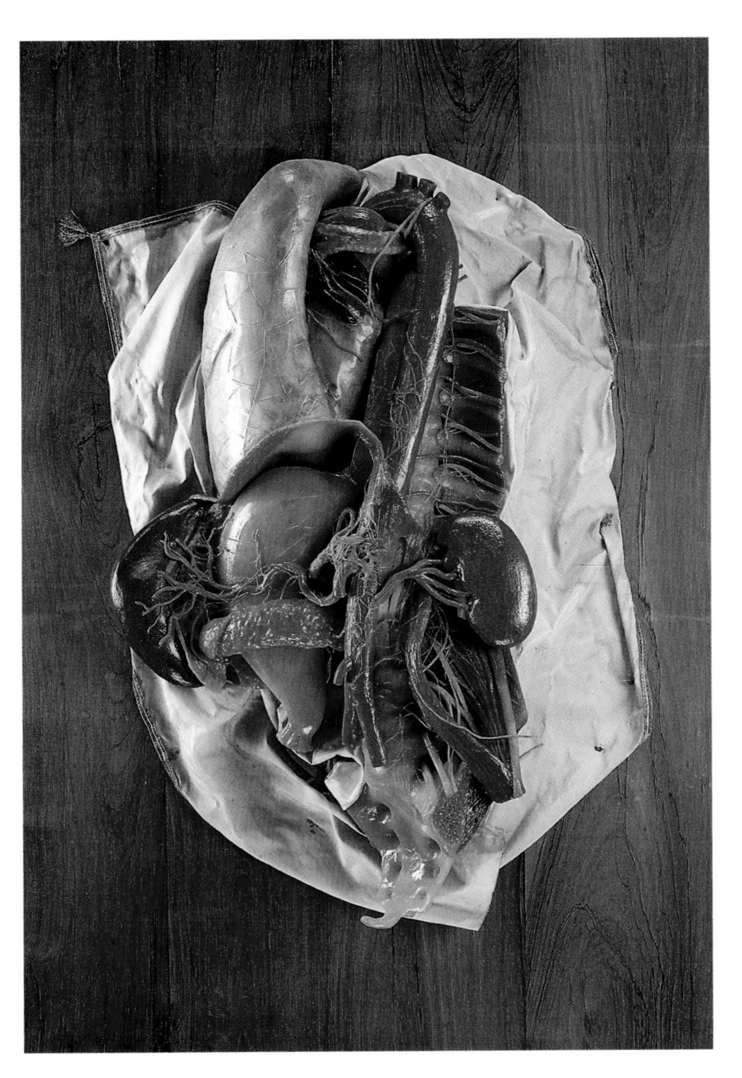

Specimen showing the viscera with the arteries and their associated autonomic nerve fiber plexuses

Eingeweidepräparat mit Arterien und begleitenden vegetativen Nervenfasergeflechten

Système nerveux végétatif autour des artères se destinant aux viscères de l'abdomen

[XXIX, 811]

Display showing the autonomic nerve plexuses near the vertebral column and associated with the abdominal arteries

Darstellung von vegetativen Nervenfasergeflechten im Bereich der seitlichen Wirbelsäule und der Baucharterien

Système nerveux végétatif dans la région latéro-vertébrale et autour des artères de l'abdomen

[XXIX, 799]

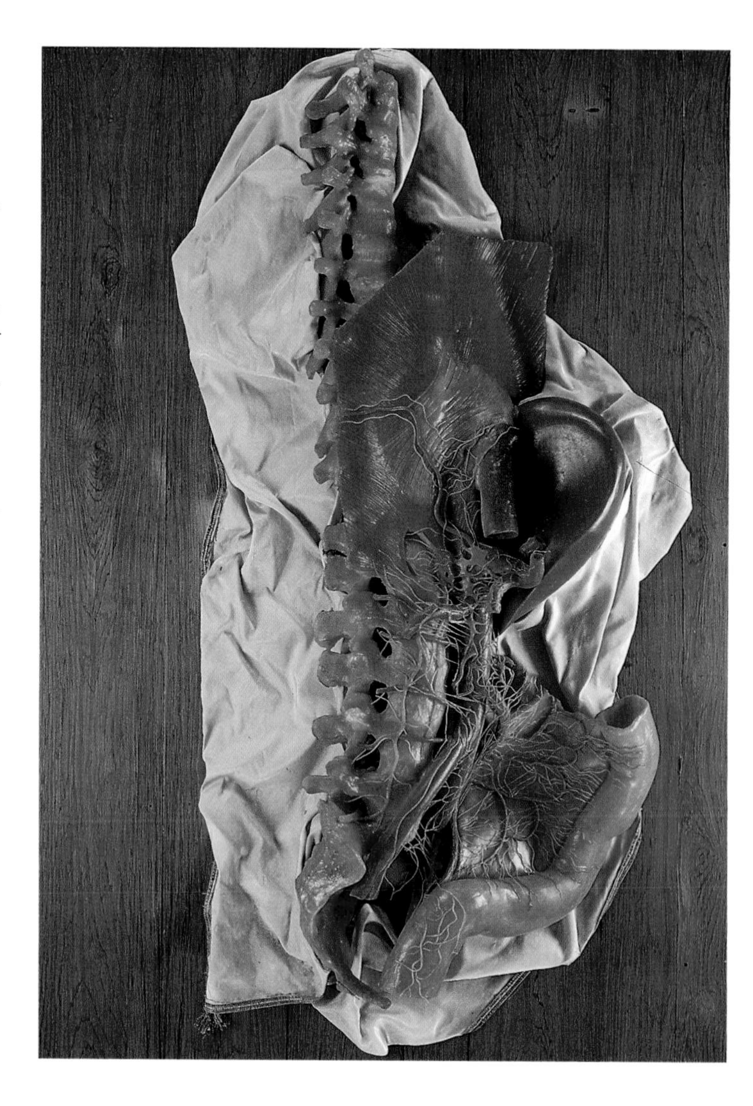

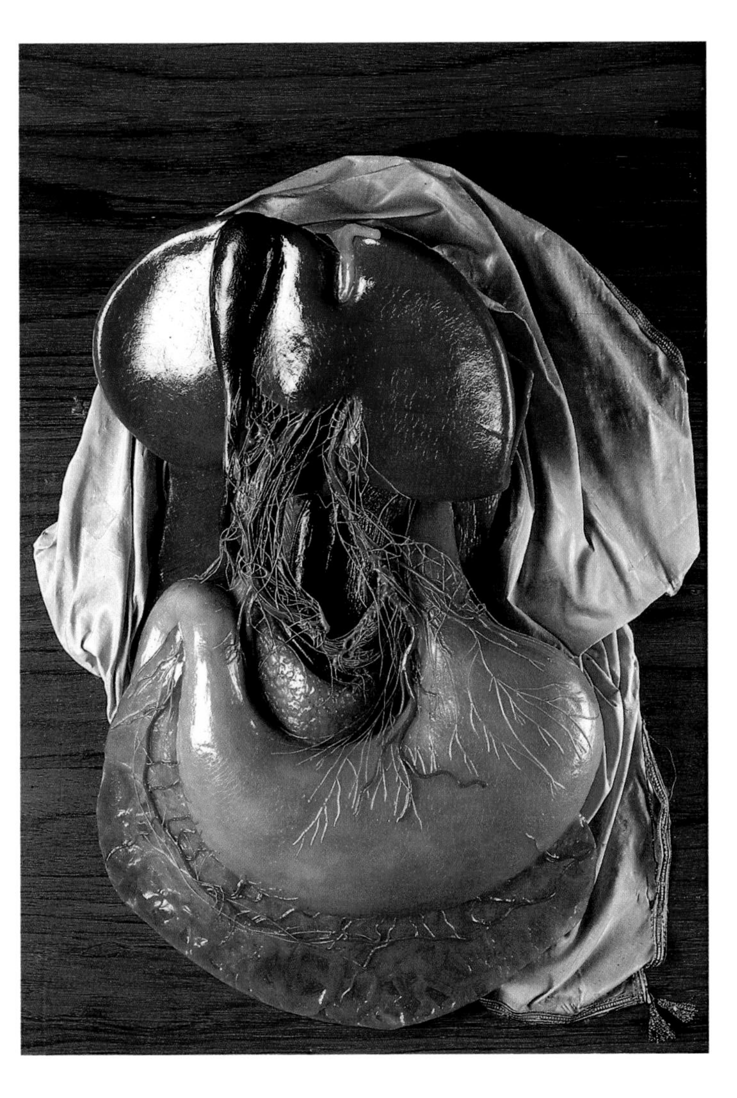

Display showing the autonomic nerve plexus associated with stomach and liver

Darstellung der vegetativen Nervenfasergeflechte an Magen und Leber

Système nerveux végétatif de l'estomac et du foie

[XXIX, 767]

Truncus sympathicus, pars cervicalis, Ganglion cervicale superius et medius, Ganglion stellatum

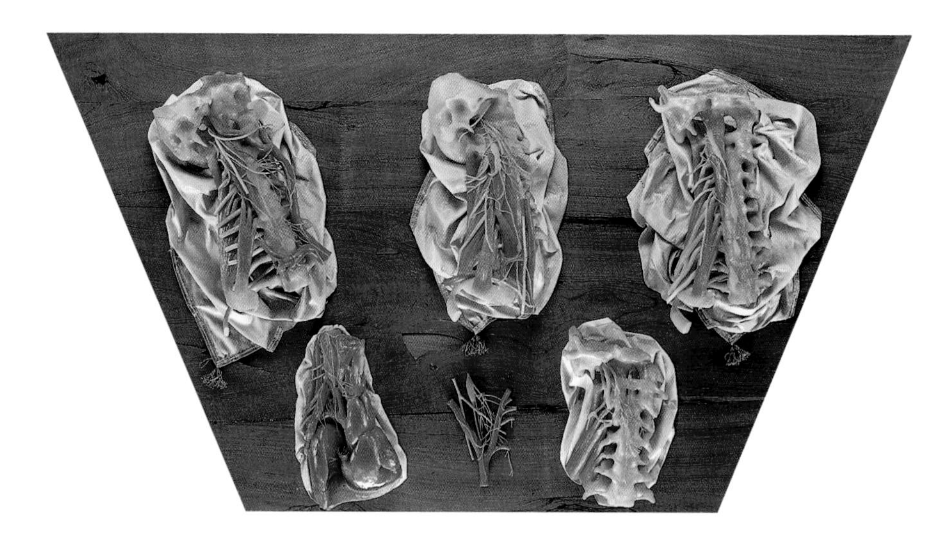

Display showing the sympathetic nerves and their ganglia in the deep region of the neck

Darstellungen sympathischer Nerven und ihrer Ganglien im Bereich der tiefen Halsregion

Nerfs et ganglions sympathiques de la région profonde du cou

[XXIX, 812]

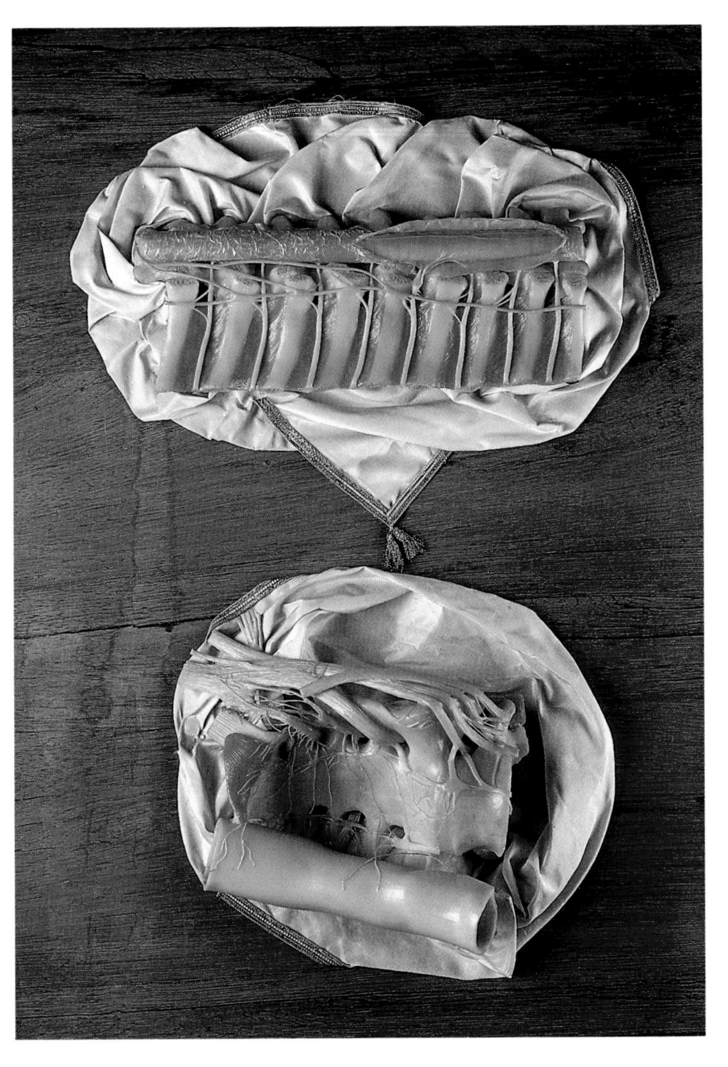

Display showing the sympathetic nerves and their ganglia in the region of the thoracic cage (upper specimen) and in the region of the sacrum (lower specimen)

Darstellung sympathischer Nerven und ihrer Ganglien im Bereich des Brustkorbs (oberes Präparat) und im Bereich des Kreuzbeines (unteres Präparat)

Nerfs et ganglions sympathiques de la région thoracique (en haut) et sacrée (en bas)

[XXIX, 780]

Systema digestorium et respiratorium

Viscera

Eingeweide

Appareil digestif, appareil respiratoire

Viscera

The recognition of surgery as a legitimate part of the medical faculty coincided in time with the collecting of these models and had a lasting effect on anatomical teaching and research. As a result of the new status acquired by surgery, accurate anatomical knowledge acquired a new significance. Particular attention was paid to the anatomical dissection of the organs of the thorax, abdomen and true pelvis. Continued developments dissection techniques led to increasingly detailed specimens representing the topographical (spatial) relationship between individual organs, and also their blood and nerve supply. In contrast to the drawings and woodcuts, the preparation of wax models of original specimens gave an entirely new impetus to the study of anatomy, allowing the complex structures to be represented in three dimensions.

A large number of the models in this collection show the organs within the thoracic and abdominal cavities in their natural positions. Beginning with these overall displays, many of the detailed models also show the complete extent of the individual organs together with their blood supply, and for this purpose adjacent organs were displaced from their natural relationship or completely removed from the body. In these specimens the arteries and veins are clearly distinguished, and colored either red or blue. Sections of the various organs are arranged to show their internal structure, including the relative internal measurements of hollow organs, as in the case of sections through the digestive tract in which the internal surfaces are also illustrated.

Pages · Seiten · Pages 474–475:

Whole body specimen showing the lymphatic vessels in the thoracic and abdominal cavities

Ganzkörperpräparat mit Darstellung der Lymphgefäße in Brust- und Bauchraum

Préparation du corps entier représentant les vaisseaux lymphatiques des cavités thoracique et abdominale [XXIX, 746]

Die Anerkennung der Chirurgie als Fach der Medizin fiel zeitlich mit der Herstellung der Modellsammlung zusammen und hatte nachhaltigen Einfluß auf die Lehr- und Forschungsinhalte der Anatomie. In der praktischen Umsetzung durch die Chirurgie erlangten die genauen Kenntnisse der Anatomie einen neuen Stellenwert. Das besondere Interesse galt der anatomischen Sektion der Organsysteme von Brust- und Bauchhöhle sowie des kleinen Beckens. Immer weiter entwickelte Sektionstechniken ermöglichten eine zunehmend detaillierte präparatorische Darstellung der topographischen (räumlichen) Beziehungen der einzelnen Organe sowie ihrer Gefäß- und Nervenversorgung. Im Gegensatz zu Zeichnungen und Holzschnitten eröffneten die vom Originalpräparat hergestellten Wachsmodelle dem anatomischen Studium eine völlig neue Dimension, indem sie die komplexen Strukturen des menschlichen Körpers dreidimensional wiedergaben.

Eine Vielzahl von Modellen dieser Sammlung zeigt die Organe des Brust- und Bauchraums in ihrer natürlichen Lage. Ausgehend von diesen Übersichtsdarstellungen sind in vielen Detailmodelle zusätzlich die Organe in ihrer gesamten Ausdehnung sowie ihre Gefäßversorgung zu sehen. Hierzu wurden benachbarte Organe aus ihrer natürlichen Position verlagert oder ganz herausgenommen. Arterien und Venen sind in diesen Präparaten zur klaren Differenzierung rot bzw. dunkelblau koloriert. Anschnitte der verschiedenen Organe zeigen deren inneren Aufbau, einschließlich der luminalen Größenverhältnisse von Hohlorganen, wie dies beispielsweise der Anschnitt des Darms mit dem Schleimhautrelief verdeutlicht.

Eingeweide

Les viscères

La chirurgie a été reconnue comme discipline médicale à part entière à l'époque où a été réalisée la collection de modèles de la Specola ; cette intégration de la chirurgie a influencé l'enseignement et la recherche en anatomie. Les connaissances précises en anatomie ont été valorisées par leur transposition dans la pratique chirurgicale.

La dissection des organes et appareils contenus dans le thorax, l'abdomen et le bassin a été l'objet d'un intérêt particulier. Le perfectionnement continu des techniques de dissection a entrainé une représentation de plus en plus détaillée des rapports topographiques entre les différents organes ainsi que la mise en évidence de leur vascularisation et de leur innervation. Contrairement aux dessins et aux gravures, le modèle en cire réalisé à partir d'une préparation originale offrait à l'étude de l'anatomie une dimension entièrement nouvelle par la représentation tridimensionnelle des structures complexes du corps humain.

Les organes du thorax ou de l'abdomen dans leur situation naturelle sont présentés par un grand nombre de modèles de cette collection. A côté de ces représentations d'ensemble, des détails des organes sont aussi représentés par de nombreux modèles comme par exemple les pédicules vasculaires. Pour permettre ces observations, les organes voisins sont alors déplacés de leur position naturelle ou entièrement enlevés. Les artères et les veines sont clairement différenciées dans ces cires par l'emploi des couleurs rouge ou bleu foncé respectivement. Les tranches de section des différents organes permettent de voir leur structure interne, ou le relief de la lumière des viscères creux comme par exemple celui de la muqueuse intestinale.

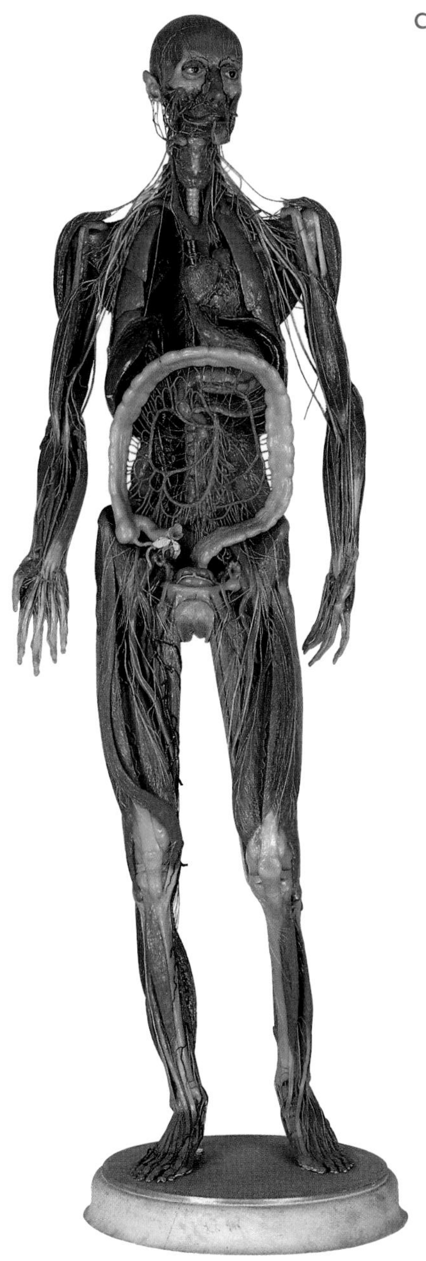

Whole body specimen with thoracic and abdominal cavities laid open. The loops of the small intestine have been removed to display the arteries, veins and nerves.

Ganzkörperpräparat mit eröffneter Brust- und Bauchhöhle. Die Dünndarmschlingen sind entfernt. Darstellung von Arterien, Venen und Nerven

Préparation du corps entier représentant des viscères du thorax et de l'abdomen, l'arbre artériel et les troncs nerveux (les anses de l'intestin grêle sont enlevées)

[XXIX, 748]

Specimen showing the viscera, including the stomach, liver and lesser and greater omentum. The liver has been lifted up to show the gallbladder.

Eingeweidepräparat mit Magen, Leber, sowie dem kleinen und großen Netz. Durch Anheben der Leber wird die Gallenblase sichtbar.

Viscères de l'abdomen : estomac, foie relevé avec vésicule biliaire, petit et grand épiploon (péritoine)

[XXX, 777]

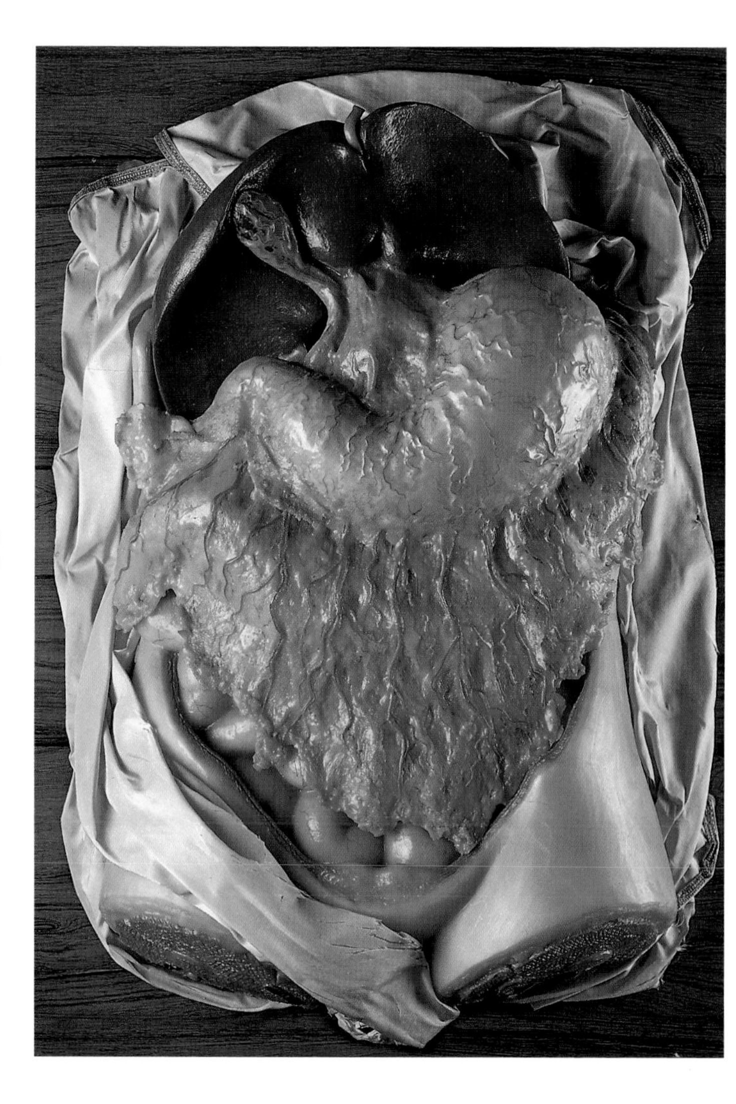

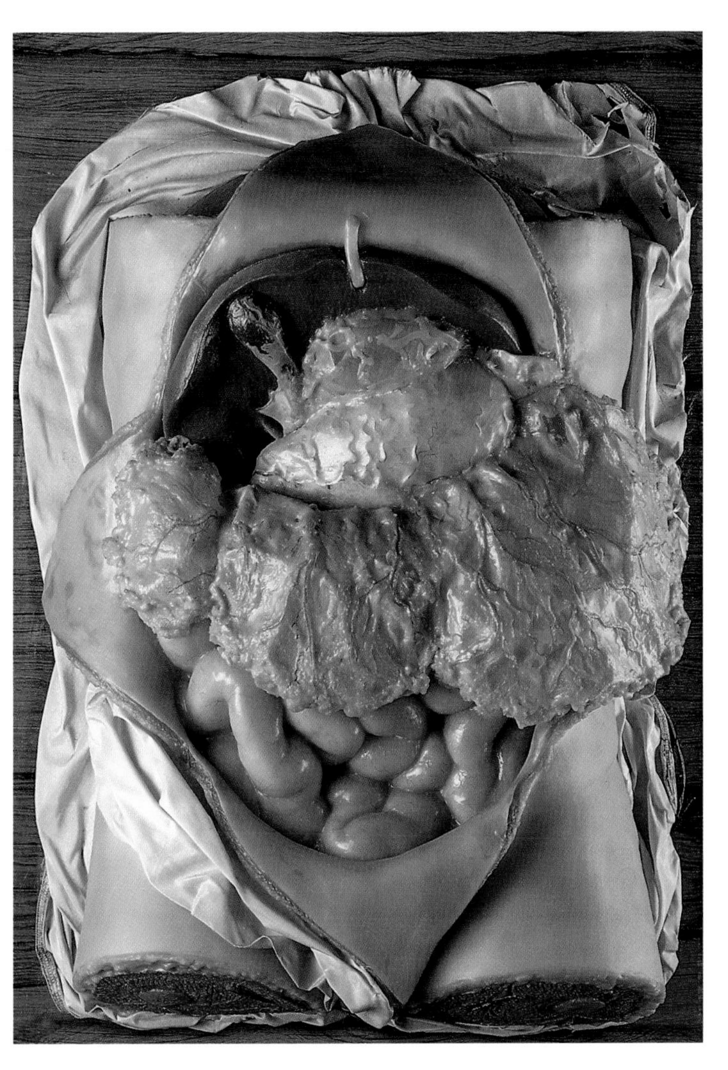

Specimen showing the viscera, including the stomach, liver and greater and lesser omentum, as well as the loops of the small intestine

Eingeweidepräparat mit Magen, Leber, großem und kleinem Netz sowie Dünndarmschlingen

Viscères de l'abdomen : estomac, foie, vésicule biliaire, anses de l'intestin grêle, petit et grand épiploon

[XXX, 837]

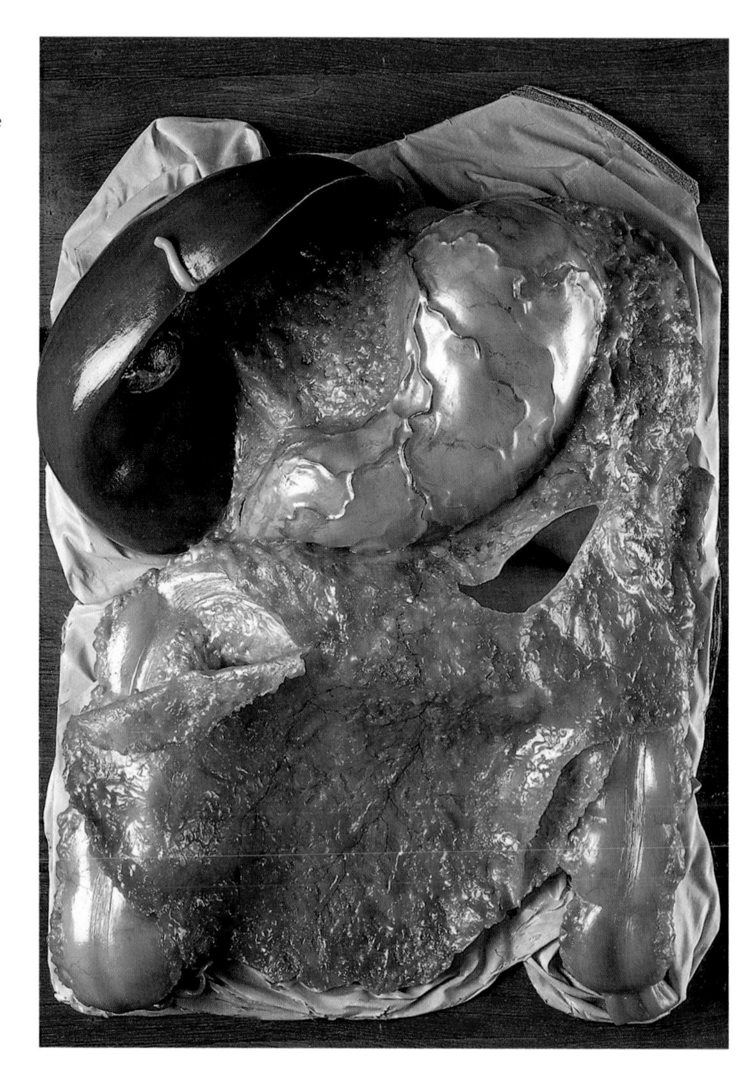

Specimen showing the viscera, including the stomach, liver, large intestine and the lesser and greater omentum

Eingeweidepräparat mit Magen, Leber, Dickdarm sowie dem kleinen und großen Netz

Viscères de l'abdomen : estomac, foie, gros intestin, petit et grand épiploon

[XXX, 778]

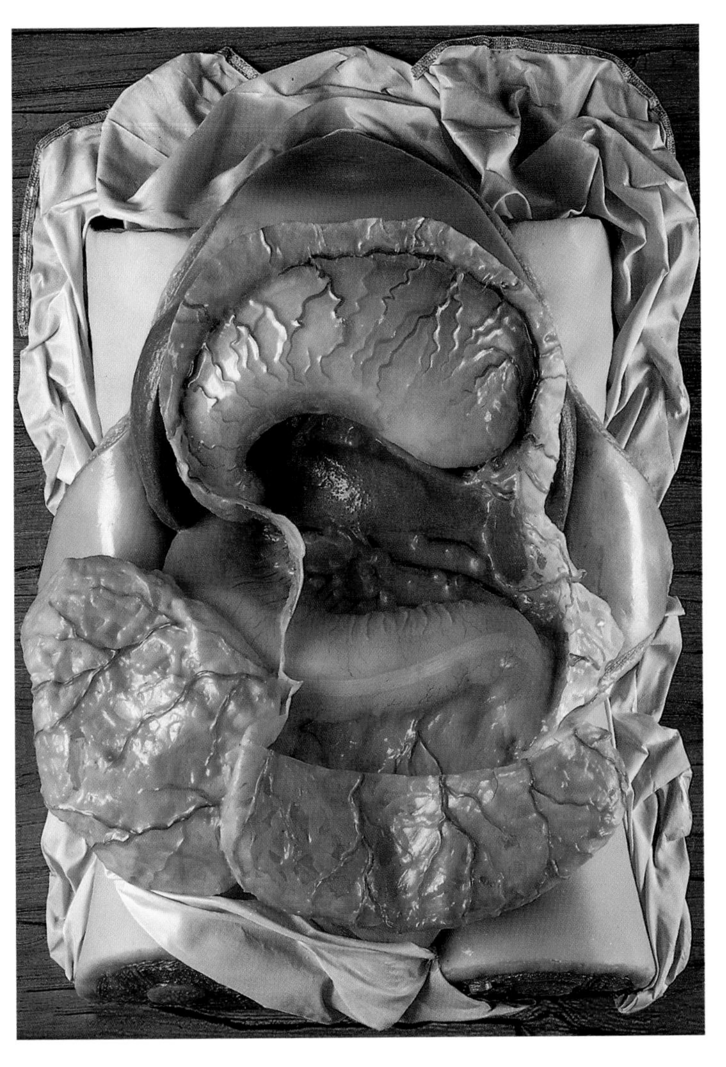

View showing the stomach bed

Blick auf den hinter dem Magen gelegenen Abschnitt der Bauchhöhle

Région de la cavité de l'abdomen située en arrière de l'estomac

[XXX, 791]

View showing the loops of the small intestine and parts of the large intestine after removal of the greater omentum

Blick auf Dünndarmschlingen und Dickdarmabschnitte nach Entfernung des großen Netzes

Viscères de l'abdomen après résection du grand épiploon : foie, estomac, intestin grêle, gros intestin

[XXX, 841]

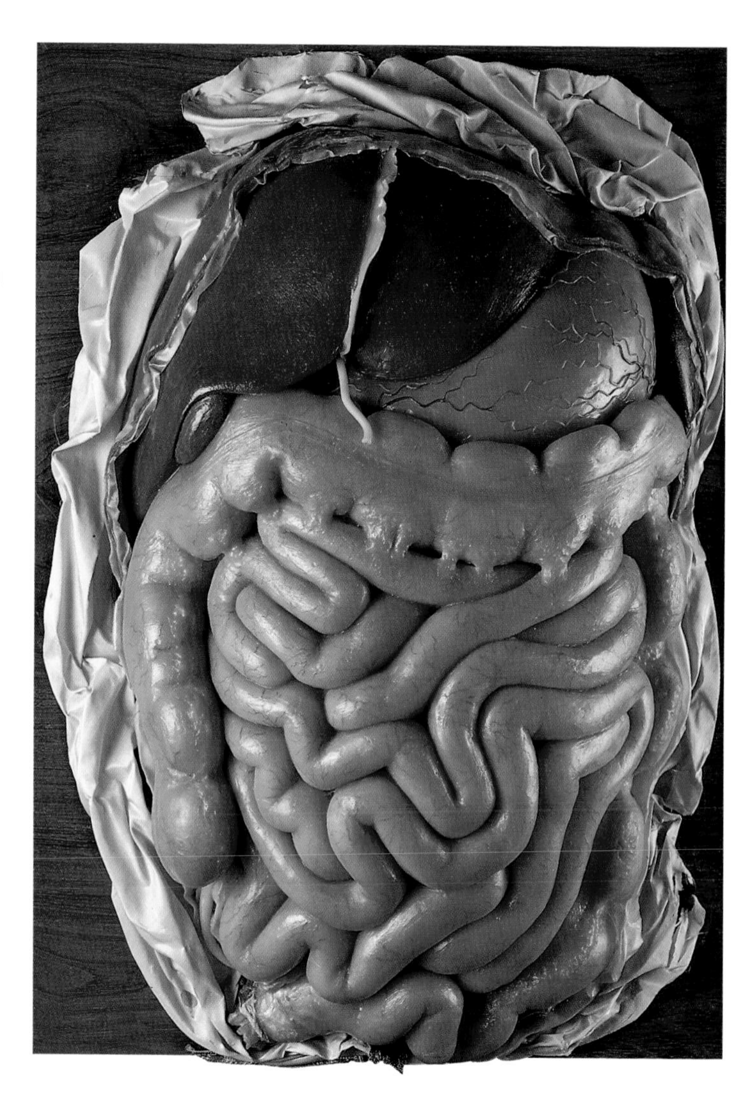

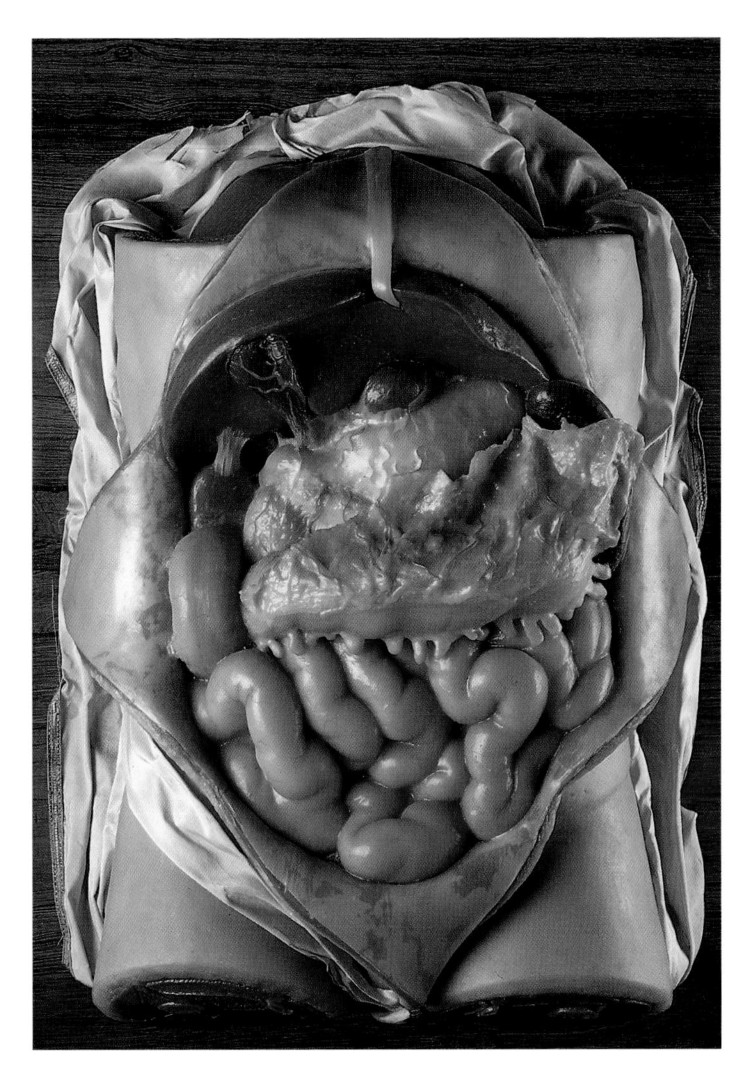

View showing the loops of the small intestine and a part of the large intestine after removal of the greater omentum

Blick auf Dünndarmschlingen und einen Dickdarmabschnitt nach Entfernung des großen Netzes

Anses de l'intestin grêle et partie du gros intestin après résection du grand épiploon

[XXX, 794]

View of the large intestine after removal of the small intestine. The vermiform appendix can be seen on the left.

Blick auf den Dickdarm nach Herausnahme des Dünndarmes. Der Wurmfortsatz ist links im Bild zu erkennen

Gros intestin (caecum, colon) après résection de l'intestin grêle, appendice visible en bas à gauche

[XXX, 842]

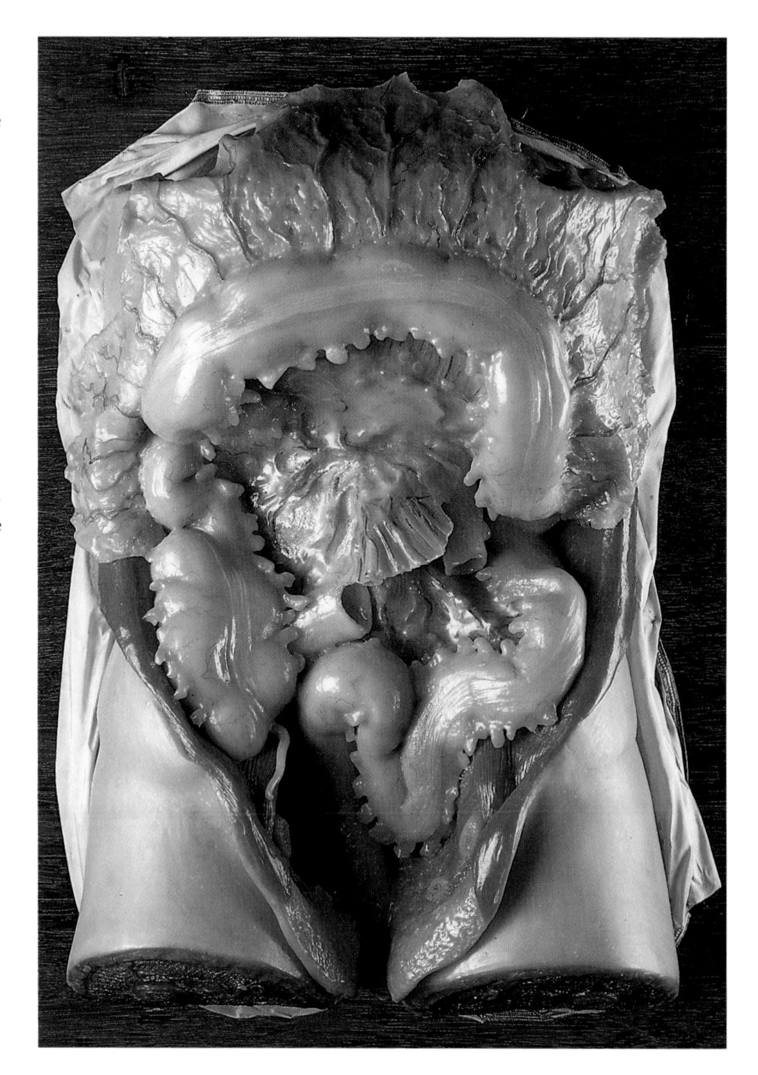

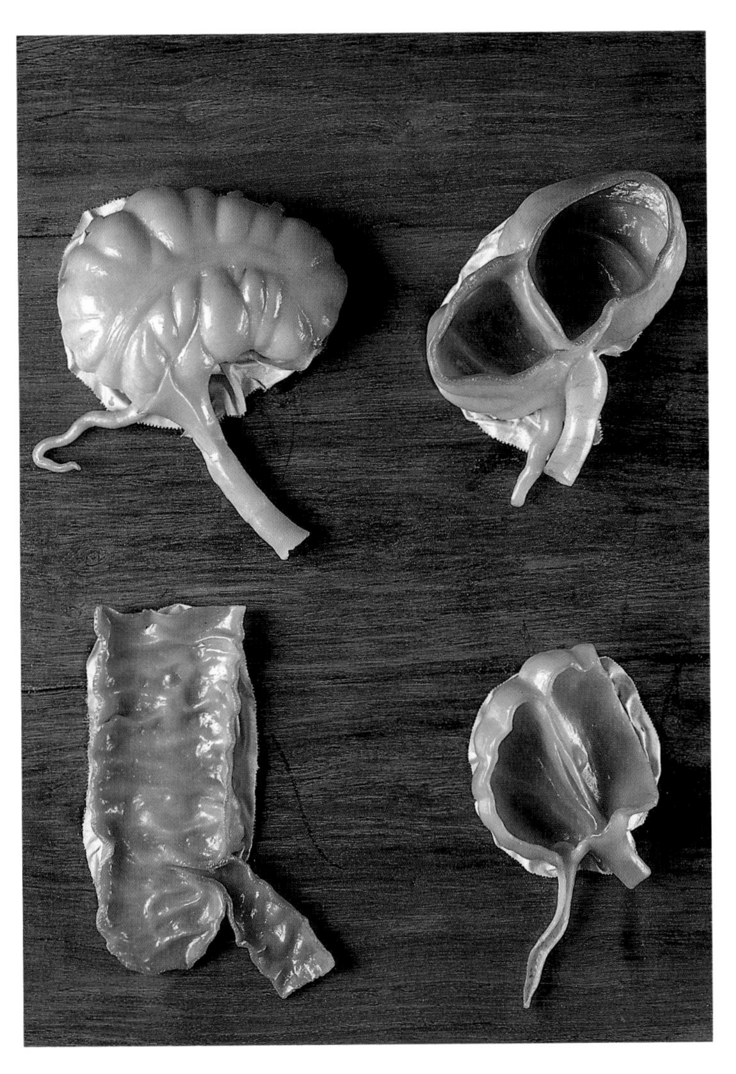

Display showing the vermiform appendix and the entry of the small into the large intestine. The opened specimens show the valve in the region of the entry.

Darstellung des Wurmfortsatzes und der Mündungsstelle des Dünndarmes in den Dickdarm. Die aufgeschnittenen Präparate zeigen die Klappe in der Mündungsregion.

Abouchement de l'intestin grêle (iléon) et de l'appendice dans le gros intestin, valve iléo-caecale

[XXX, 835]

Display showing the venous drainage of the various parts of the large and small intestines

Darstellung der venösen Abflüsse verschiedener Abschnitte des Dünn- und Dickdarmes

Réseaux veineux de l'intestin grêle et du gros intestin

[XXX, 793]

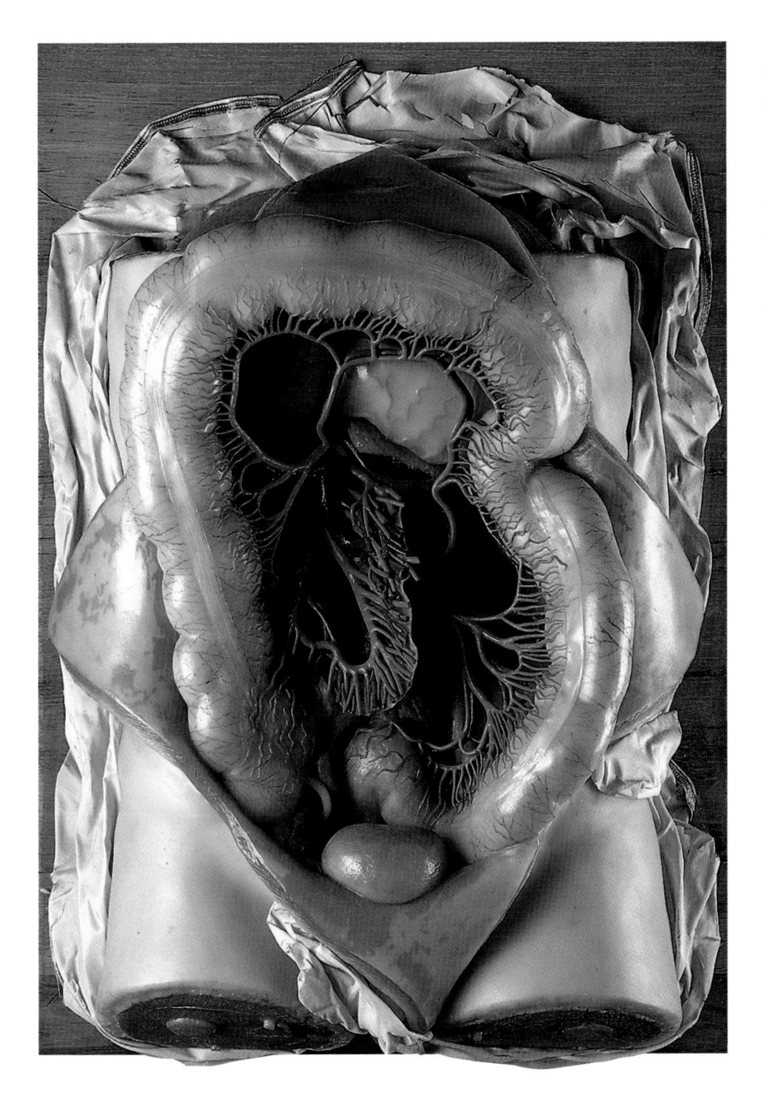

Display showing the arterial blood supply of the large intestine

Darstellung der arteriellen Versorgung des Dickdarmes

Artères du gros intestin

[XXX, 836]

Display showing the arterial blood supply of the stomach, liver, spleen and pancreas

Darstellung der arteriellen Versorgung von Magen, Leber, Milz und Bauchspeicheldrüse

Artères de l'estomac, du foie, de la rate, et du pancréas

[XXX, 840]

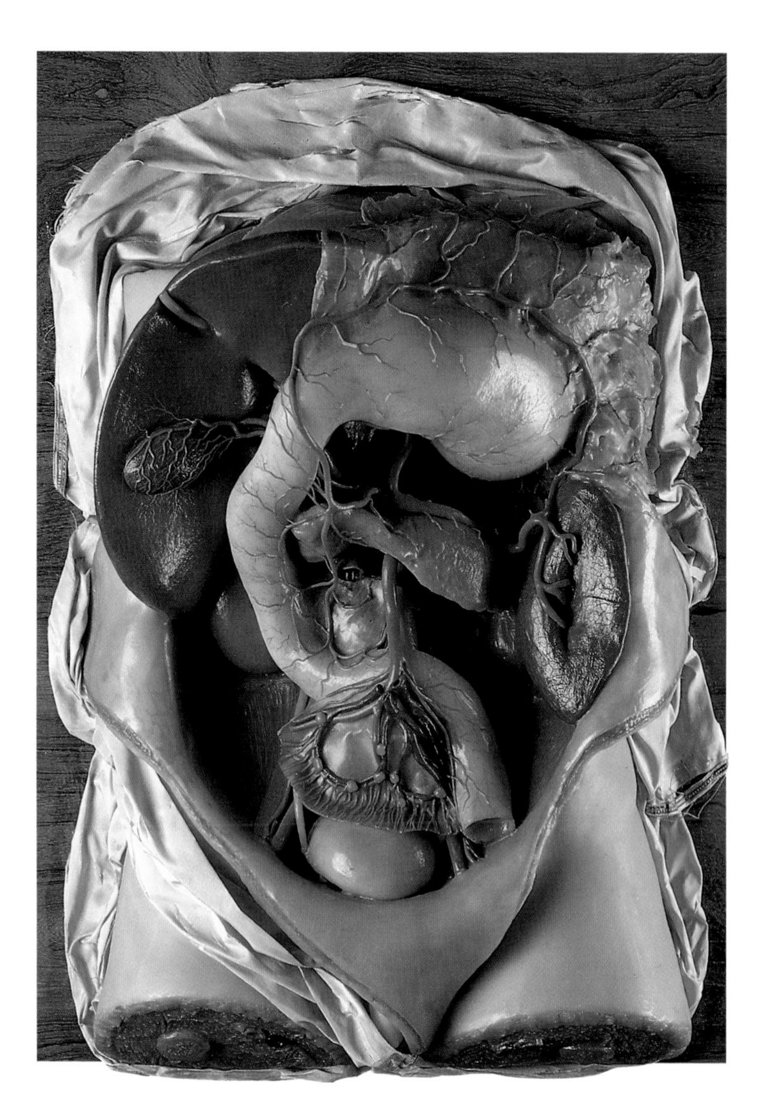

Specimens of the spleen and its vessels. The lower specimen is a bisection of the spleen.

Präparate der Milz mit ihren Gefäßen. Das untere Präparat zeigt die aufgeschnittene Milz.

Artères et veines de la rate (spléniques) ; coupe de la rate (en bas)

[XXVII, 856]

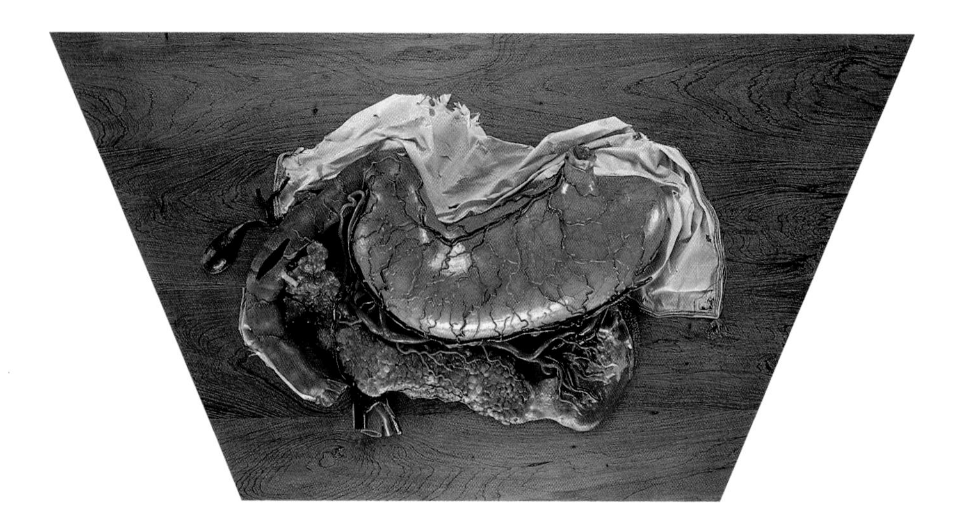

Display showing the stomach, duodenum, gallbladder and pancreas

Darstellung von Magen, Zwölffingerdarm, Gallenblase und Bauchspeicheldrüse

Estomac, duodénum, vésicule biliaire, et pancréas

[XXX, 796]

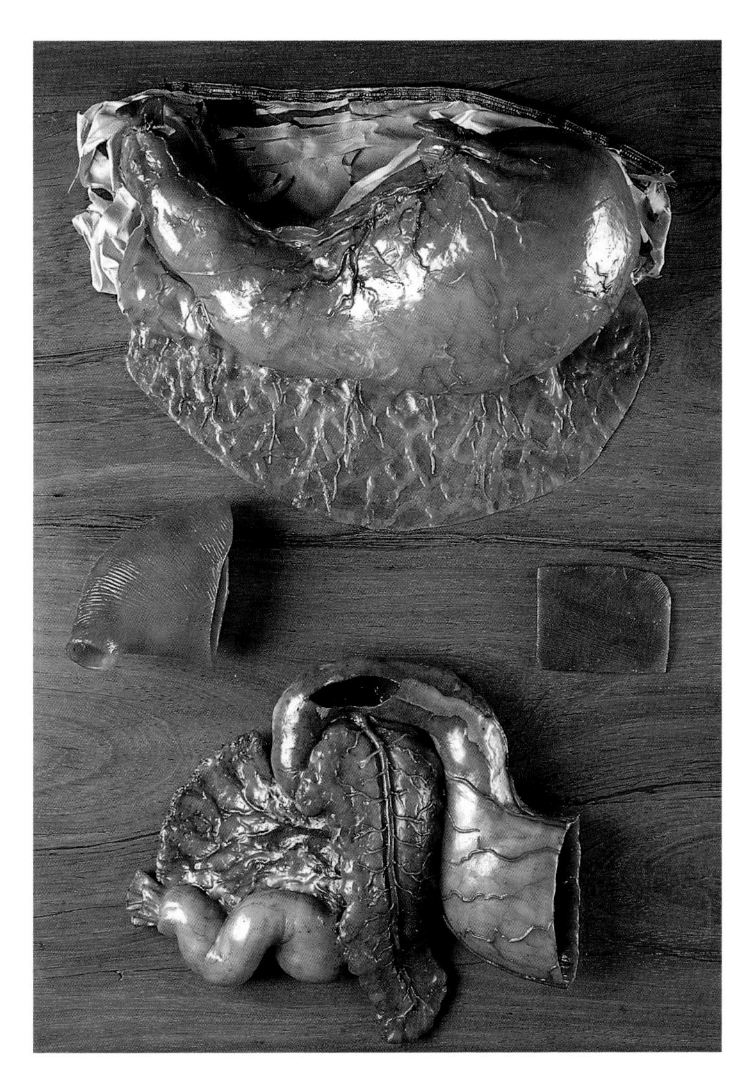

Display showing the stomach (upper specimen) and the pancreatic duct as far as its entry into the duodenum (lower specimen)

Darstellung des Magens (oberes Präparat) und des Ausführungsganges der Bauchspeicheldrüse bis zu seiner Mündung im Zwölffingerdarm (unteres Präparat)

Estomac (en haut) ; conduit excréteur du pancréas jusqu'à son abouchement dans le duodénum (en bas)

[XXX, 774]

Vena splenica

Stomach, pancreas and splenic vein

Darstellung von Magen, Bauchspeicheldrüse und Milzvene

Estomac, duodénum, pancréas, rate, et vaisseaux

[XXX, 775]

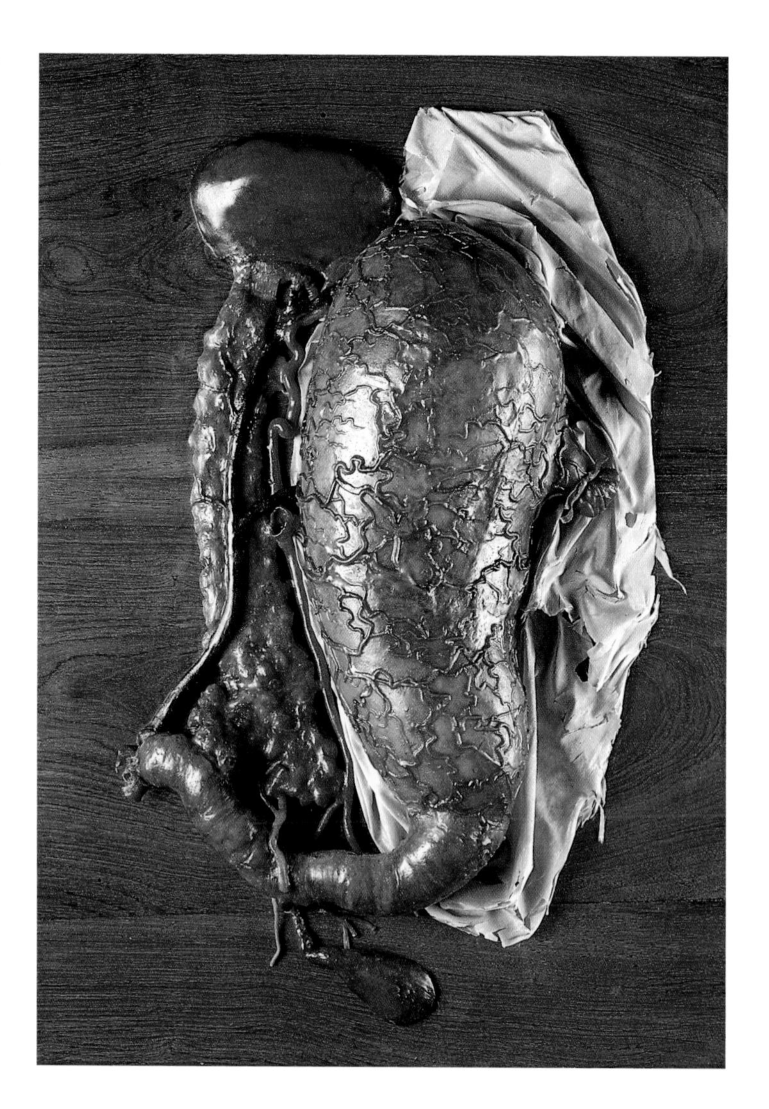

View into the opened stomach

Blick in den eröffneten Magen

Estomac ouvert : muqueuse gastrique

[XXX, 773]

Ligamentum teres hepatis, Vena umbilicalis, Ductus pancreaticus major et minor

Upper specimen: stomach, duodenum, liver and gallbladder. Lower specimen: the C-shaped form of the duodenum and the pancreatic duct

Das obere Präparat zeigt Magen, Zwölffingerdarm, Bauchspeicheldrüse sowie Leber und Gallenblase. Das untere Präparat stellt die Ringfalten des Zwölffingerdarms sowie den Ausführungsgang der Bauchspeicheldrüse dar.

Duodénum, pancréas, foie et voies biliaires (en haut) ; abouchement des voies biliaires et du conduit excréteur du pancréas dans le duodénum ouvert (en bas)

[XXX, 792]

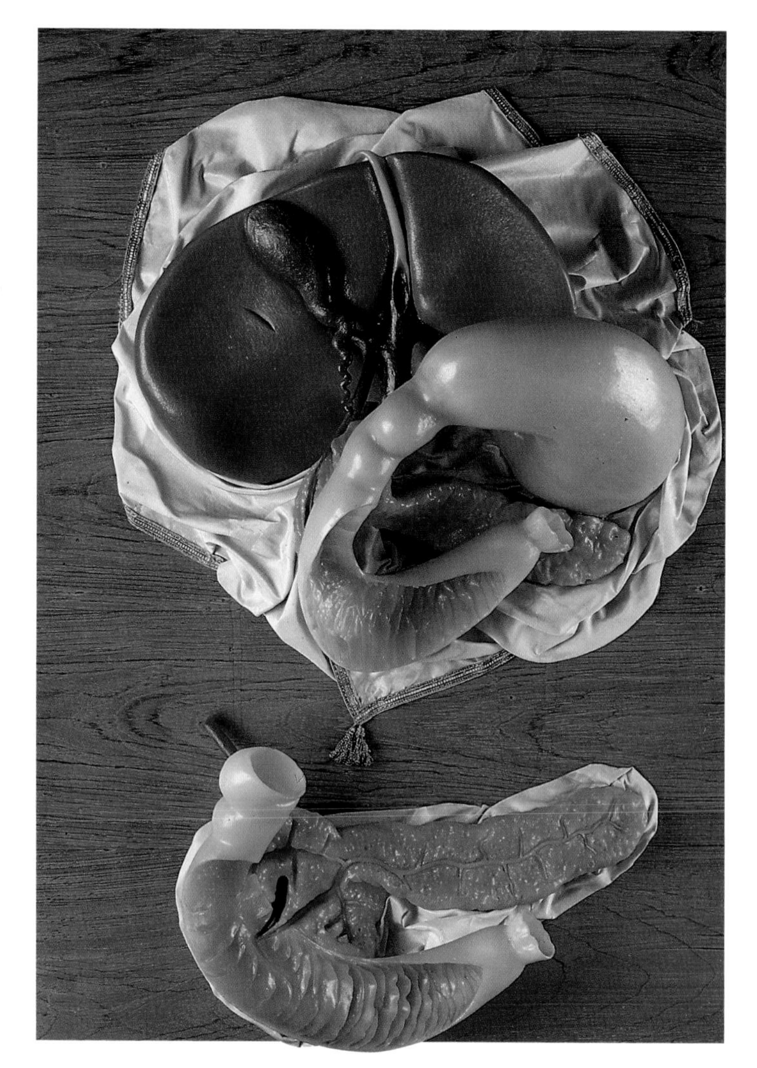

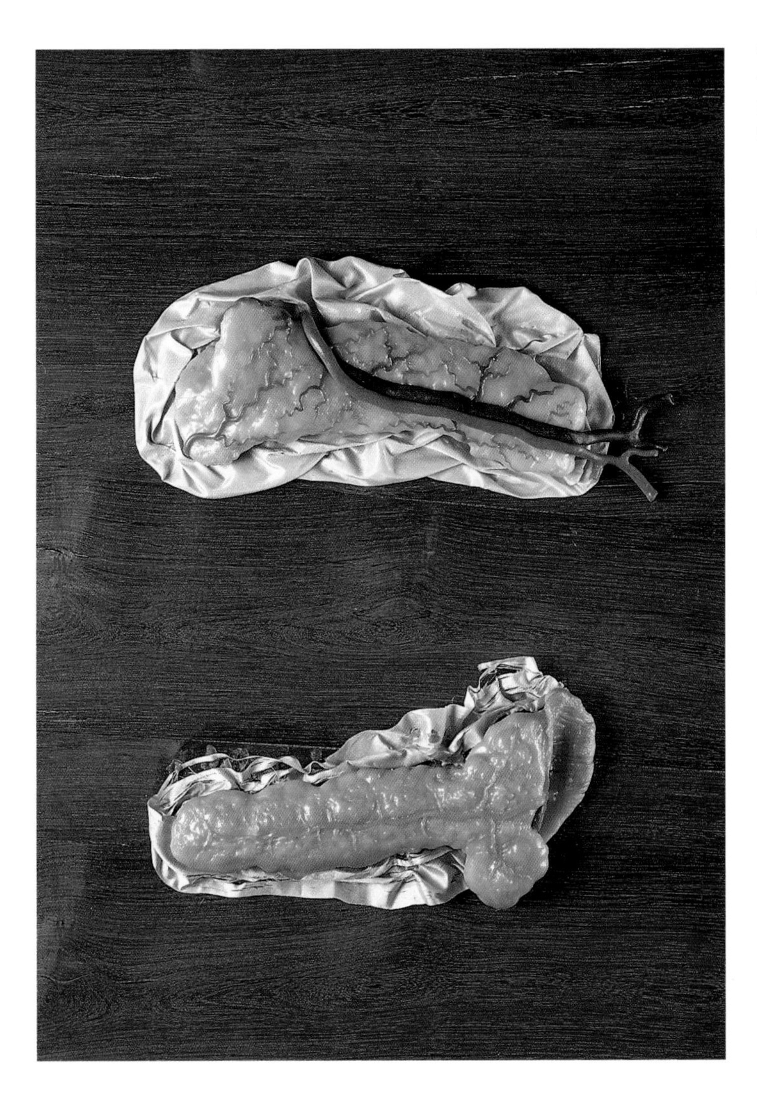

The pancreas; in the upper specimen with the splenic vessels

Darstellungen der Bauchspeicheldrüse, im oberen Präparat mit Milzgefäßen

Pancréas et vaisseaux spléniques

[XXX, 859]

Display showing the relationship of the liver to the diaphragm

Darstellung der Lagebeziehungen von Leber und Zwerchfell

Foie et ligament du péritoine le reliant au diaphragme

[XXX, 858]

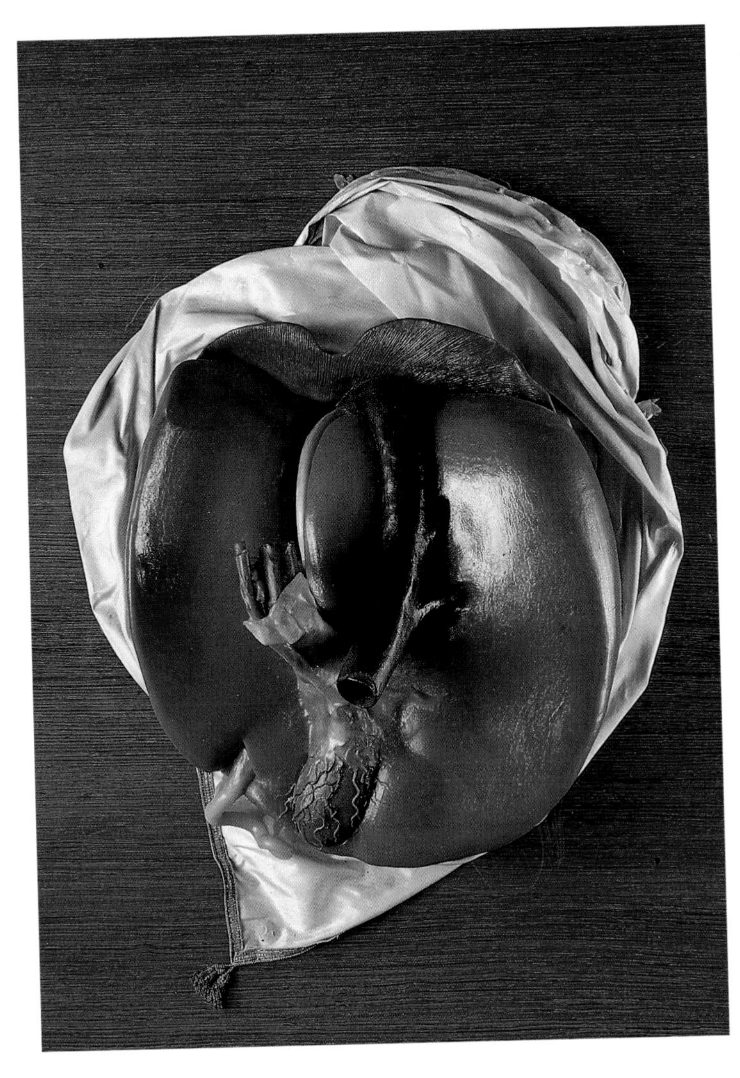

Posterior surface of the liver showing its various lobes and the related course of the inferior vena cava

Leberrückfläche mit verschiedenen Lappen und vorbeiziehender unterer Hohlvene

Face inférieure du foie : lobes hépatiques, vésicule biliaire, hile du foie, veine cave inférieure

[XXX, 857]

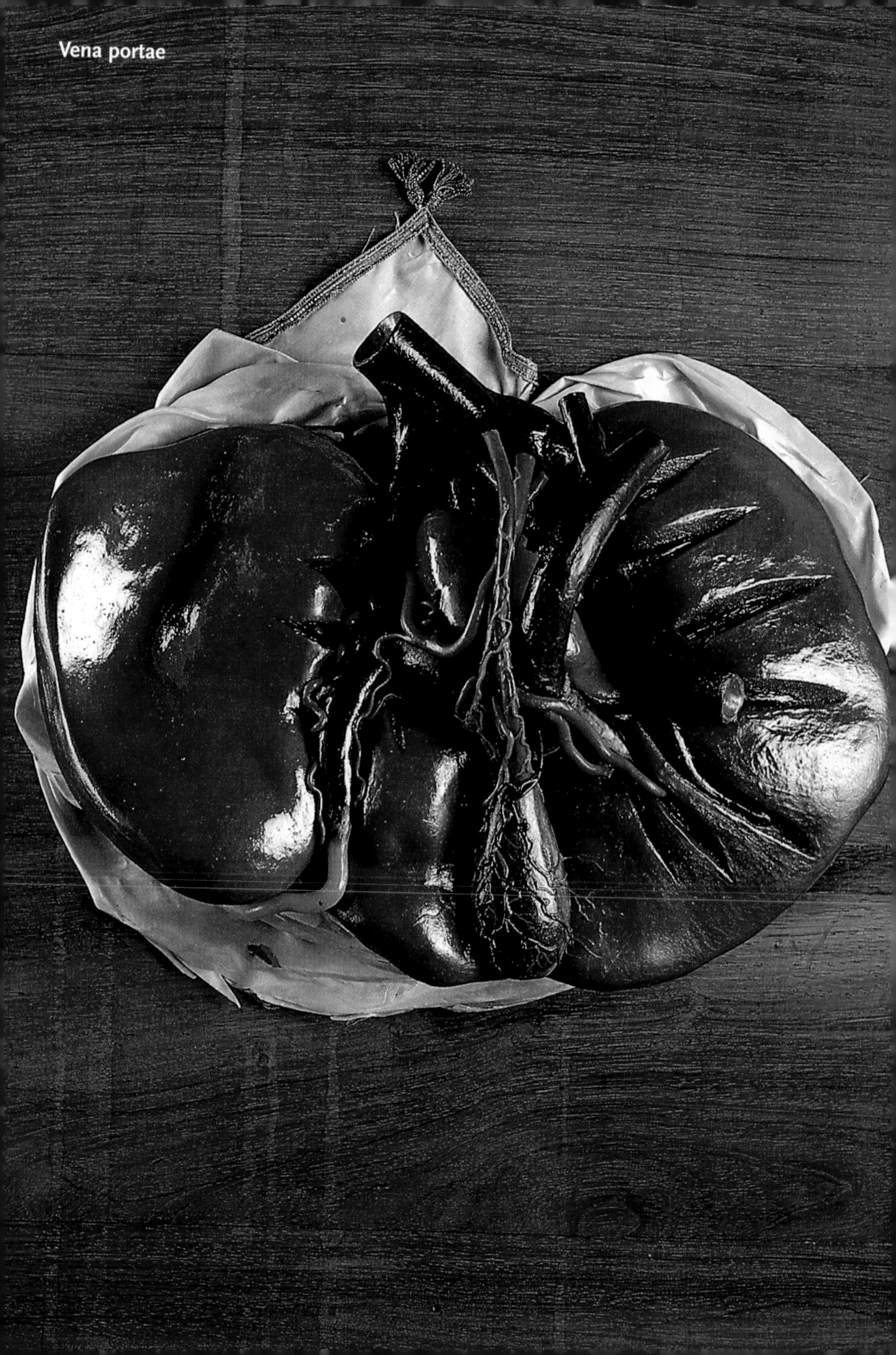
Vena portae

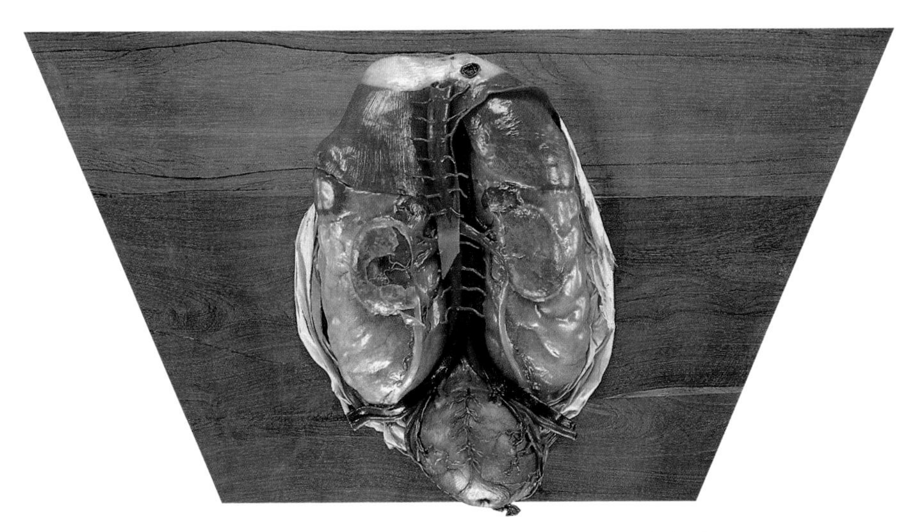

Specimen showing the diaphragm, peritoneum, kidneys and ureters, seen from behind

Blick von hinten auf ein Präparat mit Zwerchfell, Bauchfell und Nieren mit Harnleiter

Viscères de l'abdomen vus de l'arrière : péritoine pariétal, reins et uretères, portion de diaphragme conservée

[XXX, 861]

◄
Posterior surface of the liver showing the vessels of the liver

Leberrückfläche mit Darstellung der Lebergefäße

Face inférieure du foie : branches de l'artère hépatique et de la veine porte disséquées

[XXX, 838]

Systema urogenitale et genitale

Urinary and Reproductive System

Harn- und Geschlechtsorgane

Appareil urinaire, appareil génital

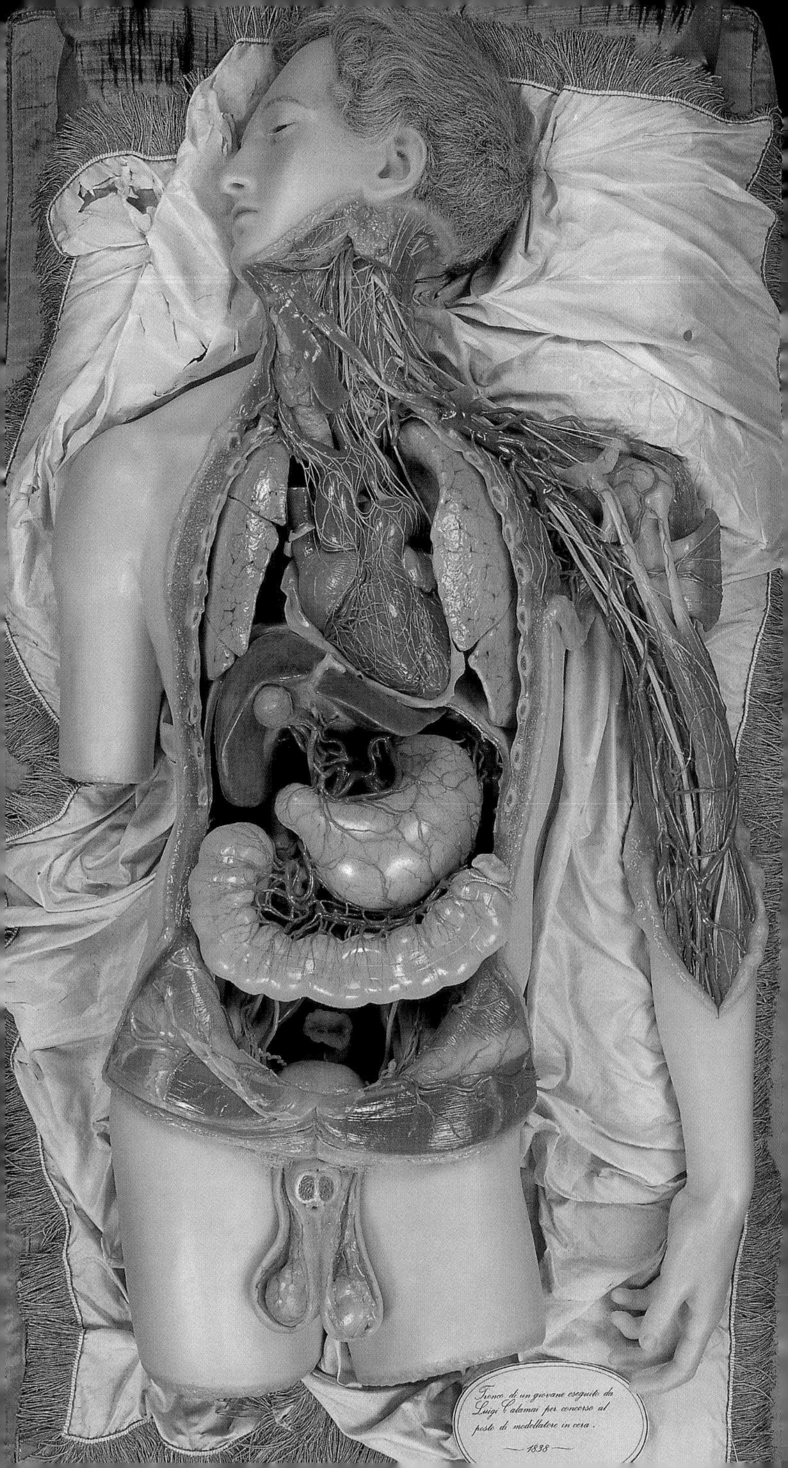
Tronco di un giovane eseguito da
Luigi Calamai per concorso al
posto di modellatore in cera.
1838

Urinary and Reproductive System

The wide range of exhibits in this section reflect very clearly the enlightened interest so characteristic of that time, and which was so far removed from the later prudery of the Victorian era.

The models are anatomically correct representations of the female and male reproductive organs shown in their natural position in the true pelvis. In addition, detailed specimens of functionally related organs are displayed from various points of view. Additional information is given by sections through organs which allow the observer a view into the wide or narrow inside space. Particular emphasis has been laid upon presentations of the uterus taken from the various stages of pregnancy.

The models of embryos are especially informative, as well as those which record the different stages of fetal development up to the time of delivery. Observing the specimens of infants in which the thoracic and abdominal organs are represented in detail, it is the difference in relative size of these organs as compared with those of the adult which is particularly impressive.

Harn- und Geschlechtsorgane

Das breite Spektrum der Exponate in diesem Kapitel spiegelt den aufgeklärten und interessierten Geist der damaligen Zeit eindrucksvoll wider, der sich deutlich von der späteren Prüderie des viktorianischen Zeitalters abhebt.

Die Modelle zeigen anatomisch korrekt die weiblichen und männlichen Geschlechtsorgane in ihrer natürlichen Lage im kleinen Becken. Zudem werden funktionell zueinander in Beziehung stehende Organe als Detailpräparate in verschiedenen Ansichten dargestellt. Zusätzliche Informationen liefern die Schnittpräparate der Organe, die dem Betrachter Einblicke in die verschiedenen weit oder eng gestalteten Innenräume erlauben. Einen thematischen Schwerpunkt bildet die Darstellung der Gebärmutter in den verschiedenen Stadien der Schwangerschaft.

Besonders informativ sind die Modelle der Embryonen sowie die Präparate, die die fötalen Entwicklungsstadien bis zum geburtsreifen Neugeborenen dokumentieren. Bei der Betrachtung der Säuglingspräparate mit Detaildarstellungen der Brust- und Bauchorgane sind besonders die – im Vergleich zur Organanatomie des Erwachsenen – anderen Größenverhältnisse der sich noch in der Entwicklung befindlichen Organe auffällig.

Page · Seite · Page 503:

Male torso with the thoracic and abdominal cavities laid open

Männlicher Torso mit Einblick in die eröffnete Brust- und Bauchhöhle

Viscères du thorax, de l'abdomen, et du bassin masculin [XXX, 447]

Le large choix des pièces concernant les organes génitaux et la grossesse reflète l'état d'esprit éclairé et intéressé qui régnait à l'époque de leur réalisation.

Des représentations anatomiques exactes des organes génitaux féminins et masculins en place dans le petit bassin sont données par plusieurs modèles en cire. A ces vues d'ensemble sont adjointes des préparations plus détaillées d'organes ayant des rapports fonctionnels mutuels, vus sous différentes incidences. Des sections et coupes apportent des précisions supplémentaires en donnant à l'observateur une vue dans les diverses cavités, plus difficilement explorables autrement.

La représentation de l'utérus aux différents stades de la grossesse est aussi un thème particulièrement étudié. Les modèles d'embryons ou les préparations démontrant les stades du développement du fœtus jusqu'à la naissance sont particulièrement instructifs. Des préparations montrent avec précision les organes thoraciques et abdominaux de nouveaux-nés ; il ressort remarquablement de leur observation les différences de proportions entre les organes du nouveau-né et ceux de l'adulte.

Appareil urinaire, appareil génital

View of the posterior abdominal wall after removal of the abdominal viscera. Kidneys, upper urinary tract, urinary bladder, aorta and inferior vena carva

Blick auf die hintere Bauchwand nach Entfernung der Bauchorgane. Darstellung der Nieren, der Harnleiter, der Blase sowie der Bauchschlagader und der unteren Hohlvene

Reins, uretères, vessie, aorte abdominale et veine cave inférieure contre la paroi postérieure de l'abdomen après résection des autres viscères

[XXX, 901]

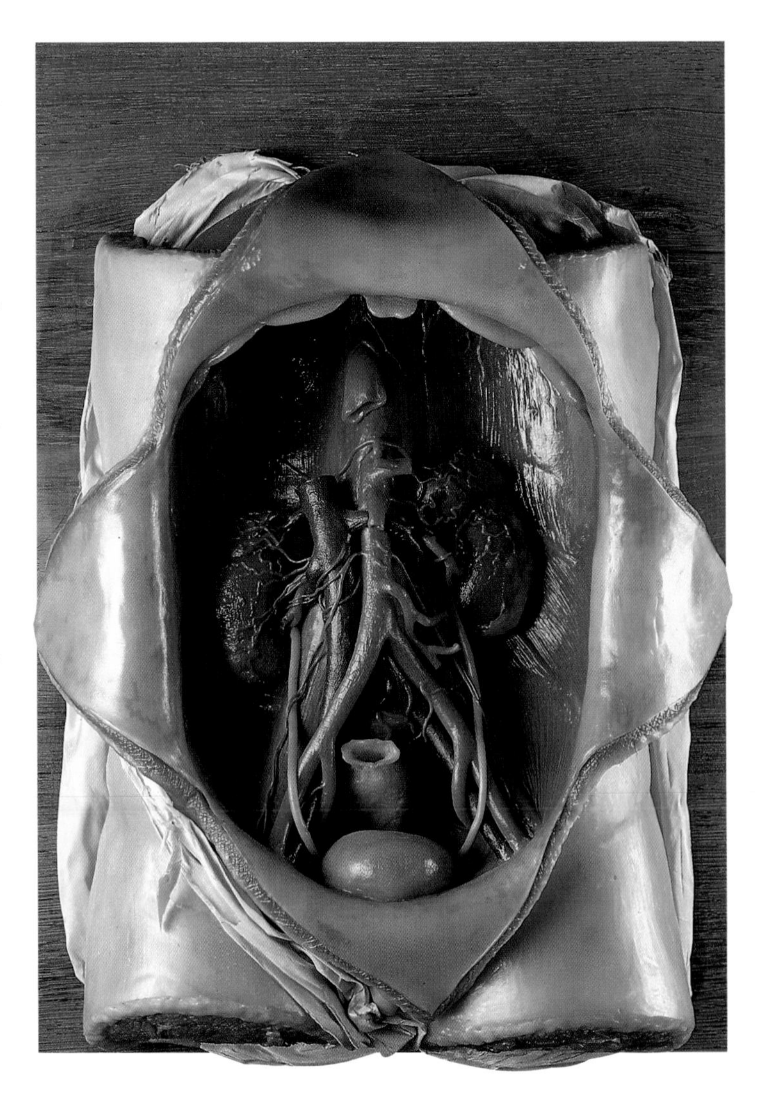

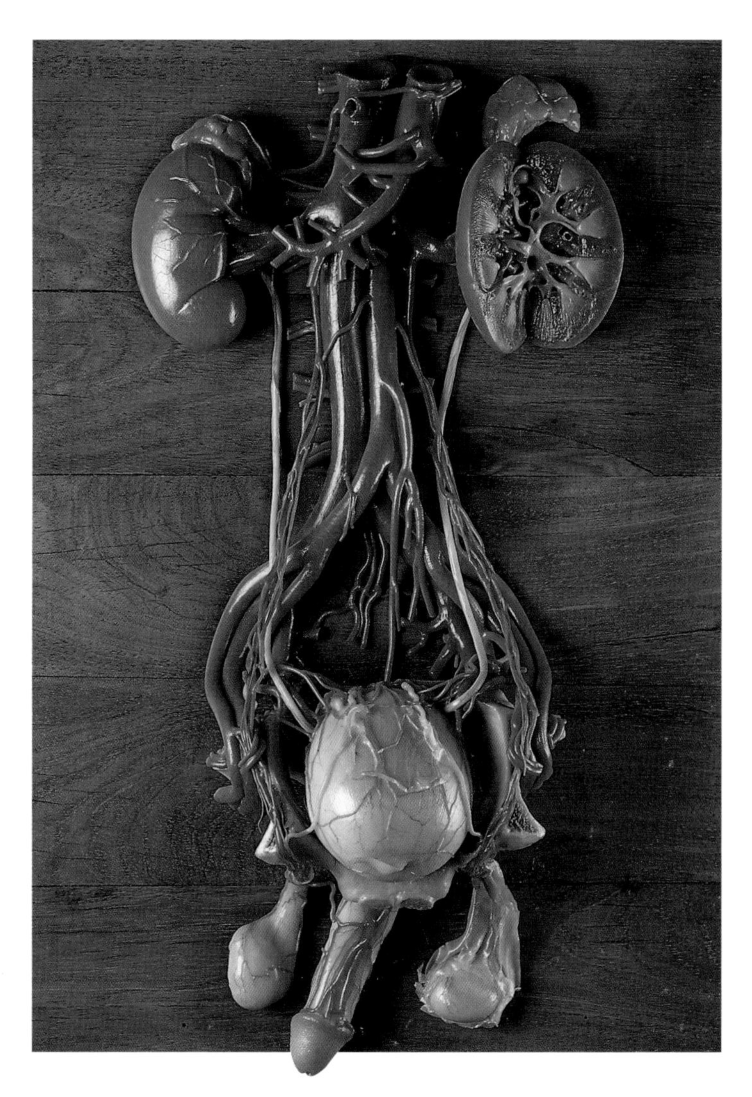

Male urogenital tract with its blood supply. The left kidney has been cut open to reveal the calyces and renal pelvis.

Darstellung des männlichen Urogenitaltraktes mit Gefäßversorgung. Die linke Niere ist aufgeschnitten zur Darstellung der Nierenkelche und des Nierenbeckens.

Appareils urinaire et génital masculins avec leur vascularisation artérielle, coupe d'un rein : calices, bassinet

[XXX, 830]

Various aspects of the kidney

Verschiedene Darstellungen der Niere

Reins : morphologie, structure, vascularisation

[XXX, 831]

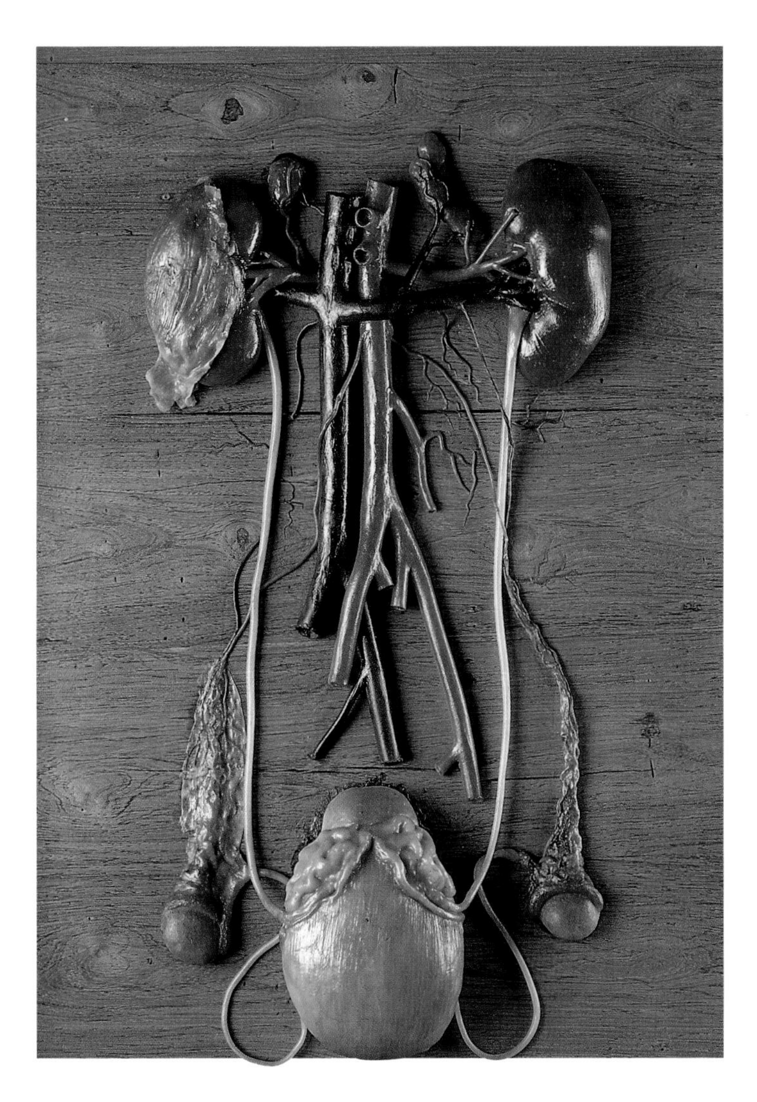

Specimen of the male urogenital tract showing bladder wall with seminal vesicles and prostate

Präparat des männlichen Urogenitaltraktes mit Blick auf die hintere Blasenwand mit Bläschen- und Vorsteherdrüse

Appareils urinaire et génital masculins : vessie vue de l'arrière, glandes séminales, prostate, testicules

[XXX, 847]

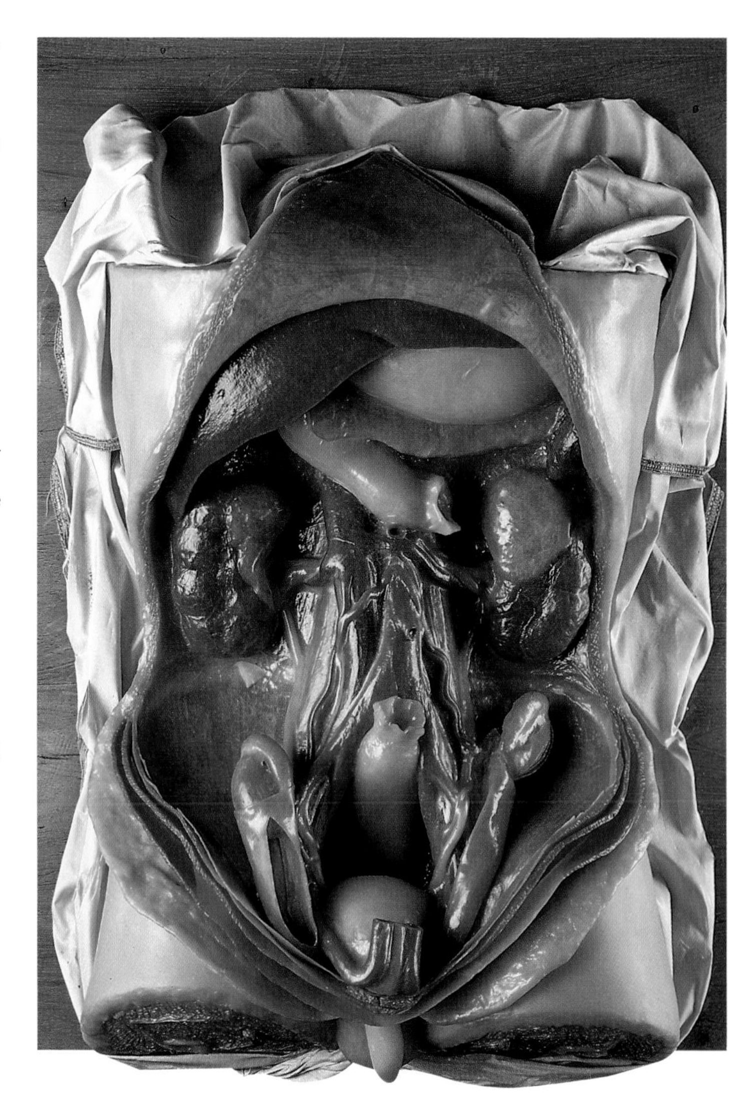

Organs on the posterior abdominal wall of a male fetus. In the true pelvis the right and left testes can be seen. At this stage of development they have not yet reached the scrotum.

Blick auf die Organe der hinteren Bauchwand eines männlichen Föten. Im Bereich des kleinen Beckens sind rechts und links die Hoden zu erkennen, die zu diesem Zeitpunkt noch nicht im Hodensack liegen.

Appareils urinaire et génital du fœtus masculin : testicules dans la cavité abdominale et non encore descendus dans les bourses

[XXX, 846]

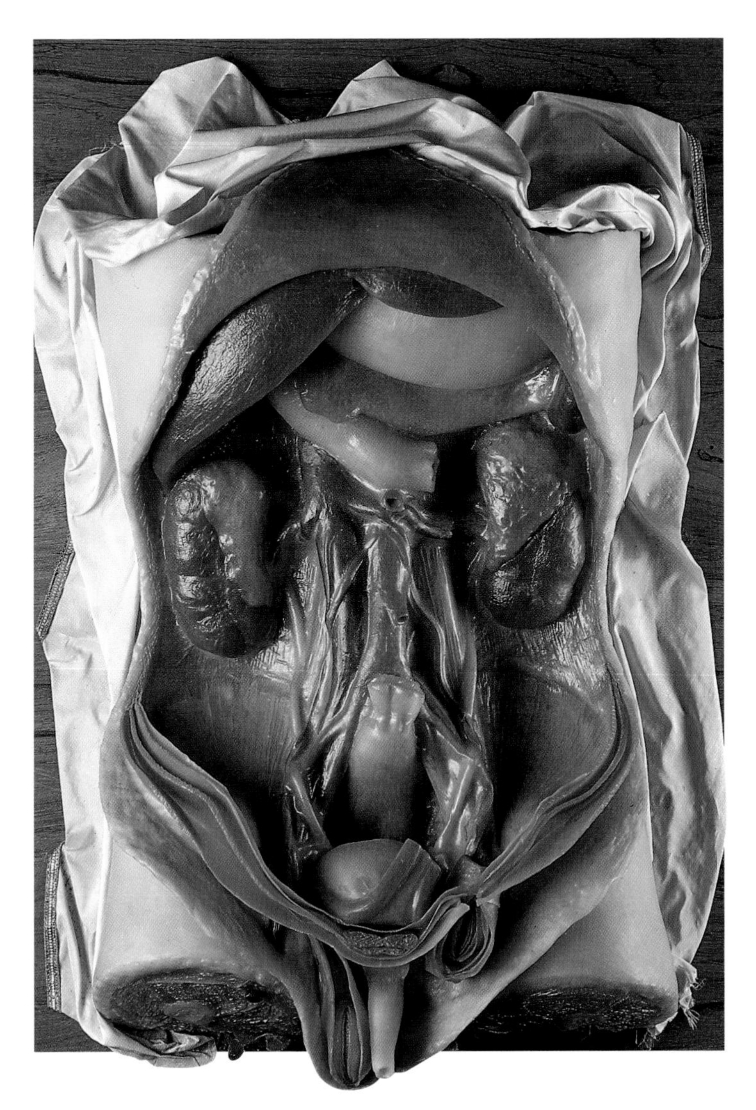

Coverings of the testis in a newborn infant

Darstellung der Hodenhüllen von einem Neugeborenen

Appareil uro-génital d'un nouveau-né : enveloppes du testicule

[XXX, 779]

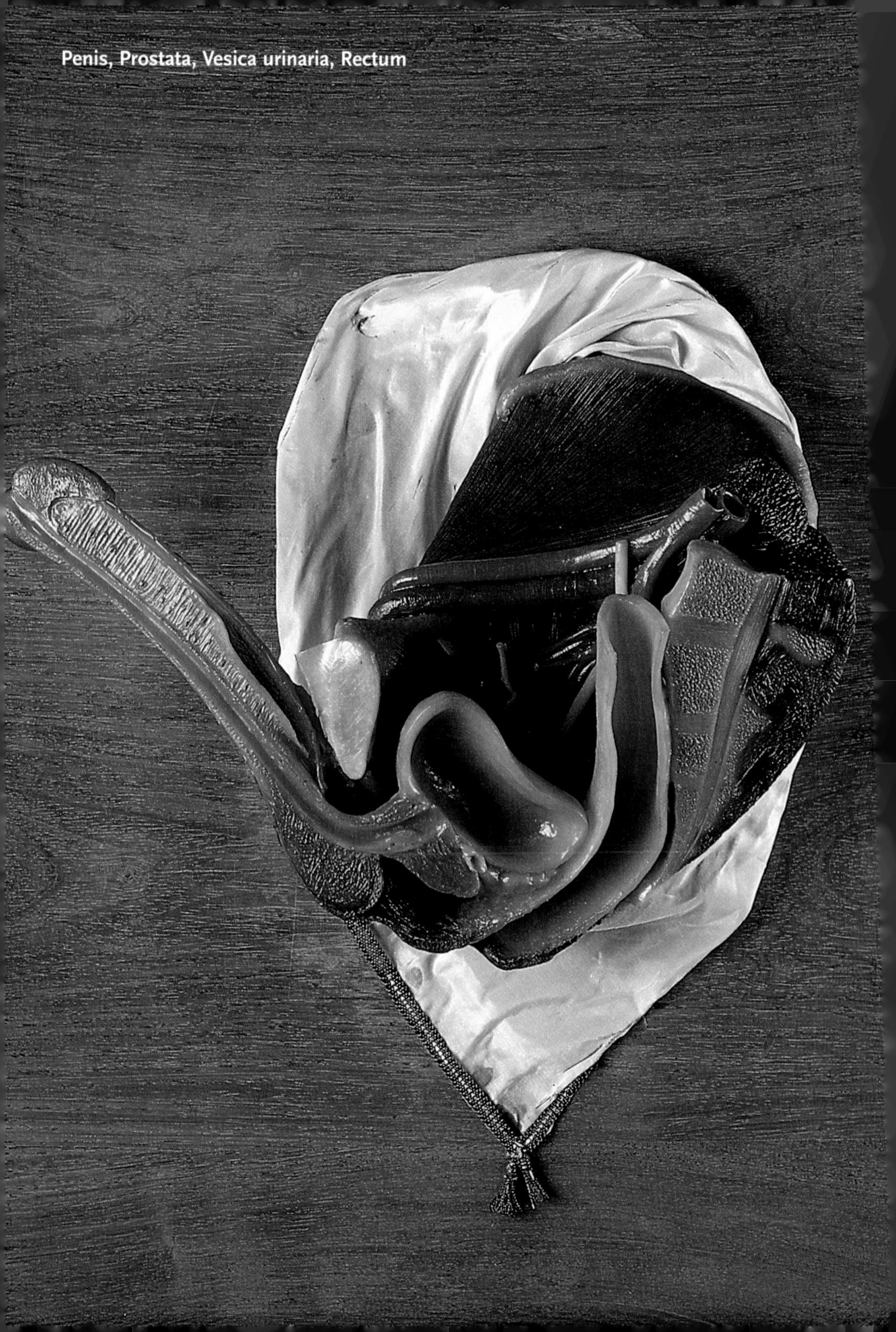

Penis, Prostata, Vesica urinaria, Rectum

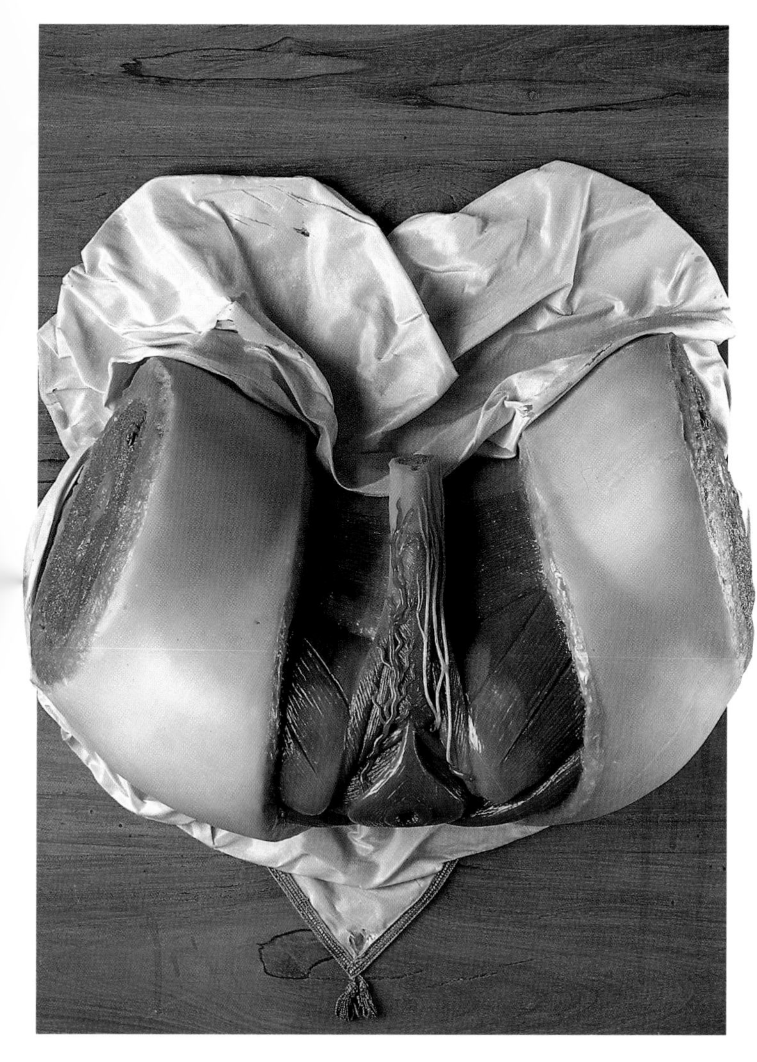

External view of the floor of the pelvis showing the root of the penis with its nerves and vessels

Präparat des äußeren Beckenbodens sowie der Peniswurzel mit Nerven und Gefäßen

Plancher du bassin masculin (périnée) : racine du pénis, artères, nerfs

[OSTETRICIA, 990]

◄

Specimen of a male pelvis which has been divided in the midline

Präparat eines median durchtrennten männlichen Beckens

Bassin masculin, coupe médiane

[OSTETRICIA, 989]

View into a hemisected male pelvis

Einblick in ein halbiertes männliches Becken

Viscères du bassin masculin : vessie, rectum

[XXXI, 400]

►
View into the male pelvis showing urinary bladder and rectum

Einblick in ein männliches Becken mit Harnblase und Mastdarm

Viscères du bassin masculin : vessie, rectum

[XXX, 401]

Vesica urinaria

Various aspects of male urinary bladder with prostate gland. The upper specimen on the left has been opened to show the urinary bladder and urethra.

Männliche Harnblase mit Vorsteherdrüse in verschiedenen Ansichten. Das obere linke Präparat zeigt die eröffnete Harnblase und Harnröhre.

Vessie chez l'homme, prostate, artères ; vessie et urèthre ouverts (en bas à droite)

[XXX, 848]

Specimen of the male genital organs. Above center: the urethra has been laid open. Below right: the urinary bladder seen from behind and the chestnut-shaped prostate gland

Präparte von männlichen Geschlechtsorganen. Im oberen zentral gelegenen Präparat ist die eröffnete Harnröhre dargestellt. Das Präparat rechts unten zeigt die Harnblase von hinten und die kastanienförmige Vorsteherdrüse

Organes uro-génitaux masculins : urèthre ouvert (au milieu) ; vessie, prostate en forme de châtaigne, glandes séminales, vus de l'arrière (en bas à droite)

[OSTETRICIA, 987]

Penis, Arteria dorsalis penis

Various aspects of the blood supply of the penis

Verschiedene Darstellungen der Blutversorgung des Penis

Pénis : artères, veines, et nerfs

[OSTETRICIA, 988]

Various aspects of testis, epididymis and their coverings

Verschiedene Darstellungen des Hodens, Nebenhodens und ihrer Hüllen

Testicules, épididyme, et enveloppes, cordon spermatique

[OSTETRICIA, 1011]

Glandula vesiculosa, Prostata, Musculus levator ani, Musculus bulbospongiosus, Musculus bulbocavernosus

View of the male pelvic floor from below

Blick von unten auf ein Präparat des männlichen Beckenbodens

Plancher du bassin masculin vu du dessous

[XXX, 827]

►
Specimens of the female breast

Präparate der weiblichen Brust

Seins, glande mammaire

[XXX, 790]

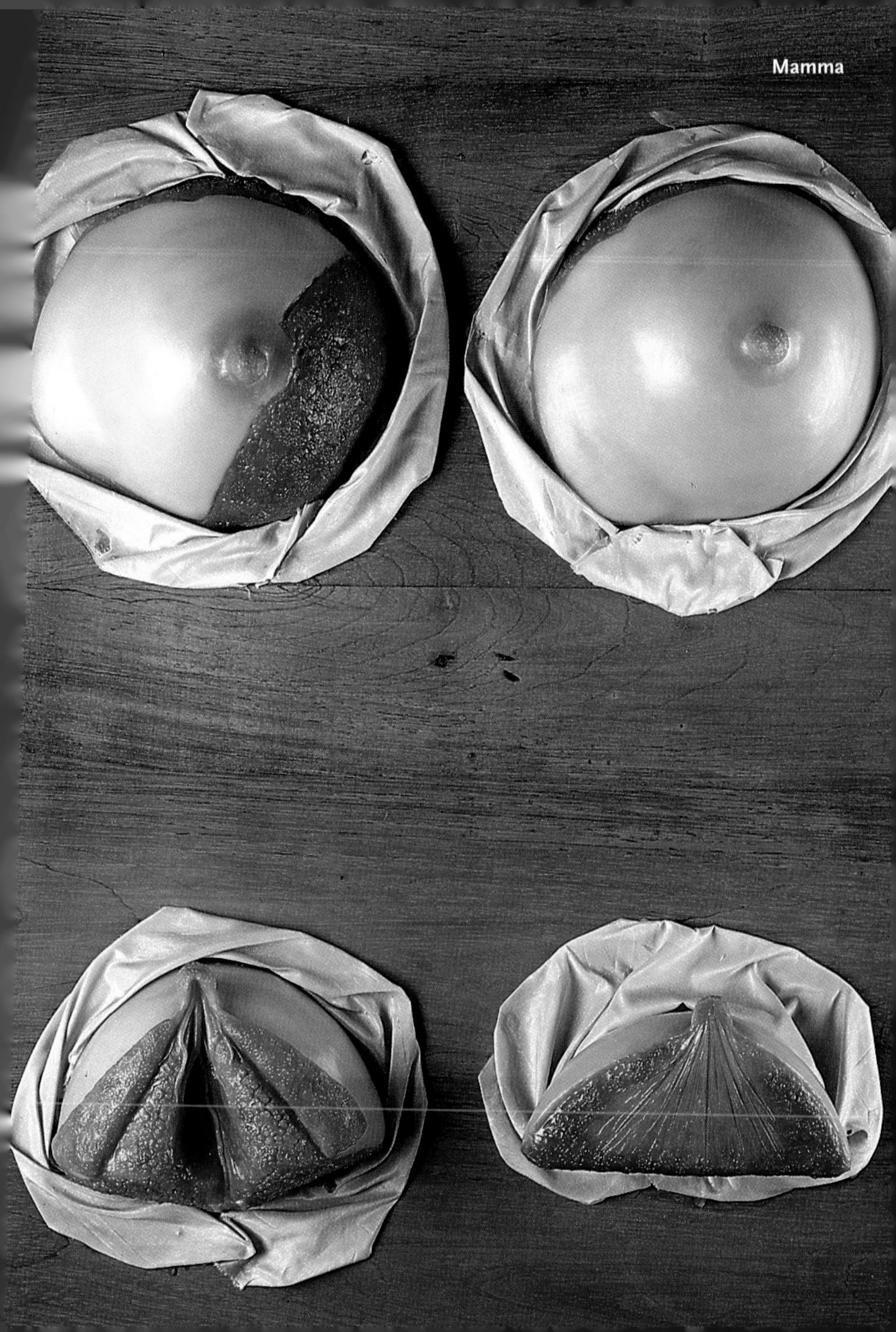

Apparatus urogenitalis femininus

Female urogenital tract with its blood supply

Darstellung des weiblichen Urogenitaltraktes mit Gefäßversorgung

Appareils urinaire et génital féminins et leur vascularisation

[XXX, 832]

View from in front onto the internal genital organs and urinary bladder of a woman. Above: uterine tube with the infundibulum for reception of the ova

Blick von vorne auf die inneren Geschlechtsorgane und die Harnblase der Frau. Das obere Präparat zeigt den Eileiter mit dem Trichter für die Aufnahme der Eizellen.

Organes génitaux internes féminins et vessie vus de l'avant (en bas) ; trompes utérines rattachées à l'utérus (en haut)

[OSTETRICIA, 982]

View from behind onto the internal genital organs and urinary bladder of a woman (lower specimen). Above: the clitoris and its muscles

Blick von hinten auf die inneren Geschlechtsorgane und die Harnblase der Frau (unteres Präparat). Das obere Präparat zeigt die Klitoris mit ihrer Muskulatur.

Organes génitaux internes féminins : utérus, trompes, vagin, et vessie vus de l'arrière (en bas) ; clitoris et ses muscles (en haut)

[OSTETRICIA, 981]

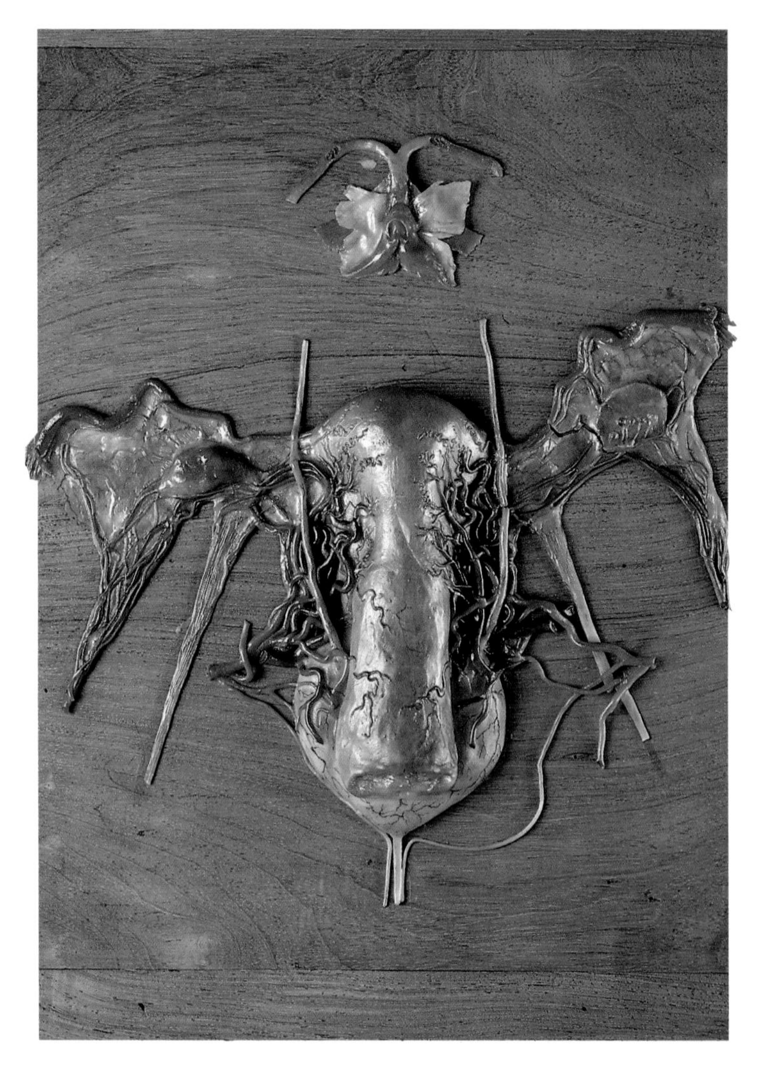

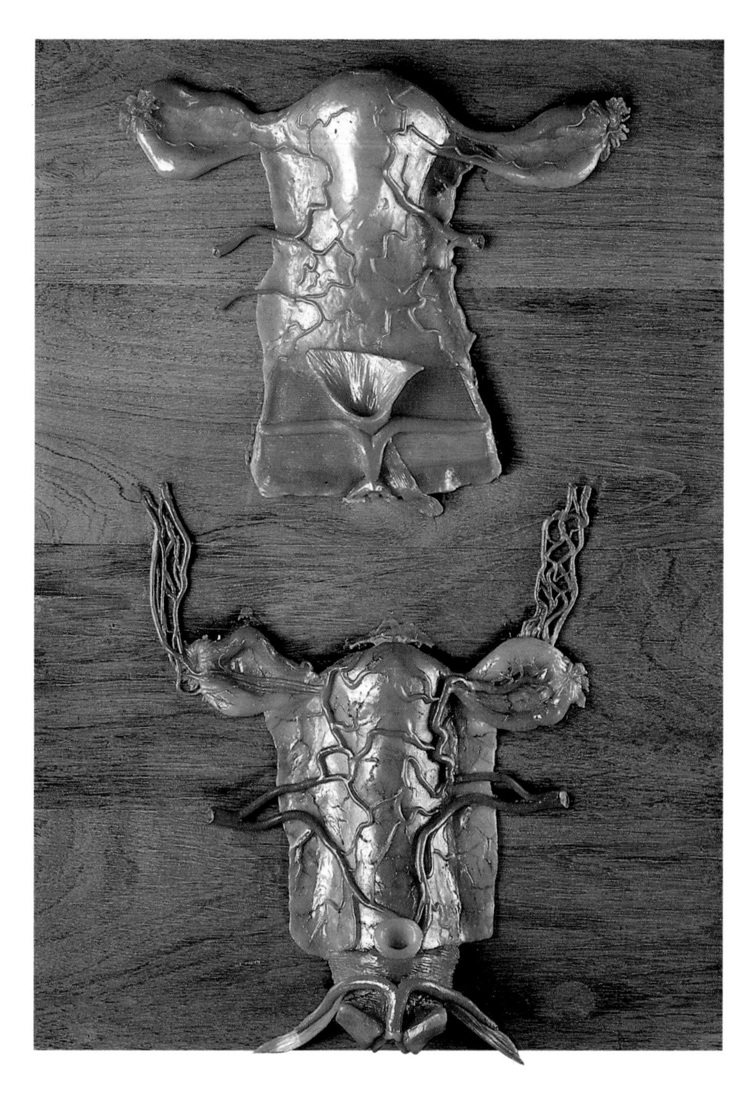

The internal reproductive organs of a woman

Darstellungen der inneren Geschlechtsorgane der Frau

Organes génitaux internes féminins, artères (en haut) et veines (en bas)

[OSTETRICIA, 984]

Pathologically affected placenta

Krankhaft veränderte Plazenta

Placenta dégénéré par une maladie

[DEP. 11]

Various specimens of the female reproductive organs with ovary, tube, uterus and vagina

Verschiedene Präparate der weiblichen Geschlechtsorgane mit Eierstock, Eileiter, Gebärmutter und Scheide

Organes génitaux féminins : ovaire, trompe utérine, utérus, vagin

[OSTETRICIA, 1009]

Specimen of a female pelvis divided in the midline

Präparat eines median durchtrennten weiblichen Beckens

Bassin féminin, coupe médiane

[OSTETRICIA, 1008]

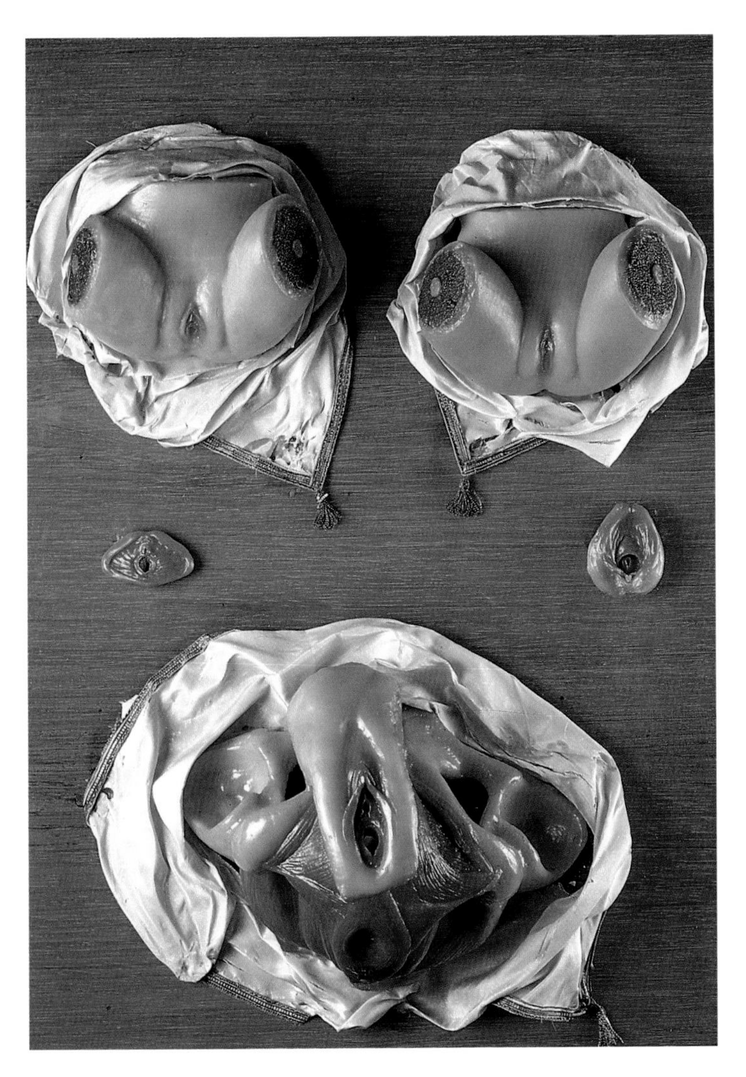

External genital organs of a female infant

Äußere Geschlechtsorgane eines weiblichen Säuglings

Organes génitaux externes féminins du nourrisson

[OSTETRICIA, 1010]

Female external genital organs

Präparate der äußeren weiblichen Geschlechtsorgane

Organes génitaux externes feminins

[OSTETRICIA, 1012]

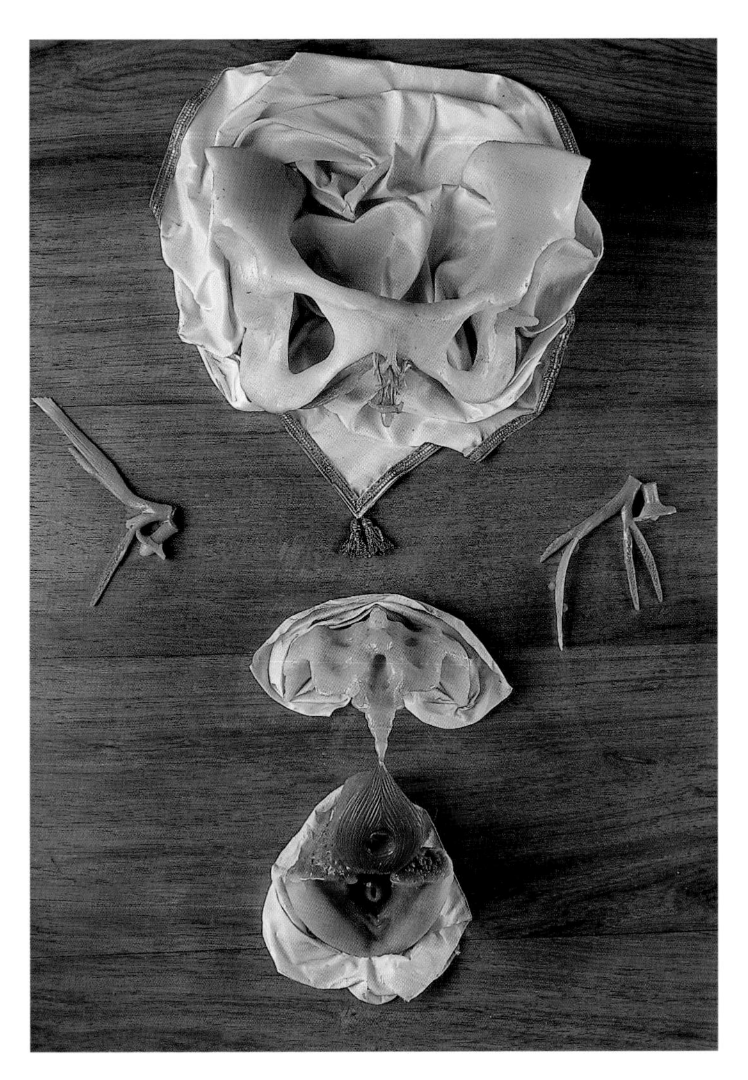

Above: female pelvis with clitoris. Below: view of the coccyx and part of the pelvic floor, seen from below

Oberes Präparat: Weibliches Becken mit Darstellung der Klitoris. Unteres Präparat: Ansicht auf das Steißbein und Teile des Beckenbodens von unten

Bassin osseux avec clitoris (en haut) ; sacrum, coccyx et partie du plancher du bassin (en bas)

[OSTETRICIA, 1007]

Female external genital organs

Darstellung der äußeren weiblichen Geschlechtsorgane

Organes génitaux externes féminins (position gynécologique)

[OSTETRICIA, 1015]

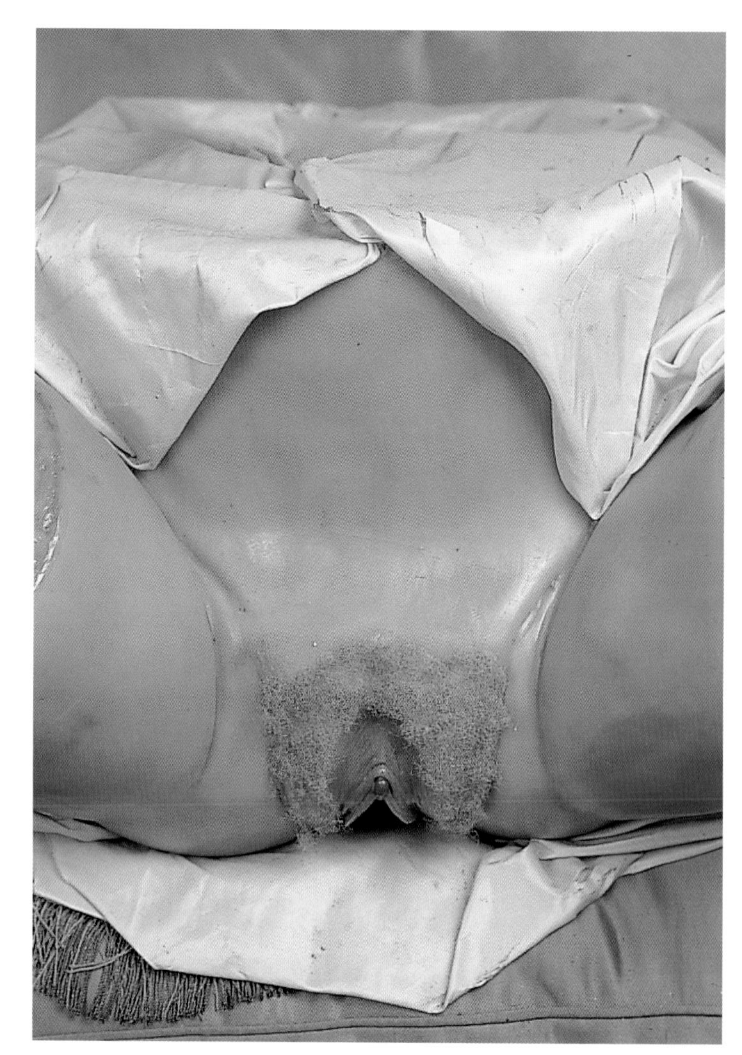

Pages · Seiten · Pages 532–533:

Whole body specimen seen from below (cf. pp. 474–475)

Ganzkörperpräparat in Untersicht (vgl. S. 474–475)

Préparation du corps entier, vu du bas (cf. p. 474–475)

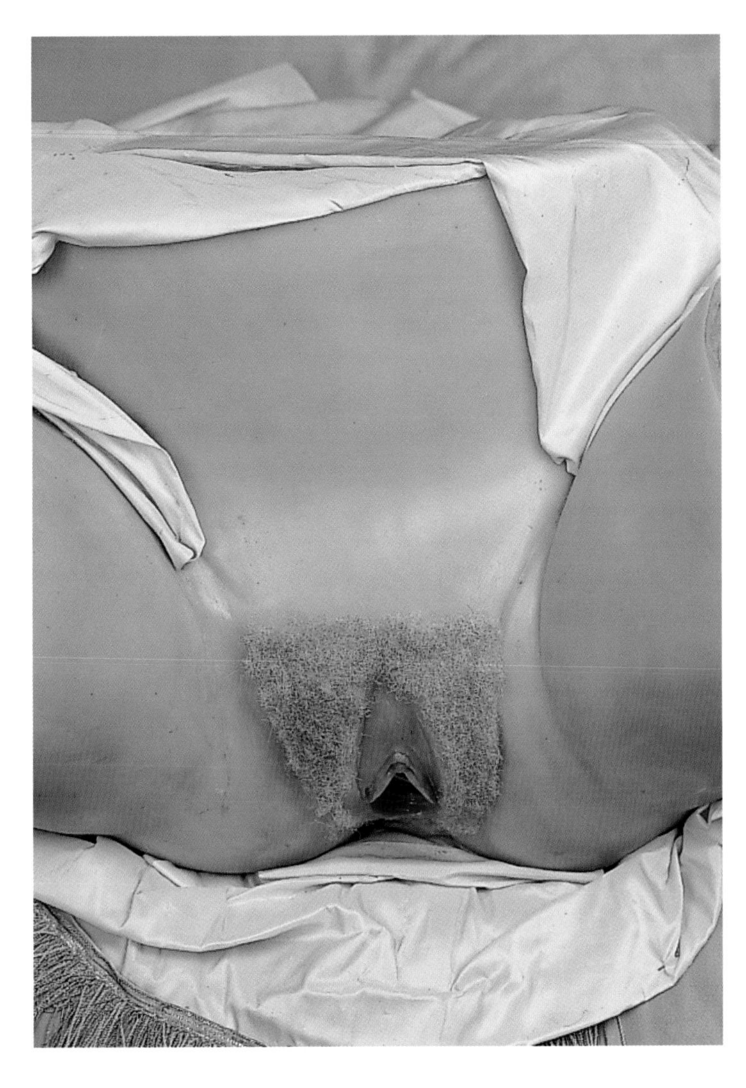

Female external genital organs

Darstellung der äußeren weiblichen Geschlechtsorgane

Organes génitaux externes féminins (position gynécologique)

[OSTETRICIA, 1015]

Lower part of the belly of a pregnant woman; abdominal cavity laid open

Präparat eines Unterleibes einer Schwangeren mit eröffneter Bauchhöhle

Utérus pendant la grossesse après ouverture de la paroi abdominale

[OSTETRICIA, 1015]

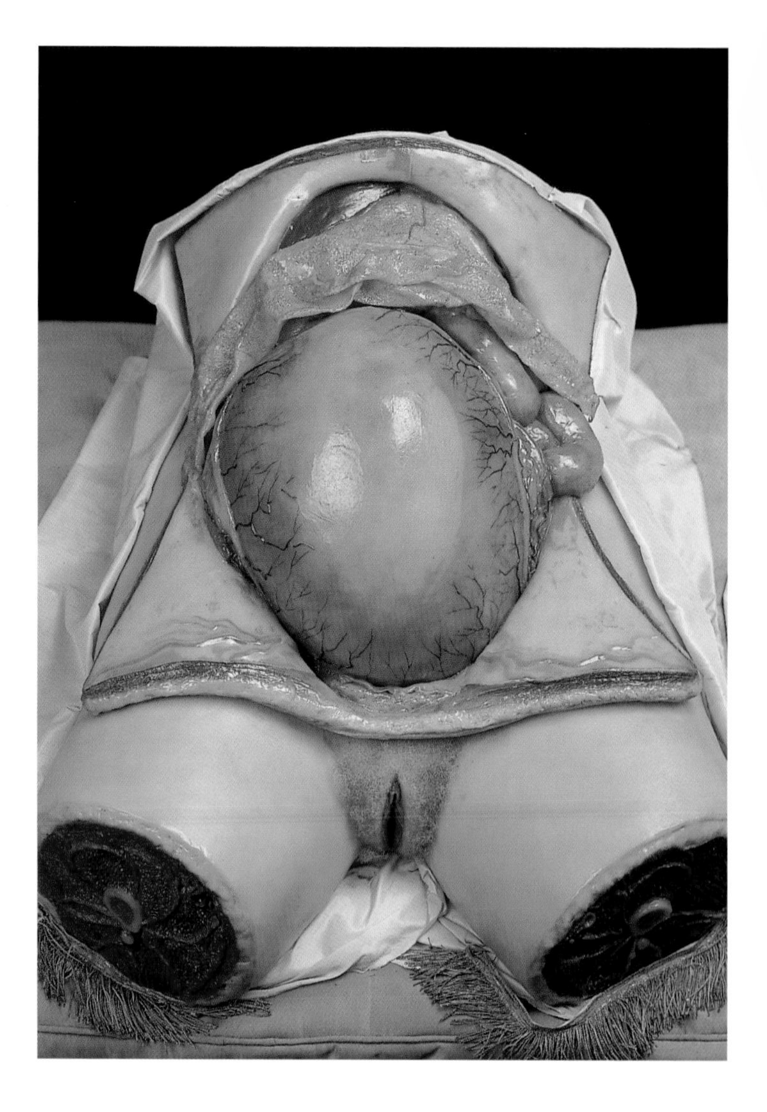

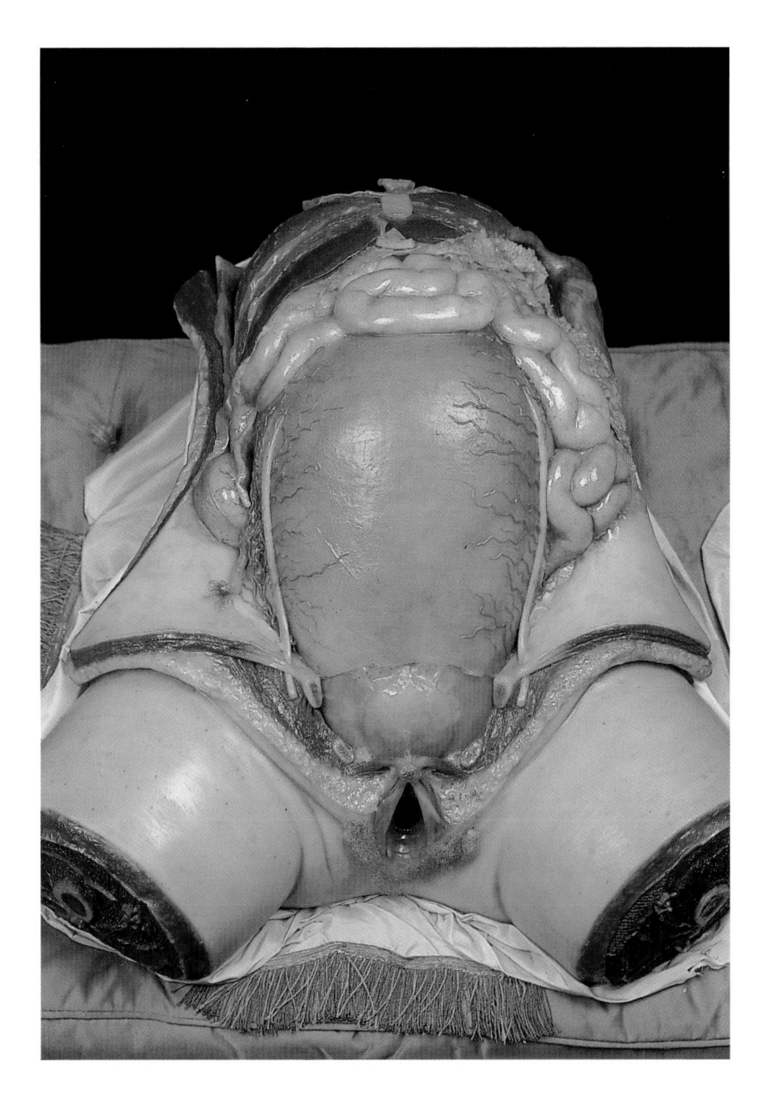

Lower part of the belly of a pregnant woman; abdominal cavity laid open

Präparat eines Unterleibes einer Schwangeren mit eröffneter Bauchhöhle

Utérus à la fin de la grossesse après ouverture de la paroi abdominale

[OSTETRICIA, 1016]

Lower part of the belly of a pregnant woman; abdominal cavity laid open. The superficial layers of the uterine wall have been removed to show the blood vessels.

Präparat eines Unterleibes einer Schwangeren mit eröffneter Bauchhöhle. Oberflächliche Schichten der Gebärmutterwand sind zur Darstellung der Blutgefäße entfernt.

Utérus à la fin de la grossesse : après ouverture de la paroi abdominale et dissection des couches superficielles de l'utérus, mise en évidence des vaisseaux utérins

[OSTETRICIA, 1016]

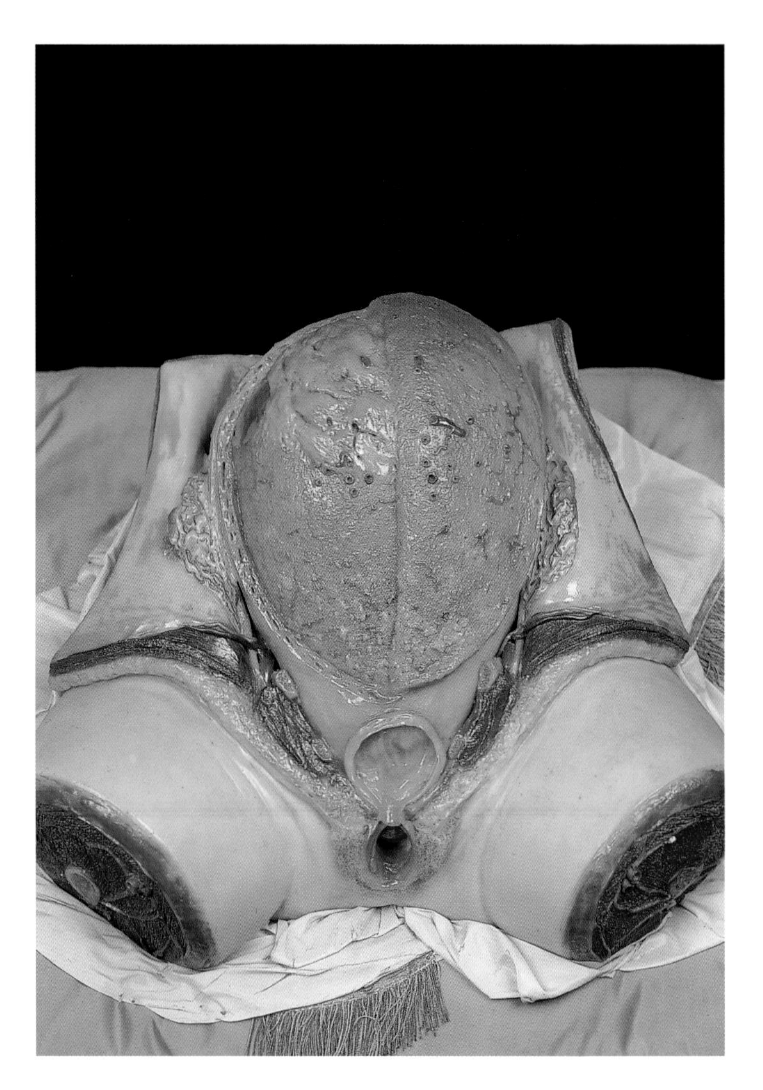

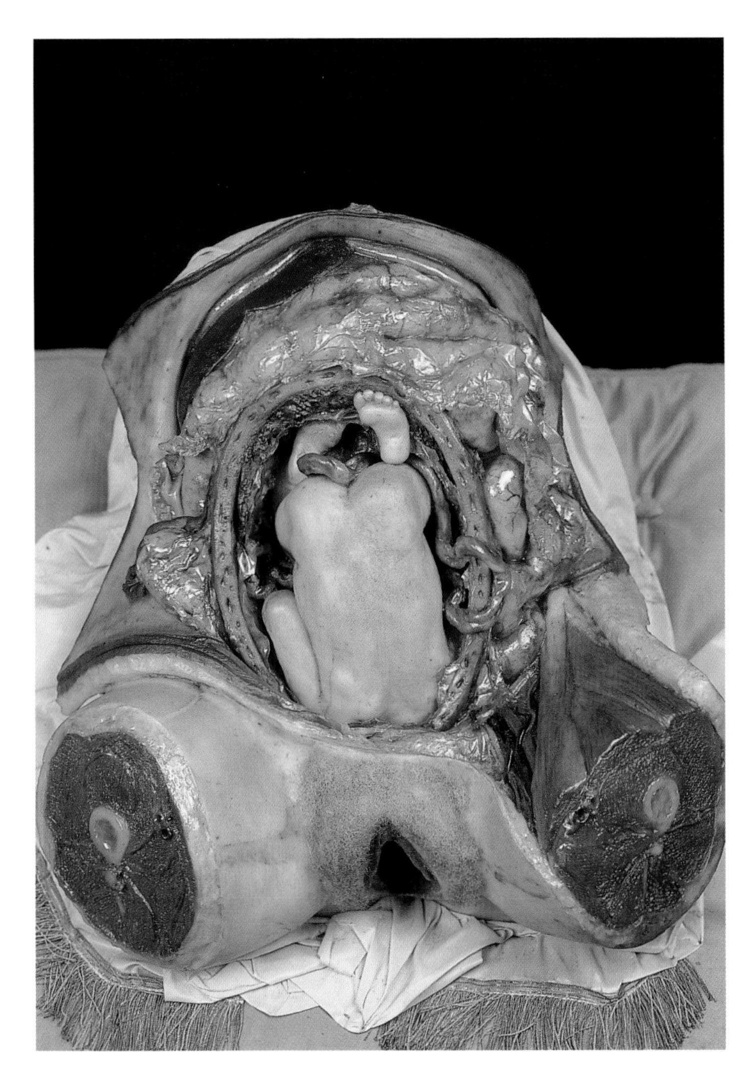

View into the opened uterus of a pregnant woman during parturition

Einblick in die eröffnete Gebärmutter einer Schwangeren während des Geburtsvorganges

Fœtus dans l'utérus ouvert

[OSTETRICIA, 1016]

Uterus of a pregnant woman laid open; dilatation of external os

Präparat einer eröffneten Gebärmutter einer Schwangeren mit sich öffnendem Muttermund

Fœtus jumeaux dans l'utérus ouvert, béance du col utérin

[OSTETRICIA, 1014]

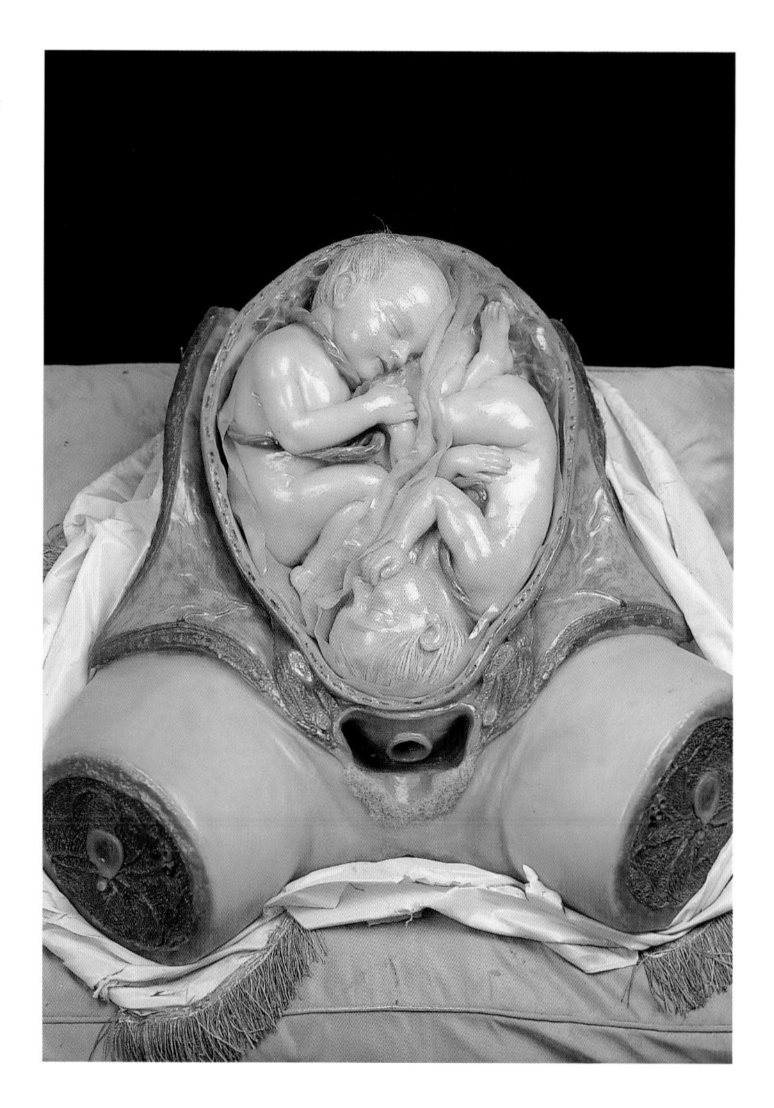

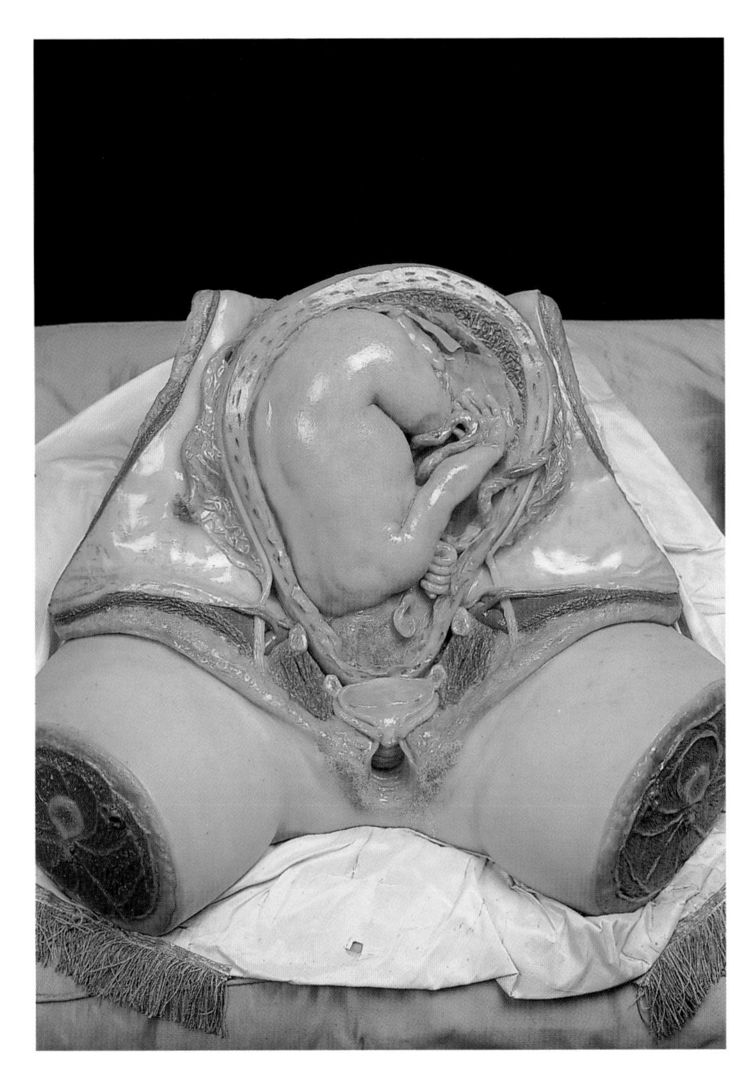

Uterus of a pregnant woman laid open

Präparat einer eröffneten Gebärmutter einer Schwangeren

Fœtus dans l'utérus ouvert

[OSTETRICIA, 1014]

Uterus after delivery of child and placenta

Präparat einer Gebärmutter nach der Geburt des Kindes und des Mutterkuchens

Utérus juste après l'accouchement et l'expulsion du placenta (coupe médiane)

[OSTETRICIA, 1014]

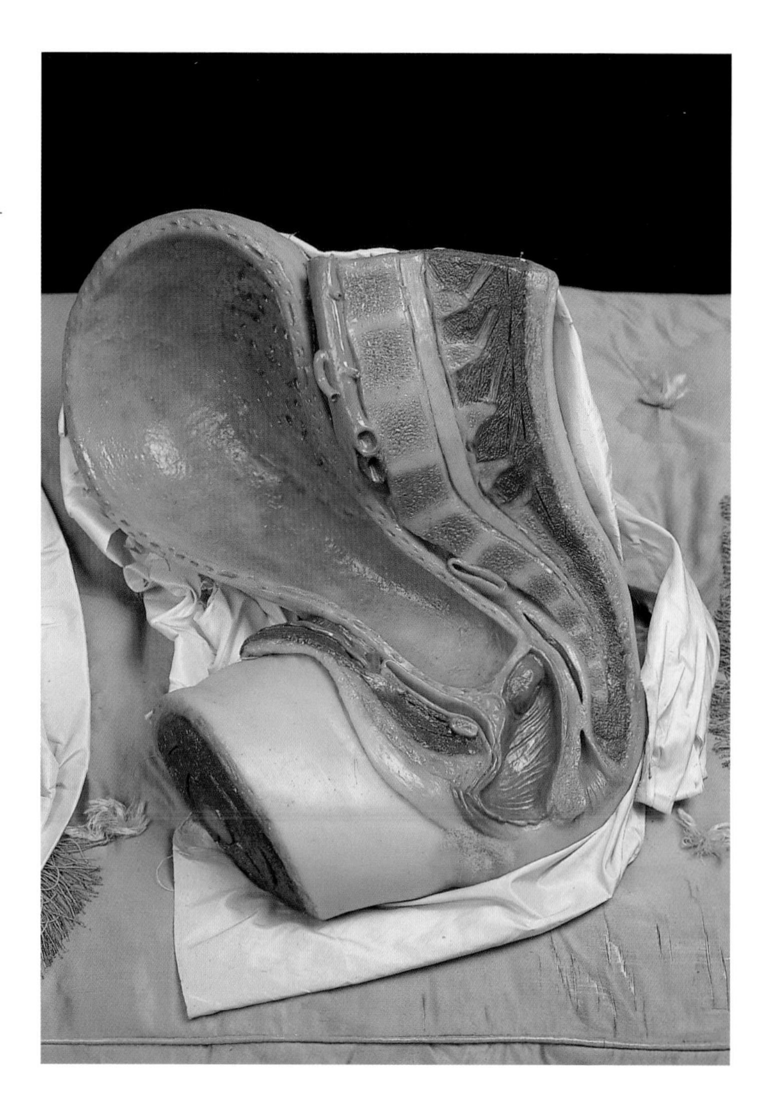

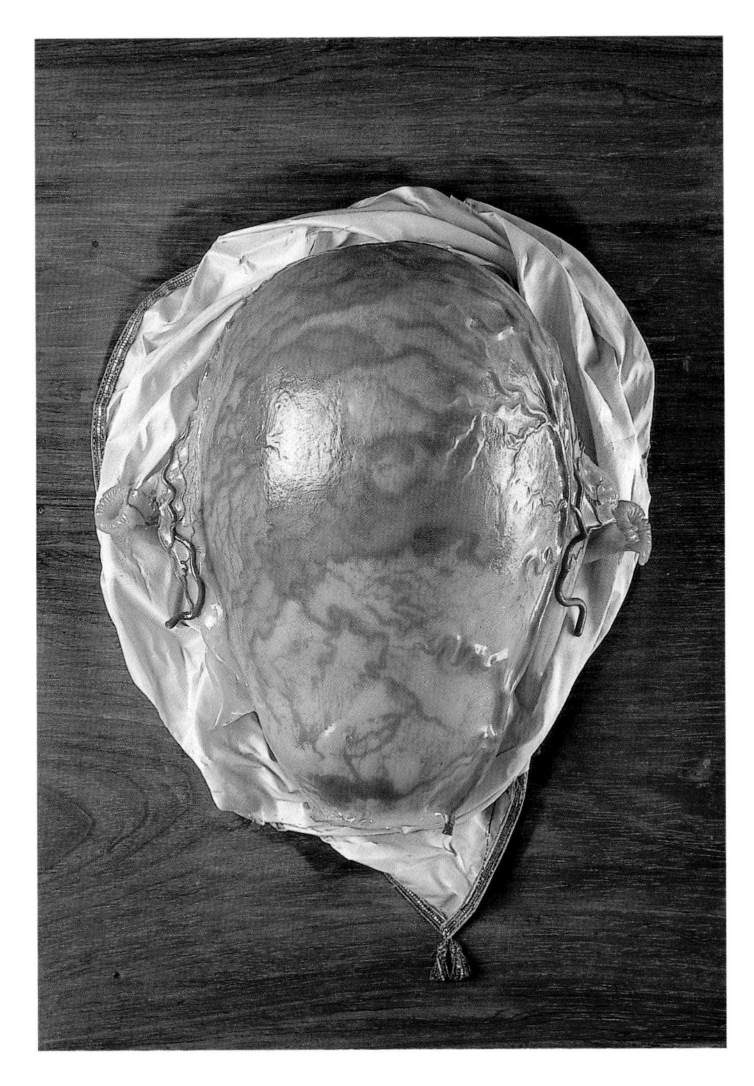

Uterus of a pregnant woman

Präparat der Gebärmutter einer Schwangeren

Utérus pendant la grossesse

[OSTETRICIA, 1005]

Uterus of a pregnant woman; deeper layer showing blood vessels

Präparat der Gebärmutter einer Schwangeren, tiefere Schicht mit Darstellung der Blutgefäße

Utérus pendant la grossesse : artères et veines

[OSTETRICIA, 1002]

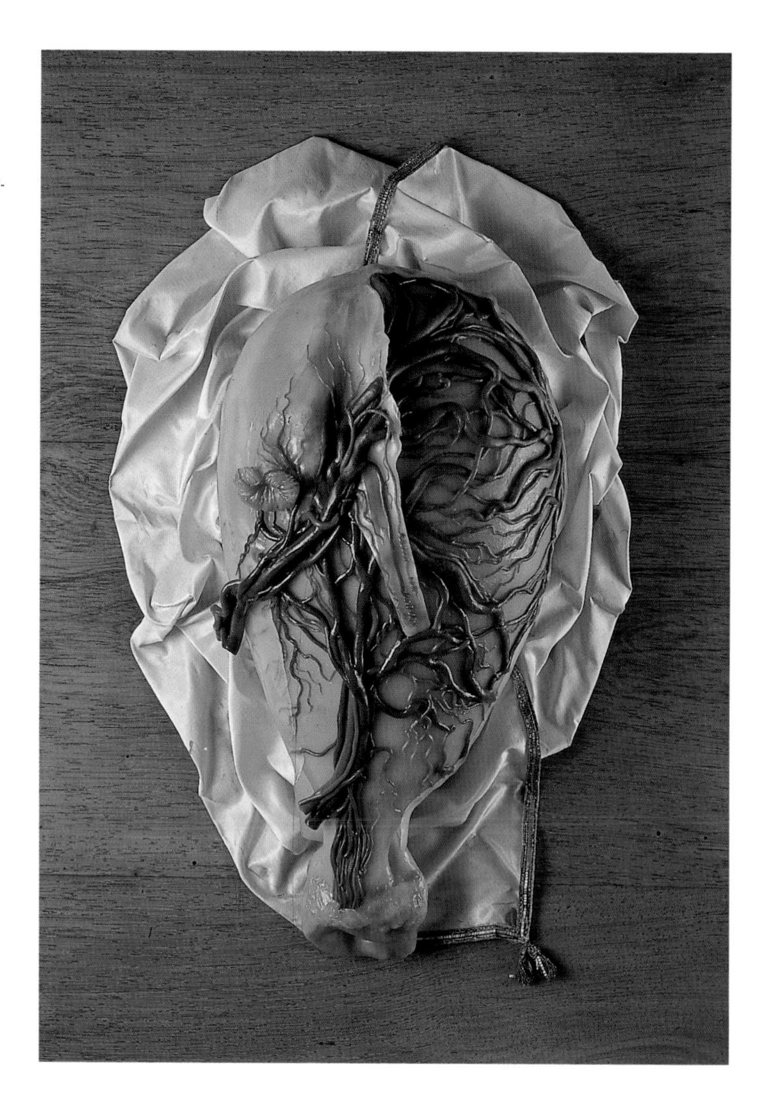

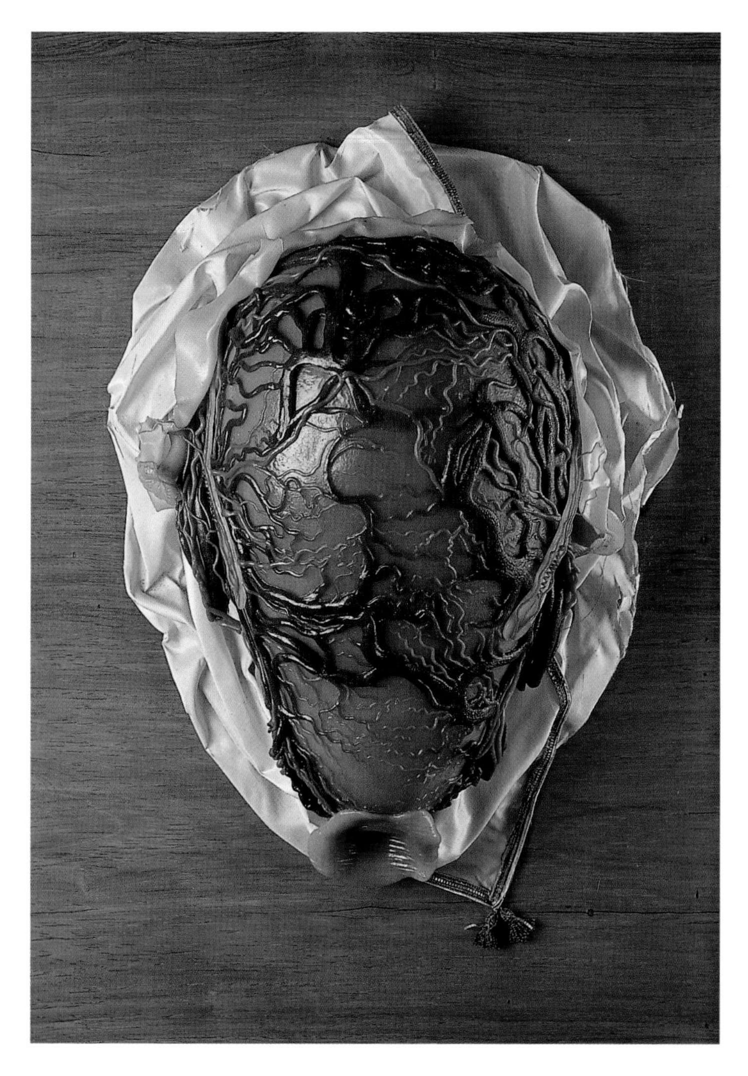

Uterus of a pregnant woman; deeper layer showing blood vessels

Präparat der Gebärmutter einer Schwangeren, tiefere Schicht mit Darstellung der Blutgefäße

Utérus pendant la grossesse : artères et veines

[OSTETRICIA, 1003]

Uterus gravidus

Uterus of a pregnant woman; deeper layer showing blood vessels

Präparat der Gebärmutter einer Schwangeren, tiefere Schicht mit Darstellung der Blutgefäße

Utérus pendant la grossesse : couches profondes, réseaux vasculaires

[OSTETRICIA, 1004]

Uterus of a pregnant woman; uterine musculature displayed

Präparat der Gebärmutter einer Schwangeren mit Darstellung der Gebärmuttermuskulatur

Utérus pendant la grossesse : organisation des fibres musculaires

[OSTETRICIA, 1000]

Uterus gravidus

Uterus of a pregnant woman; uterine musculature displayed

Präparat der Gebärmutter einer Schwangeren mit Darstellung der Gebärmuttermuskulatur

Utérus pendant la grossesse : organisation des fibres musculaires

[OSTETRICIA, 1001]

Uterus with amniotic sac

Präparat einer Gebärmutter mit Fruchtblase

Utérus ouvert et enveloppes du fœtus (sac amniotique)

[OSTETRICIA, 976]

Uterus gravidus, Fetus

Uterus with a fetus

Präparat einer Gebärmutter mit Fötus

Fœtus dans l'utérus ouvert

[OSTETRICIA, 974]

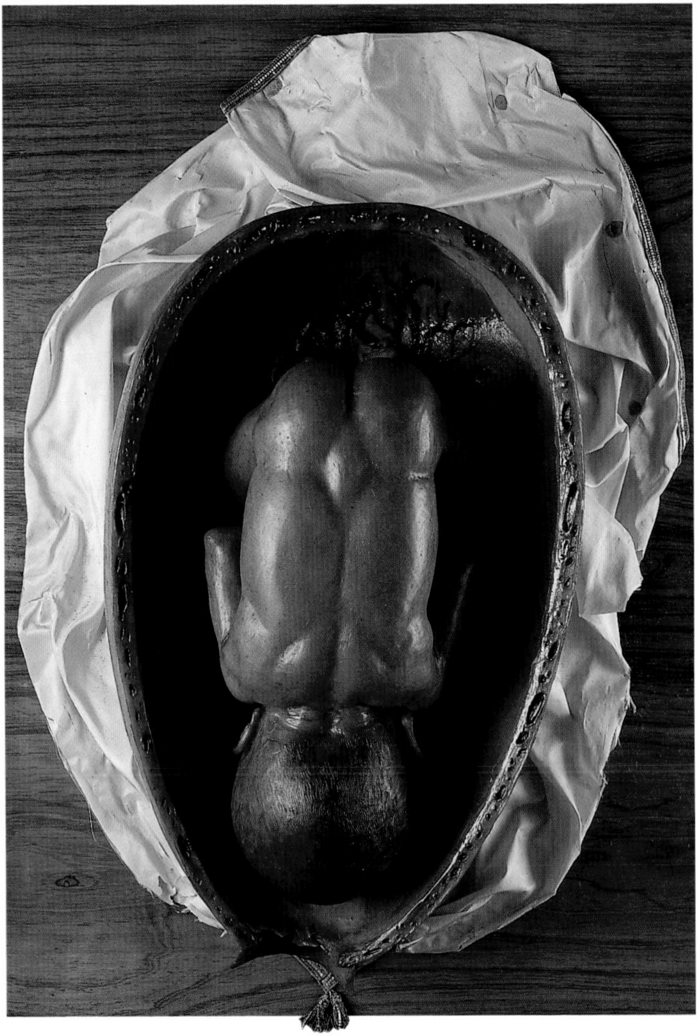

Uterus with placenta

Präparat einer Gebärmutter mit Mutterkuchen

Utérus ouvert avec le placenta

[OSTETRICIA, 979]

Uterus post partum

Uterus after delivery of the placenta

Präparat einer Gebärmutter nach der Geburt des Mutterkuchens

Utérus ouvert après l'expulsion du placenta

[OSTETRICIA, 978]

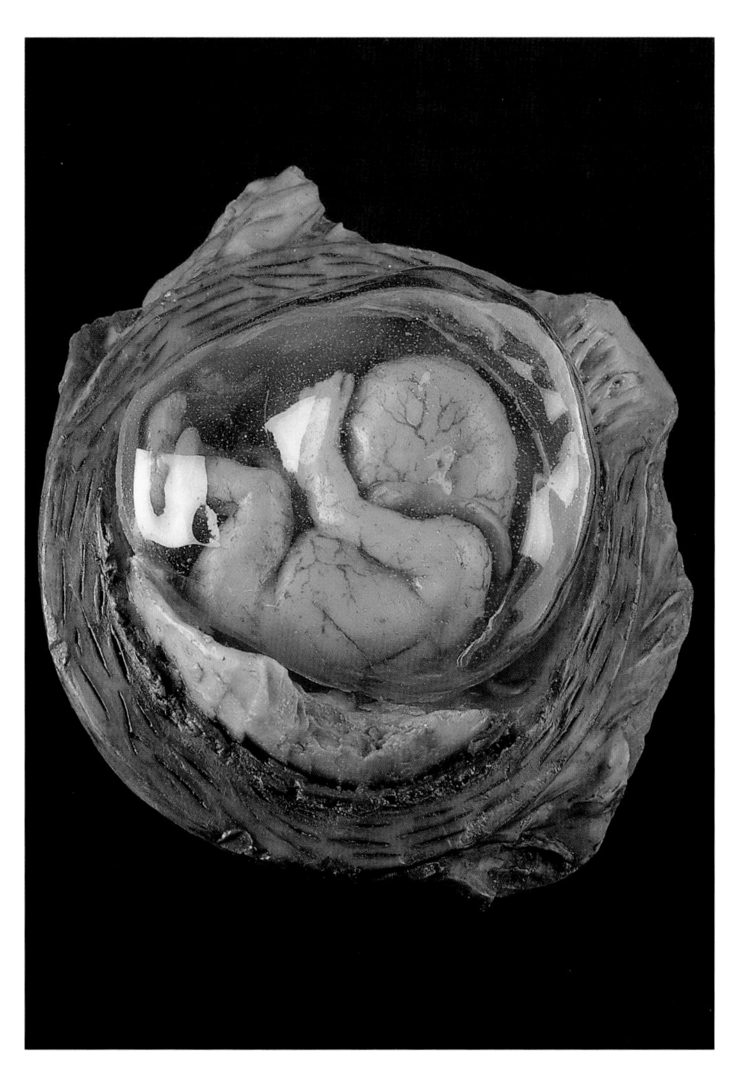

Specimen showing a fetus within the amniotic sac

Darstellung eines Fötus in der Fruchtblase

Fœtus dans le sac et le liquide amniotique dans l'utérus ouvert

[DEP. 3]

View of a fetus inside the uterus

Darstellung eines Föten in der Gebärmutter

Fœtus dans l'utérus ouvert

[DEP. 6]

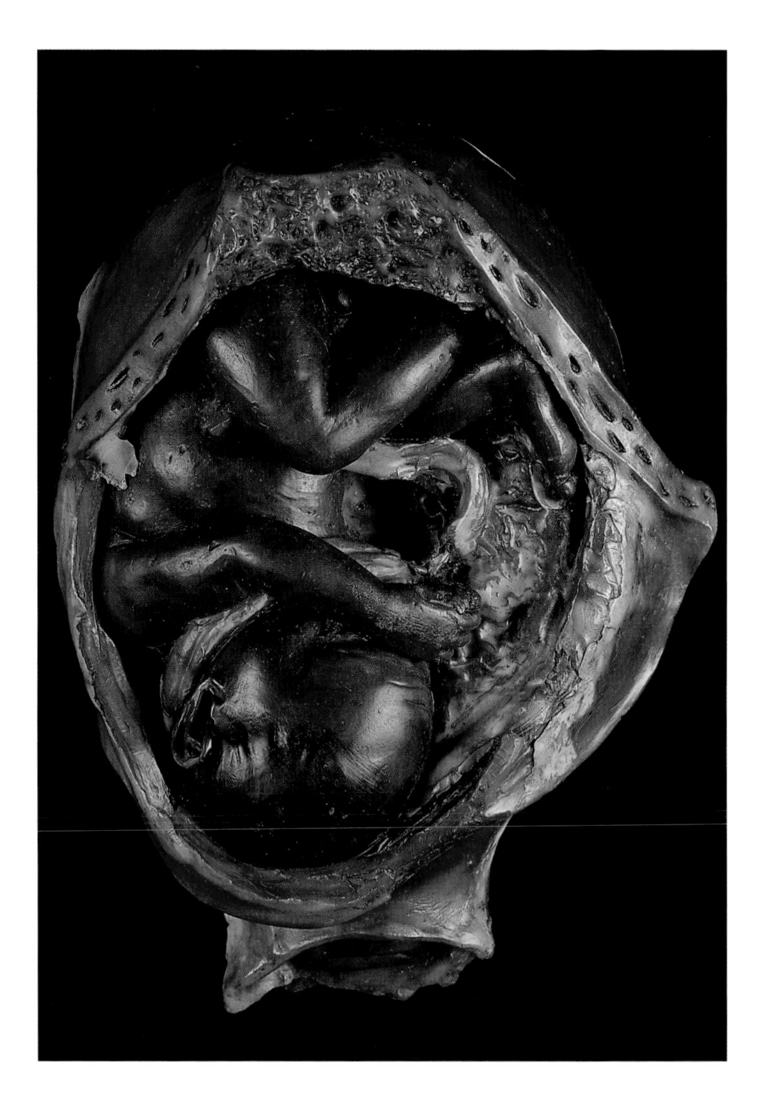

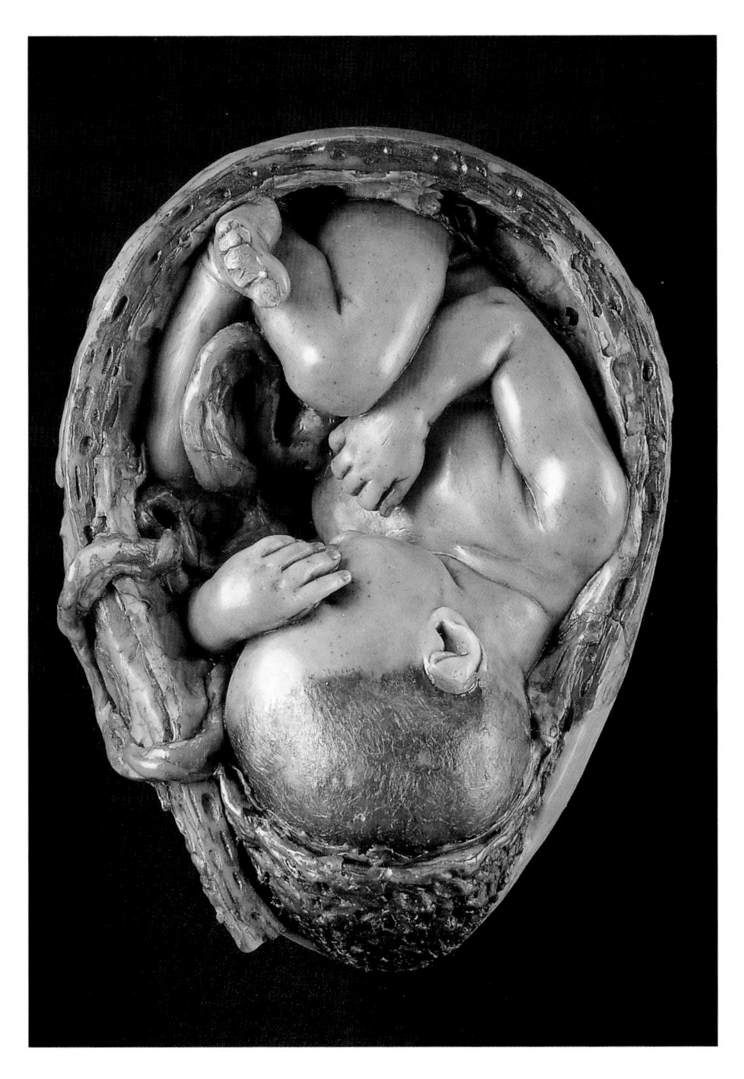

View of a fetus inside the uterus

Darstellung eines Föten in der Gebärmutter

Fœtus dans l'utérus ouvert

[DEP. 7]

View of a fetus inside the uterus

Darstellung eines Föten in der Gebärmutter

Fœtus dans l'utérus ouvert

[DEP. 8]

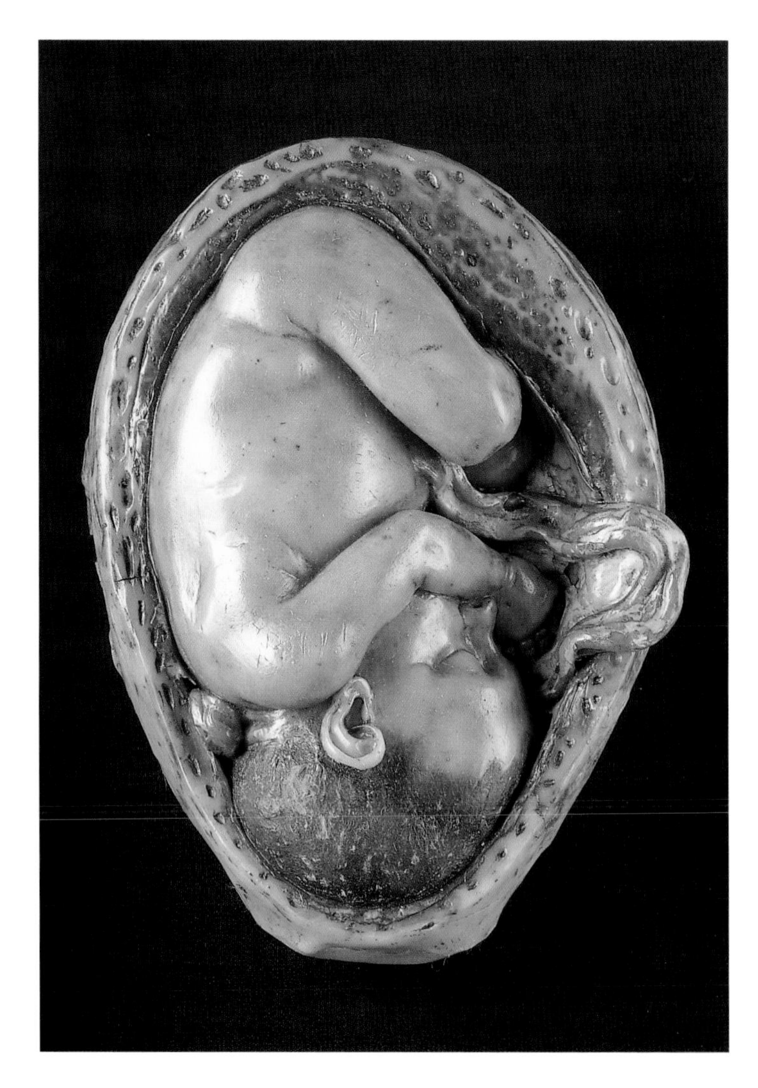

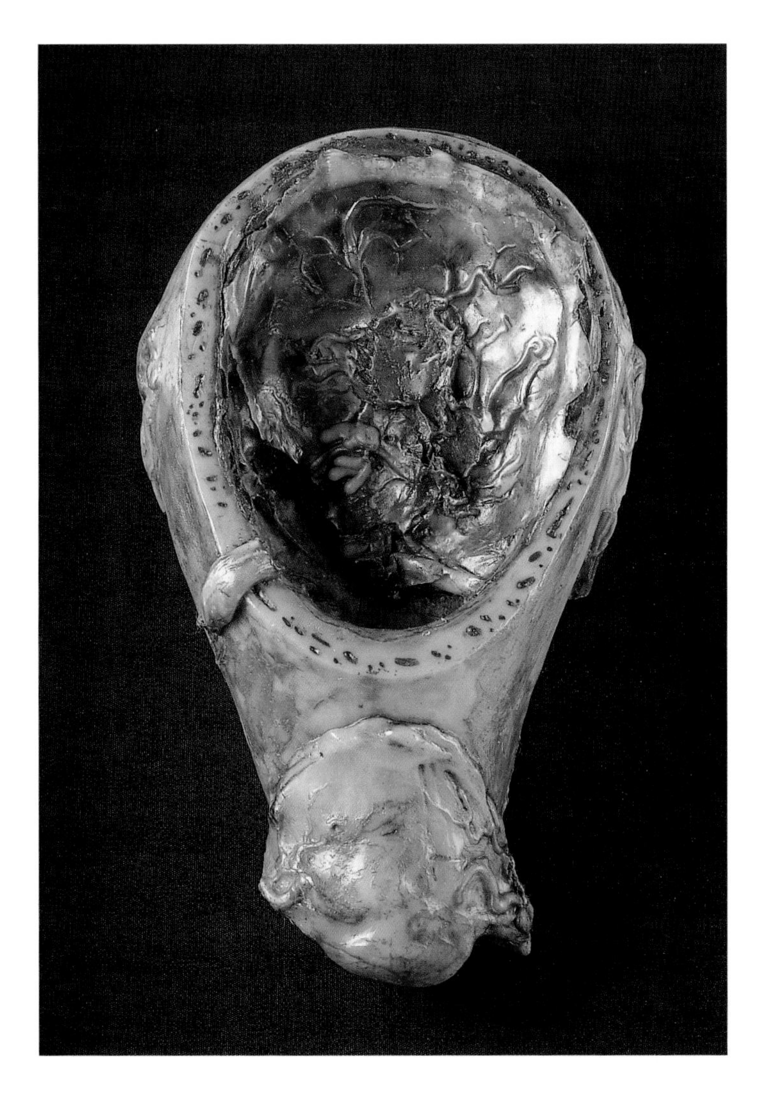

The uterus of a pregnant woman containing the placenta

Darstellung der Gebärmutter einer Schwangeren mit Mutterkuchen

Utérus ouvert avec le placenta

[DEP. 2]

The uterus of a pregnant woman with amniotic sac, fetus and placenta

Darstellung der Gebärmutter einer Schwangeren mit Fruchtblase, Fötus und Mutterkuchen

Fœtus dans le sac amniotique et placenta dans l'utérus ouvert

[DEP. 1]

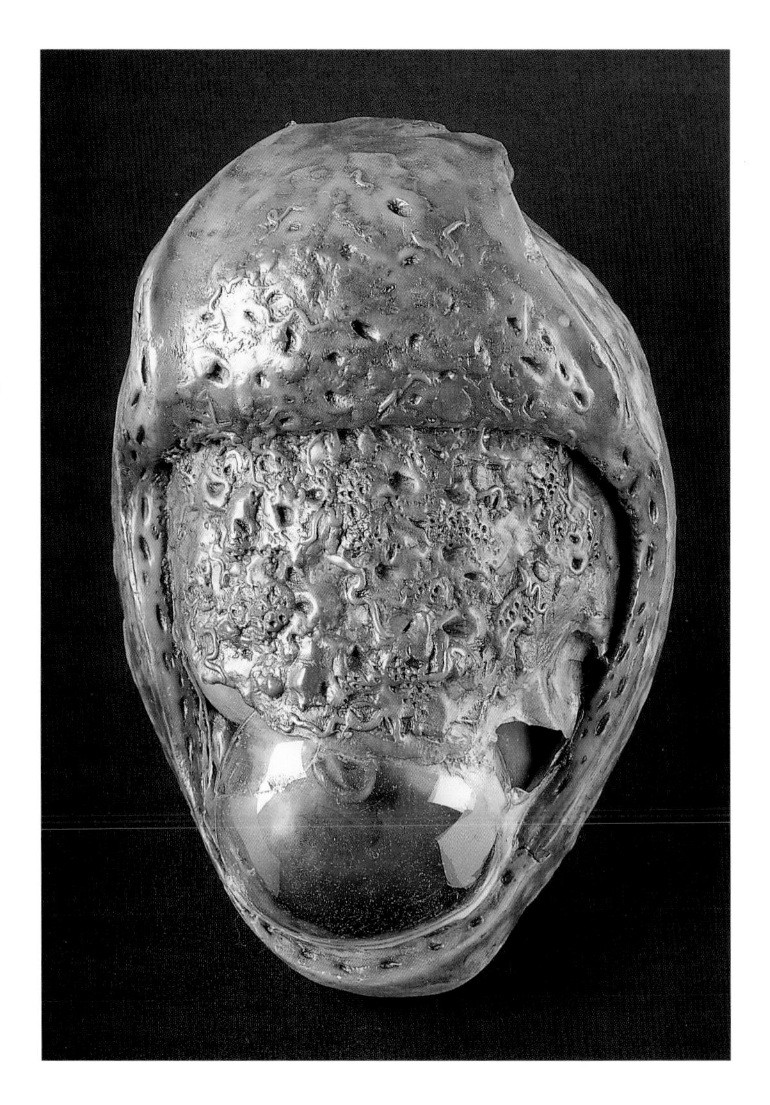

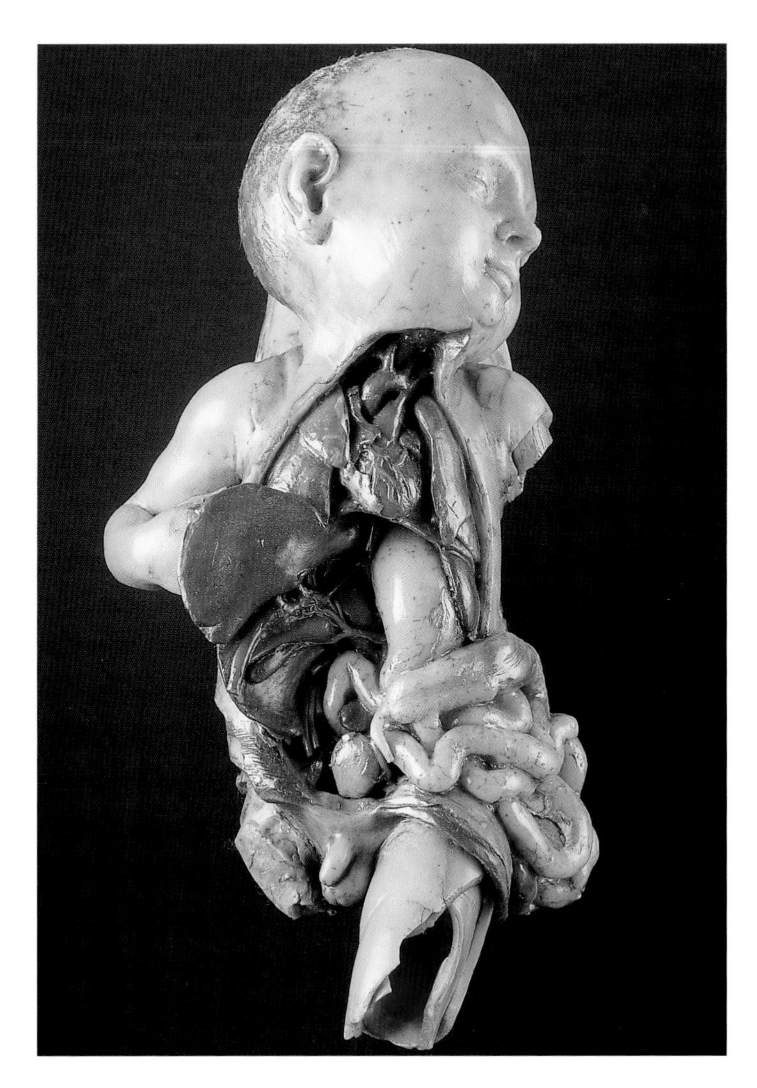

A newborn infant with the thoracic and abdominal cavities laid open. The arms and legs of the specimen have been damaged.

Darstellung eines Neugeborenen mit Einblick in den Brust-und Bauchraum. Das Präparat ist an den Armen und Beinen beschädigt.

Nouveau-né : viscères thoraciques et abdominaux (membres supérieurs et inférieurs endommagés)

[DEP. 5]

Specimen showing abdominal pregnancy

Darstellung verschiedener Typen von Bauchhöhlenschwangerschaften

Présentation de différents types de grossesses extra-utérines

[OSTETRICIA, 986]

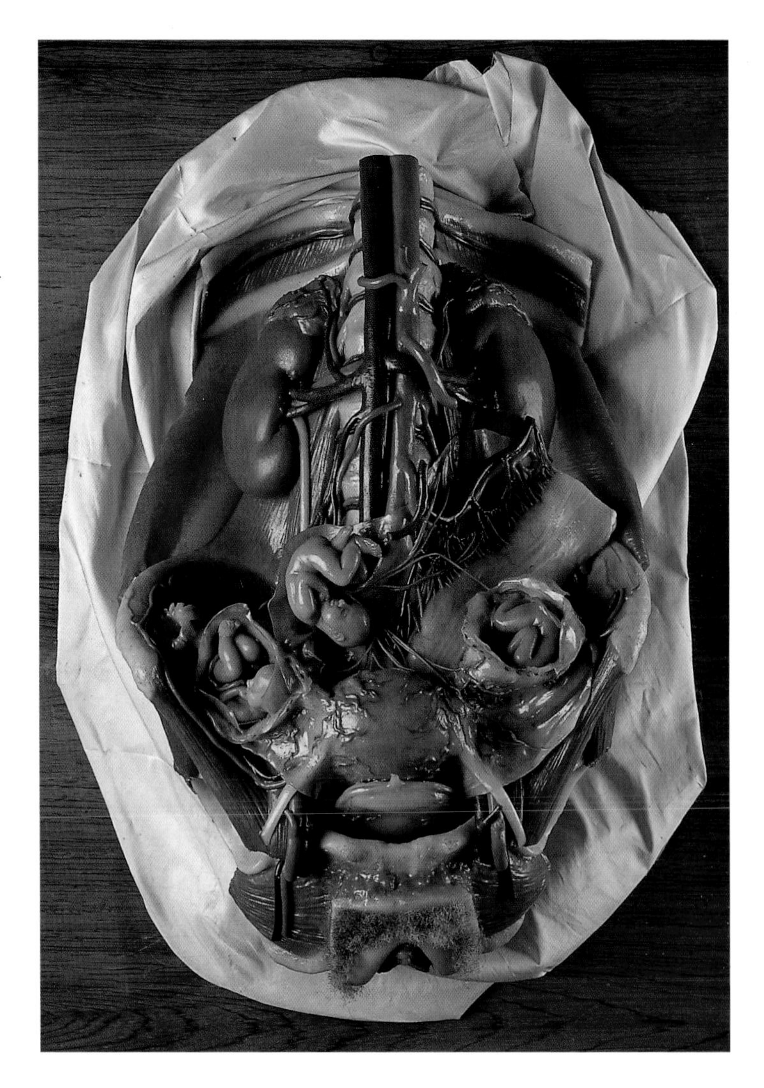

Uterus with placenta, umbilical cord and fetus

Präparat der Gebärmutter mit Mutterkuchen, Nabelschnur und Fötus

Fœtus, cordon ombilical, placenta, utérus ouvert

[OSTETRICIA, 975]

Newborn infant with umbilical cord and placenta

Präparat des Mutterkuchens mit Nabelschnur und Neugeborenem

Nouveau-né, cordon ombilical, placenta

[OSTETRICIA, 973]

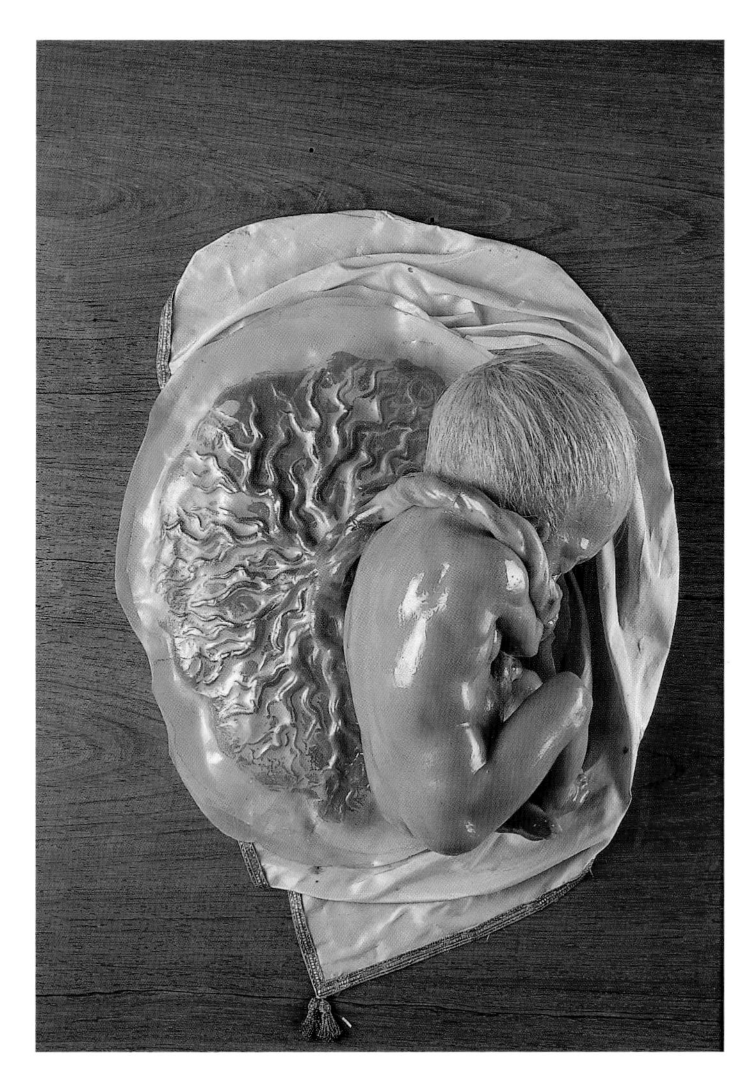

Newborn infant with umbilical cord and placenta

Präparat des Mutterkuchens mit Nabelschnur und Neugeborenem

Nouveau-né, cordon ombilical, placenta

[OSTETRICIA, 999]

Specimen to show the blood supply and circulation of a fetus

Präparat zur Darstellung der Blutversorung und des Kreislaufes des Föten

Fœtus : circulation sanguine, artères et veines ombilicales, vascularisation du placenta

[OSTETRICIA , 1013]

Placenta and umbilical cord

Präparat des Mutterkuchens und der Nabelschnur

Placenta et cordon ombilical : artères et veines

[OSTETRICIA, 977]

Fetus with abdominal cavity laid open to display the umbilical arteries and vein

Präparat eines Föten mit eröffnetem Bauchraum zur Darstellung der Nabelarterien und der Nabelvene

Fœtus : artères et veines ombilicales

[OSTETRICIA, 998]

Fetus with abdominal cavity laid open and abdominal organs removed to display the urogenital system. The left testis has descended into the scrotum, the right testis is in the abdominal cavity.

Präparat eines Föten mit eröffnetem Bauchraum. Zur Darstellung der Urogenitalorgane sind die Bauchorgane entfernt. Der linke Hoden ist in den Hodensack abgestiegen, der rechte Hoden befindet sich im Bauchraum.

Fœtus : appareils urinaire et génital ; testicule gauche descendu dans les bourses, testicule droit encore dans la cavité abdominale

[OSTETRICIA, 995]

Fetus with thoracic and abdominal cavities laid open

Präparat eines Föten mit eröffnetem Brust- und Bauchraum

Fœtus : viscères du thorax et de l'abdomen

[OSTETRICIA, 996]

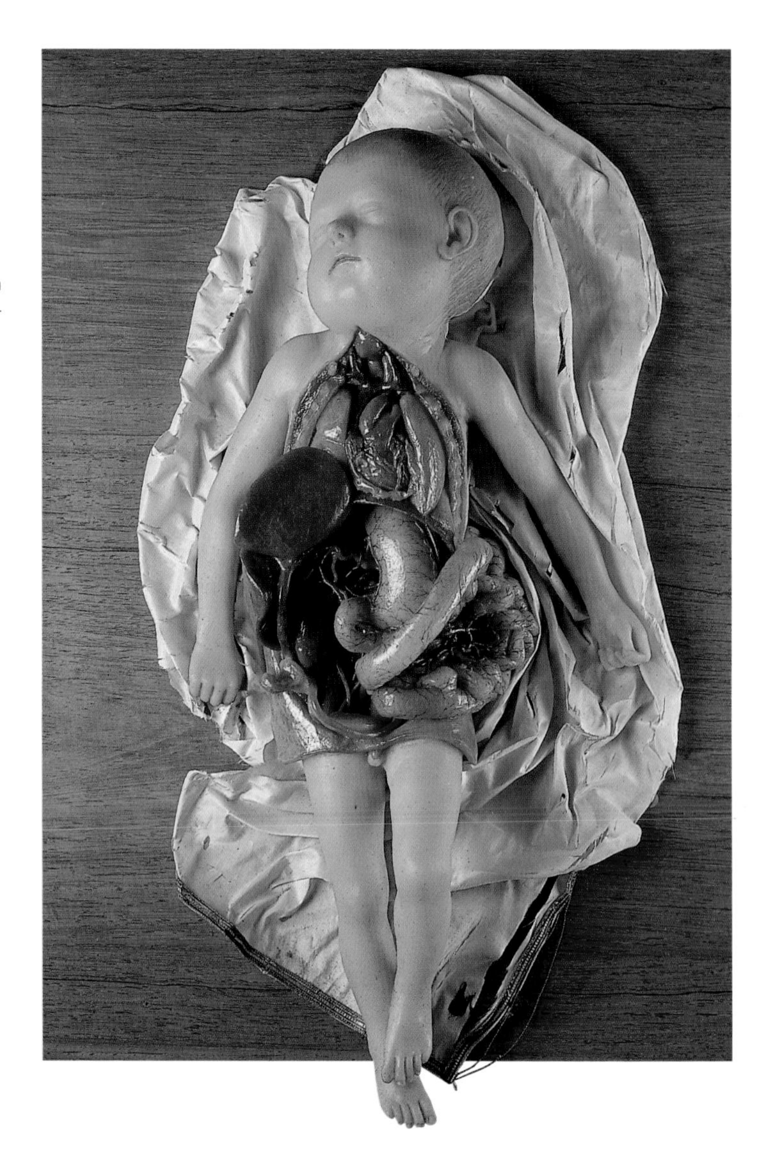

Display showing the various developmental stages of the embryo up to the third month of pregnancy

Darstellung verschiedener Stadien der Keimesentwicklung bis zum dritten Schwangerschaftsmonat

Embryons et fœtus : stades du développement jusqu'au troisième mois

[OSTETRICIA, 969]

Fetus, fourth month of pregnancy

Darstellung von Föten im vierten Schwangerschaftsmonat

Fœtus au cours du quatrième mois de la grossesse

[OSTETRICIA, 970]

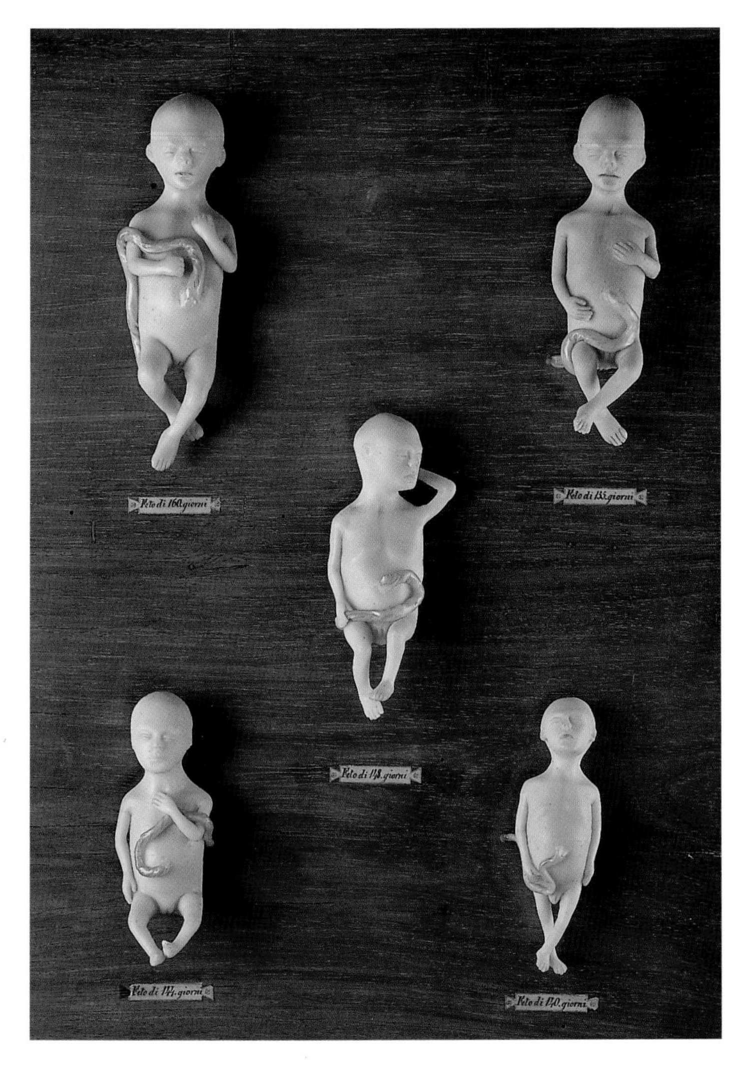

Fetus, fifth month of pregnancy

Darstellung von Föten im fünften Schwangerschafts-monat

Fœtus au cours du cinquième mois de la grossesse

[OSTETRICIA, 994]

Fetus, Mens VI

Fetus, sixth month of pregnancy

Darstellung von Föten im sechsten Schwangerschafts-monat

Fœtus au cours du sixième mois de la grossesse

[OSTETRICIA, 993]

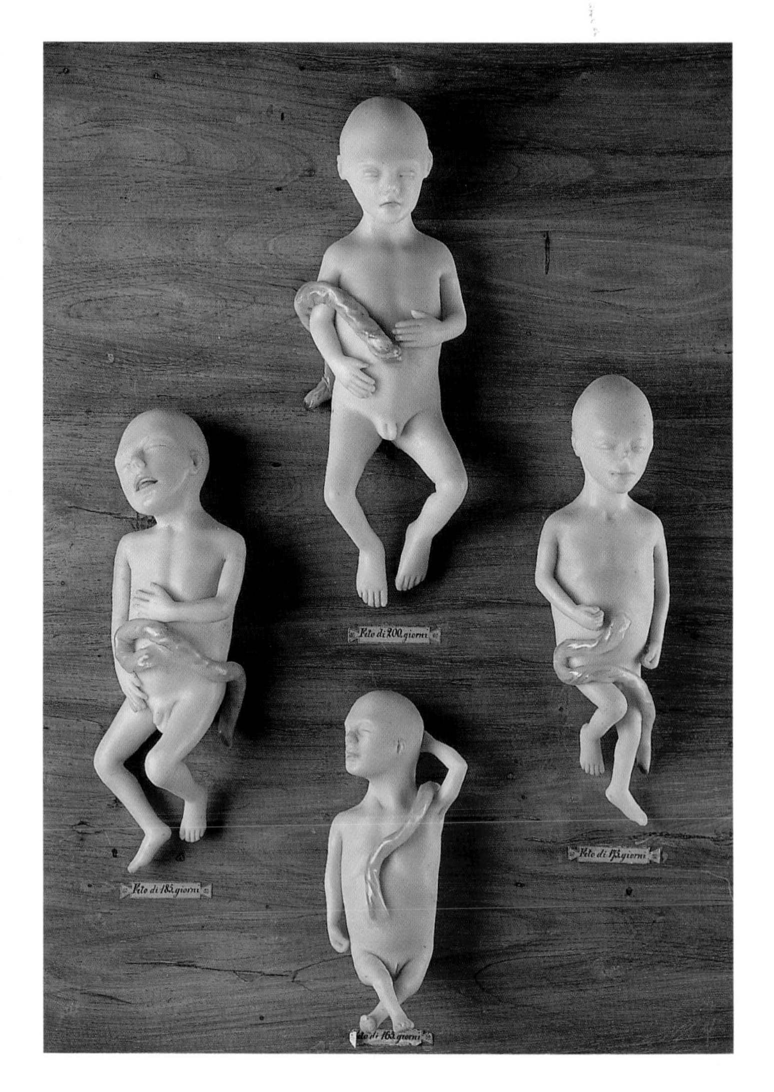